EQUINE SCIENCE

EQUINE SCIENCE

Rick Parker

Delmar Publishers

an International Thomson Publishing company I(T)P®

Albany • Bonn • Boston • Cincinnati • Detroit • London • Madrid
Melbourne • Mexico City • New York • Pacific Grove • Paris • San Francisco
Singapore • Tokyo • Toronto • Washington

NOTICE TO THE READER

Cover photo courtesy of Images © 1996 Photo Disc Inc.
Cover design: Kristina Almquist
Photo credits: Unless noted otherwise, photographs in the text were taken by Marilyn Parker or the author.

Delmar Staff:
Publisher: Tim O'Leary   Senior Project Editor: Andrea Edwards Myers
Acquisitions Editor: Cathy L. Esperti   Production Manager: Wendy A. Troeger

COPYRIGHT © 1998
By Delmar Publishers
a division of International Thomson Publishing Inc.

The ITP logo is a trademark under license

Printed in the United States of America

For more information, contact:

Delmar Publishers
3 Columbia Circle, Box 15015
Albany, New York 12212-5015

International Thomson Publishing Europe
Berkshire House 168–173
High Holborn
London, WC1V 7AA
England

Thomas Nelson Australia
102 Dodds Street
South Melbourne, 3205
Victoria, Australia

Nelson Canada
1120 Birchmount Road
Scarborough, Ontario
Canada M1K 5G4

International Thomson Editores
Campos Eliseos 385, Piso 7
Col Polanco
11560 Mexico D F Mexico

International Thomson Publishing GmbH
Königswinterer Strasse 418
53227 Bonn
Germany

International Thomson Publishing Asia
221 Henderson Road
#05–10 Henderson Building
Singapore 0315

International Thomson Publishing – Japan
Hirakawacho Kyowa Building, 3F
2-2-1 Hirakawacho
Chiyoda-ku, Tokyo 102
Japan

1 2 3 4 5 6 7 8 9 10 XXX 03 02 01 00 99 98 97

Library of Congress Cataloging-in-Publication Data

Parker, R. O.
    Equine science / Rick Parker.
      p.   cm.
    Includes index.
    ISBN 0-8273-7136-5
    1. Horses. 2. Horses—Health. I. Title.
SF285.3.P36 1997
636.1—dc20   96-35298
                         CIP

CONTENTS

Cows, pigs, and sheep serve humans by providing food, fiber and power. Over the centuries, their role has changed little. Horses also provided power for transportation and for industry and even food for some people. Now, at least in North America, the horse is thought of as an animal for recreation, sport, and companionship. Other livestock will never change their involvement with humans as completely as the horse. With this change, people want more and better information about horses, so the numbers of books, pamphlets, videos, and even Internet sites increase at a rapid pace.

Similar to other books on horses, this book starts with where the industry and horses have been and where they are now. A clear vision of the past and present is necessary to move into the future. Also, similar to other books on the horse, this book contains information about the breeds and types of horses. But this information is presented in a large table format, since so many books on breeds and types of horses are available.

Chapters on soundness, selecting and judging, and determining the age of horses could be considered quite traditional for a book on horses. However, in this book, understanding these topics receives strong support from the biology or science that is involved. Chapters on cells, tissues and organs, and functional anatomy are unique to a "horse" book. These chapters also support a student's understanding of reproduction, nutrition and digestion, feeding, health, and parasite control. A chapter specifically on genetics further enhances a student's understanding of reproduction and breeding.

While a chapter on health management is probably not unique, a complete discussion of the immune system is. For some reason, textbooks on horses have ignored this important concept. This book does not. An understanding of the immune system is essential to healthy foals, disease prevention, and vaccination programs.

Shoeing and hoof care is such an important part of successful equitation care, that it stands on its own as a chapter. Again, students' understanding of this subject is increased because they have already read about anatomy and the biomechanics of movement.

To describe all of the possible types of buildings and equipment used in the horse industry would be an impossible task. The chapter on buildings and equipment provides some solid guidelines for construction and purchase. After all, a good building starts on a solid foundation.

In the past, most people grew up around animals and they learned animal behavior by working with and observing animals every day. A textbook discussion of animal behavior was unheard of a few years ago. Now, any textbook about animals would be considered incomplete without a chapter on behavior. This is especially true for horses. Understanding their behavior is critical to a person's success as a trainer or rider.

Finally, many people who learn about horses want to be employed in the equine industry. This happens two ways—by starting a new equine business or by getting a job or career in the equine industry. The last two chapters in the book point students in these directions.

This book is packed with information. Several features of the book make the information more useful. These features include the chapter objectives, key terms, tables, charts, illustrations, pictures, exercises, a glossary, and an appendix. Each chapter also contains at least one sidebar to help stimulate the students' interest. To be useful the book must be used as a whole, since each chapter and each feature complements the others. In a complete education, nothing stands on its own; every piece of information is interrelated.

ACKNOWLEDGMENTS

An author's efforts never stand on their own merit. The genesis of this book depended on teamwork. Team members included people who were willing to go "above and beyond" to create this book. One member of the author's team, his wife, Marilyn, has been on the team for over twenty-seven years. Without her constant help, support, and encouragement this book would never have been completed. Marilyn concentrated on the details of the artwork, key terms, and questions. Another team member, Rosemary Vaughn, works with the author on a daily basis. She willingly gave up evening and weekend time to do typing and follow up on resource material. Other members of the team included the author's children who tolerated his mental absence while working at home. The author also appreciates the support, encouragement, and even the prodding of Cathy Esperti and the team at Delmar. Finally, the author and Delmar Publishers wish to express their appreciation to the following reviewers:

Ted Arthur
Perkins, Oklahoma

Glenn Boettcher
Lakeland, Minnesota

Donna Crompton
Glastonbury, Connecticut

Jim McCall
Mt. Holly, Arkansas

Ray Whelihan
Cobleskill, New York

Dedicated to
nine people who provide meaning
and happiness in my life:
my wife, Marilyn, and our
eight children—
Shaura, Sariah, Justus, Liberty,
Cole, Morgan, Spencer, and Samuel.

R.O. (Rick) Parker grew up on an irrigated farm in southern Idaho. Because of a love of agriculture he went on in his education. Starting at Brigham Young University, he received his bachclor's degree and then moved to Ames, Iowa, where he finished a Ph.D. in animal physiology at Iowa State University. After completing his Ph.D., he and his family moved to Edmonton, Alberta, Canada where he completed a post-doctorate at the University of Alberta. Next he and his wife, Marilyn, and their children moved to Laramie, Wyoming, where he was a research and teaching associate at the University of Wyoming.

The author is currently working as a division director at the College of Southern Idaho in Twin Falls. As director, he works with faculty in agriculture, computer applications, office technology, and business management. Occasionally, he teaches a computer class, agriculture class, or a writing class.

CHAPTER 1

History and Development of the Horse

Since prehistoric times, the swift and powerful horse has been domesticated by human beings for use as a draft animal, for transportation, and in warfare, and has figured notably in art and mythology. Riding horses was not practical until suitable bits and other controlling devices were invented, and the horse did not replace humans and oxen at heavy farm labor until the appearance of an efficient harness. Today, horses are used primarily for sports such as racing, show competition, rodeos, and simple riding for pleasure. Horseflesh has occasionally been consumed by humans since prehistoric times, and it is used as a pet food.

A large herbivore adapted for running, the horse, Equus caballus, *is a mammal of the family Equidae, order Perissodactyla.*

OBJECTIVES

After completing this chapter, you should be able to:

- Name the major evolutionary horse-like animals
- Identify the position of the horse in the zoological scheme
- Describe how humans eventually changed the way they used the horse
- Give the scientific name for the horse and three of its close relatives
- List the four evolutionary trends exhibited by horse fossils
- Identify the Romans' influences on the use of the horse
- Describe the affect of the Middle Ages and the Renaissance on the use of horses
- Name three horses in mythology or legend
- Name three famous horses of the films

- Describe the use and decline of horses in agriculture in the United States
- Discuss how racing started in the United States
- Identify the factors that changed the use of horses in the twentieth century
- Name four geologic time periods (epochs) used to discuss the evolution of the horse

K E Y   T E R M S

Archeohippus	Hybrids	Oligocene epoch
Calippus	*Hyracotherium*	Onagers
Centaur	*Merychippus*	Paleocene epoch
Eocene epoch	*Mesohippus*	Pleistocene epoch
Eohippus	Miocene epoch	Pliocene epoch
Evolution	Monodactyl	*Pliohippus*
Geldings	Morrill Land Grant Act	Przewalski's horse
Hackamore	Mules	Rodeo

EVOLUTION OF THE HORSE

Evolution of the horse did not occur in a straight line toward a goal, like a ladder. Rather, it was like a branching bush, with no predetermined goal. Many horse-like animals branched off the evolutionary tree and evolved along various unrelated routes, with differing numbers of toes and adaptations to different diets. Now one genus—*Equus*—is the only surviving branch of a once mighty and sprawling evolutionary bush. Of the several species within that genus, *Equus caballus* is today's true horse. Here's how it fits into the zoological scheme:

Kingdom: Animalia
Phylum: Chordata
Class: Mammalia
Order: Perissodactyla
Family: Equidae
Genus: *Equus*

Equus asinus—the true asses and donkeys of northern Africa. (The African wild asses are sometimes called *Equus africanus*.)

Equus burchelli—the Plains zebra of Africa, including "Grant's zebra," "Burchell's zebra," "Chapman's zebra," the half-striped Quagga, and other subspecies. The Plains zebra is what people usually think of as the "typical zebra," with rather wide vertical stripes, and thick horizontal stripes on the rump.

Equus caballus—the true horse, which once had several subspecies.

Equus grevyi—Grevy's zebra, the most horse-like zebra. This is the big zebra with the very narrow vertical stripes and huge ears.

Equus hemionus—the desert-adapted onagers of Asia and the Mideast, including the kiang.

Equus przewalski—the oldest living species of horse, discovered in remote Mongolia. (*Equus caballus* first appeared in Central Asia, probably as **Przewalski's horse**.)

Equus zebra—the Mountain zebra of South Africa. This is the little zebra with the dewlap and the gridiron pattern on its rump.

The Fossil Record

Eohippus.   The earliest forerunner of our present horse, ***Eohippus*** or ***Hyracotherium***, was a small, primitive horse about the size of a fox. It had an elongated skull, a moderately arched back, and a shortened tail. There were four functional toes on each front foot, but only three toes on each hind foot. The structure of its teeth suggests that it was a browser (see Figure 1-1). The earliest remains of this extinct animal are found in rocks of the late **Paleocene epoch** (about 54 million years old) in North America. More recent fossils have been found in rocks of the **Eocene epoch** (about 50 million years old) in Europe.

Mesohippus.   During the **Oligocene epoch**, about 35 million years ago, Earth's temperature and climate changed; conifers began to outnumber deciduous trees. The forest thinned, grass became more prevalent, and ***Mesohippus*** appeared. This animal was larger than *Eohippus*. Its teeth had further evolved. It had only three toes on its front feet and was better suited to outrunning its enemies. As swamp gave way to soft ground, toes became less essential. On *Mesohippus*, the lateral supporting toes decreased in size while the middle toe strengthened. The toes now ended in little hooves that still had a pad behind them. In both Europe and North America, these browsing horses became extinct about 7 million years ago (see Figure 1-2).

Merychippus.   In the **Miocene epoch**, about 20 million years ago, a totally new type of horse appeared. ***Merychippus*** evolved in North America and adapted to the hard grasses of the plains. This was the beginning of the grazing horse of today. The horse now increased in size to about 35 inches. It was increasingly gregarious and lived in herds. In order to chew the rough, hard grass, *Merychippus* developed complicated grinding teeth quite similar to those of present-day horses. Its lateral toes diminished and no longer reached the ground, while the main toe thickened and hardened for swift travel on the dry

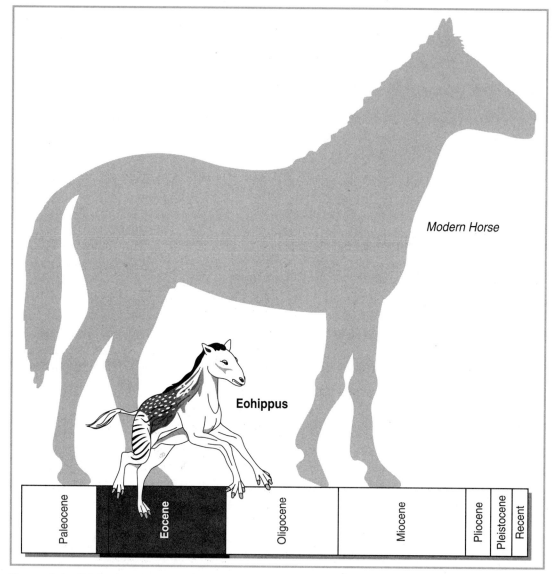

| Paleocene | Eocene | Oligocene | Miocene | Pliocene | Pleistocene | Recent |

Modern Horse

Eohippus

Figure 1-1 *Eohippus*

ground. The feet were without pads and the weight was carried on an enlarged single hoof on the central toe (see Figure 1-3).

Pliohippus. At the beginning of the **Pliocene epoch**, about 5 million years ago, one branch of horses crossed into Asia, quickly multiplied, and spread to Europe. Meanwhile, in North America, the horse developed into the final model. ***Pliohippus*** was the first true **monodactyl** (one-toed animal) of evolu-

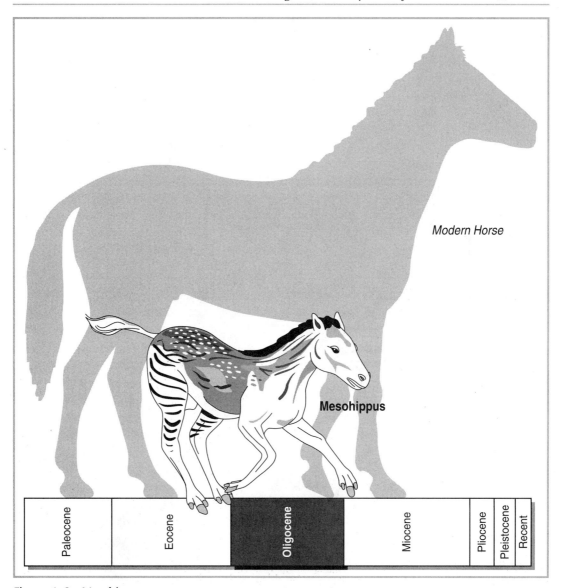

Figure 1-2 *Mesohippus*

tionary history. *Pliohippus* had increasing need for speed to outrun its enemies, so the hoof evolved from the continued over-development of the middle toe. Its teeth and limbs were the nearest approach to our present-day horses (see Figure 1-4). This horse now spread into South America, Asia, Europe, and Africa. The last two million years, from the present to the **Pleistocene epoch**, represent the final evolutionary stage of *Equus*. About eight thousand years ago, *Equus* became extinct in the Western Hemisphere and was not to return

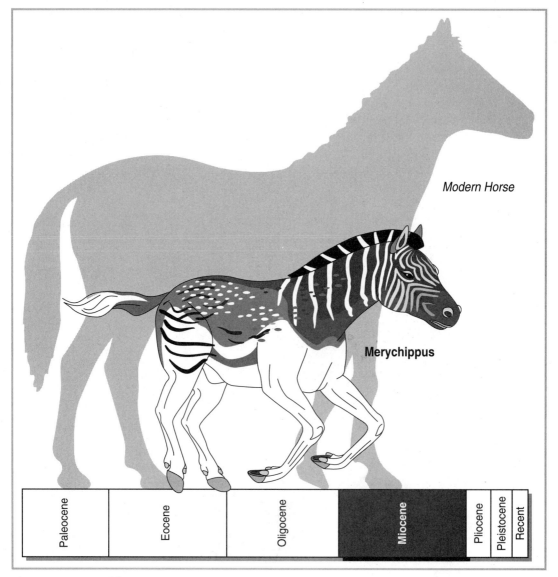

Figure 1-3 *Merychippus*

until the Spanish brought horses to the New World in the 1400s (see Figures 1-5 and 1-6).

How Evolution Works

Common evolutionary trends are not seen in all of the horse lines. On the whole, horses got progressively larger, but some (***Archeohippus***, ***Calippus***) then got smaller again. Many evolved complex facial pits, only to have some of

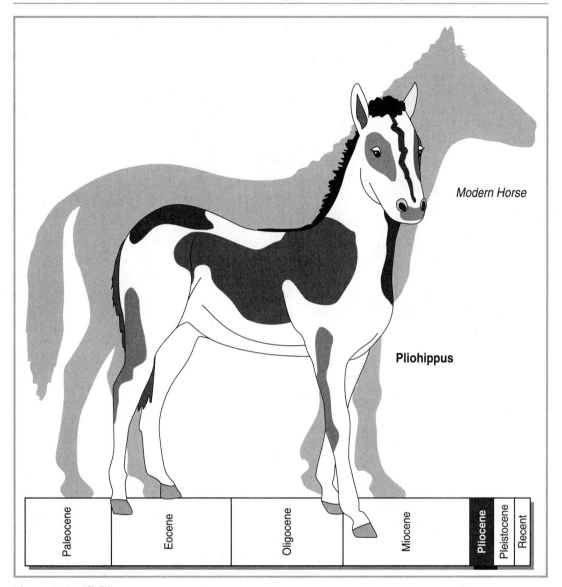

Modern Horse

Pliohippus

| Paleocene | Eocene | Oligocene | Miocene | Pliocene | Pleistocene | Recent |

Figure 1-4   *Pliohippus*

their descendants lose them again. Most of the recent (5 to 10 million years) horses were three-toed, not one-toed. One-toed animals prevailed only because all the three-toed lines became extinct.

Additionally, these traits did not necessarily evolve together, or at a steady rate. The various structural characteristics each evolved in an interrupted series of changes. For example, throughout the Eocene, feet changed little, and only the teeth evolved. Throughout the Miocene however, both feet

Figure 1-5　Horse bones found in the Hagerman, Idaho fossil beds. These have been shown at the Smithsonian Institute. *(Photo courtesy Dr. Greg McDonald)*

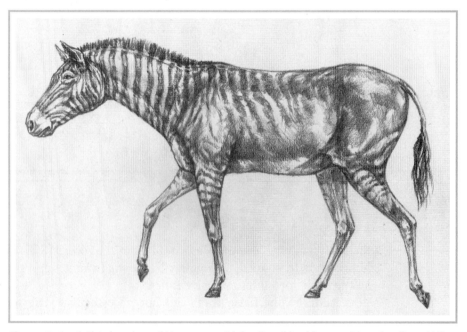

Figure 1-6　Artist drawing of Hagerman, Idaho fossil bed horse. *(Drawing from U.S. Department of Interior, National Park Service, Hagerman Fossil Beds National Monument)*

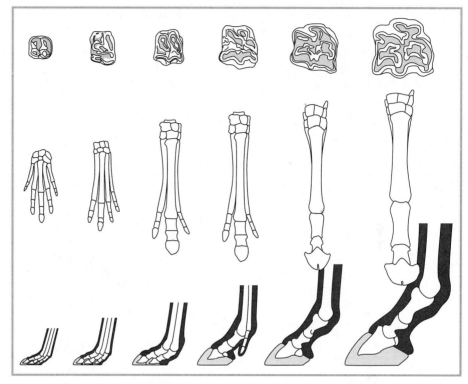

Figure 1-7 Evolution of the horse tooth and hoof

and teeth evolved rapidly. Rates of evolution depended on the ecological pressures facing the species.

Evolving along with the modern horse were other species of *Equus*, such as the ass, or donkey, the onager, and the various zebras.

Tracing a line of descent from *Eohippus* to *Equus*, fossils reveal four trends (see Figure 1-7):

- reduction in the number of toes
- increase in the size of the cheek teeth
- lengthening of the face
- increase in body size

Przewalski's Horse

The oldest species of horse still in existence is the wild Przewalski's horse (*Equus przewalski*). Ironically, it was not discovered until 1879 when the Russian Capt. Nikolai Mikailovich Przewalski sighted it in the remote valleys of Mongolia. The modern Przewalski's horse resembles many of the animals appearing in the cave paintings at Lascaux, France. It typically stands twelve to fourteen hands high, has a dun (yellowish) coloring, a light-colored muzzle, a short, upstanding mane, a dark streak along its back, and dark legs (see Figure 1-8). In its native Mongolia it feeds on tamarisk, feather grass, and the white roots of rhubarb. The former Soviet Union established a refuge for the horse in the late 1970s to ensure both its continued existence and its freedom.

Although held in captivity in many zoos around the world, Przewalski's horse has never been effectively tamed, and in fact can be vicious if threatened.

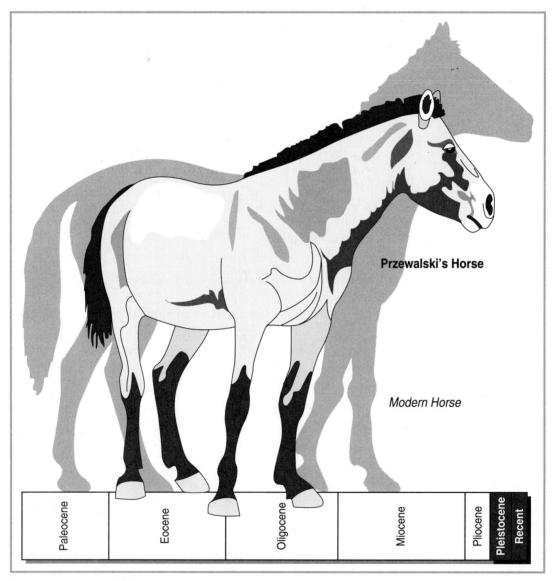

Przewalski's Horse

Modern Horse

Paleocene	Eocene	Oligocene	Miocene	Pliocene	Pleistocene	Recent

Figure 1-8 Przewalski's horse

The Hunted Horse

Our first insight into human relationship with the horse comes from Stone Age paintings on the walls of caves in Western Europe. Although frequently showing the horse as an object of prey, these prehistoric cave paintings also reveal the majesty the artists saw in the horse and show the great effort they expended to capture its beauty.

But while Cro-Magnon man admired the horse, he primarily considered it an important source of food. Lacking the speed to pursue it or a way to kill it from a distance, prehistoric hunters learned to drive the elusive prey to its death. Evidence of this is seen at Salutre in France, where the bones of some ten thousand horses dating from that period have been found at the base of a cliff.

DOMESTICATION OF THE HORSE

For perhaps half a million years, this was mankind's only contact with the horse—as a hunter in search of food. Only in the relatively recent past—between 4,000 and 3,000 B.C.—did humans begin to domesticate horses on the steppes north of the Black Sea. Oxen were already being yoked in draft in Mesopotamia and by the early third millennium B.C., asses and **onagers** (wild donkeys of central Asia) had been similarly harnessed there (see Figure 1-9). When the horse was introduced to the region in numbers during the early second millennium B.C., a tradition of driving was already well established.

Some of the history of the domestic horse is rather obscure, but our knowledge of the donkey is more certain. Artifacts reveal that donkeys were first domesticated in Egypt as early as 3,400 B.C., and by 1,000 B.C. had spread from Egypt into Asia.

From Cow of the Steppes to Pack Animal

Excavations show that the horse became progressively more important in the economy of the people of the steppes. Still considered a source of food, tame horses were evidently first kept for meat and possibly their milk. Later, as these now-docile animals began to carry the goods of nomadic tribes, their importance grew. The horse was now a worker—not just a meal on the hoof.

Role of the Wheel

Oxen were yoked to the pole of a plow probably early in the fourth millennium B.C., in the Near East. Toward the end of the millennium, they were yoked to sledges that were eventually mounted on rollers and then on wheels. Vehicles with disk wheels appeared near the beginning of the third millennium B.C., drawn by oxen, onagers, or donkey **hybrids**. The four-wheeled war wagon

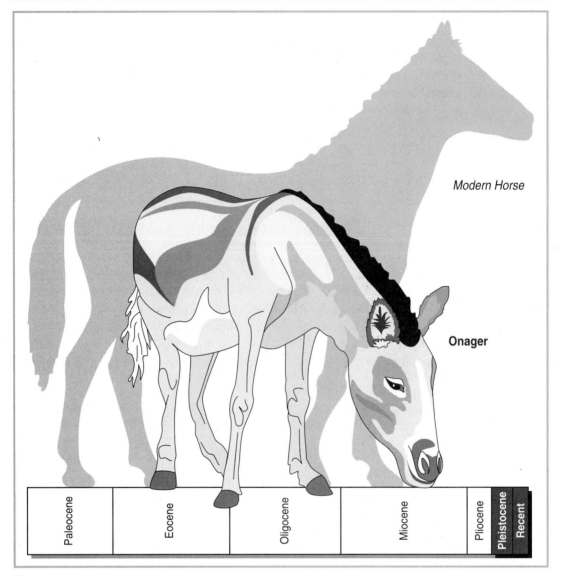

Figures in the image: Modern Horse, Onager, and timeline labels: Paleocene, Eocene, Oligocene, Miocene, Pliocene, Pleistocene, Recent

Figure 1-9 Onager

from Ur, in southern Mesopotamia, of about 2,500 B.C., was pulled by a yoked team of four donkeys with nose-ring control.

Probably imported from the steppes of southern Asia, the horse first appeared as a domesticated draft animal in the Near East between 3,000 and 2,000 B.C., and because of its speed, soon became the favorite draft animal. By the time horses were numerous in the region, a light chariot with spoke wheels had been developed for war and hunting. Yoked to it, the horse rapidly superseded its other relatives in harness for these purposes.

Learning to Control and Harness Horses

The ability to control the horse and effectively connect it to useful implements depended on developing appropriate draft systems that would allow the horse to work at its best.

The Yoke. Equine draft was preceded by and modeled after a draft system developed for oxen, that was not well-adapted to equine anatomy. Horses were harnessed in pairs, with each horse on either side of a pole and under a yoke. The yoke was secured by a strap around the throat that tended to press on the horse's windpipe. By the fifteenth century B.C. in Egypt, the yoke saddle was introduced. This was a wishbone-shaped wooden object, lashed to the yoke by its "handle" with its "legs" lying along the horse's shoulders. This design took considerable pressure off the horse's throat and allowed it to breathe more easily. The yoke saddles rested on pads and their ends were joined by crescent-shaped straps that went across the lower part of the horse's throat.

Early Bits. All-metal bits were first implemented in the Near East around 1,500 B.C. The increased use of the light chariot in warfare called for stronger and more effective control of the teams. Two types of snaffle bits appeared almost simultaneously—the plain bar snaffle and the jointed bit. Both types usually had studs on the inner surfaces of the cheekpieces to enforce directional control when one rein was pulled.

Learning to Ride

Driving came before riding in the Near East. Large chariot forces required schooled, disciplined, and highly conditioned horses. Riding was still pursued only in a casual fashion. Disciplined cavalry mounts, trained to function with their riders in formation, came only after 1,000 B.C. The state of riding before 1,000 B.C. is depicted as a scantily clad, unarmed rider, probably a groom or messenger.

At first, riders may have controlled their mounts with no more than a rope around the jaw or some sort of **hackamore**. Antler cheekpieces, which served as toggles to soft mouthpieces of rope, rawhide, or sinew, have been found at sites of the earliest domesticated horse on the steppes north of the Black Sea.

The Scythians

Emerging from a collection of scattered southern Russian steppe tribes, the Scythians unified as a group of nomadic horsemen with common customs and interests about 800 B.C. During the seventh century B.C., they invaded the Near East, riding as far south as Palestine, and occupied part of northern Iran for some forty years.

Scythians were primarily archers, skilled at using the powerful composite bow from horseback. One of the techniques they mastered was that of shooting backwards over their horses' croups as they turned away from the enemy.

The Scythian's nomadic way of life, which enabled them to burn and destroy all their property when they retreated, allowed them to survive encounters with two of the greatest invading armies of the time, those of Darius I of Persia (512 B.C.) and Alexander the Great (325 B.C.). All this was made possible by the mobility provided by vast herds of horses.

Scythian horses are the first recorded **geldings**. Horses in the Near East were not castrated at that time.

A Scythian's wealth was counted in horses, and a belief in the continuation of material life after death caused the wealthy to take quantities of horses (in one case 400) with them into the grave.

The Roman Army

During the more than four centuries of its existence, the Roman army changed from an essentially infantry formation to a predominantly cavalry-led force. The change was brought about primarily by the type of enemy it faced on the frontiers. To the east, rivals such as the Persians, who employed all-cavalry armies, had inflicted serious defeats to Roman infantry. The only way to effectively counter these armies was with more and better cavalry forces. The same was essentially true when facing the mounted Germanic tribes to the north and west, and eventually the mounted nomadic tribes of the steppe.

China

Horse drawn war chariots were first used in China during the Shang Dynasty (about 1,450 to 1,050 B.C.). But repeated invasion and devastating plunder by barbarians of the northern steppes and the Hiung-nu or Huns led to the development of a Chinese light cavalry, that provided a more effective defense against the invaders.

Despite completion of the Great Wall in 209 B.C., continued clashes with the Huns prompted China to adopt and refine their enemy's riding technique based on the use of a saddle and the bow and arrow.

Although the Chinese did not use the horse in great numbers until the third century B.C. (well after its use was common in the West), by the seventh century A.D., the T'ang emperors had huge stud farms holding as many as 300,000 horses, with each horse given seven acres of pasture. Paintings from the tenth and eleventh century reveal the Chinese as complete horsemen. Their equipment is rather modern in appearance, and they seem at ease on their mounts.

Europe After the Romans

The fall of the Roman Empire in 476 A.D. began the Middle Ages, a period that lasted some seven hundred years. The early portion of this period is sometimes called the Dark Ages since the glories of the former Roman Empire virtually vanished. In this period, learning and invention stagnated except in a few isolated monasteries. These were times of religious wars and barbarian invasion. The horse became largely a vehicle for battle or the hunt as the Roman roads, which had previously united Europe, fell into disrepair. Travel from one area to another was dangerous due to the hostile relations between kingdoms. Generally, chariots fell from use and the wagon remained a farm vehicle. Despite a decline in the quality of technological innovation in many spheres of life during the Middle Ages, the horse adapted to new roles, particularly in agriculture.

Horses were both expensive to buy, and, compared to oxen and donkeys which were foragers, expensive to keep. They required specialized feed, constant care, and good shelter. The feudal system of the Middle Ages placed the farmer of the land under the control of a lord. Since the lord had the financial means to supply his farmers with horses to work his fields, the Middle Ages saw the horse used on a large scale in agriculture for the first time in history.

In the Middle Ages, hunting deer on horseback became a very popular sport, especially in Norman France. By the time of the Norman conquest of England in 1066, the deer hunt was enjoyed by most noble Norman gentlemen and William the Conquerer brought the sport, with all its rules and traditions, with him to Britain. "Ty a Hillaut," the old Norman phrase used to warn hunters that a deer had been roused, became the "Tally-ho" familiar to the fox hunter of today (see Figure 1-10).

The Renaissance

Between 1450 and 1650, Europe experienced a cultural rebirth. (Renaissance literally means "rebirth.") Interest in the natural laws governing the world and the universe were renewed. Gutenberg's invention of the printing press introduced an age of study and creation. The Renaissance removed the veils of mystery and ignorance that had characterized the Middle Ages.

The events affecting the history of the horse in the Renaissance grew from people's zeal for discovery. The anatomy of the horse became a subject of scientific study, and the training of horses became a disciplined art. The Renaissance enriched the quality of life for humans and horse alike. Vehicle design was advanced and horses assumed a more prominent role in the transportation of goods and people.

Hungary emerged as supreme in the art of carriage making. Anne of Bohemia (circa 1380) made a great impression when she brought coaches from her native land to England. The sophistication of the coach was remarkable for

Figure 1-10 Fox hunt

its time. The body was constructed of a light wood frame with wickerwork attached. The wheels were light and the single trees, which connect the harness to the vehicle, were ingeniously secured to the rear axle. The coach, and other vehicles developed along similar lines during the Renaissance, were dramatically superior to the lumbering carts and wagons of earlier times (see Figure 1-11).

MYTHOLOGY AND THE HORSE

Ancient humans held the horse in awe and placed it among the gods and in their legends. Cultures of the ancient world evolved various mythologies, bodies of legend and belief, that reflected their values, ideals, and visions of the past. Some examples of the horse's place in these stories include:

- Poseidon creates the horse
- Pegasus the wild winged horse tamed by Bellerophon by using a golden bridle
- **Centaur** the magnificent creature who had a body that was half horse and half man (see Figure 1-12)

Figure 1-11 A carriage from the Denver Livestock Show

Figure 1-12 The Centaur—half man and half horse

Figure 1-13 A unicorn

- Epona, an ancient Gaul goddess of horses, who lovingly protected the horse and stable and also kept watch over the grooms and carters

- Horses in ancient warfare in Homer's *The Iliad*

- Trojan Horse, the wooden horse that got the Greeks inside the walls of Troy

- Unicorn, an animal with the legs of a buck, the tail of a lion, the head and body of a horse, and a single horn in the middle of its forehead (see Figure 1-13)

- Horse-drawn Chariot of the Sun used by the ancient gods of India

- Four Horsemen of the Apocalypse in the *Bible*

HISTORY OF HORSES AND MULES IN THE UNITED STATES

When the first Spanish conquistadors came to the Americas in the early 1500s, they considered themselves the explorers and colonists of a vast new world. For the many horses the Spanish brought with them, the voyage to the Americas was really a homecoming. Although the horse is believed to have originated in North America, none survived prehistoric times except those that emigrated to Asia over an ancient land bridge near modern-day Alaska.

In 1519 Coronado set out for North America with 150 horsemen, followed by DeSoto's expedition with 237 horses in 1539. By 1547, Antoni de Mendoza, the first governor of New Spain (Mexico), had eleven haciendas and over 1,500 horses. The Spanish colonization depended on horses, and they recognized the tactical value of the animal. Indians were forbidden to ride horses, unless they had the specific permission of their masters. When unconquered Indians encountered stray horses, they killed them for food rather than learn to ride them. Spanish horses eventually found their way back to their prehistoric home in the wilds of the Americas.

Colonization and Settlement

With colonization, towns and cities began showing signs of the growing importance of the horse—hitching posts, mounting blocks, water troughs, stables, and carriage houses. By the late 1800s the horse was a central element in urban life. It hauled goods, pulled omnibuses and cabs, and moved people about in carriages. The increasing prosperity of the urban population created a huge new market for horses. In turn, carriage makers, wheelwrights, harness makers, and feed merchants prospered because of the horse's increased prominence in everyday life.

The exploration and settlement of new frontier land in America also created an enormous need for the horse. Settlers saw the horse as a means of expansion and as power for taming the wilderness and cultivating the virgin soil. The sluggish but easily maintained ox had previously fulfilled the needs of the farmer. But the versatility of the horse made it an even more valuable asset to the farmer of the 1800s. Horses plowed fields, pulled wagons and carriages, and became an essential part of the rural economy. The horse population grew rapidly during the 1800s. In 1867, the rural horse population in America was estimated at nearly eight million, while the number of farm workers was well under seven million.

Mules. Changes in farm machinery also increased the demand for **mules**. Beginning in the 1830s, farm machinery such as mowing, reaping, and threshing machines, John Deere's steel plow, the corn planter, and the two-horse cultivator were invented. These inventions called for the heavier and stronger horse or mule. Mules were especially valued in coal mines where the poor working conditions defeated many horses. Typical coal mules hauled between 60 and 100 tons of coal a day in two to five cars from the mine face to the parting. Mules also had a long career in the U.S. Army. The government used them from 1775 until 1957 to transport supplies both in packs and under harness.

Commercial Uses

The draft horse played a significant role in the growth of urban America. From the end of the Civil War to the beginning of World War I, the United States was in transition from an agrarian to an urban society. As cities grew, so did the need for mass transportation. Before reasonably priced and effective horse-powered mass transit systems, most people were forced to live within walking distance of their work.

The development of horse-powered mass transit systems allowed the cities to expand into the new suburbs. In 1880, horse-car lines were operating in every city in the United States with a population of 50,000. By 1886, over 100,000 horses and mules were in use on more than 500 street railways in more than 300 American cities (see Figure 1-14).

As cities grew, so did the demand for powerful horses. Heavy horses hauled cargo unloaded at city terminals by railroads, steamships, and canal boats, and they distributed the goods produced in urban factories. The vans used for cartage were fifteen to twenty feet long and often carried loads of over ten tons. Strength and endurance were the prime considerations in selecting the horses to haul the goods. Some businesses used brightly painted delivery wagons pulled by handsomely matched teams to advertise their products. Breweries, meat packers, and dairies were particularly fond of this practice, assembling elaborate wagons, powered by four or six regally harnessed draft horses, which, by 1890, averaged 2,000 pounds apiece.

These hitches soon began to compete in the show ring, especially at the annual International Livestock Show held at the Chicago Stock Yards. Their legacy can be seen today in the famous Budweiser Clydesdales and other show hitches performing in American show rings (see Figure 1-15).

Fire Protection

The horse became an essential part of urban fire protection during the 1850s. Since the total destruction of Jamestown in 1608, one of the greatest dangers faced by urban Americans was fire. As cities grew, the magnitude of destruction from urban fires became even greater. With the introduction of heavier and more efficient steam pumpers and ladder trucks in the 1850s, horses became an integral part of urban fire departments. Speed was essential. Intricate systems were developed to hasten the harnessing of the fire horse teams. When an alarm sounded, stall doors were automatically opened and the horses were moved below their suspended harness. The harness, complete with hinged collars, was then dropped onto their backs and quickly secured by the driver. With a good crew, the entire operation could be completed in around two or three minutes. Fire horses were most always draft crosses selected for speed and strength. In New York City, the first fire horse was purchased in 1832. By 1906, their number had grown to nearly 1,500.

Figure 1-14 Horses in the big city in the late 1800s. *(USDA, Photography Division, Office of Public Affairs)*

Figure 1-15 One wagon big hitch in Ketchum, Idaho. *(Photo by Terrell Williams, Wendell, ID)*

Transportation

When roads became worn between town and farm, a lighter wagon became desirable. The "pleasure wagon" had a seat on two hickory springs, while other wagons had none at all. Even so, riding in the pleasure wagon would have been far from a pleasure. For farmers who could afford to have such a vehicle and a pair of horses, the pleasure wagon hauled the crop to town and took the family to church on Sunday.

Some folks drove a one-horse shay on local errands. The shay got its name from the Yankee rendering of the French word *chaise* meaning chair. It had two wheels, a fixed top, and a body hung on straps. The shay's had two large strong wheels could absorb some of the roughness of a road by spanning the holes, ruts, and bumps.

Long-distance travel by public stage was decidedly uncomfortable. Two big horses normally pulled the stage wagon, but in bad weather four horses were needed. Nine to twelve passengers sat three abreast on backless board seats. The wagon had roll-down curtains in case of rain, but no springs to soften the ride. Schedules were often not convenient either. The stage for Lancaster, Pennsylvania, left Philadelphia at 3:30 A.M. and didn't stop at an inn for breakfast until 9:00.

Often we need to be reminded of how far we have progressed. The following excerpt is from the *Elements of Agriculture*, published in 1914. Supposedly, it was a high school textbook. (Keep in mind, this was written over eighty years ago.)

In 1830, it required an average of three hours of time for each bushel of wheat grown; it now requires about one hour. In 1850 it took four and one-half hours to grow, harvest, and shell a bushel of corn; it now requires about one hour. This saving of time has been due to the substitution of machinery drawn by horses for human labor. According to the last census (1900), we had twenty-one million horses in the United States, or one horse to each four persons. In Great Britain there is one horse to twenty-six persons; in France, one to ten; in Germany, one to thirteen.

In America, we have gone the farthest in the substitution of brute force for human energy. Human labor is the most expensive of all labor, even if the person be a slave. One horse, properly directed can do the work of ten men, while his "board and room" on the farm cost about half as much as that of one man. The farm boy who drives a good four-horse team to a gang-plow is doing as much work as if the horses were replaced by forty men. In the West, the farmer is no longer content to use a single team in his farm operations when it is possible to use larger numbers. The four-horse gang-plow and four-horse harrow have rapidly replaced the two-horse machines. There are many parts of the country in which similar methods can be used. In this way one man can do the work that would require many men under European conditions. Because we make our labor count for so much, we are able to make farming an attractive business, rather than a peasant's drudgery.

The extensive use of horses has had a great influence on our national character and history. The boy who trains a colt gets a lot of training himself. It makes a man expand as he learns to manage a spirited horse.

How time changes the world we live in. Today horses have been replaced with tractors and only 1.8 minutes are needed to produce one bushel of wheat. The number of horses in the United States is only six to eight million compared to the twenty-one million at the time this book was written.

Today, colt training is a valuable experience for girls, boys, women, or men.

Agriculture

Throughout the eighteenth and early nineteenth centuries, horses in the United States were used primarily for riding and pulling light vehicles. Although two draft-type horses, the Conestoga Horse and the Vermont Drafter, were developed in the new nation, both were absorbed into the general horse population by 1800. Oxen were the preferred draft animal on many American farms. They cost half as much as horses, required half the feed, and could be eaten when they died or were no longer useful. Oxen, however, worked only half as fast as horses, their hooves left them virtually useless on frozen winter fields and roads, and physiologically they were unsuitable for pulling the new farm equipment developed in the nineteenth century. The revolution in agricultural technology, westward expansion, and the growth of American cities during the nineteenth century, led to the emergence of the draft horse as America's principal work animal.

In 1862, Congress passed the **Morrill Land Grant Act** that led to the establishment of state agricultural colleges. The first of the nation's veterinary colleges opened at Cornell University in 1868. As farmers became more educated, there was a corresponding improvement in the care, feeding, and breeding of horses.

The revolution in agricultural technology between 1820 and 1870, created a demand for a larger and stronger horse. New and improved farm equipment greatly increased the productivity of the American farmer. With the McCormick reaper, which both cut and tied grains into stocks, one person could do the work of thirty. New steel plows, double-width harrows and seed drills, mowers, binders, combines, and threshers decreased the need for manpower, but increased the demand for horsepower. Toward the end of the century, the typical Midwestern wheat farm had ten horses, each of which worked an average of 600 hours per year. During harvest, it was not unusual to see giant combines pulled by teams of over forty draft horses (see Figure 1-16).

With the use of new equipment and fertilizers, wheat yields increased seven times between 1850 and 1900. Better rail and steamship transportation opened new markets in the United States and in Europe. America was coming of age as a world agricultural power.

The acreage one family could cultivate increased as technology and equipment improved. The average American farm in 1790 was 100 acres. This figure more than doubled over the next sixty years. By 1910, 500-acre wheat farms were not uncommon. While oxen and light horses had been adequate for tilling the long-worked fields of Europe and the eastern United States, a stronger power source was needed to work the sticky, virgin soil of the American prairie.

As a result, the first European draft horses were imported to the United States in the late 1830s. Farm labor became scarce due to westward migration

Figure 1-16 Combine with a forty-horse hitch. *(USDA, Photography Division, Office of Public Affairs)*

and casualties from the Civil War. This created an even greater demand for the new farm equipment and the draft horses to power them. By 1900, there were over 27,000 purebred Belgians, Clydesdales, Percherons, shires, and Suffolk punches in the United States. Although the purebred draft stock was seldom used in the field, the infusion of their blood resulted in an increase of the average horse size to between 1,200 and 1,500 pounds by 1900.

The offspring of these heavy farm horses soon found additional uses as the nation moved toward the pacific. The railroads employed thousands of draft crosses, working side-by-side with mules and oxen, to carry ties, rails, and supplies to the railheads, and to haul dirt and rock from the excavation of mountain tunnels. Many of the western stagecoach lines used up to six draft crosses to haul mail and passengers over dangerous, rough roads. By century's end, large grain farms, comparable to those in the Midwest, had been established on the western prairies. These farms, like their predecessors, relied on draft horses to power their plows, threshers, and combines.

Recreation, Sports, and Shows
One of the most celebrated equestrian events in the United States is the National Horse Show, held each November at Madison Square Garden in New York City. The National Horse Show began in 1883, and was soon a major event of the social season.

Team events were begun in 1911. Although interrupted by World War I, the show again thrived in the 1920s and 1930s. In the early years of the show, military teams dominated the jumping competition, but civilians now constitute the majority of entrants. The National Horse Show includes international team jumping, competition for national hunters and jumpers, saddle seat equitation, and harness competition.

Another celebrated event that happens every four years is the Olympic Games. Equestrian events include jumping, dressage, vaulting, and endurance. This is an international event where riders and horses both benefit from sharing information and experiences with other countries (see Figure 1-17).

Rodeo.   The informal sport of cowboys became an organized event. **Rodeo** (Spanish for cattle ring) began as an amusement among cowboys who had reached the end of the long cattle-drive and had to remain with their herds until they were sold. Given a few days of freedom, it was not long before an empty cattle pen was appropriated and one cowhand challenged another to a calf-roping contest, or dared him to ride "the meanest horse between here and the Rio Grande." The popularity of these informal sports grew until the first rodeo with paid attendance was held in Prescott, Arizona, on July 4, 1886. At the turn of the century, rodeos combined with the popular Wild West Show. These events became extravaganzas, including wagon races, bull riding, and

Figure 1-17   Michael Matz on Rhum at the recent Olympics. *(Courtesy of Carol Drake)*

steer wrestling. The Wild West Show soon fell from popularity, but its influences remained in the rodeo, which steadily grew in popularity throughout the western United States and Canada. In more elaborate rodeos, even the cooks got into the act by racing their chuckwagons. Today major events in a rodeo include bareback bronco riding, steer roping, calf roping, bull riding, team roping, and barrel racing (see Figure 1-18).

Racing. Kentucky has long been recognized as a horse-breeding region. But back when Kentucky was only a remote and unknown woodland, the chief horse-breeding region of the United States was Rhode Island. At one time Rhode Island had farms with as many as 1,000 horses, predominantly Narragansett Pacers. From these Rhode Island farms, horses were shipped to all of the sea-coast colonies, as well as to the islands of the Caribbean, for use on the plantations. Rhode Island was the only New England colony that allowed horse racing, and a one-mile track was maintained at Sandy Neck Beach, South Kingston. As always, competition was the key to improved breeding and Rhode Island gathered the best stock from neighboring areas to upgrade its horses.

Many towns and cities in America have streets called "Race Street." Such streets gained their names from the habit of running horse races on them. In

Figure 1-18 American Rodeo—barrel racing. *(Courtesy of Appaloosa Horse Club, Inc, ID)*

1674, the citizens of Plymouth evidently grew tired or frightened of the races in their village, and created an ordinance forbidding racing. However, the sting of the fine or the humiliation of punishment in the stocks did not seem to discourage colonial racing enthusiasts. About a century later, Connecticut enacted a law that demanded the forfeit of a person's horse, in addition to a fine of forty shillings, for racing in the streets.

Until the mid-nineteenth century, horse racing was the principal form of organized sport in America. In colonial America, town rivalry was centered around horse racing. Often competitors and spectators travelled far to early quarter-mile race paths in the woods and placed considerable wagers on their town's horse. Typical wagers included money, tobacco, slaves, and property. Tempers frequently ran high if a start was questioned or if one rider allegedly interfered with another. Thus, the official who started the race was selected as much for his brawn and ability to defend himself as for his honesty. The race was generally started by firing a pistol, sounding a trumpet, or hitting a drum. Even after land became available for long circular tracks, the sport of quarter-mile racing remained a popular American institution.

The first Kentucky Derby was run on May 17, 1875. The Derby was sponsored by the Louisville Jockey Club and Driving Association, which owned the track now known as Churchill Downs. Col. M. Lewis Clark founded the Association after he had visited Europe to study their farms and racing regulations. Clark was particularly impressed by the English system. He called the Kentucky race a "derby" after the Epsom Derby, which was first run in 1780 under the sponsorship of the Earl of Derby.

Today, the Kentucky Derby is the most prestigious race for Thoroughbreds in the United States and the first race in the Triple Crown for three-year-olds. Each year in May, horse enthusiasts look to Churchill Downs in Louisville to see who will become the year's contender for the Triple Crown. The Kentucky Derby is the oldest continuously-run race in America. For many horse owners, winning the Kentucky Derby brings not only prize money, but entrance into an honored tradition of superior racing (see Figure 1-19).

Mining

Ponies are frequently seen grazing on the farms near coal mines. Most of these rugged little ponies are descendants of pit ponies that were used to haul coal from mines as early as the 1600s. Breeds such as the hearty Shetland ponies from northern Scotland were imported in great numbers to work in the first mines of Pennsylvania, Ohio, West Virginia, and Kentucky.

Life in the coal mines has never been easy for either man or beast. But few equines have received better care and respect than the pit pony. In some larger mines, particularly in Europe, a pony would be bred, born, and put to work without ever having seen the light of the sun.

Figure 1-19   Advertisements for horse racing appear in unexpected places.

Old West

The original cowboys were Native Americans, enslaved by Spanish conquerors who put them to work tending herds on their vast rancheros in Mexico. They wore broad sombreros to protect them from the burning sun and chaparjos (chaps) to protect their legs against cactus and mesquite. Their saddles were fashioned after the medieval Spanish saddle.

Men who came from Kentucky and Tennessee to settle Texas were the first of the American cowboys. The growing population of the eastern United States in the mid-1800s created a market for beef. When construction of the western railroad provided the means of carrying the beef to the East, the cattle business began to flourish. Cattlemen raised stock and drove them great distances to the railheads. The men who tended and drove the cattle came to be known as cowboys and were as ethnically diverse as the growing nation. With the increasing demand for beef, the cowboy's domain spread northward to Canada, and westward to the Rockies. The cowboy's life was often lonely and sometimes violent. His manners, dress, language, and amusements remain a symbol of the rugged independence and determination that characterized the American West (see Figure 1-20).

The ranges were not fenced, and the cattle had to be watched constantly for fear they might stampede during a thunderstorm or when threatened by a predator. Regular chores included cutting out calves for branding and, in the

Figure 1-20 The legendary cowboy

earlier days of the range, fighting off the Native Americans who were protecting their hunting grounds.

In the fall of each year the cattle were rounded up in preparation for the drive to market. The riding and roping skills of the cowboy and the agility and cow-sense of his horse were especially important in the round-up.

THE TWENTIETH CENTURY

By the turn of the century, at least half of the 13 million horses in the United States carried between 10 percent and 50 percent draft horse blood. More than three million of these were in use in nonfarm capacities by 1910. With the continued growth of heavy industry, and increased European immigration, American cities were experiencing unprecedented growth. New interest in public health, rising real estate values, and improvements in electric- and gasoline-powered alternatives to horse power combined to mark the rapid decline of the horse's significance in the city.

Decline of Draft Horses

Within a decade, the horse was replaced in public transportation by motorized taxies, and electric streetcars and subways. Large new gasoline-powered trucks had a similar impact on the transportation of goods. The new trucks were three times faster (ten miles an hour) than horse powered transportation, took less room to store, and eliminated the problem of manure disposal. One of the last urban uses of the horse to succumb to mechanization was the horse-drawn hearse, which continued to be utilized into the 1930s.

The market for heavy horses in agriculture went into a steady decline after World War I. The reduction in the number of domestic draft horses, an increased demand for American grain exports, and improvements in gasoline-powered tractors combined to hasten the replacement of draft horses by machines. This was especially true of pure-bred draft stock. In 1920, there were 95,000 registered draft horses in America. By 1945, this figure had dropped to under 2,000.

Particularly hard hit were the Clydesdale and the shire. Both breeds had been used primarily in the city, and were affected earlier than other draft breeds. The heavy feathering on the feet of the shire and Clydesdale was considered a maintenance problem on the farm, therefore diminishing their popularity. What remained of the draft horse market was centered primarily on the farms of the Midwest. The American farmer looked for a smaller, more economical animal. Belgian breeders responded by breeding a more compact horse, and by 1937, the Belgian was the most numerous draft breed in the United States (see Figure 1-21).

By the early 1950s, registrations for all draft breeds had dropped dramatically and many breeders went out of business. The numbers of shires and Suffolks dropped so low that in 1985 they were listed as rare by the American Minor Breeds Conservancy.

Figure 1-21 Transportation being provided at an Agricultural Equipment Show. *(Photo courtesy Robert Lowder, Twin Falls, ID)*

Personal transportation in the early 1900s also saw the transition from horsepower to gasoline-engine power. Since many families called their faithful carriage horse "Lizzie," their first car was often dubbed "Tin Lizzie." At first the car was only used in good weather, and then just for a Sunday outing. The tin Lizzie was apt to get stuck or to boil over. But soon, as roads and engines improved, it was the auto that was used every day, and the horse that was reserved for the Sunday drive.

Horses in the Military

The entry of the United States into World War I in 1917 tipped the balance in favor of an Allied victory. Long before the United States sent its men into the struggle, it had sent another resource—its horses. World War I was the twilight of the cavalry. Except for limited skirmishes in the Middle East and on the Western front, the cavalry fought mostly on foot. In previous wars the cavalry swept across a battlefield to surprise an enemy force. But now, tangles of barbed wire were not easily penetrated, and the machine gun could mow down man and horse alike with chilling efficiency. The days of the offensive horse were ended.

Black Jack

The profound grief of Americans at the death of President John F. Kennedy was accentuated by the sight of Black Jack, the riderless horse with boots reversed in the stirrups, a symbol of a fallen hero. Black Jack was the last horse issued to the Army by the quartermaster, and the last to carry the "U.S." brand common to all army horses. Like so many thousands of army horses, his breeding was unknown. He was foaled on January 19, 1947.

Black Jack was sent to the Third Infantry (The Old Guard) from Fort Reno, Oklahoma, in 1953. He was named after Gen. John J. "Black Jack" Pershing, Supreme Commander of the American Expeditionary Force in World War I. Black Jack served in ceremonial functions, participating in the funerals of Presidents Hoover, Kennedy, and Johnson, General Douglas MacArthur, and thousands of others in Arlington National Cemetery. Black Jack was semi-retired on June 1, 1973, and died February 6, 1976, at the age of 29. His ashes were placed in an urn at his monument at Fort Meyer, Virginia.

The war used horses in great numbers for noncavalry purposes also. About six million horses served, and substantial numbers of them were killed. The death of these millions of horses depleted the world's equine population. By 1914, the British had only 20,000 horses and the United States was called upon to supply the allied forces with remounts. In the four years of the war, the United States exported nearly a million horses to Europe and when the American Expeditionary Force entered the war, it took with it an additional 182,000 horses. Of these, 60,000 were killed and only a scant 200 returned to the United States.

In one year, British veterinary hospitals treated 120,000 horses for wounds or diseases. Like human combatants, horses required ambulances and field veterinary hospitals. The motorized horse van was first used as an equine ambulance on the Western Front.

Movies and Entertainment

Some of the best-loved motion picture and television stars have been horses. Beginning with the *Great Train Robbery* (1903), the West became one of film's dominant themes and depended inevitably on the horse. The horses of the screen, and later television, became as familiar as the heroes who rode them to fame and fortune. They included:

- Tom Mix and Tony
- Gene Autry and Champion
- Roy Rogers and Trigger
- The Lone Ranger and Silver
- Tonto and Scout
- Hopalong Cassidy and Topper

Motion pictures and television also produced individual equine stars, such as:

- Francis (the talking mule)
- Fury
- Flicka
- Black Beauty
- Mr. Ed (the talking horse)

Like their human counterparts, horses in the movie world can be divided into three classes: the stars, the stunt horses, and the extras. The brilliance of a particular horse is based more on training than on breed or pedigree. Both stars and stunt horses receive systematic schooling so that they will respond to their rider's or trainer's commands. They must be of even temperament since a film set is a mix of equipment and bright lights and unfamiliar sounds and faces.

Horses on the set must respond to the trainer's visual commands since verbal commands interfere with the recorded sound. Stunt horses must have the skill and courage to run over cliffs or crash to the ground from a full gallop. Stunt horses are carefully watched by members of humane societies to ensure their safety. What appears to be a bone-shattering crash on rocky ground is, in reality, a well-rehearsed fall on soft mattresses covered with plowed earth. Whether star or extra, the talented movie horse has a special place in the equine world.

Circus Horses. The circus is an exciting tradition in which the horse has played a prominent role. In the early days, the circus parade announced the coming show. All the horses and rolling stock paraded through the village streets to advertise the animals and performers on the bill.

In the early 1900s, the Barnum and Bailey Circus used 750 horses in draft and performance. Ringling Brothers had 650 horses at the turn of the century. The circus was moved almost exclusively by horses, first from town to town, and later, to and from railroad yards. The dappled gray Percheron was one of the trademarks of the Ringling Brothers Circus. By 1938, the circus was mechanized, although horses remain popular performers.

SUMMARY

Through a series of evolutionary stages involving millions of years, *Eohippus*, the small, primitive ancestor of the horse evolved. After the horse evolved into what we know today, humans began interacting with it. Humans hunted the horse and eventually different civilizations learned to use the horse for work and transportation. The horse also became and remained an important role in the history of the United States.

The 1800s enjoyed an unprecedented pace of economic growth. As new markets for manufactured goods were opened, the need for horse transportation increased dramatically. As a result, many horse-drawn vehicles were built by local carriage makers or by large wagon factories. The need for new harnesses and constant repairs on old ones created a demand for skilled harness makers. Wheelwrights, farriers, and blacksmiths were essential to the livelihood of every city and town. Other horse-related crafts and occupations included saddlers, grain farmers, feed merchants, veterinarians, grooms, coachmen, and horse breeders. The Industrial Revolution and the growth in the economy and population created a peak of interdependence between human and horse.

The twentieth century brought radical changes in the world of the horse. With the steady rise of technology, the horse was eclipsed by the internal combustion engine. In 1915, the horse population in America peaked at over 21 million. But immense numbers of horses were sent to the battlefields of Europe during World War I. This export decreased America's horse population, which steadily declined until recently, when the horse entered new arenas as a pleasure rather than work animal. Equine numbers now continue to grow rapidly. Instead of being a beast of burden, the modern horse enjoys a major role in recreation and organized competition. Many breeds of horses are now being revived, and systematic breeding is raising the quality of horses to heights unknown in the past. The future promises a continued increase in the world horse population. Perhaps the ultimate Age of the Horse is yet to come.

REVIEW

Success in any career requires knowledge. Test your knowledge of this chapter by answering these questions or solving these problems.

True or False

1. The first fossil horse, *Eohippus*, was about the size of one of today's donkeys.

2. *Equus asinus* is the scientific name for today's true horse.

3. Przewalski's horse was discovered as a fossil in 1879.

4. Prehistoric humans probably hunted and ate horses.

5. Metal bits were first used about 1810 in England.

Short Answer

6. Arrange the following geologic time periods from oldest to most recent: Pliocene, Oligocene, Paleocene, and Miocene.

7. List four trends documented in the evolution of the horse.

8. List the following ancestors of today's horse in order of their appearance on earth: *Pliohippus, Eohippus, Mesohippus,* and *Merychippus.*

9. Where and when was Przewalski's horse discovered?

10. Name three horses or horse-like animals mentioned in mythology or legend.

11. Name three horses made famous by the movies or television.

12. List the scientific name for the horse and three of its close relatives, like the donkey or zebra.

13. Name two factors that caused an increased demand for draft horses during the history of the United States.

14. Name the zoological kingdom, phylum, class, and order for the horse.

Discussion

15. Discuss how the development of the wheel influenced humans' uses of the horse.

16. How did horse racing start in the United States?

17. Describe how the Romans influenced the use of horses.

18. What twentieth-century events have changed how horses are used in this country?

19. Describe the events during the nineteenth and early twentieth century that caused the number of horses to steadily increase in the United States.

20. What effect did the Middle Ages have on the use of horses?

21. Discuss the concept that Spanish conquistadors actually reintroduced the horse to America.

STUDENT ACTIVITIES

1. Visit a virtual museum on the Internet and learn more about the fossil record and evolution of horses. Report your findings.

2. Read a legend or myth that involves a horse or horse-like animal. Retell the story in your own words.

3. Research and report on plants and other animals that were present and evolving during the same geologic time periods as the horse.

4. Prepare a presentation showing the horse in the art of specific civilizations.

5. Report on the status and populations of feral horses in the United States today, or report on the donkeys in the Grand Canyon.

6. Make a pictorial of the various modes of transportation that horses have pulled over the ages.

7. Choose a civilization and report on its use of horses in warfare, or compare the use of horses in the Civil War, World War I, and World War II.

8. Compare a piece of farm equipment pulled by draft horses to one now propelled by the gas-powered engines used in farming today.

9. Choose a race horse and track its winnings for several months.

10. Horses are still used for work in some police departments, the Forest Service, and search and rescue operations. Find out how these horses are selected and trained.

11. Discover how horses used in the movies are selected and trained. Report your findings.

12. View an old Tom Mix, Gene Autry, Roy Rogers, Lone Ranger, or Mr. Ed video and discuss the training and use of the horse used in the video.

13. Trace the history of horses and their use to transport people.

14. Use the Internet or other resources to find the average selling prices of horses used in rodeo events, racing, dressage, and so on.

15. The history of horses can be used to teach geography, even though many of the names have changed. Obtain a world map and identify the locations discussed in the chapter. Also, the history of horses can be used as a springboard to other history lessons.

ADDITIONAL RESOURCES

Books

Blakely, J. 1981. *Horses and horse sense: the practical science of horse husbandry.* Reston, VI: Reston Publishing Company, Inc.

Davidson, B., and Foster, C. 1994. *The complete book of the horse.* New York, NY: Barnes & Noble Books.

Dossenbach, M., and Dossenbach, H. D. 1994. *The noble horse.* New York, NY: Cresent Books (Random House).

Price, S. D. 1993. *The whole horse catalog.* New York, NY: Fireside.

Silver, C. 1993. *The illustrated guide to horses of the world.* Stamford, CT: Longmeadow Press.

Internet

Kentucky Horse Park, International Museum of the Horse at http://www. horseworld.com on the World Wide Web.

CHAPTER 2

Status and Future of the Horse Industry

The golden age of horses in the United States extended from the 1890s to the mechanization of agriculture. Industries associated with horses were essential parts of the national economy. In 1900, the automobile was still a rich man's toy and the truck and tractor were unknown. Then in 1908, Henry Ford produced a car to sell for $825. The truck, the tractor, and improved roads soon followed. As the automobile, truck, and tractor numbers increased, horse numbers declined.

Today few horses are ever seen on the streets of cities or towns. Horses hitched to a delivery wagon or plow are a novelty. Now the horse is popular for recreation and sport.

OBJECTIVES

After completing this chapter, you should be able to:

- Identify countries or areas with the most horses, donkeys, and mules
- Compare the population of horses, donkeys, and mules in the United States to that in the world
- Describe the rise and fall of the horse population in the United States
- Compare the total worldwide population of horses, donkeys, and mules
- Project changes in the horse population in the United States
- Identify the top ten horse-producing states
- Name four general areas of equine research and give two specific research projects in each
- Identify activities and organizations associated with the United States horse industry
- Discuss the future of the United States horse industry

K E Y   T E R M S

Bucked shins	Genetic	Protozoal
Colic	Gymkhanas	Synovitis
Dressage	Immune system	Tack
Embryo	Influenza	Triple Crown winners
Farrier	Jockeys	Viral
Food and Agriculture	Jumping	
Organization (FAO)	Laminitis	

WORLD DISTRIBUTION OF HORSES, DONKEYS, AND MULES

The majority of the world's horses, donkeys, and mules are not found in the United States. According to statistics maintained by the **Food and Agriculture Organization (FAO)** of the United Nations only about 8 percent of the world's horses and less than one percent (0.13 to 0.19 percent) of the donkeys and mules in the world are located in the United States.

Distribution of Horses

The world population of horses is about 60 million. As Figure 2-1 shows, more than half (52 percent) of the world's horses are found in the countries of Asia and South America. But Mexico, Africa, Europe, and the United States all have significant horse populations.

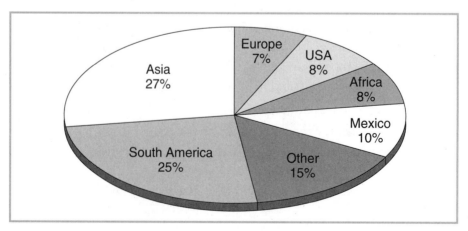

Figure 2-1   Horse populations in the world

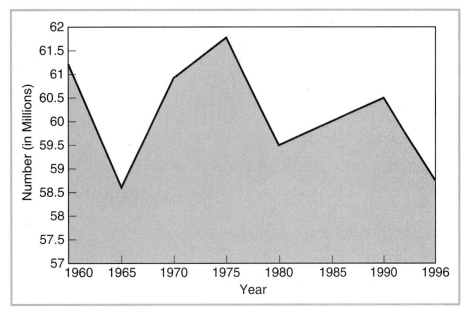

Figure 2-2 History of world horse population

Over the course of years the world population of horses has fluctuated slightly up and down from 61 million since 1960. This is shown in Figure 2-2.

Distribution of Donkeys

As Figure 2-3 indicates, most (80 percent) of the donkeys in the world are found in Africa and Asia. The worldwide population of donkeys is about 43 million. Only about 54,000 donkeys are found in the United States, while Mexico has over 3 million.

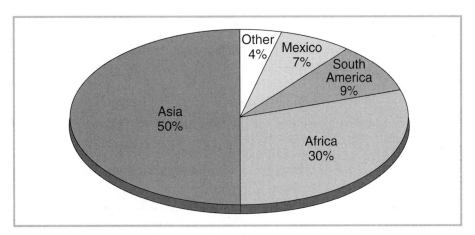

Figure 2-3 Donkey populations in the world

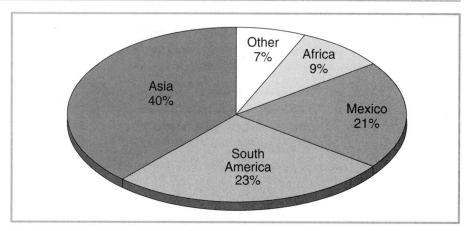

Figure 2-4 Mule populations in the world

Distribution of Mules

The worldwide population of mules is about 15 million. Figure 2-4 shows that most (84 percent) of the world's mules are located in Mexico, South America, and Asia. The United States possesses only about 28,000 mules.

GROWTH AND DECLINE OF UNITED STATES HORSE, DONKEY, AND MULE INDUSTRY

In the United States, the number of horses increased until 1915. At that time statistics showed over 21 million horses in this country. As discussed in Chapter 1, horse production expanded with the growth and development of manufacturing, commerce, and farming. But by 1960, only slightly more than 3 million horses remained in the United States—the lowest number ever recorded. After 1960, as Figure 2-5 shows, horse numbers increased slightly and declined slightly. According to several sources, the United States horse population today is estimated to be between 5 and 8 million.

Horses are found all across the United States, but the top ten horse-producing states are Texas, California, Oklahoma, Colorado, New York, Ohio, Michigan, Pennsylvania, Washington, and Kentucky.

The breeds with the highest individual registrations are the quarter horse, Arabian, and Thoroughbred.

Since 1960, according to statistics from the FAO, the donkey population in the United States steadily increased from 1960 to about 1985. It then dramatically increased and now stands at almost 60,000. Figure 2-6 shows this change.

According to the FAO, in 1961 mule numbers in the United States were about 20,000. They have increased now to about 28,000.

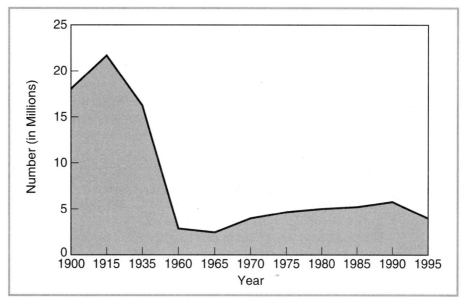

Figure 2-5 History of United States horse population

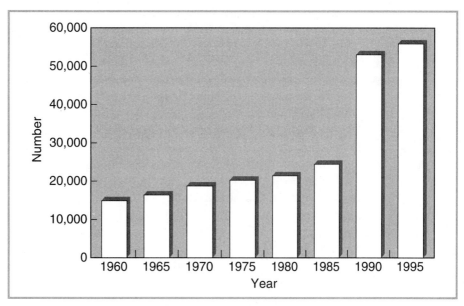

Figure 2-6 Growth of United States donkey population

STATUS OF UNITED STATES HORSE INDUSTRY

For many people, the horse industry is a business and a way of life. Even though the numbers of horses have declined in this century, the United States horse industry contributes about $16 billion annually to the economy. This represents about 16 percent of the gross national product. Of the $16 billion, horse owners spend about $13 billion for the development and maintenance of their horses. Another $3 billion is accounted for by horse-related spectator events, and about $508 million is generated in farm income by the actual sale of horses and mules.

Revenue derived directly from horses includes the actual sale of horses, stud (breeding) fees, races, shows, rodeos, and entertainment. The indirect revenue from horses includes such items as feed, training, veterinary and **farrier** services, transportation, labor, and equipment. Money from all of these revenue sources stimulates the economy.

Sports that involve horses attract more than 110 million spectators each year. Horse racing is a leading spectator sport, and, as with other sports, can lead to status and big money. The attendance at race tracks exceeds 70 million people each year and people wager over $13 billion on the races. Leading **jockeys** can win millions of dollars a year and include great professionals such as Willie Shoemaker, Braulio Baeza, Laffit Pincay, Jr., Chris McCarron, and Angel Cordero, Jr.

Many small race tracks operate throughout the United States and racing is a part of county fairs or other annual events, contributing to about 14,000 racing days each year. The three most famous races are the

- Kentucky Derby at Churchill Downs, started in 1875; a distance of 1.25 miles

- Preakness at Pimlico in Baltimore, started in 1873; a distance of $1\frac{3}{16}$ miles

- Belmont Stakes at Elmont, New York, started in 1867; a distance of 1.5 miles

Three-year-old horses winning all three of these races in one season are called **Triple Crown winners**. These horses, their owners, trainers, and jockeys make their way into sports history. Winners of the Triple Crown include such outstanding horses as War Admiral, Whirlaway, Citation, Secretariat, and Seattle Slew.

About 800 rodeos representing over 2,200 performances are held each year in the United States and the number continues to grow. The ten major rodeos based on prize money are given in Table 2-1.

People ride horses for pleasure more than ever before. More than 27 million people ride horses each year. In the national forests, horseback riding is the third most popular activity, involving about 24 million visitor-days each year. The number of 4-H club horse and pony projects is about double the number of 4-H beef cattle projects. As family pets, horses rank fourth behind the dog, cat, and pet bird.

Table 2-1   Major Rodeos in North America[1]

Location	Month	Location	Month
Las Vegas, NV	December	Reno, NV	June
Houston, TX	February	Calgary, Canada	July
Scottsdale, AZ	October	Denver, CO	January
Pocatello, ID	March	Fort Worth, TX	January
Cheyenne, WY	July	San Antonio, TX	February

1. Las Vegas is by far the biggest rodeo with prize money over $2 million. Prize money at the other rodeos range between $200,000 and $100,000.

Horse shows have also increased in size and number. In the past 20 years their number has more than doubled.

Despite all the mechanization in today's world, some jobs are still better suited to horses. For example, the Forest Service uses horses; remote areas inacessessable to vehicles require horses for packing and travel; and law enforcement agencies have found that mounted patrols are the most effective ways to handle crowds and riots (see Figure 2-7).

Figure 2-7   A police horse at the Denver Livestock Show.

Horses for the Canadian Mounted Police

On May 23, 1873, the Canadian Parliament authorized the establishment of the North West Mounted Police Force. The force's immediate objectives were to stop the liquor trade among Native Americans, halt tribal warfare and attacks on white settlers, collect customs fees, and perform normal police duties. Their vast area of responsibility was roughly composed of the current provinces of Manitoba, Saskatchewan, Alberta, and the Northwest Territories. In the fall of 1874, the first post was established on the banks of Old Man River and was named for the Force's Assistant Commissioner, James Macleod.

One of the first problems of the newly formed North West Mounted Police was to obtain and train horses suitable for the rigors of their western duties. The sleek black horses of today's Royal Canadian Mounted Police (RCMP) bear little resemblance to the tough work animals originally used to patrol the vast Canadian wilderness. Initially two types of horse were selected: the tough western bronco and a primarily Standardbred-type purchased in Ontario. Neither of these horses proved totally satisfactory and in 1875, the NWMP first began breeding their own horses. This proved to be too expensive and was turned over to private ranchers. By 1889 specifications for the force's mounts called for "a horse standing from 14.3 to 15.2 hands, fine clean-cut head, long neck, high chest, broad round quarters with plenty of good flat bone, and strong feet."

Throughout the remainder of the nineteenth century the North West Mounted Police continued to bring law and order to the vast Canadian wilderness. Recognition of their outstanding contributions came in 1904 when King Edward VII proclaimed that the prefix "Royal" be added to their name.

By the outbreak of World War I, the Royal North West Mounted Police had grown to 1,268 officers and men. In 1920 the RNWMP began enforcing all Dominion statutes and its name was changed to the Royal Canadian Mounted Police. Between 1928 and 1950, the RCMP assumed the Provincial Police duties for all Canadian provinces with the exception of Ontario and Quebec. Today's RCMP is essentially a federal police force with responsibilities similar to those of the Federal Bureau of Investigation in the United States.

As the equestrian (as opposed to law-enforcement) duties of the Royal Canadian Mounted Police shifted from practical to ceremonial, a need was identified to provide a different type of mount. The result was the reestablishment in 1943 of the RCMP's breeding program to supply the primarily Thoroughbred-type mares and geldings used in their famous exhibition, the musical ride.

Since the breeding program began, the force has continued to experiment, introducing the blood of Clydesdales, Percherons, Hanoverians, and Trakehners in an effort to develop a heavier boned, well-mannered, Thoroughbred-type horse. Today's RCMP mount should be black in color, approximately 16 hands in height, and weigh between 1,200 and 1,300 pounds.

The Musical Ride. The origins of the famous musical ride can be traced to the intricate Prussian Cavalry drills of the eighteenth century. The first recorded riding exhibition performed by the NWMP was at Ft. Macleod in 1876, and the first performance accompanied by music was presented in 1887. Since 1966 the only RCMP members to receive equestrian training are those associated with the ride. Today's musical ride consists of 32 horses and riders performing numerous intricate figures, always ending in an exciting charge.

FUTURE OF THE UNITED STATES HORSE INDUSTRY

Even though horse numbers in the United States may never again match those they reached at the beginning of the twentieth century, horses have never been so popular. More people are enjoying a greater variety of equestrian activities than ever before. For example, the art of coachman, almost lost, has made an exciting return to competitive performance trials, and side-saddle riding, showing, hunting, **jumping**, and **dressage** all are attracting more devotees. Riding has never enjoyed so much popularity and this popularity continues to grow among all ages.

A short list of some of the myriad equestrian activities provides some idea of the popularity, diversity, and promising future of the United States horse industry.

Popular Equestrian Activities

- Horse shows
  - Hunter division
  - Jumper division
  - Saddle-horse division
  - Harness division
  - Western division
  - Equitation division
  - Breed divisions
- Dressage
- Rodeos
- Cutting
- Polo
- Combined training
- Fox hunting
- Driving
- **Gymkhanas** (games for horses and riders)
- Distance riding
- Riding for the handicapped
- Holidays on horseback
  - Summer camps
  - Dude ranching
  - Pack trips
  - Cross-country riding
- Draft horse demonstrations

Another indication of the popularity and future of horses is the number of publications, videos, and organizations that support the various equestrian activities. Membership in equine organizations has grown in the past decade. Table A-17 in the Appendix lists more than fifty international horse organizations with their addresses, including:

- American Horse Shows Association
- United States Trotting Association
- United States Dressage Federation
- United States Pony Clubs
- United States Combined Training Association
- National Cutting Horse Association
- International Pro Rodeo Association
- American Driving Society
- National Steeplechase Association

Figure 2-8 Browsing the world of horse literature at the Denver Livestock Show.

Many of the breed registries have their own magazine or newsletter. Almost every equestrian activity has an organization and some type of publication. Also, the popularity of horses caused a proliferation of books and videos on every imaginable equine-related topic (see Figure 2-8). Finally, the Internet and the World Wide Web contain numerous home pages and interest groups dedicated to equestrian activities. Based on the increasing activity in all areas of publication and communication, horses in the United States have a bright future.

The making and selling of **tack** or virtually anything a horse or rider wears is a growing multimillion-dollar industry. For each type of horse-related activity, catalogs or stores sell needed equipment and apparel.

People have more money to spend and more leisure time than at any other period in history. A shorter work week, increased automation, and more suburban and rural living contribute to more free time. Equestrian activities will play a larger role in physical fitness and well being.

RESEARCH IN THE HORSE INDUSTRY

Research in the horse industry involves four general areas—unsoundness and injury, breeding and reproduction, nutrition, and disease prevention and control. Research is slow and costly but if the growing horse industry is to take advantage of science and technology, research is necessary.

Current Areas of Research

Unsoundness and Injury

- Safe anesthetization of horses after strenuous exercise
- Relationship of training to the occurrence of **bucked shins**
- Molecular mechanism in **synovitis**
- Blood clots and **laminitis**
- Investigations into the skeletal muscles and the "tying up" syndrome
- Fatal muscular and skeletal injuries
- Horseshoeing and shoes associated with injury
- Types and management of surfaces for prevention of injury
- Identification, prevention, and treatment of injuries
- Best techniques for training

Breeding and Reproduction

- Pregnancy diagnosis
- Estrous cycle of the mare
- Causes of reproductive failure
- **Embryo** transfer
- **Genetic** mapping

Nutrition

- Reduction and understanding of **colic**
- Grazing methods and pasture types
- Factors influencing nutritional requirements
- Interactions of nutrients
- Nutrient requirements: energy, protein, mineral, and vitamins
- Characteristics and suitability of feeds

Disease Prevention and Control

- More effective immunizations against herpesvirus, equine **influenza**, equine **viral** arteritis, and pneumonia caused by *Rhodococcus equi*
- Isolation of the agent causing equine **protozoal** myeloencephalitis
- Prevention and treatment of equine influenza
- Better avoidance and cure for most all diseases
- Improvement of resistance to disease by stimulating response of the **immune system**
- Improved control of parasites

Research solves many problems and increases our understanding. Good research also generates new questions and the need for more research. This progress will help move the horse industry successfully into the twenty-first century.

Equine research is expensive. Organizations such as the American Horse Shows Association Equine Health Research Fund, the Bolshoi Colic Research Program, the Grayson-Jockey Club Research Foundation, Inc., the Morris Animal Foundation, and the University of Kentucky Equine Research Foundation fund important equine research projects.

SUMMARY

Even though the number of horses has declined drastically from the early 1900s, the horse industry is a major agriculturally based industry that combines business, sport, and recreation into a program of economic impact that involves millions of people. Opportunities for expansion and participation by even more people seem unlimited. The industry involves many types of horses and horse programs and events. It may be divided into three major segments: racing, showing, and recreation. These segments involve numerous support industries including feed, veterinary, education, insurance, tack, farriery, and so on. Racing and shows and other events also involve the general public as spectators. In the United States today, around 80 percent of all horses are kept for recreation and 40 percent of all horses are used in youth programs. Horse owners represent a wide socio-economic group and reside in rural, urban, and suburban areas.

Opportunities are unlimited to expand the user and spectator base. Youth programs of all kinds form the foundation for sport and recreation and have wide appeal. Growing markets include new owners, suburban horse owners, senior citizens, and amateurs. Horses provide for a life-long sport and interest for all people.

Challenges facing the industry include the need for improved marketing and more educational programs, recognition by traditional public service bureaucracies, lack of adequate census and economic analysis, expanding youth programs, the long-term agenda of animal rights extremists, lack of adequate trails and riding areas, zoning and related environmental concerns, increasing costs and profitability for the business segments, insurance issues, and the overall lack of unity and communication between the various parts of a very large and complex industry.

Meeting challenges will become more and more complex. The industry needs to enhance unity and communication, and to expand advertising, promotional, marketing, and educational programs. This includes promoting the generic horse with the public, expanding support for youth programs, implementing cost-effective management, expanding trail development, and increasing riding instruction and opportunities for more people.

REVIEW

Success in any career requires knowledge. Test your knowledge of this chapter by answering these questions or solving these problems.

True or False

1. The United States has more horses, mules, and donkeys than any other country in the world.

2. New York is one of the top ten horse-producing states.

3. Horse shows have decreased in size and number over the past twenty years.

4. A gymkhana is an equestrian activity.

5. Genetic mapping is an area of equine research.

Short Answer

6. List five of the top ten horse producing states.

7. According to the Food and Agriculture Organization of the United Nations, what percentage of the world's horses are located in the United States?

8. List five equine organizations.

9. List five equine activities.

10. Name four general areas of equine research.

11. Where are most of the donkeys in the world?

Discussion

12. Briefly describe the changes in the United States horse population from 1900 to the present.

13. Describe four trends that suggest the United States horse industry has a bright future.

14. Using the statistics from the Food and Agriculture Organization of the United Nations, compare the worldwide populations of horses, donkeys, and mules.

15. Why does the number of equestrian activities and membership in equestrian organizations give some idea of the popularity and future of the United States horse industry?

16. Compare the United States population of horses, donkeys, and mules to that of the world.

STUDENT ACTIVITIES

1. Contact your state Department of Agriculture or Cooperative Extension Service for current statistics on horses, donkeys, and mules in your state.

2. Locate and attend an equine event. Or, instead of attending the event, check the schedule of ESPN or PHN (Premier Horse Network) and view several types of events on television.

3. Write to one of the equine organizations listed in Appendix Table A-17 and ask how members of the organization have benefitted from equine research. Also, ask what type of research the organization supports.

4. Develop a report, written or oral, on one of the subjects of equine research. Describe the problem and the current progress being made through scientific research.

5. Write a report on the importance and use of horses, donkeys, or mules in other areas of the world, for example, Asia, Mexico, South America, Africa, or Europe.

6. Investigate how science has changed horse racing. Report your findings.

7. Research the Internet for information about equine associations or equine events. Report your findings.

ADDITIONAL RESOURCES

Books

Davidson, B., and Foster, C. 1994. *The complete book of the horse.* New York, NY: Barnes & Noble Books.

Dossenbach, M., and Dossenbach, H. D. 1994. *The noble horse.* New York, NY: Cresent Books (Random House).

Price, S. D. 1993. *The whole horse catalog.* New York, NY: Fireside.

U.S. Department of Agriculture. 1960. *Power to produce: the yearbook of agriculture 1960.* Washington, DC: U.S. Government Printing Office.

Articles

Wildermuth, P. 1995. Trends at a glance. *Equus*, November: 28–31.

Kilby, E. 1995. Are we better yet? *Equus*, November: 33–40.

Videos

The following videos are available from Equine Research, Inc., Grand Prarie, TX:

Life & times of Secretariat

Great American trotters

Thoroughbred heroes

Fast track to the hall of fame

Breeds, Types, and Classes of Horses

Through selection, inbreeding, and outcrossing, humans created horses for speed, strength, endurance, size, good nature, hardiness, beauty, and athletic ability. Today, over 300 breeds exist. These breeds represent numerous types and classes. The various breeds and types of horses are also bred to donkeys to produce different types of mules.

This chapter acquaints the reader with the breeds of horses and the methods and terms used to group the breeds.

O B J E C T I V E S

After completing this chapter, you should be able to:

- Describe how horse breeds started with foundation stallions
- Understand the concept of breed, types, and classifications
- Describe the common height measurement for horses
- Define the terms warmblood, coldblood, cob, and hack
- Name ten common breeds of light horses and their origin
- Name five common breeds of draft horses and their origin
- Name five common breeds of ponies and their origin
- List five color breeds of horses
- Name five lesser-known breeds of horses or ponies and their origin
- Explain the origin of feral horses
- Describe how mules are produced
- Identify the common breeds of donkeys

- List ten uses for horses
- Describe some of the uses for the miniature donkeys and horses
- List six uses for mules

K E Y   T E R M S

Breed	Draft horses	Miniature
Breeding true	Feral	Mustangs
Breed registry	Foundation sires	Pony
Cob	Hack	Roadsters
Coldblood	Hand	Warmblood
Color breed	Hinny	
Conformation	Light horses	

BREEDS

Through selective breeding, people learned to develop specific desirable characteristics in a group of horses. After a few generations of selective breeding a **breed** of horse was born.

A breed of horses is a group of horses with a common ancestry that breed true to produce common characteristics such as function, conformation, and color. **Breeding true** means that the offspring will almost always possess the same characteristics as the parents.

Recognized breeds of horses have an association with a stud book and breeding records. Many recognized breeds have certain **foundation sires** and all registered foals must trace their ancestry back to these stallions. For example, the three foundation stallions of the Thoroughbred are the Darley Arabian, the Byerly Turk, and the Godolphin Arabian. Justin Morgan is the foundation sire of the Morgan horse breed. Allen F-1, a Morgan stallion, is the foundation sire of the Tennessee walking horse. Morgan horse stallions also contributed to the development of the Standardbred, quarter horse, American albino, and the palomino breeds.

People who found particular colors appealing established registries with color requirements. Some of these registries require only color for registration, but others have conformation standards as well. The Palomino Horse Association of California was the first **color breed** registration. Other color breed registries now include the Appaloosas (see Figure 3-1), albinos, paints, pintos, buckskins, whites, cremes, and spotted. Color breeds do not breed true. Table A-16 in the Appendix lists the names and addresses of many breed registries.

Figure 3-1 Appaloosa colt from Sheldak Ranch, Sheldon, North Dakota. *(Photo courtesy K. Burman, Appaloosa Journal, Moscow, ID)*

CLASSIFICATIONS AND TYPES

After breeds, horses can be classified several different ways. For example, horses can be grouped as light, draft, or **pony**, according to size, weight, build, and use. Within these groupings horses can be divided by use. For example riding, racing, driving, jumping, or utility. They can also be classified as **warmblood**, **coldblood**, or ponies.

Horse classifications depend on the height and weight of the horse. The common measurement of horse height is the **hand**. The height of a horse is measured from the top of the withers to the ground. A hand is equal to 4 inches. So a horse that is 15 hands is 60 inches. A horse that is 15.2 (15 hands 2 inches) is 62 inches tall from the top of the withers to the ground.

Light Horses

Light horses are 14.2 to 17.2 hands high (hh) and weigh 900 to 1,400 pounds. They are used primarily for riding, driving, showing, racing, or utility on a farm or ranch. Light horses are capable of more action and greater speed than **draft horses** (see Figure 3-2).

Draft Horses

Draft horses are 14.2 to 17.2 hands high and weigh 1,400 pounds or more. They are primarily used for heavy work or pulling loads. Historically, when draft horses were bought and sold for work, they were classified according to their use as draft, wagon, farm chucks, or southerners (see Figure 3-3).

Figure 3-2   Arabian horse. *(Photo courtesy Jerry Sparagowski, Dunromin Arabians)*

Figure 3-3   A Clydesdale herd sire. Owner James and Betty Groves and family of Pecatonica, IL. *(Photo courtesy Maureen Blaneg)*

Figure 3-4 A Welsh pony stallion. Owner Jean du Pont. *(Photo courtesy Welsh Pony and Cob Society of America, Winchester, VA)*

Ponies

Ponies stand under 14.2 hands high and weigh 500 to 900 pounds. Ponies possess a distinct conformation on a reduced scale. They are either draft, heavy harness, or saddle type (see Figure 3-4).

Warmblood

Warmblood does not relate to horses with a certain blood temperature. It refers to the overall temperament of light-to-medium horse breeds. Warmblood horses are fine-boned and suitable for riding. In some countries, the warmblood is distinguished as a horse having a strain of Arab breeding. Some groupings classify all light horses as warmbloods. According to some, all breeds that are not definitely Thoroughbred, draft, or pony are classified as warmblood (see Figure 3-5).

Figure 3-5   A Trakehner mare owned by Lancaster Oakes. *(Photo courtesy Lancaster Oaks/American Trakehner Association)*

Coldblood

Coldblood horses are heavy, solid, strong horses with a calm temperament. This term is probably best thought of as another way of describing draft horses.

Types and Uses

Types of light horses include riding, racing, showing, driving, all-purpose, and **miniature**. Riding horses are generally thought of as the gaited horses (three- and five-gait), stock horses, horses for equine sports, and ponies for riding and driving. Racing horses are running racehorses, pacing/trotting racehorses, quarter racehorses, and harness racehorses (see Figure 3-6). Driving horses include the heavy and fine harness horses, ponies, and the **roadsters**. All-purpose horses and ponies are used for family enjoyment, showing, ranch work, etc. Miniature horses and donkeys are used for driving and as pets.

Obviously, some breeds fit better into some of these types than other breeds.

The terms **cob** and **hack** are also used to describe types of horses. A cob is a sturdy, placid horse. It stands 14.2 to 15.2 hands high and is not heavy or coarse enough to be classified as a draft animal. A hack is an enjoyable, good riding or driving horse, sometimes considered a small Thoroughbred in Europe or a Saddlebred in America.

Figure 3-6 1993 Fine Harness World's Grand Champion. Owned by Linda Dake, Wilford Ranch, Santa Fe, California. Trained and shown by Tom Moore. *(Photo courtesy Jamie Donaldson)*

COMMON BREEDS OF HORSES

Table 3-1 briefly describes some of the more common breeds of horses, their origin, classification, and height. Table 3-2 lists some of the less well-known breeds of warmblood or light horses and their origin. Table 3-3 lists some other breeds of draft or coldblood horses and their origin, while Table 3-4 provides the name and origin of some lesser-known breeds of ponies and their origin.

MINIATURE HORSES

Miniature horses are scaled-down versions of a full-size horse and are not dwarfs. Miniatures are not a breed but can be registered with the Miniature Horse Registry. The maximum height for registration is 34 inches at the withers.

Miniatures are often kept as pets. Some are exhibited as driving horses in single pleasure and roadster driving classes. Also, some people exhibit miniature horses in multiple hitches pulling miniature wagons, stagecoaches, and carriages. Because of their size, only a small child can ride them (see Figure 3-7, page 67).

Table 3-1 Well-Known Breeds of Horses

Name	Origin	Classification	Height	Color	Comments
Akhal–Teke	Turkmenstan	Light	15–15.2 hh	Gold with metallic sheen; also, bay, cream, chestnut	Less than 2,000 purebreds in world; Marco Polo said foundation sire was Alexander the Great's horse, Bucephalus
Albino	United States	Light	No height requirements	White only	Foals born white
Alter–Real	Portugal	Light	15–15.2 hh	Mostly bay or brown; some chestnuts and gray	High-strung temperament; does well in dressage; carries Andalusian breeding
American Buckskin	United States	Light	14 hh	Four color patterns accepted	Descendants of Norwegian Dun and Spanish Sorraia
American Creme	United States	Light	Varied from 12.2–17 hh	Three variations of creme accepted	Color breed
American Quarter Horse	United States	Light	15.2–16.1 hh	Any solid color; mostly chestnut	Oldest of American breeds; most versatile horse in the world; largest equine registry in the world; natural cow sense
American Saddlebred	United States	Light	15–16 hh	Black, bay, brown; white markings on face and legs	Formerly Kentucky saddler, amiable; can perform several gaits; very showy
American Standardbred	United States	Light	14–16 hh	Any solid color, mostly brown, bay, black, chestnut	Developed as trotter/pacer; direct line can be traced to one male, Messenger

(continued)

Table 3-1 Well-Known Breeds of Horses (*continued*)

Name	Origin	Classification	Height	Color	Comments
American Warmblood	United States	Light	Varies	Any color	Relatively new breed; common crosses are Thoroughbred/draft or Thoroughbred/warmblood
American White	United States	Light	Varied from 12.2–17 hh	Snow or milk white hair, pink skin, brown, black, or hazel eyes	Color breed; not true albino
Andalusian	Spain	Light	15–16.2 hh	Gray, born dark and becoming lighter over years	Oldest and purest of all horses after Arabian; breed founded in 710 AD; almost became extinct in 1830s; saved by monks; used by mounted bullfighters in Spain; no arab blood used in development
Appaloosa	Spain, United States	Light	14–15.3 hh	White schelera, striped hooves, mottled skin and coat pattern	Bred by Nez Perce Indians; third largest breed registry in world; popular in U.S. and Australia
Arabian or Arab	Arabia	Light	14.3–16 hh	Bay, brown, chestnut, gray, black	Large nostrils and long eyelashes adapted for desert conditions; one less vertebra than any other breed; has influenced the foundation of all light breeds; can carry more weight over longer distance than Thoroughbred or Quarter Horse; characteristic dished face; oldest purebred, 5,000 BC

(continued)

Table 3-1 Well-Known Breeds of Horses *(continued)*

Name	Origin	Classification	Height	Color	Comments
Barb	North Africa	Light	14–15 hh	Dark brown, bay, chestnut, black, gray	One of great foundation horses; used to strengthen other breeds; considered forerunner of Thoroughbred
Bashkir Curly	Russia	Light	13.2 average	All colors	Noted for long, curly coat of hair, milking ability, cold-hardy
Belgian	Belgium	Draft	Up to 17 hh	Mostly roan with black points, chestnut, sometimes bay, brown, dun, gray	Descendent of medieval great horses; magnificent animal; one of most powerful of horse breeds
Chickasaw	United States	Light	13.2–14.7 hh	Bay, black, chestnut, gray, roan, sorrel, palamino	Developed by Native Americans of Tennessee, North Carolina, Oklahoma; used as cow ponies
Cleveland Bay	England	Light	16–16.2 hh	Bay, mahogany with black points main, tail; feet blue in color	Very versatile and hardy; easy keepers
Clydesdale	Scotland	Draft	16.2–18 hh	Bay, brown, black, roan, much white on face and legs and sometimes body	Displays action; popular in big hitches much feathering on foot; regularly exported from Britain to wherever horses are needed for over 100 yrs.
Connemara	Ireland	Pony	13–14.2 hh	Gray, black, brown dun	Hardy, sure-footed

(continued)

Table 3-1 Well-Known Breeds of Horses (*continued*)

Name	Origin	Classification	Height	Color	Comments
Dales	England	Pony	14.2 hh maximum	Mostly black	Very hardy; good for children; used as work horse
Dartmoor	British Isles	Pony	11.2–11.3 hh	Bay, brown, black	Used for pack ponies in the mines of England; good for children
Dutch Warmblood	Holland	Warmblood	16 hh	Any color	Mix of Groningen and Gelderland breeds; willing temperament
Fell Pony	England	Pony	14.2 hh maximum	Black, brown, bay, gray, no white markings	Very hardy, all-purpose horse
French Saddle Horse or Selle Français	France	Light	15.2–16.3 hh	Usually bay or chestnut	Descended from Anglo-Norman studs; developed as a competition horse
Friesian (West Friesian)	Holland	Draft	15 hh	Black	Used by knights of old; have heavily feathered legs; breed lighted for carriage and sport horse; tail and mane may touch ground
Gotland (or Skogsruss)	Sweden	Light	12–14 hh	Dun, black, brown, bay, chestnut, palomino	One of the oldest breeds; excellent youth horse, jumper, trotter
Hackney	England	Pony	14 hh maximum	Dark brown, black, bay, chestnut	Trotting horse; good carriage horse

(*continued*)

Table 3-1   Well-Known Breeds of Horses *(continued)*

Name	Origin	Classification	Height	Color	Comments
Hackney	England	Light	14.2–16 hh some taller	Black, brown, chestnut, bay	Flamboyant pacers, usually used in shows and harness; distinctive trotting action
Haflinger	Austria	Pony	14.2 hh	Chestnut, white main and tail	All of today's Haflinger's are traced back to foundation sire; 249 Folic; family horse
Hanoverian	Germany	Warmblood	16–17.2 hh	Any solid color	Dominate in international competition; stable and willing temperament
Highland (or Garron)	Scotland	Pony	14.2 hh maximum	Various shades of dun; dorsal eel stripe, black points or silver hair in tail and mane; also, gray, chestnut, bay, black	Very versatile; sturdy, sure-footed
Holstein	Germany	Warmblood	15.3–16.2 hh	Any solid color, mostly black, brown, bay	One of oldest warmbloods from great horse types; competes well; good carriage horse
Lipizzaner	Austria	Light	15–16 hh	Mostly gray	Most famous horses from Spanish Riding School of Vienna performing haute ecole riding; great athletic ability, performing airs–above–the–ground
Lustiano	Portugal	Light	15–16 hh	Usually gray	Bred from Andalusian stock; used for bull fighting

(continued)

Table 3-1 Well-Known Breeds of Horses *(continued)*

Name	Origin	Classification	Height	Color	Comments
Missouri Fox Trotter	United States	Light	14–17 hh	Any color, usually sorrel	Natural ability for specialized gaits; comfortable ride
Morab	United States	Light	14.3–15.2 hh	Usually solid	Cross of Morgan and Arabian breeds only
Morgan	United States	Light	14–15.2 hh	Bay, brown, black, chestnut	One common foundation sire, Justin Morgan of Massachusetts; works very well under harness or saddle
Mustang	United States	Light	14–15 hh	All colors	Original cow-pony; feral horses of American West; small but tough; Native Americans used extensively
New Forest	England	Pony	12–14.2 hh	Any color except piebald or skewbald	Very hardy; good family pony; allowed to run wild most of year; easy to train
Norwegian Fjord Pony	Norway	Pony	13–14 hh	Dun with black eel stripe down center of back; Zebra stripes on legs	Primitive-looking horse, resembling Przewalski's horse; hardy and sure-footed; still used as farm ponies in Norway; gentle
Oldenburg	Germany	Warmblood	16.2–17.2 hh	Any solid color	Tallest and heaviest of German warm-bloods; based on Friesian breeding
Paint	United States	Light	Variable	Black and white in bold patches all over body	All Paint horses must be sired by a registered Paint, Quarter Horse, or Thoroughbred

(continued)

Table 3-1 Well-Known Breeds of Horses *(continued)*

Name	Origin	Classification	Height	Color	Comments
Palomino	United States	Light	14.2–15.3 hh	Gold coat; white mane and tail; no markings	Not possible to breed true to color; first color registry
Paso Fino	Caribbean, Puerto Rico, South America	Light	14.3 hh	All colors	Shows the natural lateral 4-beat gaits
Percheron	France	Draft	15.2–17 hh	Gray, black	Most popular cart horse in world; slight arab features in face
Peruvian Paso	Peru	Light	14–15.2 hh	Mostly bay and chestnut	Has unique gait; can carry rider long distances not becoming too tired
Pinto	Spain, United States	Light	Variable	Black and white in bold patches all over body	Associated with Native Americans
Pony of the Americas (POA)	United States	Pony	11.2–13 hh	Appaloosa color pattern	Cross between Appaloosa and Shetland; one of newest breeds; very good for young riders
Shetland	England	Pony	11.2 hh maximum	Variable	Popular with children; very hardy, gentle
Shire	England	Draft	17 hh average	Bay and brown most common with white markings	Very docile; can be trusted with a child; tallest horse in the world; heavy feathering on foot; descended from great horses; popular as team horse
Spanish Barb	Spain, United States	Light	13.3–14.1 hh	Varied	Three strains recognized—scarface, rawhide, buckshot

(continued)

Table 3-1 Well-Known Breeds of Horses *(concluded)*

Name	Origin	Classification	Height	Color	Comments
Spotted Saddle	United States	Light	14–16 hh	Spotted coloring	Good all-around horse; good disposition
Suffolk (or Suffolk Punch)	England	Draft	15.2–16.2 hh	Chestnut	Developed as work horse; not as big as other draft breeds
Swedish Warmblood or Halfbred	Sweden	Warmblood	16.2 hh	Usually chestnut, bay, brown, gray	Outstanding saddle horse; competes very well in dressage
Tennessee Walking Horse	United States	Light	15–16 hh	All solid colors	Well known for two unique gaits—flat walk and running walk; bred for comfortable ride; good for beginners
Thoroughbred	England	Light	14.2–17 hh	Any solid color, white markings allowed	Bred mainly for racing; must be handled carefully
Trakehner	Germany, Poland	Warmblood	16–16.2 hh	Any solid color	Very versatile, considered most handsome of all German warmbloods; competes well in all sports
Welsh Pony (Sections A, B, C, D)	Wales	Pony	13.2 hh maximum (height determines which section)	Any solid color	Very hardy; very good trotting ability; good jumper; influenced trotters all over the world
Württemberg	Germany	Warmblood	16 hh average	Black, bay, chestnut, brown	Developed to do work on small mountain farms

Table 3-2 Lesser-Known Breeds of Warmblood and Light Horses			
Name	**Origin**	**Name**	**Origin**
American Remounts	USA	Furioso North Star	Hungary
Anglo Arab	Britain, France, Poland, Hungary	Gelderland	Holland
		German Trotter	Germany
Anglo-Argentine	Argentina	Gessian	Germany
Bavarian Warmblood	Germany	Gidran	Hungary
Beberbeck	Germany	Groningen	Holland
Brandenburg	Germany	Hispano (Spanish Anglo-Arab)	Spain
Brumby	Australia		
Budyonny	Russia	Iberian	Iberian Peninsula
Calabrese	Italy	Iomud	Central Asia
Campolina	Brazil	Irish Hunter	Ireland
Charollais Halfbred	France	Jaf	Iran/Kurdistan
Criollo	South America	Kabardin	Russia
Dølegudbrandsdal	Norway	Karabair	Uzbekistan
Danubian	Bulgaria	Karabakh	Azerbaidzhan
Darashouri	Iran	Kladruber	Czechoslovakia
Don	Central Asia	Knabstrup	Denmark
East Bulgarian	Bulgaria	Kustanair	Kazakhstan
East Friesian	Germany	Latvian Harness Horse	Latvia
Einsiedler	Switzerland	Libyan Barb	Libya
European Trotter	France, USA, Russia	Limousin Halfbred	France
Fox Trotting Horse	Ozarks	Lokai	Uzebikistan
Frederiksborg	Denmark	Malapolski	Poland
Freiburger Saddle Horse	Switzerland	Mangalarga	Brazil
		Maremmana	Italy
French Trotter	France		*(continued)*

Table 3-2   Lesser-Known Breeds of Warmblood and Light Horses *(concluded)*

Name	Origin	Name	Origin
Masuren	Poland	Sardinian	Sardinia
Mecklenburg	Germany	Shagya Arab	Hungary
Metis Trotter	Russia	Sokolsky	Poland/Russia
Murgese	Italy	Spotted Saddle Horse	USA
Native Mexican	Mexico		
New Kirgiz	Kirgiz/Kazakhstan	Tchenaran	Iran
Nonius	Hungary	Tersky	Russia
Novokirghiz	Central Asia	Toric	Estonia
Orlov Trotter	Russia	Waler	Australia
Plateau Persian	Iran	Westfalen	Germany
Pleven	Bulgaria	Wielkopolski	Poland
Rhinelander	Germany	Yorkshire Coach	Ireland
Salerno	Italy	Zweibrucker	Germany

Figure 3-7   Miniature horse pulling a cart at the Boise Horse Show.

Table 3-3 Lesser-Known Breeds of Coldblooded or Draft Horses

Name	Origin	Name	Origin
Ardennais	France/Belgium	North Swedish	Sweden
Auxios	France	North Swedish Trotter	Sweden
Boulonnais	France	Poitevin	France
Breton	France	Rhineland Heavy Draught	Germany
Comtois	France		
Døle Trotter	Norway	Russian Heavy Draught	Ukraine
Dutch Draught	Holland		
Finnish	Finland	Schleswig Heavy Draught	Germany
Irish Draught	Ireland		
Italian Heavy Draught	Italy	Schwarzwälder	Germany
Jutland	Denmark	Soviet Heavy Draught	Russia
Lithuanian Heavy Draught	Baltic States		
		Swedish Ardennes	Sweden
Mulassier	French	Trait du Nord	France
Murakov	Hungary	Vladimir Heavy Draught	Russia
Noriker Pinzgauer (Oberländer, South German)	Austria/Germany		
		Woronesh	Russia

FERAL HORSES

Horses that were once domesticated and have become wild are called **feral** horses. No one knows for sure where, when, and how the first horses escaped from or were stolen from the Spaniards in America. During the 1700s and 1800s the number of feral horses in America could have been two to five million. Most of these were located in the Southwest.

Currently, habitats for feral horses are found in California, Colorado, Idaho, Montana, Nevada, New Mexico, Utah, and Wyoming. These habitats are public lands administered by the U.S. Bureau of Land Management and the U.S. Forest Service. Some horses on these lands have been feral for many generations but others have been recently released.

Table 3-4   Lesser-Known Ponies of the World

Name	Origin	Name	Origin
Acchetta	Sardinia	Huzule	Romania
Ariège	France	Icelandic	Iceland
Assateague	USA	Java	Indonesia
Australian	Australia	Kathiawari	India
Avelignese	Italy	Kazakh	Kazakhstan
Balearic	Balearic Islands	Konik	Poland
Bali	Indonesia	Landis	France
Bashkirsky	Russia	Leopard Spotted	England
Basuto	South Africa	Macedonian	Yugoslavia
Batak (Deli)	Indonesia	Manipur	Assam-Manipur
Bhutia	India	Marwari	India
Bosnian	Yugoslavia/Bosnia-Herzegovina	Merens	France
		Mongolian	Mongolia
Burma (Shan)	Burma	Native Turkish	Turkey
Camarguais	France	Peneia	Greece
Caspian	Iran	Pindos	Greece
China	China		
Chincoteague	USA	Sable Island	Canada
Costeno	Spain, Peru	Sandalwood	Indonesia
Dülmen	Germany	Skyros	Greece
Exmoor	England	Spiti	India
Falabella	Argentina	Sumba	Indonesia
Fjord (Westlands)	Norway	Sumbawa	Indonesia
Fjord-Huzule	Czechoslovakia	Tarpan	Eastern Europe
Galiceño	Mexico	Tibetan (Nanfan)	Tibet
Garrano (Minho)	Portugal	Timor	Indonesia
Gayoe	Indonesia	Viatka	Russia
Huçul	Poland	Zemaituka	Russia

Man o' War was not just another racehorse. Man o' War was one of 1,680 Thoroughbreds foaled in 1917. He was born on March 29 at the Kentucky Nursery Stud owned by August Belmont II. Man o' War's sire was Fair Play, by Hastings. To provide balance to Fair Play's temper, Belmont bred him to Mahubah, by Rock Sand, who had won the British Triple Crown. Mrs. Belmont named the foal "My Man o' War," since he was a war baby. Belmont had to serve in the Army in 1918, so he ordered his entire crop of yearlings sold at Saratoga.

At Saratoga, Man o' War was bought by Pennsylvania horseman Samuel Riddle for $5,000 and shipped to Riddle's training farm, "Glen Riddle," in Maryland. Man o' War was trained by Louis Feustel who had trained Mahubah, Fair Play, and Hastings. When he was sent off to his first race at Belmont Park, a retired hunter named Major Treat accompanied him and would continue to travel with Man o' War throughout his racing career.

On June 6, 1919, ridden by Johnny Loftus, Man o' War won his first race by six lengths, crossing the finish line at a canter. He showed his desire to be a front runner and never liked to have any other horse in front of him. He ran only in expensive stake races for the remainder of his career. As his wins built up, so did the weight he was required to carry. By his fourth race, Man o' War was carrying 130 pounds.

His sixth race was the Sanford Memorial and the only defeat of his career. A bad start left him with a ten-length deficit and once he caught the pack he was boxed in. He was beaten by a horse ironically named Upset. Man o' War had beaten Upset on six other occasions. In spite of this lone defeat, he was selected Horse of the Year at the end of his two-year-old season.

Public concern for the plight of feral horses led to the passage of two federal laws to protect them—Public Laws 86-234 and 92-195. Feral horses are also called **mustangs** (see Figure 3-8).

DONKEYS

The breeds registered by the American Donkey and Mule Society, which was founded in 1968, are the mammoth (or American standard) jack, large standard donkey (Spanish donkey), standard donkey (burro), miniature Mediterranean donkey, and American Spotted Ass.

Man o' War went undefeated as a three-year-old in 1920, and reduced the American record for the mile by two-fifths of a second, to 1:35⅘; and even at that, his jockey, Clarence Kummer, had held him back. Later, in the Belmont Stakes, he set a record that stood for 50 years. By the time of the Dwyer Stakes at Aqueduct, Man o' War could find only one opponent, John P. Grier, a horse from the Whitney Stables. In this race, Grier challenged Man o' War, and Kummer used his whip for the first time. Man o' War dashed to victory, and set a new American record at 1:45⅕.

The crowning event of Man o' War's career came in a match race against the celebrated Canadian horse Sir Barton, the first winner of the Triple Crown. On October 12, 1920, Sir Barton and Man o' War met in Windsor, Ontario. Man o' War won by seven lengths! In his career he won 20 of 21 races.

Man o' War's stud career was just as distinguished as his career on the track. At stud on Hinata Stock Farm in Lexington, Kentucky, he sired 13 foals his first season, the most notable of which

was American Flag. Riddle did not allow Man o' War to breed many mares besides his own. Although Man o' War became history's leading sire in terms of his off-spring's winnings, his stud career might have been even greater had he been bred to better mares. Among Man o' War's most important get were Triple Crown winner War Admiral, Crusader, Blockade, War Hero, War Relic, Clyde Van Deusen, and Battleship. In all, he sired 379 foals that won 1,286 races.

Under the close care of his groom, Will Harbut, Man o' War was visited by thousands each year at Riddle's Faraway Farm. In 1947, Will Harbut died of a heart attack; barely a month later he was followed by Man o' War also victim of a heart attack at the age of 30. Man o' War's burial was a time of national mourning; his funeral was broadcast over radio and covered by the press from all over the world. Samuel Riddle had commissioned Herbert Hazeltine to sculpt a memorial statue of Man o' War while he lived, to be placed on his grave. In 1977 the remains of Man o' War and his famous statue were brought to the Kentucky Horse Park in Lexington, Kentucky.

The mammoth breed is a blend of several breeds of jack stock first imported into the United States in the 1800s from southern Europe. It is the largest of the asses, with the jacks being 56 inches or more high. The foundation sire was a jack named Mammoth. His name was given to the breed.

The large standard donkey (Spanish donkey) is between 48 and 56 inches high, while the standard donkey (the burro) is between 36 and 48 inches high. The miniature Mediterranean donkey, originally imported from Sicily and Sardinia, must be under 36 inches (down from the original 38 inches) to qualify for registration. The height restriction is the only requirement for registration by the American Donkey and Mule Society (see Figure 3-9).

Figure 3-8 Feral horses from the Wyoming Red Desert penned up at the University of Wyoming, Laramie, dairy farm. Note height of fence.

Figure 3-9 Two wild jacks registered as standard donkey/wild burros. Owned by Elmer Zeiss of Valley, Nebraska. *(Photo courtesy American Donkey and Mule Society, Denton, TX)*

The American Spotted Ass is a trademark of the American Council of Spotted Asses, founded in 1967. It can be registered as either white with colored spots or colored with white spots. However, the spots have to be above the knees and hocks, and behind the throat latch. Stockings and face markings do not qualify.

Miniature Donkey

The Miniature Donkey Registry of the United States, founded in 1958, is currently governed by the American Donkey and Mule Society. Color, and other considerations, such as ancestry, do not define the miniature donkey. The only requirement is that it be 36 inches or less in height.

The original imported donkeys had the typical gray-dun color, in which the hairs are all gray and not mixed with white hairs. All shades of brown are also common, and black, white, roan, and spots are possible. True gray is extremely rare in donkeys of any size, and is distinguished from the gray-dun because the true gray donkeys are born with a dark coat that lightens to almost white over the years. One other characteristic of the donkey is the cross, consisting of a dorsal stripe from mane to tail, and a cross stripe between the withers. In black animals the cross marking may be difficult to detect.

The miniature donkey with good conformation should give the impression of being small, compact, and well-rounded, with four straight strong legs, and all parts in symmetry and balance. The coat of the miniature donkey is not as thick in winter as the coat of larger donkeys, probably because of its ancestry from climates in the Mediterranean.

Although the most obvious use of these little donkeys is as pets, they can also be used as companions to foals at weaning time to relieve foal stress. Their calm also serves when they are used as companions for nervous horses or horses recovering from surgery. They do not take up much room in the stall, but have a great calming effect.

MULES

A cross between a donkey and a horse is called a **mule** or a **hinny**. A mule is the offspring of a male donkey (jack), and a female horse (mare). It is like the horse in size and body shape but has the shorter, thicker head, long ears, and braying voice of the donkey. Mules also lack, as does the donkey, the horse's calluses, or chestnuts, on the hind legs (see Figure 3-10).

The reverse cross, between a male horse (stallion), and a female donkey (called a jennet or jenny) is a hinny, sometimes also called a jennet. A hinny is similar to the mule in appearance but is smaller and more horselike, with shorter ears and a longer head. It has the stripe or other color patterns of the

Figure 3-10 Meredith Hodges cross country jumping with her mule. *(Photo courtesy Meredith Hodges and American Donkey and Mule Society, Denton, TX)*

donkey. Hinnies are more difficult to produce than mules. A fertile male mule or hinny may display normal sex drives. Rarely, a female mule or hinny may come into heat and produce a foal. Mules and hinnies have 63 chromosomes and donkeys, 62, while horses have 64 chromosomes (32 pairs).

Classifications of Mules

Historically, mules were classified as draft, sugar, farm, cotton, and pack and mining.

Draft and sugar mules were the largest being 17.2 hh (hands high) to 16 hh and 1,600 to 1,150 pounds. Farm and cotton mules were intermediate in size (16 hh to 13.2 hh and 1,250 to 750 pounds). Pack and mining mules were smaller, but could range from 16 hh to 12 hh and 1,350 to 600 pounds.

Today mules are classified as draft, pack/work, saddle, driving, jumping, or miniature. The type of mule produced depends on the breed or type of horse and breed or type of donkey used to produce the mule.

SUMMARY

Worldwide, about 300 breeds of horses exist. They range in size from the gentle giant draft horses at almost 6 feet in height to the miniature horses at barely three feet in height. People have bred and selected horses for specific, common characteristics such as function, conformation, and color. Horses breeding true or with a common ancestry are registered in **breed registry** associations. These horses meet the standards defined by the registry. Besides breeds, horses are classified by type, such as light, draft, and pony, and by use such as riding, driving, harness, sport, gaited, stock, and all-purpose. Some breeds have specific purposes while other breeds serve a variety of uses.

Five breeds of donkeys are recognized. Donkeys are crossed with horses to produce mules. The type of mule that results depends on the breed and type of donkey and horse used in the cross. Both donkeys and horses have miniatures. These miniatures are used for pets and exhibition hitches and as companions to sick or nervous horses.

REVIEW

Success in any career requires knowledge. Test your knowledge of this chapter by answering these questions or solving these problems.

True or False

1. Feral horses were commonly bred by the early Spaniards.

2. Coldblood horses and draft horses are similar classifications.

3. Warmblood horses exhibit a body temperature three degrees above normal.

4. A mule is the offspring of a stallion bred to a female donkey or jennet.

5. Mules are ridden in contests.

Short Answer

6. Name the three foundation stallions of the Thoroughbred breed.

7. Name five color breeds of horses.

8. Name five common breeds of light horses and give their place of origin.

9. How long is the measurement of one hand?

10. Name five common breeds of draft horses and give their place of origin.

11. How many chromosomes do horses, donkeys, and mules possess?

12. Name two common breeds of donkeys.

13. List five less well-known breeds of horses and give their classification and country of origin.

14. List six uses for mules.

Discussion

15. Define a breed.

16. Define the terms light, draft, pony, warmblood, and coldblood and explain the relationships among any of the terms.

17. Describe ten uses for horses.

18. Discuss some of the uses for the miniature donkeys and horses.

19. Compare light horses to draft horses.

20. Compare a mule to a horse.

STUDENT ACTIVITIES

1. Choose a competitive event such as racing, driving, dressage, or riding. Research the breed of horse most commonly used for this event and explain why the breed is appropriate for the event.

2. Write to a breed registry association listed in Appendix Table A-16 and request more information and pictures of a breed of light horse, draft horse, or pony.

3. Use the Internet to discover more information about five horse breeds of your choice. Write a report comparing the five breeds.

4. Develop a report on the chromosome numbers of horses, donkeys, and mules. Which chromosome is missing in the mule when compared to the chromosomes of a horse? Why are mules generally sterile?

5. Construct a family tree for a famous Thoroughbred showing how this horse's ancestry can be traced to the foundation stallions.

6. Some horse breeds have their own magazine or newsletter. Select two common breeds from Table 3-1 and obtain sample copies of their newsletter or magazine. Next, read an article of your choice in the magazine or newsletter and write a summary.

7. Create a poster showing the color markings of the Appaloosa, the paint, the pinto, and the buckskin horse. Describe how horses are bred to produce these color breeds.

8. Explain why the process of blood typing could be important to breed registration, and diagram how blood typing is done.

ADDITIONAL RESOURCES

Books

American Youth Horse Council. 1993. *Horse industry handbook: a guide to equine care and management.* Lexington, KY: American Youth Horse Council, Inc.

Davidson, B., and Foster, C. 1994. *The complete book of the horse.* New York, NY: Barnes & Noble Books.

Dossenbach, M., and Dossenbach, H. D. 1994. *The noble horse.* New York, NY: Cresent Books (Random House).

Ensminger, M. E. 1990. *Horses and horsemanship.* 6th ed. Danville, IL: Interstate Publishers, Inc.

Evans, J. W. 1989. *Horses: a guide to selection, care, and enjoyment.* 2nd ed. New York, NY: W. H. Freeman and Company.

Knight, L. W. 1902. *The breeding and rearing of jacks, jennets and mules.* Nashville, TN: The Cumberland Press.

Mills, F. C. 1971. *History of American jacks and mules.* Hutchinson, KS: Hutch-Line, Inc.

Silver, C. 1993. *The illustrated guide to horses of the world.* Stamford, CT: Longmeadow Press.

Associations and Registries

Any of the associations or registries in Appendix Table A-16 can be contacted for more information about a specific breed.

Magazines

Magazines such as *Horse Illustrated*, *Horse & Rider*, *Western Horseman*, and *Horse and Horseman* often feature articles on a single breed.

Internet

Information about many of the breeds can be found on Internet sites by using a good search engine like Web Crawler or Yahoo. Use search words like horse breeds or a specific breed, for example Arabian, Thoroughbred, or paint. Some good World Wide Web sites include the following:

http://horseworlddata.com/index1.html

http://www.ansi.okstate.edu/breeds/horses

http://www.freerein.com/haynet/

http://www.horseworld.com/imh/kyhp/hp1.html

CHAPTER 4

Cells, Tissues, and Organs

Functional anatomy begins with the cell, an understanding of cell types and cell processes. Depending on their location, cells develop specialized functions. Next, groups of cells form tissues and then organs. These organs become parts of systems that function together as a heathy, productive animal.

OBJECTIVES

After completing this chapter, you should be able to:

- Explain the importance of cells and their function to the horse

- Identify the parts and organelles of animal cells

- List and describe the functions of each of the major types of specialized animal cells

- List the cell organelles and the functions of each part

- Describe how specialized cells are organized to form a tissue type

- List and describe the six types of specialized animal tissues and their individual functions

- Describe the difference between meiosis and mitosis

- Describe blood and its function

78

KEY TERMS

Adenosine triphosphate (ATP)	Granules	Organelles
Adipose	Histogenesis	Ovum
Axon	Interphase	Oxidative phosphorylation
Blastula	Lysosomes	Plasma membrane
Cell	Meiotic cycle	Receptors
Centrioles	Metabolites	Ribonucleic acid (RNA)
Chromatids	Microfilaments	Ribosomes
Cytoplasm	Microtubules	Sarcolemma
Epithelial	Mitochondria	Spermatozoa
Fertilization	Mitosis	Stimuli
Gametes	Morphogenesis	Synapses
Gastrulation	Myelin	Tissue
Golgi apparatus	Myofibrils	Vacuoles
	Nucleus	

CELLS

All living material is made of cells or the chemical products of cells. Understanding the **cell** as the fundamental unit of life is the basis for an understanding of living organisms such as the horse.

Modern cellular biology makes six assumptions:

1. All living material is made up of cells or the products of cells.

2. All cells are derived from previously existing cells; most cells arise by cell division, but in sexual organisms they may be formed by the fusion of a sperm and an egg.

3. A cell is the most elementary unit of life.

4. Every cell is bounded by a **plasma membrane**, an extremely thin skin separating it from the environment and from other cells.

5. All cells have strong biochemical similarities.

6. Most cells are small, about 0.001 cm (0.00004 inches) in length.

The three general functions of most cells include:

- maintenance
- synthesis of cell products
- cell division

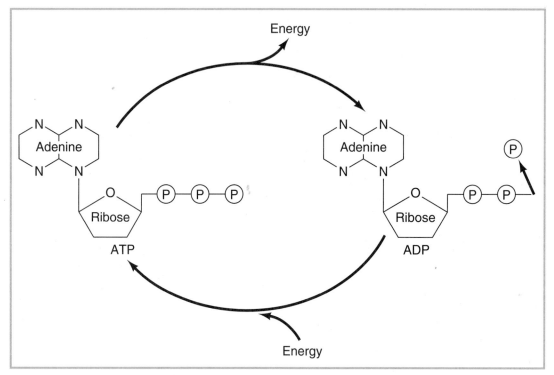

Figure 4-1 ATP molecules are continually rebuilt from ADP (adenosine diphosphate) molecules, phosphates, and chemical energy. Thus ATP acts as an energy carrier for those reactions in the cell that release energy and those reactions that consume energy.

These functions require the cell to take in nutrients and excrete waste products. The nutrients are used either as building blocks in synthesizing large molecules, or they are oxidized—burned—producing energy for powering the cell's activities. Because synthesis, maintenance, and mechanical and electrical activity all require energy, a major chemical activity in nearly all types of cells is the energy-linked conversion of **metabolites**. **Adenosine triphosphate (ATP)** is the universal energy-transfer molecule. ATP is constantly used and regenerated by the energy-yielding chemical reactions shown in Figure 4-1.

Components of Cells

A cell is enclosed by a cell membrane. The material known as the **cytoplasm** lies within the cell membrane and contains several **organelles** and **granules** in suspension (see Figure 4-2). Major components of the cell include:

- plasma membrane
- nucleus
- ribosomes
- endoplasmic reticulum

Cell Structure

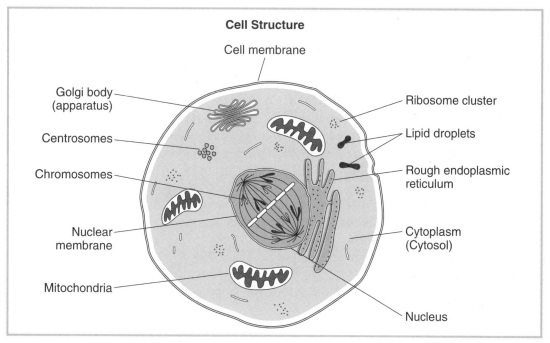

Figure 4-2   A cell and its components

- Golgi apparatus
- centrioles
- microfilaments, microtubules, lysosomes, and storage particles

Plasma Membrane. Cells are surrounded by a thin membrane of lipid (fat) and protein. This membrane controls the transport of molecules in and out of the cell and thus serves as a boundary between the cell and the surrounding **tissue**. Other membranes also occur in a cell's interior, for example, as part of the endoplasmic reticulum, nucleus, and **mitochondria**. The exterior portions of the cell surfaces determine cell-to-cell interactions, so they are important in the formation and control of tissues. The extracellular material also acts as a glue that holds cells together in tissues.

Nucleus. Most cells have a single **nucleus** enclosed by a nuclear envelope, or membrane, with pores. Pores provide continuity between the nucleus and the cytoplasm. The nucleus contains one or more discrete structures, known as nucleoli, which are sites of ribosomal **ribonucleic acid (RNA)** synthesis. Hereditary information is carried in the DNA contained within the chromosomes in the nucleus. In the nucleus, this information is transcribed into the RNA, which serves as a messenger. The messenger then moves outside of the nucleus to the **ribosomes** where it guides the synthesis of proteins. Thus, the nucleus directs the activity of the cell.

Ribosomes. Ribosomes are tiny particles within the cell. Made of RNA and protein, they are present in large numbers in most cells and are the site of protein synthesis (the manufacture of large protein molecules from amino acid subunits).

Endoplasmic Reticulum. Within most cells is a complex set of membranous structures. When viewed in the electron microscope, the membranes are either rough—covered with granules or ribosomes—or smooth. Generally, the rough endoplasmic reticulum is highly developed in cells that make large amounts of protein.

Golgi Apparatus. A special type of membrane mixture is often found near the nucleus. This collection of membranes is called the **Golgi apparatus**. In cells that synthesize and secrete products, the Golgi apparatus is the site of the material that is accumulated.

Mitochondria. Mitochondria are composed of an outer membrane and a winding inner membrane. A series of chemical reactions that occur on the inner membrane convert the energy of oxidation into the chemical energy of ATP. In this process, called **oxidative phosphorylation**, the predominant energy transfer molecule is ATP. Almost all of the energy passes through this molecule before being used in cell function. Cells with high rates of metabolism usually have a large number of mitochondria.

Centrioles. Most cells have two cylindrical bodies, called **centrioles**, located near the nucleus. The centrioles appear as sets of triple tubules. Centrioles play a part in cell division.

Other Organelles. The material containing the organelles is called ground substance, or cytoplasm. It contains proteins, small molecules, and a group of entities organized as **microfilaments** and microtubules. Microfilaments are long, thin, contractile rods that appear to be responsible for the movement of cells, both external and internal.

Microtubules are hollow, cylindrical groupings of tubelike structures that help give the cell shape and form. They are also involved in other cell processes.

Lysosomes are small bodies where large numbers of enzymes are stored. Some cells may also have particularly large liquid-filled areas known as **vacuoles**. The vacuoles are believed to be involved in digestion or excretion, or both.

Storage particles comprise a diverse group of structures and contain lipid droplets and glycogen granules whose function is the long-term storage of energy.

Morphogenesis

All organisms, regardless of their complexity, begin as a single cell. By repeated cell growth and **mitosis**, or division, the organism eventually develops into an adult containing thousands of billions of cells. This process of development is called **morphogenesis**. Since many different types of cells exist in fully grown animals, morphogenesis involves not only cell growth but differentiation into specialized types of cells (see Figure 4-3). This differentiation is controlled by the genes. The information needed to program and guide the growth is contained within the chromosomes. Size, shape, and chemical activity of the cells are governed to some extent by the function of the tissue in which they are found.

Each cell contains the same total genetic information that was present in the fertilized egg. The cells are not identical because in different types of cells, groups of genes are controlled—switched on and off—by various biochemical processes. Each cell manufactures the proteins and structures needed for it to function. Blood cells make hemoglobin to carry oxygen. Sperm cells make flagella, and so forth. On average, only about 10 percent of the genes of any cell are functional—which genes, in particular, varies with the type of cell. Although morphogenesis has been scientifically described in great detail for a number of organisms, all of the processes involved at the cellular level are still not understood.

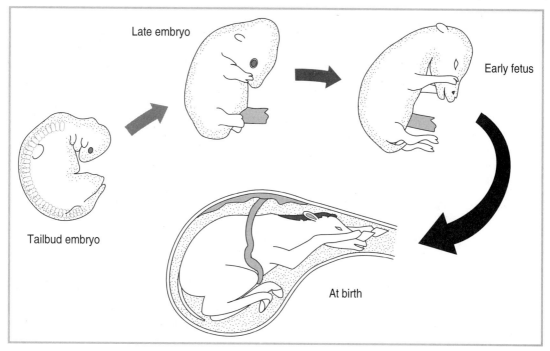

Figure 4-3 Changes of horse from an embryo to fetus to horse

Metabolism is the sum of all the chemical reactions in the living cell. These reactions produce useful work and synthesize cell constituents. Almost all cellular reactions are catalyzed by complex protein molecules called enzymes that speed reaction rates by a factor of hundreds to millions.

Many structures in the living cell must be periodically replaced. This process of building new molecules is called anabolism. Structures that are worn out or no longer needed are broken down into smaller molecules and either reused or excreted. This process is called catabolism. Great quantities of energy are required not only to produce the work needed for pumping the heart, for muscular contraction, and for nerve conduction, but also to provide the chemical work needed to make the large molecules characteristic of living cells. Anabolism and catabolism are aspects of overall metabolism. They occur interdependently and continuously.

In the combustion of feed, oxygen is used and carbon dioxide is given off. The rate of oxygen consumption indicates the energy expenditure of a horse, or its metabolic rate. The metabolic rate of a horse at any given time is highly variable and is influenced by many factors, including muscular activity, diet, digestion, lactation, pregnancy, time of day or year, sexual activity, and stress.

To fix a point of reference, a convention has been adopted to serve as the standard metabolic rate. The ideal standard established is the metabolism of an animal under the least physiologically demanding conditions. In the case of humans and livestock, this minimal-rate-of-energy metabolism is called the basal metabolic rate (BMR). It is defined as the rate of metabolism of a fasting animal at rest and under no heat stress.

Cell Division

The nucleus controls cells division; cell division depends on two events:

1. The replication (copying) of the DNA molecules that make up the basic genetic material of all cells

2. The orderly separation of the products of this replication.

To survive, each new cell must have the same genetic code as its parent cell. Cells must reproduce so the division of the nucleus precedes division of the cytoplasm, and both are necessary for cell division.

Mitotic Cycle

Mitosis is part of a more complex cycle that includes a long phase, called **interphase**. As a result of the syntheses that take place during interphase, each chromosome consists of two sister chromosomes, called **chromatids**, that are identical in their structural and genetic organization and joined at the kinetochore. Chromatids become visible when mitosis sets in; the remainder of the mitotic cycle involves their separation into two offspring nuclei (see Figure 4-4). Mitosis depends on four major events—coiling, orientation, movement, and uncoiling. The six essential stages of the mitotic cycle are:

- Prophase
- Prometaphase
- Metaphase
- Anaphase
- Telophase
- Interphase

Meiotic Cycle

Mitosis rarely lasts more than two hours, but the **meiotic cycle** that produces **gametes**, or sex cells, may take days or weeks, since it involves two successive

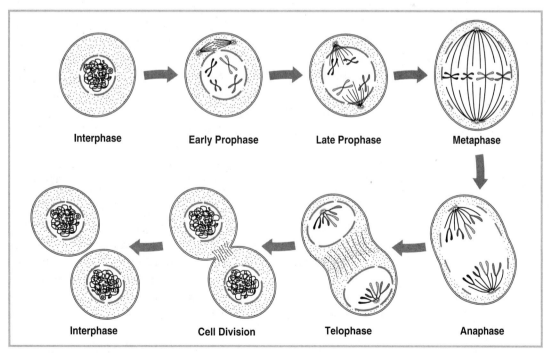

Figure 4-4   Diagram of stages of mitosis

sequences of cell division and a reduction in the number of chromosomes. Each cell resulting from meiosis has one-half the number of chromosomes as the parent cell or other body cells. Also, the chromosomes in the resulting cells are sorted. The stages of the meiotic cycle include:

- Prophase with five substages
- Prometaphase
- Metaphase
- Anaphase
- Interphase

Types of Cells

During morphogenesis, several major types of animal cells form, including absorptive, secretory, nerve, sensory, muscle, and reproductive cells. All must arise during morphogenesis from cells that are less differentiated.

Absorptive Cells. Absorptive cells often occur as continuous sheets on surfaces where material is transported to the cells. For example, the single layer of **epithelial** cells lining the surface of the small intestine selectively absorbs food molecules from the gut into the bloodstream. These cells have a free surface that faces the digestive tract and a base surface that is in contact with the capillaries. The free surface is covered with many projections called microvilli, which vastly increase the area available for molecular flow (see Figure 4-5).

Similar cells are found in the kidney, where a large surface area is needed for the absorption of protein, water, salts, and other materials. The microvilli are an example of a cell structure fitted to the function of the cell. Because an absorptive cell needs maximum area for transport, the shape of the cell surface is altered to achieve the optimum transfer of molecules.

Secretory Cells. Secretory cells produce products that are subsequently deposited in either the bloodstream or a special duct to an organ, where they are used. The pancreas and pituitary are glands that have large numbers of secretory cells. Proteins and other cell products are synthesized throughout the cytoplasm of these cells and transported to the Golgi apparatus, where they are packaged in membrane-bounded vesicles that come to a cell's surface and discharge the secretion outside the cell. Secretory cells in the spleen, lymph nodes, and other sites synthesize antibodies for the recognition and destruction of foreign molecules.

Nerve Cells. A nerve cell consists of a main cell body and a long thin structure known as an **axon** (see Figure 4-6). The function of nerve cells is to

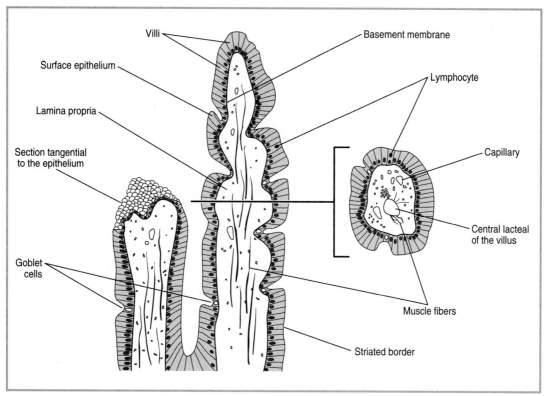

Figure 4-5   Diagram of absorptive cells

transmit electrical messages from one part of the cell body to another. These cells function like telephone transmission lines. The connections between nerve cells are called **synapses**. When these structures are combined, they form an electrical network known as the nervous system. The processes that occur at the synapses are both electrical and chemical. The axon is covered with a layer of insulation called **myelin**. Axons carry electrical signals called nerve impulses.

Sensory Cells. Sensory cells respond to impulses by emitting electrical signals. An example is the rod cell of the eye, in which the central cell body has two long, thin appendages. One appendage has an outer segment consisting of specialized stacked membranes for the reception of light. At the other end is a long, thin connection to a nerve cell that leads to the optic nerve fiber. About half of the material in the outer segment consists of rhodopsin, the pigment used in detecting light. Other sensory **receptors** include free nerve endings, pacinian corpuscles, ruffini corpuscles, taste buds, smell receptors, and hearing receptors (see Figure 4-7).

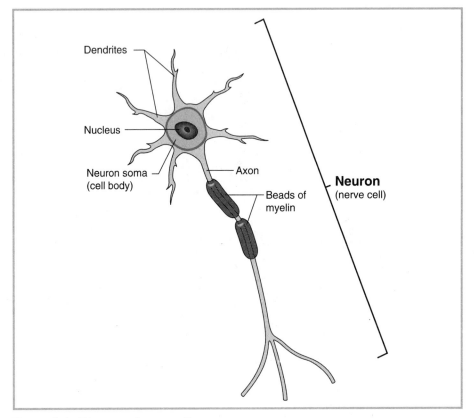

Figure 4-6 Diagram of nerve cell

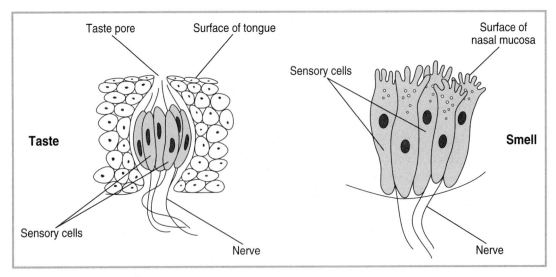

Figure 4-7 Diagram of sensory receptors

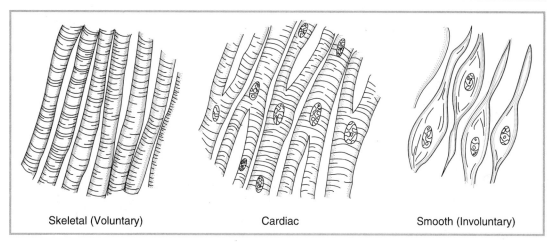

| Skeletal (Voluntary) | Cardiac | Smooth (Involuntary) |

Figure 4-8   Diagram of three types of muscle cells

Muscle Cells.   Muscles are of three types—skeletal, cardiac, and smooth (see Figure 4-8). Contraction of muscle fibers generates a mechanical force. The skeletal muscle is a multinucleate structure that has an outer envelope known as the **sarcolemma**. This system does not fit the definition of a cell. Skeletal muscle cells are actually a tissue in which the cells have merged. Most of the interior consists of long, thin **myofibrils** that are actually the contractile elements.

Reproductive Cells.   Gametes are formed after completion of the process of meiosis, which halves the number of chromosomes in each cell. Stallion gametes or **spermatoza** are motile, whereas a mare's gamete or **ovum** (egg) is larger and stationary. **Fertilization** occurs when a sperm is fused with an egg. This stage is followed by morphogenesis (see Figure 4-9).

TISSUES

Tissue are structured groupings of cells specialized to perform a common function necessary for the survival of the horse—a multicellular animal. The process of tissue formation or **histogenesis** evolves from the earlier process of cell differentiation.

The fertilized ovum or egg, a single cell, divides to form the **blastula**, in which tissues are not yet defined (see Figure 4-10). As growth continues, the cells of the blastula begin to form three germ layers—the ectoderm, mesoderm, and endoderm—through the process of **gastrulation**. Cell differentiation during gastrulation begins the process of histogenesis and continues into the formation of organs.

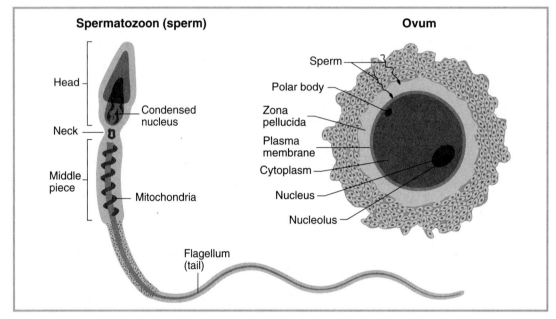

Figure 4-9 Diagram of equine sperm and egg cells

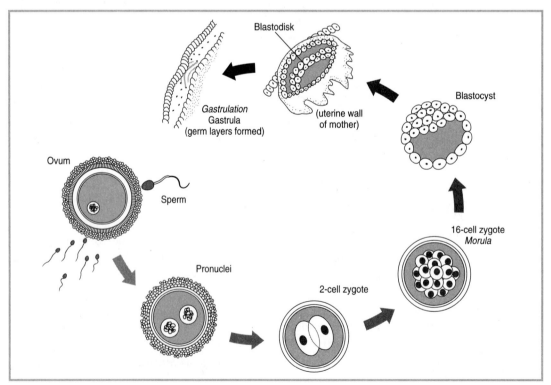

Figure 4-10 The early stages of embryology

The cells in a tissue look more or less alike and contribute the same type of service. Five general classifications of tissues are:

- nerve
- epithelial
- muscle
- connective
- fluid

Nerve Tissue

Nerve tissues consist of extraordinarily complex cells called neurons that respond in a specific way to a variety of **stimuli** so as to transfer information from one part of the body to another (see Figure 4-11).

Epithelial Tissue

Epithelial tissue consists of a layer of cells covering the external surfaces of an animal and lining its internal tubes for digestion, respiration, circulation, reproduction, and excretion. This layer controls what is absorbed into and lost from the organism. The epithelium is composed of continuous sheets of adjacent cells. Outgrowths and ingrowths of epithelium form the sensitive surfaces of sensory organs, glands, hair and nails, and other structures (see Figure 4-11).

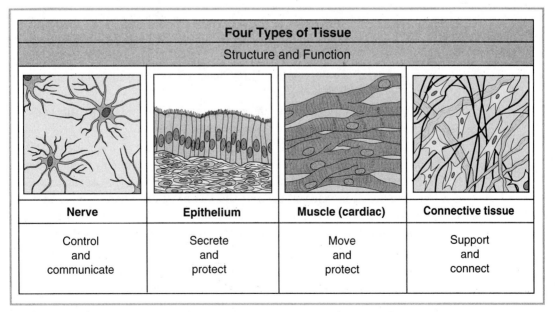

Four Types of Tissue			
Structure and Function			
Nerve	**Epithelium**	**Muscle (cardiac)**	**Connective tissue**
Control and communicate	Secrete and protect	Move and protect	Support and connect

Figure 4-11   Example of nerve tissue, epithelial tissue, muscle tissue, and connective tissue

Muscle Tissue

The ability to contract and relax and thus provide movement is characteristic of muscle tissue. Muscle tissue is of three types. Smooth muscle is activated by the autonomic nervous system. Skeletal muscle is controlled by the central nervous system and, to a certain extent, by the will. Cardiac muscle is characterized by its ability to contract rhythmically (see Figure 4-11).

Connective Tissue

Connective tissues contain large amounts of extracellular material modified into different types. They are varied in structure to permit them to support the entire body and to connect its parts. Connective tissue includes fibrous tissue found in tendons and ligaments; elastic tissue found in ligaments between the vertebrae, arterial walls, and trachea; cartilaginous tissue found in joints and in the development of bone; and **adipose** tissue, which, with its fat deposits, cushions and supports vital organs and stores excess food (see Figure 4-11).

Fluid Tissue

Fluid tissues are the blood and lymph. These tissues function to distribute food and oxygen to other tissues, carry waste products from the tissues to the kidneys and lungs, and carry defensive cells and other substances to destroy disease-producing agents (see Figure 4-12).

ORGANS

Groups of specialized tissues performing a specific function are called organs. The stomach is an organ of digestion. The uterus is an organ of reproduction. A group of organs working together is know as a system. For example, the stomach is only one of the organs in the digestive system and the uterus is only one organ in the reproductive system.

Chapter 5 discusses the systems formed by the organs in the body of the horse.

SUMMARY

All living material is made of cells or the chemical products of cells. Most of the cells in the body of the horse carry on the processes of maintenance, synthesis, and cell division. As cells carry on these life processes they require energy and produce waste products.

All cells contain the same genetic information, but through morphogenesis cells grow and develop into specialized cells. Tissues are structured groupings of cells specialized to perform a common function necessary to the survival of the horse. The five basic tissue types include nerve, epithelial,

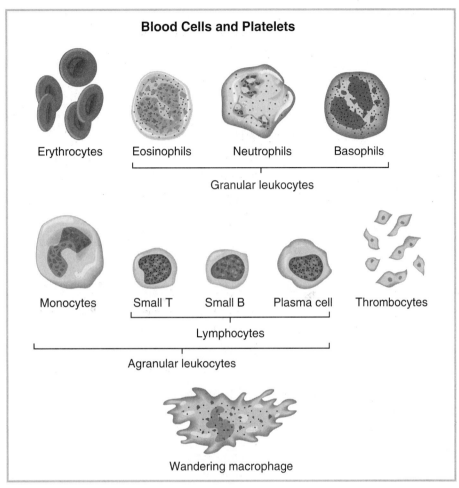

Blood Cells and Platelets

Erythrocytes Eosinophils Neutrophils Basophils

Granular leukocytes

Monocytes Small T Small B Plasma cell Thrombocytes

Lymphocytes

Agranular leukocytes

Wandering macrophage

Figure 4-12 Example of blood cells and platelets

muscle, connective, and fluid. Tissues combine to form organs and organs group to form functional body systems.

REVIEW

Success in any career requires knowledge. Test your knowledge of this chapter by answering these questions or solving these problems.

True or False

1. Prophase occurs in both mitosis and meiosis.

2. Axons are a part of absorptive cells.

3. Groups of specialized cells are called organs.

4. An end product of mitosis is sperm cells or an egg.

5. The pancreas and pituitary are glands that have large numbers of secretory cells.

Short Answer

6. List five of the ten major components of a cell.

7. Name three general functions of all cells.

8. Name two gametes.

9. Identify five general types of tissues.

10. What general type of tissue are fibrous, elastic, cartilaginous and adipose tissue?

11. Give an example of a fluid tissue.

12. What type of information is carried in DNA?

Discussion

13. Describe six assumptions that make the cell the fundamental unit of life.

14. Discuss the concept of morphogenesis.

15. Explain the difference between mitosis and meiosis.

16. Discuss the differences between skeletal, smooth, and cardiac muscle.

17. Describe the two events that cell division depends upon.

18. What is the importance of ATP?

STUDENT ACTIVITIES

1. From a biological supply company, obtain prepared microscope slides of epithelial tissue, muscle tissue, nerve tissue, connective tissue, and blood. View these through a microscope and make sketches of each type. As an alternative, search the Internet for sites with photographs of tissue cross sections and histological (tissue) discussions.

2. Draw and label your own diagram of an animal cell. Label all of the components of the cell.

3. Develop a report on the production of ATP (adenosine triphosphate) in the body. How is it produced and where is it used? Briefly, outline the biochemistry involved in producing and using ATP.

4. Obtain a blood smear and staining kit. Using your own blood, make and stain some blood smears and observe these through the microscope. Make drawings of your observations.

ADDITIONAL RESOURCES

Books

Asimov, I. 1954. *The chemicals of life.* New York, NY: New American Library.

Frandson, R. D., and Spurgeon, T. L. 1992. *Anatomy and physiology of farm animals.* 5th ed. Philadelphia, PA: Lea & Febiger.

Fraser, C. M., ed. 1991. *The veterinary manual.* 7th ed. Rahway, NJ: Merck & Co.

Hafez, E. S. E. 1993. *Reproduction in farm animals.* 6th ed. Philadelphia, PA: Lea & Febiger.

Thomas, L. 1974. *The lives of a cell: notes of a biology watcher.* New York, NY: The Viking Press.

Videos

Hawkhill Associates, Inc. 1996. *The gene.* Madison, WI: Hawkhill Associates, Inc.

Hawkhill Associates, Inc. 1996. *The cell.* Madison, WI: Hawkhill Associates, Inc.

University of Kentucky. 1995. *Growth and bone development.* VAS-0661. Agricultural Distribution Center.

Equipment and Supplies

Carolina Biological Supply Company, Carolina Science and Math Catalog 66, 2700 York Road, Burlington, NC 27215-3398

NASCO Agricultural Sciences, 901 Janesville Ave., Fort Atkinson, WI 53533-0901

Nebraska Scientific, 3823 Leavenworth Street, Omaha, NE 68105-1180

Fisher-EMD, 4901 W. LeMoyne Street, Chicago, IL 60651

CHAPTER 5

Functional Anatomy

A basic understanding of the functional anatomy of the horse is essential before discussing growth, aging, movement, selection, nutrition, health, breeding, behavior, management, and even facilities. An understanding of the functional anatomy provides a basis for the "why" of everything else involved in modern scientific equine production and equitation.

OBJECTIVES

After completing this chapter, you should be able to:

- List the nine systems of animals and the major organs that make up each system

- Explain the functions of the skeletal, muscular, digestive, urinary, respiratory, circulatory, nervous, reproductive, and endocrine systems

- Identify the components of the skeletal, muscular, digestive, urinary, respiratory, circulatory, nervous, reproductive, and endocrine systems

- List the five divisions of the vertebral column

- Name the bones in the foreleg and hindleg

- Describe three types of joints

- Identify three types of muscles and their locations in the body

- Trace the circulation of blood through the body

- Identify the endocrine glands and the hormones they secrete

KEY TERMS

Adrenal cortex	Homeostasis	Peristalsis
Adrenal medulla	Hormone	Pharynx
Alveoli	Hypercalcemia	Pineal gland
Androgens	Hypothalamus	Pituitary gland
Anterior	Hypothyroid	Plasma
Anus	Ileum	Premaxilla
Aorta	Inspiration	Pulse
Arteries	Insulin	Rectum
Articulation	Interdental space	Red blood cells
Assimilation	Internal respiration	Ridgling
Bronchi	Jejunum	Saliva
Capillary	Joint	Seminiferous tubules
Carnivores	Joint capsule	Small colon
Cartilage	Large colon	Small intestine
Caudal	Large intestine	Spinal cord
Cecum	Larynx	Sternum
Cerebellum	Ligament	Steroids
Cerebrum	Lymph	Synovial membrane
Cervix	Lymph nodes	Tendons
Coronary vessels	Lymph vessels	Testes
Cranial	Mandible	Thyroid
Cryptorchid	Maxilla	Trachea
Dorsal	Mesentery	Ureters
Duodenum	Molar	Urethra
Esophagus	Nasal	Urine
Estrus	Nephrons	Uterus
Expiration	Nerves	Vagina
Extensor	Omnivores	Veins
External respiration	Orbital	Venae cavae
Fertilization	Organs	Ventral
Flexor	Ovaries	Vertebrae
Foramen magnum	Oviducts	Visceral
Ganglia	Parotid	White blood cells
Gastric juice	Penis	Zygote
Goiter	Pericardium	
Herbivores	Periosteum	

ANIMAL SURFACES AND BODY SYSTEMS

Any discussion of the structure and function of animals begins with an understanding of the terms **dorsal**, **ventral**, **cranial** or **anterior** and **caudal** or **posterior**. Dorsal pertains to the upper surface of an animal. Ventral relates to the lower or abdominal surface. Anterior or cranial applies to the front or head of an animal. Posterior or caudal pertains to the tail or rear of an animal (see Figure 5-1).

Physiology, or the life functions of horses, occurs in body systems. Nine body systems are found in animals, including horses. These systems are:

1. Skeletal
2. Muscular
3. Digestive
4. Urinary
5. Respiratory
6. Circulatory
7. Nervous
8. Reproductive
9. Endocrine

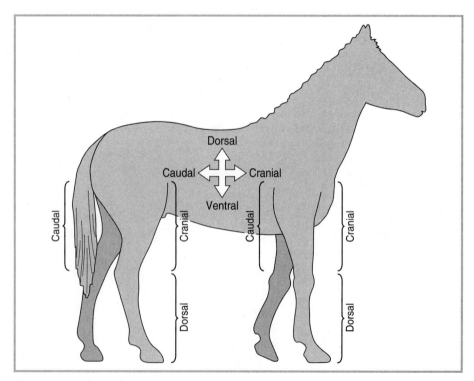

Figure 5-1   The surfaces or planes of a horse

THE SKELETAL SYSTEM

The skeletal system is the rigid framework giving the body shape and protecting the **internal organs**. It is composed of bone and **cartilage**. Bones are composed of about one part organic matter and two parts inorganic matter. The inorganic matter is mineral, mainly lime salts. The surface of each bone is covered by a dense connective tissue called the **periosteum**. A union of two bones is called an **articulation** or **joint**. **Ligament**, **tendons**, and a tough fibrous capsule provide stability or tightness to the joint. Tissues and organs attach to the skeleton.

The bones and joints together compose a complex system of levers and pulleys, which, combined with the muscular system, give the body the power of motion. The skeleton also stores up needed minerals, mainly calcium and phosphorus, and acts as a factory for the manufacture of blood cells and, in the adult animal, stores fat in the limb bones.

The relative size and position of the bones determine the form (or conformation) of the horse and its efficiency for any particular work. Bones are classified by their shape as long, short, flat, and irregular.

- Long bones are found in the limbs. They support the body weight and act as the levers of propulsion.

- Short bones occur chiefly in the knee and hock and aid in the dissipation of concussion (the shock of impact).

- Flat bones, such as the ribs, scapula, and some of the bones of the skull, help to enclose cavities containing vital organs.

- Irregular bones are unpaired bones, such as the vertebrae and some of the bones of the skull.

All bones, except at their points of articulation, are covered with a thin, tough adherent membrane, called the periosteum, that protects the bone and to an extent influences its growth. This later function is of particular interest since we know that an injury to this membrane often results in an abnormal bony growth, called exostosis, at the point of injury. Other bony growths, such as splints, spavins, and ringbone, are often the result of some injury to the periosteum. The bone is nourished partially through blood vessels in the periosteum, which also contains many nerve endings.

The articular or joint surfaces of bones are covered with a dense, very smooth bluish-colored substance known as cartilage. The cartilage diminishes the effects of concussion and provides a smooth joint surface that minimizes frictional resistance to movement.

The two main divisions of the skeleton are the trunk, or axial skeleton, and the limbs, or appendicular skeleton.

Axial Skeleton

The axial skeleton consists of the skull, spine (or vertebral column), ribs and breastbone, pelvis, and tail (see Figure 5-2).

Bones of the Skull. The skull is divided into two parts—the cranium surrounding the brain and the face enclosing the entrances to the digestive and respiratory systems. The skull is attached to the first vertebra of the spine and has a large opening, the **foramen magnum**, through which the **spinal cord** passes.

The bones of the cranium are flat or irregular bones surrounding the cranial cavity, which houses the brain. These bones join each other at immovable joints. The bone forming the poll (head area) has an articulating surface where the head joins the vertebral or spinal column. Together with the bones of the face, the cranial bones form the **orbital** (eye) and **nasal** cavities.

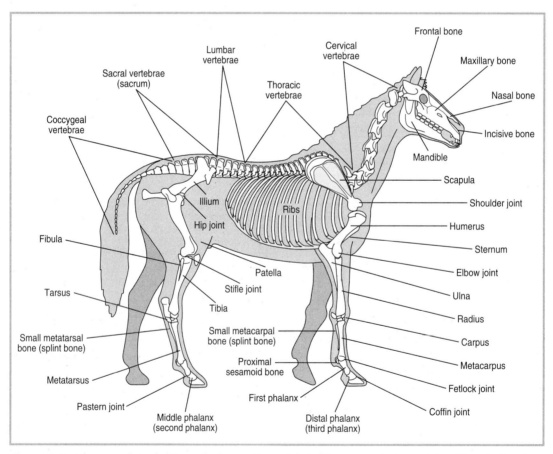

Figure 5-2   The complete skeleton of a horse. The axial skeleton, as noted, consists of the skull, spine, ribs, breastbones, pelvis, and tail.

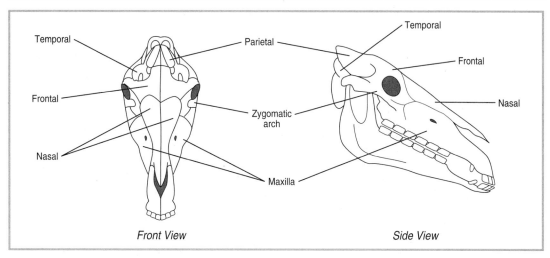

Temporal
Parietal
Temporal
Frontal
Frontal
Nasal
Nasal
Zygomatic arch
Maxilla

Front View

Side View

Figure 5-3   Bones of the horse skull

The bones of the face form the framework of the mouth and nasal cavities, and include the more important bones of the upper and lower jaws, known as the maxillae and **mandible**, respectively. Each **maxilla** has six irregular cavities for the cheek or **molar** teeth. From the maxillae forward, the face becomes narrower and terminates in the **premaxilla**, which contains cavities for the six upper incisor teeth. Enclosed in each maxilla is a cavity known as the maxillary sinus, which opens into the nasal passages. This sinus contains the roots of the three back molar teeth and may become infected from diseased teeth.

The mandible, or lower jaw, is hinged to the cranium on either side by a freely movable joint in front of and below the base of the ear. The mandible has cavities for the six lower incisors. Behind the incisors and ahead of the six lower molars in each branch of the mandible is a space known as the **interdental space**. In this space, injuries to the periosteum or possible fracture of the mandible may occur from rough usage of a bit (see Figure 5-3).

Vertebral or Spinal Column.   The spine is a flexible column of small bones called **vertebrae**. The vertebral column may be thought of as the basis of the skeleton from which all of the internal organs and passageways are suspended. It is composed of irregularly shaped bones bound together with ligaments and cartilage that form a column of bones similar to an elastic suspension bridge. An elastic pad or cushion separates each vertebra along the length of the column, from the base of the skull to the tip of the tail. Through the length of this column runs an elongated cavity or passageway, called the neural canal or spinal canal, that contains the main trunk line of **nerves** to the brain—the spinal cord. The bones of the vertebral column (see Figure 5-4) are divided into five groups:

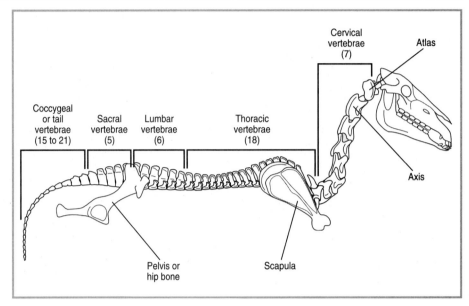

Figure 5-4   The bones of the vertebral column of the horse

- cervical: 7 vertebrae

- thoracic: 18 vertebrae

- lumbar: 6 vertebrae (sometimes 5)

- sacral: 5 vertebrae (fused together to form the sacrum)

- coccygeal or tail: 15 to 21 vertebrae

The hip bones are two large flat paired bones that form the pelvis or pelvic girdle. Each hip bone is firmly attached to the spine at the sacrum and circles around to meet at the midline below the sacrum and enclose the pelvic cavity. Each hip bone contains a cavity on its outside where the femur, or first bone of the hind leg, forms a joint. The upper front angle, together with the sacrum, forms the point summit of the croup. The back angle of the hip bone is the point of the rump.

The chest is the large cavity formed by the thoracic vertebrae, the ribs on the sides, and the **sternum**, or breastbone, on the bottom or floor. This cavity contains the heart, lungs, large blood vessels and nerves, and part of the **trachea** and **esophagus**. The eighteen pairs of ribs, all jointed to the thoracic vertebrae at their upper ends, determine the contour of the chest by their shape and length.

The sternum is a canoe-shaped prominence in the midline of the breast consisting of seven or eight bony segments connected by cartilage. The sternum forms the floor of the thorax.

Appendicular Skeleton

The appendicular skeleton of the horse consists of the forelegs and hind legs and is used for locomotion, grooming, and, to some extent, for defense and feeding. The forelimbs have no skeletal attachment to the axial skeleton, or trunk, of the horse. The connection is made only by muscles.

The bones of the foreleg of the horse (see Figure 5-5), named from the top downward, include:

- scapula
- humerus
- ulna and radius
- carpal bones
- splint bones
- cannon
- sesamoids
- first phalanx
- second phalanx
- coffin

Bones of the hind leg (see Figure 5-6), named from the top downward, include:

- femur
- patella
- tibia and fibula
- tarsals
- splint bones
- cannon
- sesamoids
- first phalanx
- second phalanx
- coffin

The hind limbs are attached to the bony pelvis at the hip joint, unlike the fore limbs, which have no bony connection to the trunk.

Functional anatomy of the hoof is discussed in more detail in Chapter 17.

Joints or Articulations

A joint or articulation is the union of two or more bones or cartilages. Joints are classified into three types according to their structure and movability—immovable, slightly movable, and freely movable.

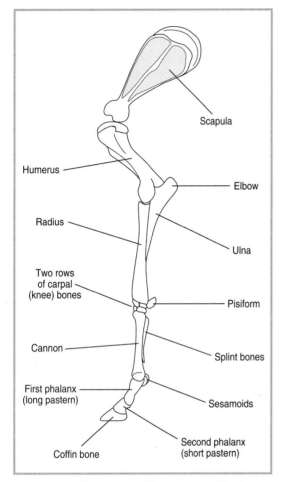

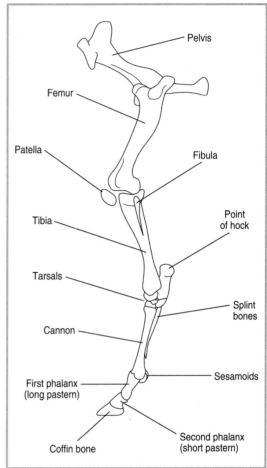

Figure 5-5 The bones of the foreleg of the horse   **Figure 5-6** The bones of the hind leg of the horse

Immovable joints are those in which the opposed surfaces of bone are directly united by connective tissue or fused bone and permit no movement, such as the bones of the cranium.

Slightly movable joints are those where a pad of cartilage, adhering to both bones, allows only slight movement due to the elasticity of the interposed cartilage. Many of the joints of the vertebrae are of this nature.

Freely movable joints are those where a joint cavity exists between the two opposed surfaces, such as the joints of the legs. The freely movable joints are the truest examples of joints. The ends of the bones entering into a freely movable joint are held in opposition by strong bands of tissue, called ligaments, which pass from one bone to the other. Ligaments possess only a slight degree of elasticity and have a limited supply of blood, which accounts for the fact that they heal slowly and often imperfectly following injury.

In freely movable joints, the ends of the bones are covered with a smooth cartilage that absorbs concussion and provides a smooth bearing surface. The entire joint is enclosed in a fibrous sac known as a **joint capsule** that assists the ligament in holding the bones in position. The inner surface of this sac is lined with a thin secreting membrane, called the **synovial membrane**, that secretes a fluid known as synovia or joint water. Synovia is a clear, slightly yellowish fluid with the appearance and consistency of the white of a watery egg. It lubricates the joint in the same manner as oil lubricates a mechanical bearing.

The joints of the foreleg, in order from the top downward include:

- shoulder, formed by the scapula and humerus
- elbow, formed by the humerus, radius, and ulna
- knee, formed by the radius, carpal bones, and the three metacarpal bones—splints and cannon bones
- fetlock, formed by the cannon, two sesamoid bones, and the first phalanx, or long pastern
- the pastern, formed by the first and second phalanges or long and short pasterns
- coffin, formed by the second phalange or coffin bone and the navicular bone

The joints of the hind leg, named in order from the top downward, are:

- hip, formed by the hip bone and the femur
- stifle, formed by the femur, patella, and tibia
- hock, formed by the tibia, tarsal or hock bones, and the metatarsal bones—splint and cannon bone
- fetlock, pastern, and coffin joints are named and formed the same as the corresponding joints of the foreleg

In addition to the ligaments that form a part of the joints, there are certain other important suspensory and check ligaments. The suspensory ligament of the foreleg is a very strong, flat ligament running from the back of the knee and upper end of the cannon bone down the back of the leg in a groove between the splint bones. Just above the fetlock, this ligament divides into two diverging rounded branches that are attached to the upper and outer part of the corresponding sesamoid bone and pass downward and forward around to the front of the long pastern bone to join at a point of union with the **extensor** tendon attached to the front of the coffin bone. From the lower part of the sesamoids, bands of ligaments pass downward and attach to the backs of the long and short pastern bones.

The check ligament is a short, strong ligament running from the back side of the upper end of the suspensory ligament at a point just below the knee downward and backward for a short distance to attach to the deep **flexor** tendon, which in turn passes down the back of the leg to a point of attachment on the under surface of the coffin bone. When the suspensory ligament is relaxed, the check ligament converts the tendon below it into a functional ligament to assist the general action of the suspensory ligament. The suspensory ligament is considerably more elastic than the binding ligaments of the joints and its supporting springlike action absorbs a great deal of concussion. This ligament is the most frequently injured in horses that do a great deal of their work at the gallop. In the hind leg, this suspensory ligament is very similar to that in the foreleg, but the check ligament in the hind leg is less perfectly developed.

The plantar ligament is a strong band of ligamentous tissue on the back of the hock bones. It extends from the point of the hock to the upper end of the metatarsus or cannon bone and, because of its strong attachment to the small hock bones, braces the hock against the strong pull of the tendon of Achilles.

The ligamentum nuchae or ligament of the neck is a fan-shaped ligament of very elastic tissue extending from the poll and upper surfaces of the cervical vertebrae downward and backward to attach to the longest spines of the thoracic vertebrae or withers. It assists the muscles of the neck in holding the head and neck in position.

THE MUSCULAR SYSTEM

The muscular system provides movement both internally and externally. Muscles are the active organs of motion and are characterized by their property of contracting or changing shape when stimulated. Each muscle is supplied by one or more nerves that not only bring commands from the brain to make it contract, but also carry back to the brain impulses that tell of the degree of contraction. This correlation results in smooth, even movements instead of jerky or staggering movements. Muscles are red flesh or lean meat and compose about 50 percent of the total body weight.

The muscular system is made up of three types of muscles:

- Smooth or involuntary muscle
- Cardiac or involuntary striated muscle
- Striated or skeletal muscle

Smooth Muscle

Smooth muscle is sometime called **visceral** muscle. It is found in the digestive syste and in the **uterus** of females. Smooth muscles are capable of prolonged periods of activity before becoming fatigued. The visceral muscles of the digestive system perform wavelike contractions called **peristalsis** for many successive hours. Contraction of smooth muscle is involuntary (see Figure 5-7).

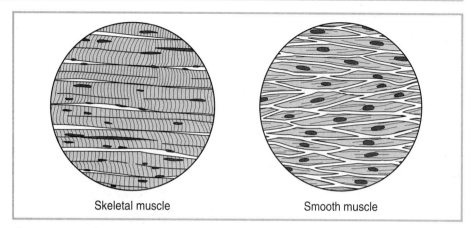

Skeletal muscle   Smooth muscle

Figure 5-7   A diagram of typical smooth muscle fibers

Cardiac Muscle

This muscle is found only in the heart. Contraction of the cardiac muscle is inherent and rhythmic, requiring no nerve stimulus. The rate of contraction is controlled by the autonomic nervous system and requires no conscious control. The heart must continue to ceaselessly contract throughout the entire lifetime of the horse with only split-second intervals of rest.

Striated or Skeletal Muscle

Skeletal muscles are usually attached to the bony levers of the skeleton and move the body voluntarily under the direct control of the will. Skeletal muscle may attach its fleshy fibers directly to a bone, but usually the main part, or belly, of the muscle terminates at either or both ends in a strong cordlike structure called a tendon that transmits the pull of the muscle as it contracts. This tendon arrangement avoids inefficient and bulky thickenings at knees, hocks, and fetlocks and also permits several large muscles to attach on one small area of bone.

Skeletal muscles are generally arranged in opposing sets, one set bending the limb or body, the other set straightening it. Usually both sets are active at the same time but to different degrees, one acting as a brake on the other. Voluntary muscles can contract for only short periods of time before becoming fatigued and requiring rest.

The contractile portion, or belly, of voluntary muscles consists of many elongated muscle cells side-by-side lengthwise of the muscle, which, when stimulated, become shorter and thicker. The tendon of a muscle is quite similar to a ligament in structure and transmits the power of the muscle to some definite point of movement. The contractile portion of a muscle has a large supply of blood, but the supply to the denser tendons is rather limited. The body of most muscles is attached to some bone at a point called the origin. The tendon of the muscle may pass one or more joints and attach (or insert) to some other bone.

The extensor and flexor muscles of the legs are an example of muscles in sets, one group having a certain general action and the other group the exact opposite action. A muscle is an extensor when its action is to extend a joint and bring the bones into alignment. A muscle is a flexor when its action is to bend the joint. Some muscles, if their points of origin and insertion are separated by two or more joints, may act as a flexor of one joint and an extensor of another joint. Except to establish fixation and rigidity of a part, such opposed muscles do not act simultaneously in opposition to each other, but act successively (see Figure 5-8).

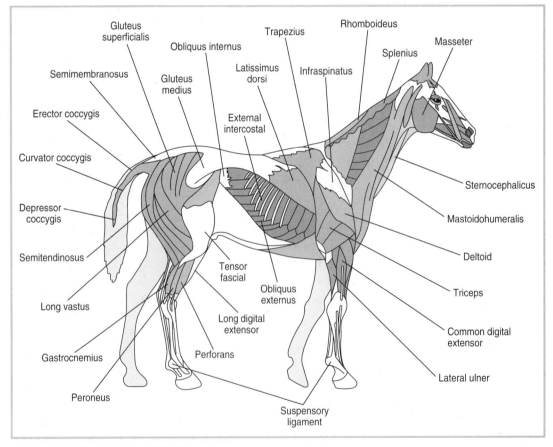

Figure 5-8 The muscles of the horse

Study of the horse's body in response to exercise is a relatively new scientific field. Equine exercise physiology includes the study of skeletal muscle, the blood and its circulation, and the cardio-respiratory system—the heart and lungs. This field of study also involves understanding concepts such as metabolic specificity, muscle fiber-type distribution, and VO_2 Max.

Metabolic specificity means that a horse trains for the type of competition it will be competing in—fast runners train by running fast. For example, if a horse runs a mile today at a 4-minute pace, his body will respond by storing fuel and rebuilding tissue so he can run a 4-minute mile tomorrow with greater ease. However, he will not be fit to run a mile at a 2-minute pace. The exercise he performed in training specifically prepared him for running a 4-minute mile.

Muscle fiber types are either fast twitch or slow twitch depending on their contraction speed and ability to use oxygen. Slow twitch fibers are best suited to low-intensity, long-duration exercise. Fast-twitch fibers are best suited to high-intensity, short-duration exercise. Muscle fiber-type distribution is highly heritable in horses and humans.

VO_2 Max is also measured in exercising horses. VO_2 Max is simply a measure of an individual's ability to use oxygen to sustain aerobic work. The V stands for volume, the O_2 stands for oxygen, and the Max stands for the maximum. Translated, VO_2 Max means the maximum volume of oxygen that an individual can consume. It is usually expressed in relation to body weight so that individuals of varying weight can be compared. An individual with a high VO_2 Max is usually better suited to long-distance types of exercise while the individual with a low VO_2 Max is usually better suited to high-intensity, short-duration types of exercise. VO_2 Max in humans is highly heritable.

Modern horse trainers will need to learn and use equine exercise physiology if they are to stay competitive.

Tendons, Sheaths, and Bursae

Many muscles, especially those of the legs, have long tendons that pass one or more joints and undergo changes of direction or pass over bony prominences before reaching their point of insertion. Tendon sheaths and tendon bursae at various points of friction along the length of the tendon eliminate undue friction to allow the muscle to act more efficiently. A tendon sheath is a synovial sac through which a tendon passes. This sheath secretes synovia to lubricate the tendon. A tendon bursa is a synovial sac interposed between the tendon and the surface over which it passes in a change of direction. It serves the same purpose as a tendon sheath but differs from a sheath in that the tendon is not surrounded by the synovial sac. Tendon sheaths and bursae are found chiefly near joints. The synovial membrane and the synovia secreted in these sacs are the same as those found in the joints.

THE DIGESTIVE SYSTEM

The digestive system converts feed into a form that can be used by the body for maintenance, growth, and reproduction. It consists of all the parts of an organism involved in taking food into the body and preparing it for **assimilation**, incorporation into the body. In its simplest form, the digestive system is a tube extending from the mouth to the **anus** with associated organs. This includes the mouth, esophagus, stomach, intestines, anus, and other associated organs like the liver, teeth, pancreas, and salivary glands. Digestive systems vary according to whether the animals are **herbivores** (eating only plants), **carnivores** (eating only animals), or **omnivores** (eating plants and animals). Horses are herbivores.

The entire digestive tract of a mature light horse is approximately 100 feet long. This length is coiled and looped many times, but is usually very small in diameter and has a total capacity of about forty to fifty gallons.

The stomach of the adult horse makes up less than 10 percent of the total capacity of the digestive tract; the small intestine, the site of most nutrient absorption, makes up only 30 percent. About 65 percent of the capacity of the digestive system is in the **cecum** and colon, which digest the forages consumed by the horse (see Figure 5-9).

Rate of feed passage through the stomach and small intestine is very rapid. Particle size affects the rate of passage; grinding or chopping increases the rate of passage and decreases absorption of nutrients by the horse. Any feed not digested and absorbed in the small intestine is passed on to the cecum and colon within two to four hours. Because of this relatively low volume capacity and rapid rate of passage through the upper gut, it is easy to overwhelm the digestive capacity of the horse's stomach and small intestine.

The horse's cecum and colon contain large microbial populations allowing for digestion of fibrous feeds. If large amounts of concentrates reach the cecum, they will become fermented very rapidly and may produce excessive gas or lactic acid and cause colic or founder.

Mouth

The mouth extends from the lips to the **pharynx** and is bounded on the sides by the cheeks, above by the hard palate, and below by the tongue. Separating

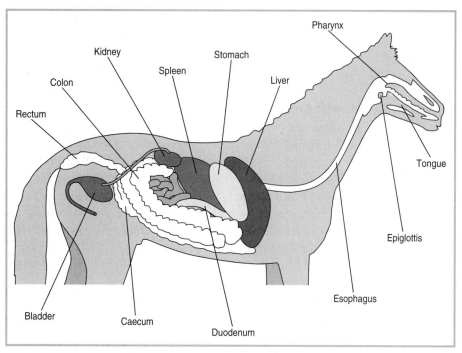

Figure 5-9 The digestive system of the horse

the mouth from the pharynx is the soft palate, a fleshy curtain suspended from the back part of the hard palate, which permits the passage of food and water from the mouth to the pharynx but prevents passage in the opposite direction. The lips pick up loose feed, which is passed into the mouth by the action of the tongue. When horses graze, they grasp food with their incisor teeth. The food is masticated, or ground up, between the molar or cheek teeth and mixed with **saliva**. The saliva is secreted into the mouth by the salivary glands, the largest of which is the **parotid** lying below the ear and back of the jaw.

Saliva moistens and lubricates the mass of food for swallowing and, as a digestive juice, acts on the starches and sugars in the feed. The ball of masticated food is forced past the soft palate into the pharynx by the base of the tongue.

Horses drink by drawing the tongue backward in the mouth and using it like the piston of a suction pump. A horse usually swallows slightly less than a half-pint at each gulp. The ears are drawn forward at each swallow and drop back during the interval between swallows.

Pharynx

The pharynx is a short, somewhat funnel-shaped, muscular tube between the mouth and the esophagus that also serves as an air passage between the nasal cavities and the **larynx**. The muscular action of the pharynx forces food into the esophagus. Food or water, after entering the pharynx, cannot return to the mouth because of the traplike action of the soft palate. (For the same reason, a horse cannot breathe through the mouth.) Food or water returned from the pharynx passes out through the nostrils.

Esophagus

The esophagus is a muscular tube extending from the pharynx down the left side of the neck and through the thoracic cavity and diaphragm to the stomach. Food and water are forced down the esophagus to the stomach by a progressive wave of constriction of the circular muscles of the organ. In the horse, this wave of constriction cannot move in the reverse direction, so vomiting is not possible. The return of food or water through the nostrils is an almost certain indication that the horse has choked because the esophagus has been blocked by a mass of food or a foreign object.

Stomach

The stomach is a U-shaped muscular sac in the front part of the abdominal cavity close to the diaphragm. Food entering the stomach is arranged in layers, the end next to the small intestine filling up first. The contents of the stomach are squeezed and pressed by the muscular activity of the organ. The digestive juice secreted by the walls of the stomach is known as the **gastric juice**.

Small Intestine

Extending from the stomach to the cecum, the **small intestine** is a tube about seventy feet in length that holds about twelve gallons and is composed of three parts, the **duodenum**, the **jejunum**, and the **ileum.** It is about two inches in diameter. After leaving the stomach, it is arranged in a distinct U-shaped curve. The small intestine lies in folds and coils near the left flank, being suspended from the region of the loin by an extensive fan-shaped membrane called the **mesentery.**

Large Intestine

The **large intestine** is divided into the cecum, **large colon**, **small colon**, **rectum**, and anus. The horse, unlike humans or dogs, consumes large quantities of cellulose in its diet. The usual digestive enzymes are not effective against cellulose, so the horse must rely upon bacteria to break down the cellulose into substances it can absorb into its body. In order to give the bacteria time to act on the cellulose, the cecum and the large colon in the horse have been greatly enlarged so that the food moves slowly through this part of the digestive tract.

    The cecum is an elongated sac extending from high in the right flank downward and forward to the region of the diaphragm. The openings from the small intestine and to the large colon are close together in the upper end of this organ. The contents of the cecum are always liquid. The cecum is about four feet long with a capacity of about eight gallons.

The large colon is about twelve feet in length, has a diameter of ten or twelve inches, and holds about twenty gallons. It extends from the cecum to the small colon and is usually distended with food. Bacterial action and some digestion of food takes place here also.

The small colon is about ten feet in length and four inches in diameter. It extends from the large colon to the rectum. The contents of the small colon are usually solid; here the balls of dung are formed. Most of the moisture in the food is reabsorbed in this portion of the large intestine.

The rectum is that part of the digestive tract, about twelve inches in length, that extends from the small colon through the pelvic cavity to the anus, which is the terminal part of the digestive tract.

THE URINARY SYSTEM

Life processes produce waste products. The urinary system is composed of the kidneys, **ureters**, bladder, and **urethra**. The kidneys are paired organs located on each side of the backbone opposite the eighteenth ribs. The chief function of the kidneys is to maintain water and mineral balance and excrete the wastes of metabolism. The urinary bladder holds the wastes until they are excreted.

The kidneys are from six to seven inches in length, four to six inches wide, and about two inches in thickness. The right kidney is roughly triangular with the corners rounded, but the left is more bean shaped and longer and narrower. In the course of a day, all the blood in the body of the horse passes through the two kidneys more than 400 times and is filtered of nitrogenous wastes each time (see Figure 5-10).

Nephrons, the tiny functional units of the kidneys, filter the blood received by the kidneys (see Figure 5-11). The outer portion or cortex of each kidney has several million tiny nephrons that filter approximately 200 gallons of liquid a day, rejecting blood cells and proteins but permitting fluid salts and other chemicals, including nitrogenous wastes, to pass through them. The kidneys also act in reverse and return to the blood stream such valuable substances as the salts, sugars, and most of the fluids—all but about 2 gallons of the 200 gallons of fluid are returned to the blood.

The urine, containing the nitrogenous waste and any excess salts or sugars not required by the body, is collected in the inner portion of the kidney, the renal pelvis, and then drained from the kidney drop by drop through the ureter to the bladder. The bladder is a sort of muscular balloon, which in the horse can expand greatly without bursting. As urine collects in the bladder, nerve endings signal that the bladder needs to be emptied. At this time the urine flows from the bladder to the outside environment through the urethra.

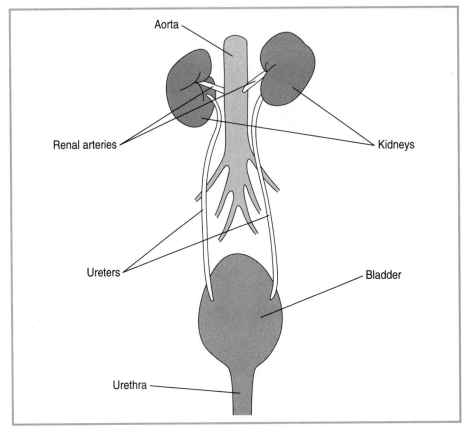

Figure 5-10   A diagram of the urinary system of the horse

In mares the urethra is short and wide. In males the urethra is long and narrow since it travels the length of the **penis**.

THE RESPIRATORY SYSTEM

The respiratory system takes in oxygen from the environment and delivers it to the tissues and cells of the body; it also picks up carbon dioxide from the tissues and cells and delivers it to the environment. The organs of respiration consist of the nasal cavity, pharynx, larynx, trachea, **bronchi**, and lungs. The lungs are the essential organs of respiration; the other parts are simply passages carrying air to and from the lungs. Air is taken into the lungs where oxygen is removed by diffusion into the blood (see Figure 5-12).

The pharynx is common to both the respiratory and digestive tracts. The larynx is a short tubelike organ between the pharynx and the trachea commonly known as the voice box. Stretched vertically within this cartilaginous box are the vocal cords, two folds of elastic tissue. By contracting the lungs and

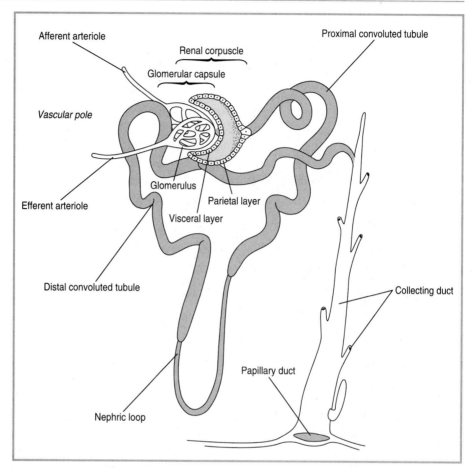

Figure 5-11 A diagram of a nephron

forcing air past these folds of tissue, the horse sets them into vibration and produces the sound known as neighing, whinnying, or nickering. The larynx also regulates the amount of air passing into or out of the lungs and aids in preventing the inhalation of foreign objects.

The trachea is a long tube connecting the larynx with the lungs and is located in the lower median border of the neck. It is composed of a series of cartilaginous rings held together by elastic fibrous material.

Bronchi are branches of the trachea that connect the trachea with each lung. Each bronchus in turn divides into a number of minute tubes that penetrate every part of the lung tissue (see Figure 5-13). The branching bronchi end in groups of minute air sacs similar to bunches of grapes, called **alveoli,** in which the gaseous exchange of carbon dioxide and oxygen takes place between the circulating blood and the air.

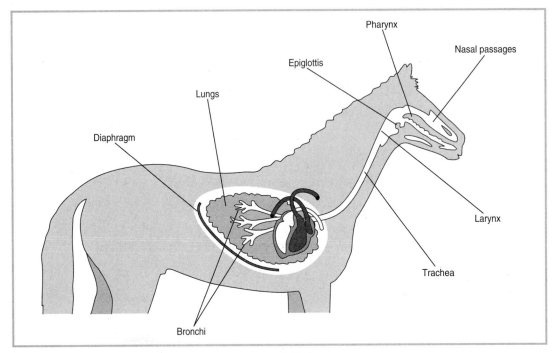

Figure 5-12 A diagram of the respiratory system of the horse

Physiology of Respiration

Respiration is the act of breathing. It consists of the exchange of oxygen in the air for carbon dioxide in the blood, and the interchange of these gases between the blood and the body tissues. The former is known as **external respiration** and the latter as **internal respiration**. External respiration consists of two movements—**inspiration** and **expiration**. Inspiration is brought about by a contraction of the diaphragm and an outward rotation of the ribs. The diaphragm bulges into the airtight thoracic cavity as a dome-shaped muscle. It works like a piston, drawing air into the lungs.

Expiration is effected by a relaxation of these muscles and a contraction of rib and abdominal muscles to force air out of the lungs. Abdominal muscles are used extensively in labored breathing. Since the diaphragm plays such an important part in respiration, it follows that the distention of the digestive tract with bulky food material interferes with normal breathing, especially when the horse is being worked at fast gaits.

The lungs of the average horse contain about 1½ cubic feet of air. The normal horse at rest breathes at the rate of eight to sixteen times a minute, and inhales at each respiration approximately 250 cubic inches of air. The amount of air required by the horse depends upon the extent of muscular work being performed.

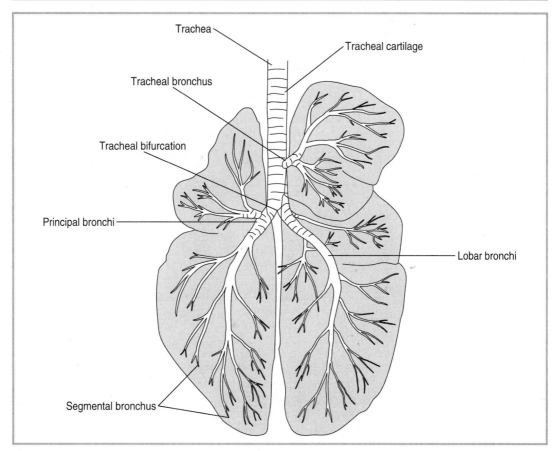

Figure 5-13   A diagram of the bronchial tree of the horse

THE CIRCULATORY SYSTEM

The circulatory system distributes blood throughout the body. This system consists of the heart, **veins**, and **arteries**. Pumping action of the heart causes blood to flow through the arteries to the lungs where it picks up oxygen and carries it to the rest of the body. Oxygen is necessary for all cells of the body. As the blood delivers oxygen to the cells of the body, it picks up carbon dioxide, a waste product, which is carried in the blood back through the veins to the heart and lungs. The lungs release the carbon dioxide to the environment and pick up more oxygen. The blood also carries food substances and waste products (see Figure 5-14).

Heart

The heart is situated in the left half of the thorax, between the lungs and opposite the third to sixth ribs. In the ordinary sized horse, the heart weighs

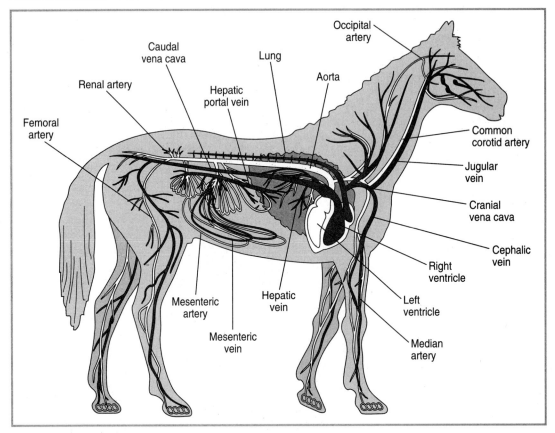

Figure 5-14   A diagram of the circulatory system of the horse

from seven to eight pounds. It is enclosed in a sac called the **pericardium**. The heart is a muscular pump composed of four chambers:

- right atrium
- right ventricle
- left atrium
- left ventricle

Right and left sides of the heart are separated by a muscular wall. Four valves in the heart keep the blood flowing in one direction.

Blood pumped out of the left ventricle into the **aorta** passes through arteries of progressively smaller diameters until it reaches the **capillary** beds of the skin, muscles, brain, and internal organs. Here oxygen and nutrients are exchanged for carbon dioxide and water. The blood is then conducted back to the heart through veins of progressively larger diameter. Finally the blood reaches the right atrium through the vena cavae.

Blood next passes into the right ventricle, which pumps it out to the pulmonary circulation and finally into capillaries around the air sacs in the lungs. Here the carbon dioxide is exchanged for oxygen and the blood returns to the left side of heart by the pulmonary vein and then to the left ventricle where the cycle begins again.

Beating of the heart is controlled internally, but the force and rate of the heartbeat is influenced by the nervous system and the endocrine system. The heart rate speeds up when a horse exercises, becomes excited, runs a fever, overheats, or experiences any circumstance when more blood is needed by the tissues.

The heart is a muscle and, as such, requires its own blood supply. The blood vessels that provide this nourishment encircle the heart like a crown at the juncture of the atria and the ventricles, sending branches to both these structures, and are known as the **coronary vessels** (see Figure 5-15).

Blood

The fluid tissue of the body—blood—carries food substances and oxygen to each cell of the body and takes waste products formed there away from the cells. Blood is a red alkaline fluid composed of blood **plasma** and red and white blood cells. It clots almost immediately upon exposure to air. The total amount

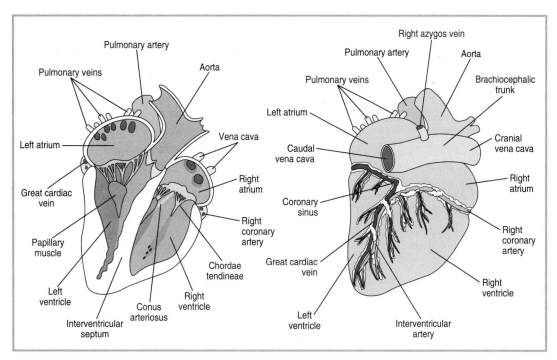

Figure 5-15 A diagram of the horse heart

is about one-fourteenth the weight of the body. The **white blood cells** are the active agents in combating disease germs in the body. **Red blood cells** originate in the red bone marrow, liver, and spleen, and carry oxygen from the lungs to the tissues and carbon dioxide from the tissues back to the lungs.

The blood is the regulator of the body, carrying food to the tissues, waste products away from the tissues, distributing heat, assisting in regulating the temperature, and neutralizing or destroying bacterial and viral invaders.

Vessels and Lymphatics

The arteries have rather thick elastic walls and carry the blood from the heart to the tissues of the body. When the blood is forced into the arteries by the heart, they expand and, in returning to their unexpanded state, force the blood onward. The expansion and contraction of the arteries is the **pulse**.

The veins have much thinner walls and, in many cases, are equipped with one-way valves at frequent intervals, opening toward the heart. The veins of the legs of the horse have such valves. The veins carry the fluid from the tissues to the heart. Veins are located between muscle masses and, as the horse moves, the veins are squeezed and the blood, having to go somewhere, is directed back to the heart by way of the **venae cavae,** the great veins from the front and back portions of the horse.

Capillaries are microscopic in size and function as numerous connecting tubes between the arteries carrying blood to the cells and the veins carrying blood away from the cells. It is through the walls of the capillaries that the exchange of food and oxygen for waste products of the body takes place.

The **lymph vessels** and **lymph nodes**, or lymphatics, consist of numerous well-defined groups of lymph nodes and connecting vessels. The vessels all converge to form one large duct that lies parallel to the aorta, the main artery from the heart, and empties into one of the large veins near the heart. Lymph glands are located at strategic places along the main vessels and act as filters for the **lymph**, which assists in carrying food from the digestive tract to the tissues and waste products back to the blood stream.

Physiology of Circulation

The heart movements are controlled by an intricate group of nerves. The heartbeat is a combined cycle of contraction and relaxation of the organ. In the normal horse at rest, the heart beats from thirty-eight to forty times a minute. The pulse rate is determined by counting the rate of pulsation in some artery that is easily palpitated, for example, the one at the angle of the lower jaw.

The pressure and rate of flow in the veins, compared with the arteries, is very low. The movement of blood in the veins is also aided by respiration movements and muscular contraction, so good circulation is made possible by exercise. The heart, however, is the main pump of the circulatory system.

THE NERVOUS SYSTEM

The nervous system supplies the body with information about its internal and external environment. This system conveys sensation impulses—electrical-chemical changes—back and forth between the brain or **spinal cord** and other parts of the body. It is a complex system consisting of the brain, spinal cord, many nerve fibers, and sensory receptors.

The nervous system is divided into two main portions: the autonomic (automatic) nervous system, and the central nervous system, each concerned with control over different functions of the body. The autonomic nervous system is concerned with control over the respiratory and digestive systems, eyes, heart and blood vessels, glandular products, and other automatic functions directed by the brain stem. The central nervous system controls the conscious or voluntary actions of the body and is again divided into two parts: the spinal cord and brain, and the peripheral nerves. In general, the nervous system is the communication system of the body and made up of the brain, spinal cord, **ganglia**, and nerves (see Figure 5-16).

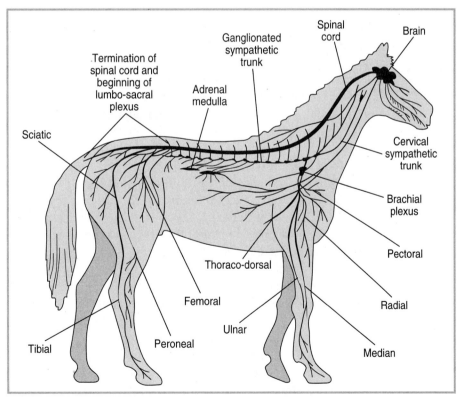

Figure 5-16   A diagram of the nervous system of the horse

The brain and spinal cord are the more important parts of the central nervous system. The brain lies in the cranial cavity of the skull and, considering the size of the horse, the brain is small when compared with the relative brain size of other animals. The size of the brain relative to body size cannot be considered an absolute indication of the degree of reasoning intelligence; however, there is a distinct correlation. The horse is considered to occupy the mid-position in the scale of intelligence of domesticated animals.

The brain is divided into three major portions: the brain stem, the **cerebrum**, and the **cerebellum**. The brain stem—the primitive brain—is the slightly expanded cranial end of the spinal cord that contains the specific nerve centers absolutely essential for the life of the animal, such as the centers controlling the heart beat, respiration, temperature, and a few others. The cerebrum, what is normally thought of as the brain, performs the functions of memory, intelligence, and emotional responses. The cerebellum controls muscular coordination, balance, and equilibrium; it is smaller than the cerebrum and is situated under the caudal part of the cerebrum.

The sense organs or receptors receive stimuli and convey them via electrical impulses over sensory nerve fibers to the brain. The brain analyzes this information and sends commands back via the spinal cord, usually over the same peripheral nerve trunks along motor nerve fibers to motor or effector nerve endings, usually located in the muscles.

Ganglia—secondary nerve centers located chiefly along the spinal cord—act almost like a sub-exchange in a telephone system. They receive and dispatch nerve impulses that do not have to reach the brain, including such stimuli as heat, pain, excessive pressure, and others. These impulses are immediately switched over to motor filaments and cause certain muscles to react instantaneously. For example, if a horse steps on a nail, the whole leg is pulled away immediately, before the brain becomes aware of the action, in an effort known as a reflex.

Nerves are bands of white tissue emanating from the central nervous system and ganglia and extending to all parts of the body. These are the peripheral nerves or nerve trunks consisting of thousands of tiny filaments or wires insulated one from the other by a myelin sheath and ending in tiny specialized knobs, coils, knots, and sprays distributed widely inside as well as on the surface of the body.

There are two kinds of nerves: one sending impulses to the brain over sensory fibers and the other carrying commands back from the brain over motor fibers. Those nerve endings receiving stimuli from the outside are called sense organs or receptors. General sense organs are responsive to pain, touch, and temperature. Special sense organs are concerned with smell, sight, taste, and hearing. In general, nerves follow the courses of the arteries and may be compared with telephone wires, the larger nerves, like telephone cable, containing many separate lines in a bundle.

THE REPRODUCTIVE SYSTEM

Sexual reproduction is the process of creating new organisms of the same species through the union of the male and female sex cells—sperm and eggs. Males and females exist in most species. **Testes** in the males produce sperm. **Ovaries** in the females produce eggs or ova. **Fertilization** occurs when the sperm unites with the egg forming a **zygote**. During a period of pregnancy the zygote develops into a fetus and eventually a new organism. An understanding of the reproductive process is important to the success of horse breeding.

Mare. The reproductive organs of the mare are shown in Figure 5-17. The ovaries produce eggs that unite with the sperm to start the new individual. They also secrete the **hormone**, estrogen, which induces **estrus**, or heat, and progesterone which conditions the reproductive tract for implantation and maintenance of the fetus.

The Fallopian tubes or **oviducts** are the customary site of fertilization of the ovum (egg) by the sperm and serve as a connecting link between the ovary and uterus.

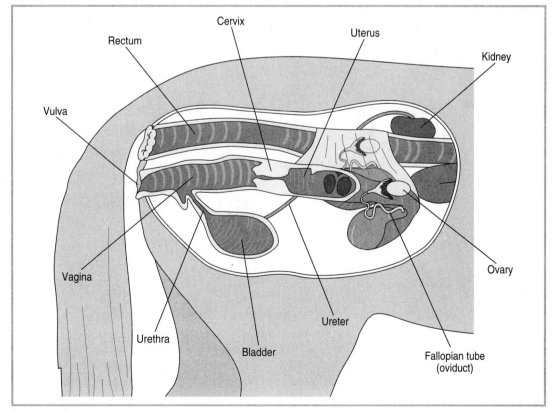

Figure 5-17 The reproductive tract of a mare

The uterus consists of a body, **cervix**, and two horns, one of which receives the fertilized ovum for development.

The **vagina** receives the sperm during mating and functions as a passageway during parturition, or birth.

All reproductive functions in the mare are directed by hormones produced in the glands of her endocrine system; hormonal balance controls all phases of reproductive tract stimulation and inhibition. The mare's reproductive cycle is discussed in detail in Chapter 11.

Stallion

The reproductive organs of the stallion are shown in Figure 5-18. The male reproductive system consists of two testes, three accessory sex glands, and a series of tubules through which spermatozoa are transported to the female reproductive tract.

Spermatozoa are produced in small coiled **seminiferous tubules** in the testes that can be extended 400 to 500 feet in length. Since these developing cells cannot live at body temperature, heat regulation of the testes is critical. Scrotal muscles contract and expand in the normal process of regulating the temperature of the testes. "**Ridgling**" or **cryptorchid** horses are those in

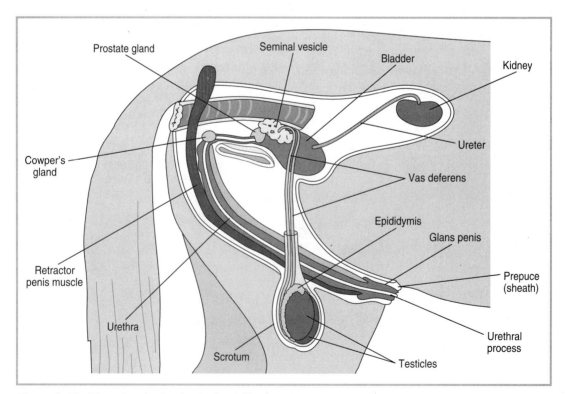

Figure 5-18   The reproductive tract of a stallion

which one or both testes have not descended into the scrotum. The testis maintained in the body cavity is sterile, but the suspended testis is fertile. Since this condition is hereditary, it should not be propagated, as castration of a cryptorchid horse is usually a serious operation.

The accessory sex glands are the seminal vesicles, prostate, and bulbourethral gland. These furnish alkaline fluid secretions to transport and neutralize the urethra in which spermatozoa are transported from the epididymis through the urethra, which terminates at the end of the penis.

THE ENDOCRINE SYSTEM

The glands producing internal secretions, or the endocrines, form a system of ductless glands that influence the vital functions of the horse from before birth until death. Endocrine secretions control the events leading up to and including conception, gestation (pregnancy), parturition (birth), digestion, metabolism, growth, puberty, aging, and many other physiologic functions. **Homeostasis**, or balance, in the horse is largely under the control of the endocrine system.

Secretions of the endocrines are called hormones. Hormones are secreted without a duct directly into the circulatory system where they travel to their target organ or tissue to influence its function.

Recent discoveries in endocrinology have blurred the lines between hormones and enzymes, and the definition of a hormone is being broadened as scientists gain a better understanding of endocrinology. The major components of the endocrine system of the horse are shown in Figure 5-19.

Hormones and Their Actions

Hormones aid in the integration of body processes by stimulating or inhibiting target organs. Although the time lapse between release and effect is longer than for the nervous system, the complementary function of the two systems provides for full coordination of body responses of horses. The ultimate purpose of hormones is to provide a means of adaptation between the body and its external or internal environment.

Hormones may be classified into two categories by their chemical composition. Steroid hormones are secreted by the **adrenal cortex** and the gonads. Protein or protein-like hormones are secreted from the **pituitary gland**, the **thyroid**, the pancreas, and the **adrenal medulla**.

Hormones regulate bodily reactions through their effects on target organs. They do not cause a reaction or event that could not otherwise occur; they merely modify the rate at which target organs perform functions. Hormones function at extremely small levels in the body, with the rate of secretion varying according to the level of stimulation required.

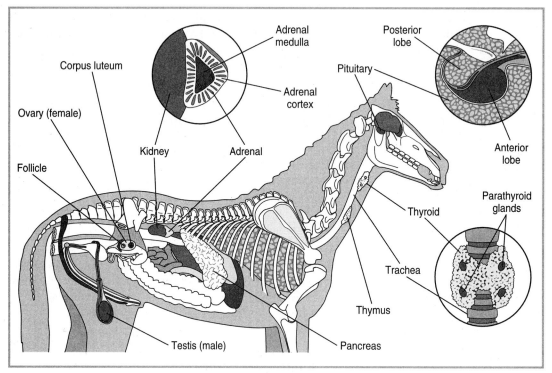

Figure 5-19 The major endocrine organs of the horse

Hormonal output is often controlled through a feedback system from the target organ. This is most evident through the interaction of the anterior pituitary gland with other endocrine glands. Hormonals released by the anterior pituitary control the level of activity of several other endocrine glands (adrenal cortex, thyroid, gonads). Increased hormone production by these glands serves as a negative feedback on the pituitary, causing in it a reduced rate of secretion of the stimulatory hormone.

The pituitary and the **hypothalamus** work together as a functional unit to coordinate the endocrine and nervous systems in their actions. The hypothalamus is the "center" of the autonomic nervous system and "master" of the pituitary. Through direct nervous connection and the releasing of hormones (factors), the hypothalamus controls the pituitary.

Posterior Pituitary

The hormones of the posterior pituitary (neurohypophysis) differ from the other pituitary hormones in that they do not originate in the pituitary, but are only stored there until needed. The two hormones, vasopressin (antidiuretic hormone, or ADH), and oxytocin (milk let-down hormone), are actually produced in the hypothalamus. Their method of transfer from the hypothalamus to the pituitary is unique because it is not through the vascular system, but along the axons of the nervous system.

ADH. Vasopressin, or antidiuretic hormone (ADH), is a small polypeptide (chain of amino acids). ADH does not always function under everyday events. Hemorrhaging, trauma, pain, anxiety, and some drugs will trigger its release, and low environmental temperatures will inhibit it. ADH exerts its effects upon the distal tubules and collecting ducts of the loops of Henle of the kidney, resulting in increased water absorption.

Oxytocin. Oxytocin controls lactation and reproductive phases of the mare. A neural stimulus, such as suckling causes the hypothalamus to stimulate the posterior pituitary into releasing oxytocin, which is circulated through the blood until it comes into contact with the myoepithelial cells surrounding the alveoli of the mammary gland. The oxytocin causes the myoepithelial cells to contract, effectively squeezing the milk out of the secreting alveoli and releasing it into the milk ducts, cistern, and teats of the mammary gland.

Oxytocin also plays a role in reproductive processes. During the estrus cycle, oxytocin stimulates uterine contractions that facilitate the transport of sperm to the oviduct at estrus; during the late stages of gestation, it aids parturition.

Anterior Pituitary

The hormones of the anterior pituitary (adenohypophysis) are produced within the pituitary gland itself. They consist of the follicle stimulating hormone (FSH), luteinizing hormone (LH), prolactin, adrenocorticotropic hormone (ACTH), thyroid-stimulating hormone (TSH), and growth hormone.

FSH and LH. The two pituitary gonadotropins, FSH and LH, are necessary for the maintenance of gonadal functioning. FSH in the mare stimulates overall follicular growth. Follicle maturation is achieved through the combined actions of FSH, LH, and the female sex hormones, which are discussed in more detail later in this chapter.

The action of LH on a follicle is to increase the growth rate and stimulate the secretion of estrogen. Ovulation (the release of an egg) is triggered by this process. The conversion of the follicle to a corpus luteum (a gland formed on the ovary after ovulation that produces progesterone) is the result of LH activity. The continued secretion of progesterone from the corpus luteum is controlled by LH. Progesterone maintains pregnancy by keeping FSH and estrogen in check.

The actions of these hormones in stallions are analogous to those in mares. FSH in the male stimulates the formation of sperm by exerting its effect on small tubules in the testes. Full sperm production cannot be accomplished without the joint effort of LH, known as interstitial cell stimulating hormone

(ICSH) in the male, and certain levels of testosterone. ICSH facilitates the production of testosterone from specialized cells of the testes.

Prolactin.   Prolactin, the lactogenic or luteotropic hormone (LTH), is vital for the proper development of lactation in horses. It cannot initiate the secretory process, and requires estrogen and progesterone to "prime" the mammary system. Prolactin does not seem to be as necessary for the continuation of lactation as it is for its initial development, and for stimulating the corpus luteum. Prolactin has not been demonstrated to have specific effects in male reproduction so far. Figure 5-20 illustrates the location in the brain of the thalamus, third ventricle, hypothalamus, pituitary gland, and infundibulum.

ACTH.   Adrenocorticotropic hormone (ACTH) secreted from the anterior pituitary causes several events to occur. Of primary importance is the release of adrenocorticoid **steroids** from the adrenal cortex into the bloodstream. Other effects include a reduction of lipid levels from the adrenocortical cells, a lowered concentration of adrenal cholesterol and ascorbic acid, a general increase in adrenal cell size and number, and an increase in adrenal blood flow. ACTH promotes the secretion of aldosterone, especially following body stress, such as loss of blood. (The hormones produced by the adrenal cortex are discussed later.) ACTH also influences processes not related to adrenal func-

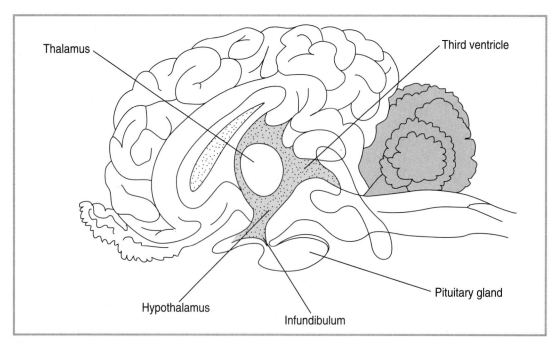

Figure 5-20   The organs that are affected by the hormones of the pituitary and hypothalamus

tion, including movement of fatty acids and neutral fats from fat deposits, ketogenesis, muscle glycogen levels, hypoglycemia, and amino acid levels of the blood.

TSH. The thyroid stimulating hormone (TSH) promotes the release of thyroxin from the thyroid gland. It also increases the rate of binding of iodine within the thyroid. The release of thyroxin serves as a general metabolic control, with higher levels of thyroxin producing an increased metabolic rate.

STH. The basic function of the growth or somatotropic hormone (STH) is to stimulate an increase in body size. Growth hormone, along with other pituitary hormones, is important in protein synthesis providing high intracellular concentrations of amino acids. It exerts its effects on bone, muscle, kidney, liver, and adipose (fat) tissues in bones; in particular, the epiphyseal plates—long bone growth sites—are sensitive to it. Growth hormone regulates, along with the thyroid hormone, the filtration rate and bloodflow through the kidney.

Growth hormone mobilizes fat from adipose tissue, resulting in increased blood levels of ketone bodies, together with stimulation of the alpha cells of the pancreatic islets, causing glucagon secretion. Growth hormone also exerts a stimulating influence on milk production in lactating mares, either partly or entirely due to an increased amount of mammary gland tissue.

Pineal

The **pineal gland** in horses and most other mammals is responsible for melatonin synthesis. It functions on a photo-receptive basis, causing different levels of melatonin production depending on light intensity. The pineal also affects the development and function of the gonads.

Thyroid

The thyroid gland secretes thyroxin. This hormone controls the rate of metabolism. Another hormone, calcitonin, also produced by the thyroid, aids in the metabolism of calcium, and is essential for general bone development. The thyroid is interrelated to other endocrine glands, the adrenals and the gonads through the pituitary.

Thyroxin. The structure of the thyroid hormone, thyroxin, is unique because the element iodine is essential for biological activity and release of the hormone from the gland. Thyroxin is necessary for the maturing of animals. While growth hormone is responsible for physical growth, thyroxin is necessary for the proper differentiation of body structures. Growth and eruption of the teeth of horses is under thyroid control. Even the skin and hair are affected by thyroid changes. A lack of thyroxin will cause a thinner coat of hair, with individual hairs being more coarse and brittle.

Reproductive failures and deficiencies in both sexes may be at least partly attributed to a lack of thyroxin, causing a variety of problems from abortions and stillbirths in mares to impaired sperm production and lowered libido in stallions.

The thyroid hormone has an impact on temperature-regulating processes. By increasing the general rate of oxygen consumption at the cellular level, heat production is increased. Thyroxin stimulates general nervous functions at all levels, decreases the threshold of sensitivity to many stimuli, shortens reflex time, and increases neuromuscular irritability.

Low levels of thyroxin during developmental stages have detrimental effects on the nervous system.

Goiter, the enlargement of the thyroid area, can be brought about by either hyperthyroid or **hypothyroid** conditions. The most common cause in animals is a deficiency of iodine making the animal hypothyroid. Many feedstuffs have goitrogenic effects (goiter producing), and inhibit the activity of the thyroid. Vegetables such as cabbage, soybeans, lentils, linseed, peas, peanuts, and all of the mustard-like plants possess goitrogens. These interfere with the process of trapping iodine by the thyroid.

Parathyroid

The parathyroid gland is located dorsal to the thyroid in horses, is responsible for the maintenance of proper calcium levels in the blood and extracellular fluids. Parathormone, the secretion of the parathyroid, increases calcium levels in the blood and affects calcium and phosphate levels of the bones and kidneys.

Thyrocalcitonin from the thyroid has the opposite effect, causing a decrease in blood serum levels of calcium during events of **hypercalcemia**. Parathormone affects bones directly by mobilizing calcium from the bones into the bloodstream. Parathormone also lowers the ability of the kidney to excrete calcium, thereby increasing calcium retention. Parathormone and vitamin D work together on calcium release from bone and in increased absorption of calcium from the intestine.

Pancreas

The pancreas is primarily an organ of digestive secretions, although there are functionally different groups of cells mixed throughout the pancreas, known as the Isles of Langerhans. These cells have rich blood supplies and consist of so-called alpha and beta cells. Beta cells are the most common and they produce the hormone **insulin**. Insulin lowers the blood glucose and gets glucose across the cell membrane where it can be metabolized. The alpha cells are responsible for the production of glucagon. Glucagon acts the opposite of insulin.

Adrenal Cortex

The adrenal cortex is the outside layer of the adrenal glands, which are located near the kidneys. The adrenal cortex produces steroid hormones. These hormones bear some structural resemblance to cholesterol. Adrenal cortical hormones include glucocorticoids, mineralocorticoids, and **androgens**. The secretion of the glucocorticoids from the adrenal cortex is stimulated by ACTH. The glucocorticoids influence metabolic functions while the mineralocorticoids influence metabolism of minerals like sodium and potassium. Androgens are masculinizing sex hormones.

Deficiencies in glucocorticoid levels have detrimental effects on general body metabolism. A primary function of the glucocorticoids is as a catalyst in the gluconeogenic process—the formation of glucose from proteins and fats. They also help regulate water metabolism together with the mineralocorticoids.

Increase in the size of the adrenals can be observed in animals that are involved in stress situations. The stress of crowding is a major factor in adrenal enlargement, and adrenal weights of wild animals are used as a measure of population density. Over-activity of the adrenals produces androgens that inhibit the production of gonadotropins and thereby lower reproductive performance.

Other sources of steroid hormones, besides the adrenal cortex, are the ovaries, testicles, and placenta. Steroids are inactivated by their target organs and in the liver and kidney. These inactivated hormonal substances are water soluble and are readily eliminated through the urine.

Adrenal Medulla

The adrenal medulla is located at the center of the adrenal glands and it produces two hormones—epinephrine and norepinephrine. Epinephrine (also known as adrenaline) helps the horse adjust to stress situations and activates the fight or flight mechanism. Norepinephrine helps maintain the tone of the vessels in the circulatory system. Release of these two hormones is controlled by nerves that enter the adrenal medulla.

Gonads

Sex hormones are primarily secreted by the ovaries and testes, and also, to some extent, by nongonadal organs such as the adrenals and the placenta. There are four types of hormones: androgens, estrogens, progesterone, and relaxin. The first three types are steroids while the fourth is a protein.

The strongest and most predominant of the androgens is testosterone, which is produced by the interstitial or Leydig cells of the testicles. Testosterone and related hormones are responsible for male secondary sex characteristics of stallions, body conformation, muscular development, and libido. They are also responsible for the growth and development of secondary sex glands of the male, as well as maintaining the viability of the spermatozoa and stimulating penile growth. Testosterone is rapidly used by target organs or degraded by the liver and kidneys.

The ovaries produce two steroid hormones, estradiol and progesterone, and another protein hormone, relaxin. Estrogen comes from the Graafian (mature) follicles of the ovary. Progesterone comes from the corpus luteum on the ovary. A mature follicle ruptures at ovulation to release an egg. This ruptured follicle then develops into a second endocrine structure, the corpus luteum and primary production shifts from estrogen to progesterone. The function of progesterone is to prepare the uterus for implantation and maintenance of pregnancy. Progesterone also suppresses the formation of new follicles, new estrus, and prepares the mare for lactation through increased mammary development.

Relaxin is a hormone related specifically to the birth process, and does not appear until late in pregnancy, just before parturition. It acts upon the ligaments and musculature of the pelvis, cervix, and vagina. The precise site of formation of this hormone is not known, yet it is speculated that production may occur in the cells that are located in the boundary region of the cortex and medulla of the ovaries.

During pregnancy, the uterus itself takes on hormonal functions through the production of placental hormones: pregnant mare serum gonadotropin, estrogens, and progesterone. These hormones serve to maintain the uterus in a way that is favorable for the continued growth and development of the mammary gland.

Gastrointestinal Tract

All hormones secreted by the gastrointestinal mucosa and small intestine are related to the digestive process. Five of these have been chemically identified, with the possibility of more existing, making the small intestine a major site of hormonal production, second only to the pituitary.

One hormone, secretin, is responsible for stimulating pancreatic bile, and small intestine secretions. While causing an increase in fluid levels of the intestine, secretin has no effect on actual enzymatic increases. It also seems to have negative effects on the activity of the stomach.

A second hormone, enterokinin, causes an increased rate of secretion of digestive juices and enzymes of the small intestine.

Enterogastrone and cholecystokinin are two hormones that are related to fat levels in the diet. Enterogastrone inhibits rates of gastric secretion; in response to feed fat in the intestine, it slows down rate of feed passage so that more time can be spent in the digestion of feed.

Table 5-1 summarizes the hormones of the horse and their origin and functions.

Table 5-1 Major Endocrine Glands and Hormones

Gland	Hormone	Function
Hypothalamus	Releasing hormones	Control the pituitary gland
Posterior pituitary	Oxytocin	Stimulates uterine contractions and milk let down
	Vasopressin or ADH	Increases water absorption in kidney
Anterior pituitary	Growth hormone (STH)	Promotes growth of most tissues
	Prolactin (LTH)	Promotes lactation; stimulates corpus luteum
	Adrenocorticotropic hormone (ACTH)	Stimulates adrenal cortex
	Thyroid stimulating hormone (TSH)	Stimulates thyroid gland
	Follicle stimulating hormone (FSH)	Stimulates follicle growth on the ovaries and sperm production in the male
	Luteinizing hormone (LH)/ Interstitial cell stimulating hormone (ICSH)	LH stimulates ovulation, corpus luteum function, secretion of progesterone, and secretion of estrogen in the female; ICSH facilitates production of testosterone in the male
Pineal	Melatonin	Adaptation to light-dark cycles
Thyroid	Thyroxine	Controls metabolism and affects growth, reproduction, and nutrient assimilation
	Thyrocalcitonin	Decreases blood serum levels of calcium
Parathyroid	Parathormone	Regulates metabolism of calcium and phosphorus
Pancreas	Insulin and Glucagon	Regulate glucose metabolism
Adrenal cortex	Glucocorticoids	Stimulate conversion of proteins to carbohydrates for energy; decrease inflammation and immune response
	Androgens	Regulate masculine secondary sexual characteristics
	Mineralocorticoids	Regulate sodium and potassium metabolism
Adrenal medulla	Epinephrine and Norepinephrine	Prepares animal for emergencies; mobilizes energy
Testes	Testosterone	Develops and maintains accessory sex glands; stimulates secondary sexual characteristics, sexual behavior, and sperm production
Ovary	Estrogen	Promotes female sexual behavior; stimulates secondary sexual characteristics, growth of reproductive tract, mammary growth, and feedback control
	Progesterone	Prepares uterus, maintains pregnancy and prepares mammary glands for lactation, and provides feedback control
	Relaxin	Facilitates dilation of birth canal
Gastrointestinal tract	Secretin, Enterokinin, Cholecystokinin, Enterogastrone	Control secretions and motility of digestive tract

SUMMARY

The nine body systems of the horse are: skeletal, muscular, digestive, urinary, respiratory, circulatory, nervous, reproductive, and endocrine. Proper function and control of each of these systems is essential to the survival, growth, and health of the horse. While the systems are generally discussed individually, they are interrelated and function in concert with each other.

For individuals working with horses, a basic understanding of the functional anatomy of the horse is essential before discussing growth, aging, movement, selection, nutrition, health, breeding, behavior, management, or even facilities.

REVIEW

Success in any career requires knowledge. Test your knowledge of this chapter by answering these questions or solving these problems.

True or False

1. The digestive system provides a large store of calcium and phosphorus.

2. The mouth is not part of the digestive system.

3. Food passes from the mouth through the trachea to the stomach.

4. Blood carries carbon dioxide and oxygen.

5. Capillaries are the largest of the blood vessels.

6. The pituitary produces the steroid hormone testosterone.

Short Answer

7. List the bones in the foreleg of the horse from the shoulder joint down to the hoof.

8. Name six types of cells that form during morphogenesis.

9. List the nine body systems.

10. Identify the four surfaces of an animal.

11. What are the two major divisions of the skeletal system?

12. List the five divisions of the vertebral column.

13. What organ transports food from the mouth to the stomach?

14. What organ filters the waste products out of the blood and helps maintain water and mineral balance?

15. Name the two movements of external respiration.

16. What is the name for the air sacs at the end of branching bronchi in the lungs?

17. List the two main divisions of the nervous system.

18. List five reproductive organs in the mare.

19. List five reproductive organs in the stallion.

20. Name three accessory sex glands in the stallion.

21. Identify the hormones from each the following: posterior pituitary, anterior pituitary, thyroid, pancreas, adrenal, testes, and ovaries.

Discussion

22. Identify the four classifications of bones according to their shape and describe their location and function based on shape.

23. Describe three types of joints.

24. Explain the concept of extensor and flexor muscles.

25. Describe one cycle of external respiration.

26. Briefly, outline the circulation of blood through the body of the horse including the heart and lungs.

27. Define a hormone.

28. Describe the relationship of the anterior pituitary to the other endocrine glands.

29. Why is the nervous system like a communication system?

STUDENT ACTIVITIES

1. Dissect a fresh or preserved heart. Ideally, this should be from a horse, but one from another livestock species will work. Identify all of the parts of the heart and trace the flow of blood through the heart.

2. Construct a model of the visible horse. Hobby shops often sell a model called the visible horse. (Contact a hobby shop or mail order source for a model.) This model reinforces understanding of the structure of many of the systems.

3. Find a mounted skeleton of a horse or some other species. Identify the bone shapes and joint types. Or, instead of the whole skeleton, obtain a model of the front or hind leg and carefully study the relationship of each bone and the joints formed.

4. From a biological supply company, obtain a three-dimensional model of the kidney to study for a better understanding of its function. As an alternative, dissect a fresh or preserved kidney from any of the livestock species.

5. Draw and label your own diagram of the reproductive tract of the mare and stallion.

6. Develop a report on the senses: sight, smell, hearing, touch, and taste. Describe how these sensations are transmitted to the brain and interpreted. How is pain sensed and interpreted? In the report draw diagrams of the various sensory receptors.

7. Create a crossword puzzle of the various hormones, using their site of origin and action as the hints.

8. Construct a model of the lungs using a bottle, some tubing, and balloons. Demonstrate how the movement of the diaphragm fills the lungs. Details can be found in a variety of old laboratory manuals.

ADDITIONAL RESOURCES

Books

American Youth Horse Council. 1993. *Horse industry handbook: a guide to equine care and management.* Lexington, KY: American Youth Horse Council, Inc.

Asimov, I. 1954. *The chemicals of life.* New York, NY: New American Library.

Evans, J. W. 1989. *Horses: a guide to selection, care, and enjoyment.* 2nd ed. New York, NY: W. H. Freeman and Company.

Frandson, R. D., and Spurgeon, T. L. 1992. *Anatomy and physiology of farm animals.* 5th ed. Philadelphia, PA: Lea & Febiger.

Fraser, C. M., ed. 1991. *The veterinary manual.* 7th ed. Rahway, NJ: Merck & Co.

Griffin, J. M., and Gore, T. 1989. *Horse owner's veterinary handbook.* New York, NY: Howell Book House.

Hafez, E. S. E. 1993. *Reproduction in farm animals.* 6th ed. Philadelphia, PA: Lea & Febiger.

Kainer, R. A., and McCracken, T. O. 1994. *The coloring atlas of horse anatomy.* Loveland, CO: Alpine Publications, Inc.

McKinnon, A. O., and Voss, J. L. 1993. *Equine reproduction.* Philadelphia, PA: Lea & Febiger.

Videos

University of Kentucky. 1995. *Anatomy of lower leg.* VAS-0675. Agricultural Distribution Center.

University of Kentucky. 1995. *Biology of breeding.* VAS-0665. Agricultural Distribution Center.

University of Kentucky. 1995. *Management of mare's estrous cycle.* VAS-0667. Agricultural Distribution Center.

University of Kentucky. 1995. *Muscle physiology.* VAS-0672. Agricultural Distribution Center.

Equipment and Supplies

Carolina Biological Supply Company, Carolina Science and Math Catalog 66, 2700 York Road, Burlington, NC 27215-3398

NASCO Agricultural Sciences, 901 Janesville Ave., Fort Atkinson, WI 53533-0901

Nebraska Scientific, 3823 Leavenworth Street, Omaha, NE 68105-1180

Fisher-EMD, 4901 W. LeMoyne Street, Chicago, IL 60651

Software

Compendia! on Horses. Diskettes or CD-ROM, Woodland, CA: Compendia! Inc.

1996 Annual Horse Chronicles. CD-ROM, Dover, NH: Orion Publishing, LLC (Internet address http://www.horse.chron.com).

CHAPTER 6

Biomechanics of Movement

Movement of a horse requires the complex integration of several physiological systems. The bones and joints together compose a complex system of levers and pulleys which, combined with the muscular system, imparts the power of motion to the body. Nerves and sensory organs control the movement. Movement is affected by a horse's conformation, or structure.

OBJECTIVES

After completing this chapter, you should be able to:

- Describe muscle contraction
- Describe the nervous control of muscle contraction
- List four functional groups of muscles
- Explain why the heat generated by muscular contraction affects performance
- Contrast aerobic to anaerobic metabolism during muscular contraction
- Name three types of muscle fibers and identify their function
- Name three extensor and three flexor muscles on the hind and front leg
- Describe the two phases of a stride
- Name three factors of a gait that determine a horse's speed
- Define height, directness, spring, regularity, and balance as they relate to gaits
- Describe the walk, trot, gallop, rack, and canter
- Explain the role of conformation in the movement or performance of a horse
- List and describe six common defects in a horse's way of going
- Describe how the center of gravity may affect the movement of a horse

K E Y   T E R M S

Abductors	Electrolytes	Run
Actin	Fatigue	Running walk
Adductors	Forging	Scalping
Aerobic	Gait	Speedy-cutting
Afferent	Gallop	Spring
Amble	Interfering	Stamina
Anaerobic	Lactic Acid	Step
Back	Metabolism	Stepping pace
Balance	Myocin	Stride
Beat	Neurotransmitter	Stride stance
Canter	Pace	Stride suspension
Catalyzed	Paddling	Sweating
Center of gravity	Pointing	Swing
Collected	Pounding	Trappy
Contraction	Proprioceptors	Troponin
Cross-firing	Rack	Trot
Directness	Reflexes	Walk
Dwelling	Relaxation	Way of going
Efferent	Rhythmic	Winding
Endurance fibers	Rolling	Winging outward

NERVOUS SYSTEM CONTROL

A walk, trot, gallop, or any other gait requires the simultaneous **contraction** and **relaxation** of muscles. Muscular contraction is a complex interaction of many parts of the nervous system and the muscular system. As Figure 6-1 shows, muscle action starts in the brain, where information received through a variety of sensory inputs is processed. For example, the eyes may sense a jump, the ears hear a cluck, or the sides feel a nudge from the heels of the rider. The horse's brain interprets this information along with internal sensory organs like the joint **proprioceptors**, which give the horse a sense of the positions of its limbs. Next, the brain determines the appropriate muscles to contract or relax. This information is sent down the spinal cord and then to **efferent** nerves that end on muscle cells. The muscle contracts or relaxes and the bone and joint respond to produce the action.

After this, the cycle starts over again. The **afferent** nerves send information from the joint proprioceptors back to the brain. As before, the sensory information to the brain is interpreted and another signal is sent back down

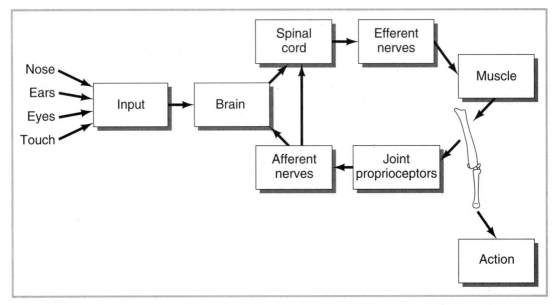

Figure 6-1 The process of how information received by the brain produces action

the spinal cord and efferent nerves, to the muscle producing movement in the bone and joint. Of course this process occurs many times and very rapidly for every movement.

Some signals traveling on the afferent nerves never reach the brain. Instead they go directly to the spinal cord and then back to the efferent nerves and the muscles. These signals are called **reflexes**. An example is a kick in response to a surprise or a twitch of the skin in response to an insect.

HOW MUSCLES CONTRACT

Figure 6-2 illustrates how muscles are organized starting with the muscle, to the muscle bundles, the muscle fibers, and finally the myofibrils. Muscular contraction occurs at the myofibril level.

Figure 6-2 also shows a muscle filament, or myofibril, in cross-section at various states of contraction. Each muscle is made up of thousands of these filaments. Contraction or relaxation is controlled by the nervous impulses received by the muscle cells.

When a muscle contracts, a **neurotransmitter** called acetylcholine (ACH) excites the muscle cells. This causes the release of calcium ions, which bind to a special protein called **troponin**. In turn, two other muscle proteins, **actin** and **myosin**, are free to bind to each other. This causes the muscle to contract. When the calcium concentration drops and the muscle is no longer excited by ACH, actin and myocin no longer bind and the muscle relaxes (see Figure 6-2).

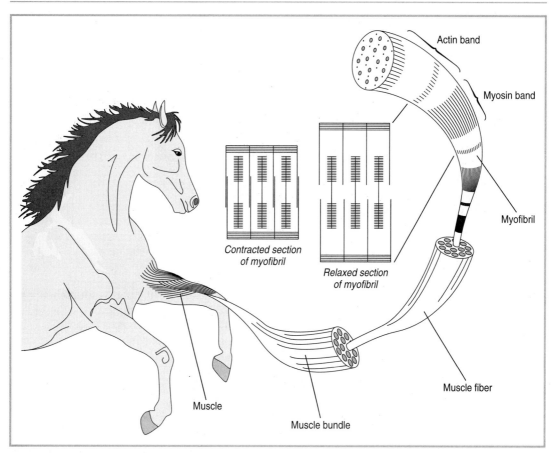

Figure 6-2 The parts of a muscle

Muscular contraction requires energy. This energy is derived from metabolic processes that produce adenosine triphosphate (ATP). ATP is produced from fats, carbohydrates such as glucose or glycogen, and protein. Oxygen from respiration (breathing) is required to produce ATP. As long as sufficient oxygen is available to produce ATP, muscular contraction is called **aerobic**. When muscular contraction is of such high intensity or long duration that adequate oxygen is not available, the products of **metabolism** are converted to **lactic acid** to produce ATP. This type of muscular work is called **anaerobic**. Exercise and training can alter the efficiency of the muscles by increasing the animal's ability to deliver oxygen to the tissues.

Figure 6-3 presents very simplified chemical reactions for aerobic metabolism and anaerobic metabolism. In the horse's body each of these reactions involves numerous series of reactions all linked together and **catalyzed** by enzymes.

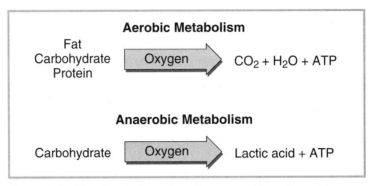

Figure 6-3   The chemical reactions for anaerobic and aerobic metabolism

MUSCLE FIBERS

Muscle fibers require nutrients to contract. Different energy sources can be used by horses performing different types of activity depending on the type of muscle fiber involved in the activity. Three different muscle fiber types are associated with the athletic horse:

- Type I (slow-twitch fibers, aerobic)
- Type IIa (fast-twitch fibers, aerobic)
- Type IIb (fast-twitch fibers, anaerobic)

Type I fibers are in use during relatively slow or light activity and use carbohydrates, fat, or protein. Type IIa fibers are the **stamina** or **endurance fibers** used during periods of aerobic work such as jogging or long-distance riding. These fibers can use carbohydrates, fat, or protein for energy. Type IIb fibers are the speed or power fibers used for periods of strenuous anaerobic work such as sprinting, jumping, or cutting. These fibers use carbohydrates only.

For example, the quarter horse is born with a relatively large proportion of type IIb fibers and does best on a diet of carbohydrates—hay and grain. The endurance horse, such as the Arab, is born with a higher proportion of type I and IIa fibers and does best on a diet of both carbohydrates and fat—hay, grain, and oil.

FATIGUE OF MUSCLES

Fatigue of muscles follows continued work, principally due to the accumulation of waste products in the muscle cells. Recovery requires removal of the accumulated waste products by the blood and lymph, and a fresh supply of nutrition brought to the muscles. Hand rubbing of the legs of a horse after exercise stimulates the blood and lymph vessels in the removal of waste products. It also causes the blood to circulate more freely. Fatigue may also be overcome in part by feeding easily digested carbohydrates for a maximum of energy.

An untrained horse, one not accustomed to steady work, fatigues more easily than a trained horse, mainly because the muscles, respiration, and circulation do not operate as efficiently. There is a limit to the amount of continued muscular effort a horse can expend; harmful fatigue can be avoided by only working the horse at a moderate rate in order to maintain the proper balance between the products of muscular activity and the ability of the blood to remove waste material. An animal should never be worked until exhausted.

HEAT

Heat is a by-product of muscle contraction. To prevent an excessive increase in core body temperature, heat must be dissipated. In the horse, heat is dissipated through **sweating** (evaporation) and by air movement across the body. The blood transports the heat from the working muscles and the core to the skin where it is cooled.

During exercise in hot environments, the need to control body temperature causes a large shift in blood flow to the skin. This may adversely affect the exercising horse by decreasing the blood flow to the muscles. Fluid losses during exercise in hot environments can also significantly decrease plasma (blood) volume. This too may negatively impact the horse, making it harder to maintain adequate blood flow to the muscles. Finally, exercise in hot environments increases the amount of **electrolytes** lost in the sweat. These electrolytes are important for fluid balance, acid-base balance, muscle contraction and nerve function.

MUSCLES INVOLVED IN GAITS

Muscles that the horse uses to execute the various gaits form four functional groups (see Figure 6-4):

1. Flexors
2. Extensors
3. Abductors
4. Adductors

Contraction and relaxation of these groups in the limbs and the attachment of the limbs to the body create the gaits of a horse and other movements. **Flexors** decrease the angle of a joint, while the **extensors** increase the angle of a joint. **Abductors** move a limb away from the center plane of the horse, while the **adductors** pull a limb toward the center plane on the horse.

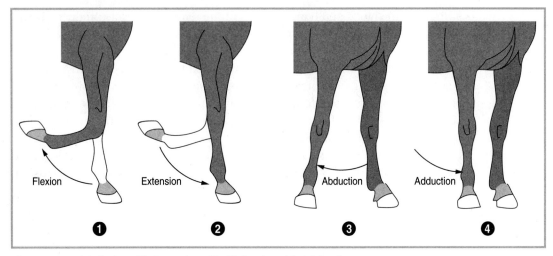

Figure 6-4   (1) Flexion, (2) Extension, (3) Abduction, (4) Adduction

Extensor muscles of the front leg include:

- Brachiocephalicus
- Supraspinatus
- Triceps brachii
- Oblique carpal extensor
- Lateral digital extensor
- Common digital extensor

Flexors of the front leg include:

- Teres major
- Latissimus dorsi
- Biceps brachii
- Flexor carpi radialis
- Flexor carpi ulnaris
- Deep digital flexor

Adductors of the front legs are the pectoral muscles. The abductor of the front leg is the deltoideus.

On the hind leg, the extensors include:

- Biceps femoris
- Semitendinosus
- Semimembranosus
- Gluteus medius

- Quadriceps femoris
- Gastrocnemius
- Long digital extensor
- Lateral digital extensor

Flexors of the hind leg include:

- Iliacus
- Popliteus
- Deep digital flexor
- Superficial digital flexor

For a review of the muscular system, the reader should refer back to Chapter 5.

GAITS AND ACTION

A **gait** may be defined as a horse's **way of going** or the way of moving its legs during progression. The horse is more versatile in selecting gaits than any other four-legged animal and it uses several gaits unique to the species. A gait is characterized by distinctive features, regularly executed. Action refers to flexion of the knees and hocks, the height the horse lifts his feet from the ground, the speed or rate of movement, and the length of the stride.

An understanding of gaits is important to detect lameness, to train a performance horse, or to signal a horse for a specific gait. Some gaits of a horse are natural while others are learned or artificial. Most horses must be trained to execute the artificial gaits.

When describing the various gaits, a **beat** refers to the time when a foot—or two feet simultaneously—strikes the ground. Beats may or may not be evenly spaced in time. A **step** is the distance between imprints of the two front legs or the two back legs. A **stride** is the distance between successive imprints of the same foot.

Components of a Stride

The stride has two phases—**stride stance** and **stride suspension**. Stride stance is the weight-bearing phase, while the stride suspension or **swing** is the nonweight-bearing phase. The speed of a horse is determined by:

- Length of stride
- Rapidity or frequency of stride
- Overlap time or the time on the ground versus time off the ground

For example, the famous race horse Secretariat ran faster because he spent less time with his legs in the stance and overlap phases. In other words,

Horses Had Role in Development of Moving Pictures

Because of his fame, his success at publicizing his activities, and his habit of patenting machines before actually inventing them, Thomas Edison received most of the credit for inventing the motion picture. As early as 1887, he patented a motion picture camera, even though it could not produce images.

Actually, many inventors contributed to the development of moving pictures, and horses helped too. Perhaps the first important contribution was the series of motion photographs made by Eadweard Muybridge between 1872 and 1877. He was hired by the governor of California, Leland Stanford, to capture on film the movement of a racehorse. Stanford had bet someone $25,000 that when a horse is at a fast trot, all four of its feet were off the ground. To prove his point, he contracted Muybridge to make a photographic study documenting animal motion. At an elaborately designed experiment station on Stanford's farm (later site of Stanford University), Muybridge set up a series of stereoscopic cameras.

Muybridge tied a series of wires across the track and connected each one to the shutter of a still camera. The running horse tripped the wires and exposed a series of still photographs, which Muybridge then mounted on a stroboscopic disk and projected with a magic lantern to reproduce an image of the horse in motion. Stanford won the bet, and Muybridge continued his research into various forms of animal locomotion, from crawling infants to elephants.

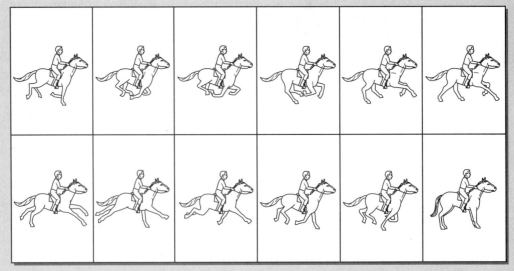

* Editor's Note: Historically, safety helmets were not used at the time this experiment took place. It is advised that whenever a rider is on a horse, an approved helmet is used.

146

Secretariat's legs completed their ground contact quicker and more time was spent in the airborne (suspension) phase.

Other terms used to describe a horse's gait include:

- **Directness**, or trueness, which is the line in which the foot is carried forward during the stride. A horse that paddles does not carry its feet straight forward during the stride (see Figure 6-5).
- Power or the pulling force exerted to create the stride.
- **Height**, which is indicated by the radius of the arc created from the point of the foot's take off to the point of the foot's contact again with the ground (see Figure 6-6).
- **Spring**, or the manner in which weight settles back on the supporting leg at the completion of the stride.
- Regularity, or the **rhythmic** precision of each stride.
- **Balance**, which is the ability of a horse to coordinate action, go composed, and in form.

Common Gaits

Historically, six gaits were considered as natural for the horse—walk, trot, pace, canter, run and back. Now all horses are considered to have four natural

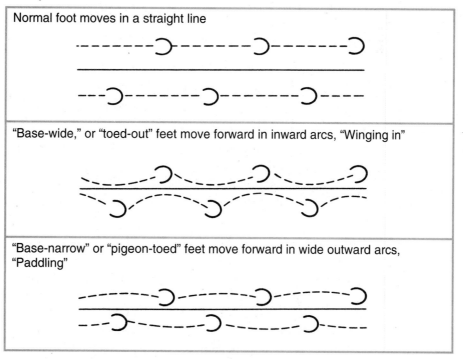

Figure 6-5 Directness vs. Winging in and Paddling

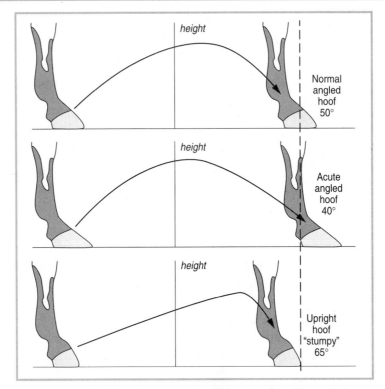

Figure 6-6 The desired arc is created when the hoof is angled correctly

gaits—walk, trot, canter and gallop (or run). Any gait that a horse will execute without training is natural. Some of the common gaits are described briefly below (see Figure 6-7).

Walk. This is a slow, even, four-beat gait. The sequence of hoof beats for the **walk** is: (1) left hind, (2) left fore, (3) right hind, and (4) right fore.

This sequence of beats is considered lateral because both feet on one side strike the ground before the feet on the opposite side strike the ground (see Figure 6-7).

Trot. The **trot** is a two-beat gait with the diagonal fore and hind leg acting together. A period of suspension in which all four feet are off the ground occurs between each beat. The road horse trot is a fast-stepping trot characterized by length and rapidity and executed with extreme degree of extension, or length of stride. Heavy harness trot and hackney trot are high-stepping gaits with a high and springy stride, very **collected** (controlled), and executed with each step showing extreme flexion and precision (see Figure 6-7).

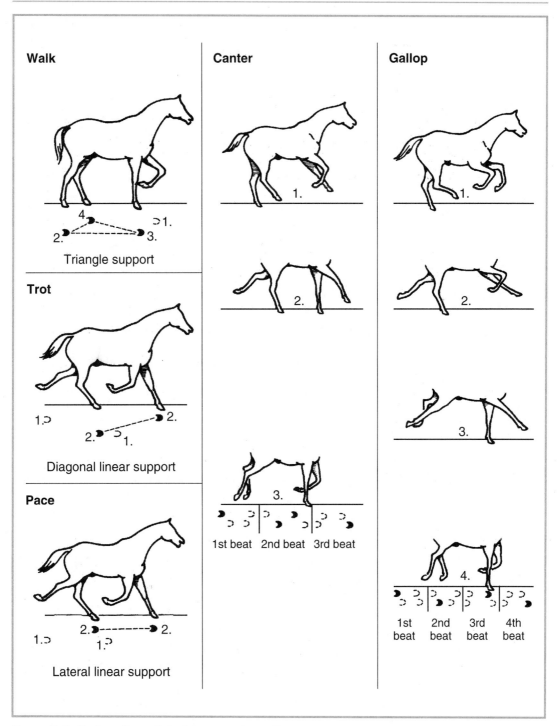

Figure 6-7 Basic gaits: walk, trot, pace, canter, gallop

Canter. The **canter** is a three-beat collected gait. The sequence of beats is: (1) the right rear hoof, (2) left rear and right front hoofs striking simultaneously, and (3) the left front hoof (see Figure 6-7). When cantering, the horse carries more weight on its haunches, or rear quarter. The gait is executed in a slow, animated, collected, rhythmic way in which the lead changes on command. If moving to the left, the horse should lead with its left leg and vice versa. If a horse is cantering to the right and leading with its left foot, the horse is exhibiting what is known as a counter canter.

Gallop. The **gallop** or **run** is a fast, four-beat gait (see Figure 6-7). One hind foot makes the first beat, followed by the other hind foot. The diagonal forefoot is the third beat, and the remaining forefoot is the fourth. A period of suspension follows the four beats. If the horse changes leads it will do so in the period of suspension. The run is the gait of a racehorse.

Pace. This is a two-beat, lateral (side-to-side) gait with fore and hind leg on the same side moving together (see Figure 6-7). A period of suspension occurs between each beat. Since the horse is shifting its weight from side to side, the gait has a rolling motion. It requires a smooth, hard footing and a minimum of draft. Trotting downhill will cause some trotters to pace; pacing uphill will cause some pacers to trot. The **pace** is a speed gait. The **amble** is a lateral gait distinguished from the pace by being slower and more broken in cadence. It is not a show gait.

Slow Gait. The slow gait or **stepping pace** is a show gait. This is a lateral, four-beat gait done under restraint in showy, animated fashion with the front foot on the right followed by the hind foot on the right (see Figure 6-7). In the stepping pace, the break in rhythm is between the lateral fore and rear foot.

Rack. The **rack** is an even, fast, flashy four-beat lateral gait. It is sometimes called a single foot and is characterized by quite a display of knee action and speed. The rack is hard on the horse, but easy on the rider. The excessive leg movement increases the amount of concussion and trauma to the forelegs.

Running Walk. The **running walk** is the fast walk of the Tennessee walking horse. It is faster than the ordinary or flat-foot walk. It is a single-foot or four-beat lateral gait with a break in the impact or rhythm occurring between the lateral fore and hind feet. The horse travels with a gliding motion because it extends the hind leg forward to overstep or overreach the forefoot print.

Back. When a horse **backs** it is actually trotting in reverse. Backing is a two-beat gait in which the diagonal pairs of legs work together.

CONFORMATION AND ACTION

Conformation, the form or structure of a horse, has a bearing on how well it functions or performs. While Figure 6-8 illustrates desired traits, irrespective of breed, this does not mean the illustration is a true representation of all breeds.

How a horse stands is indicative of how it will move. The normal stance, with width between the legs in proportion to the width of the chest, and feet placed straight, results in the legs and feet moving in a straight line. A base-wide horse, particularly if it also toes out, wings inward or moves its feet and legs in with each stride. If the condition is severe, the horse is apt to strike one leg with the other (interfering) resulting in injury and even unsoundness. Base-narrow, with toes pointing in, results in a horse that paddles. This is unsightly and results in excessive hoof wear on the outside quarters and excessive strain on the knee, fetlocks, and tendons.

If a horse stands straight, it is likely to move straight and true. If the legs are set properly, it is better able to move with collected action. A crooked-legged

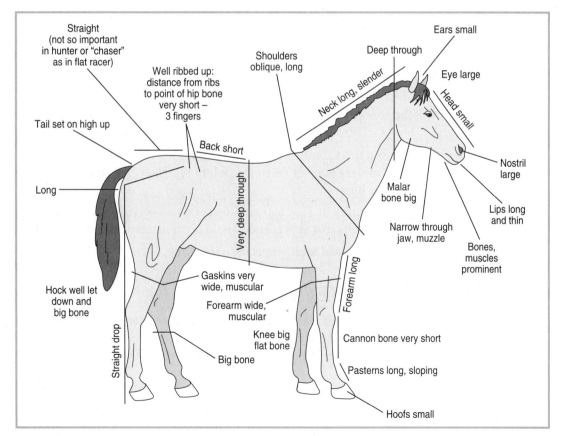

Figure 6-8 Desired horse traits

horse cannot move true. Regardless of a horse's excellent head, neck, shoulder, top, and general balance and conformation, if it is crooked on its legs, it is not a top horse.

Unsoundnesses in the pasterns, cannon bones, knees, and especially the hocks also affect movement.

For more information on conformation refer to Chapters 7 and 8.

The following conformation features affect action and gaits and may predispose an animal to certain unsoundnesses:

- A long forearm contributes to a long stride.

- Sloping shoulders and pasterns are associated with a springy stride.

- A calf-kneed (back at the knees) posture is associated with hard concussion or a pounding gait; it predisposes a horse to bone chips.

- Low rounding withers are associated with a defective gait called forging. A horse with low withers commonly hangs in the bridle, moves with its head low, and handles its front feet awkwardly.

- A pigeon-toed horse will paddle or wing out. Conversely, a splay-footed (heels in, toes out) horse will wing in and the striding leg may actually strike the supporting leg. In addition, its hooves will wear unevenly (see Figure 6-5).

- Short, steep ankles and pasterns result in a stilted stride, hard concussion, and a tendency to cocked ankles and unsoundness.

- Front legs out at the corner or legs set too far apart in front are a structural defect associated with a rolling motion when the horse moves.

- A short, thick, bulky neck too often goes with a straight shoulder and reduces neck suppleness and mobility and the rider's ease in controlling the horse.

- A short straight shoulder and forearm, accompanied by steep pasterns, results in a short stride and a tendency toward sidebones.

- Buck-knees and long toes cause stumbling.

COMMON GAIT DEFECTS

How the horse moves its feet and/or legs while executing the gaits may be considered defects. Some defects cause limb interference and may be severe enough to cause injury. Other defects are not serious but they prevent top performance from the horse. Defects and peculiarities in the gait include forging, interfering, brushing, striking, paddling, winding, scalping, speed-cutting, cross-firing, pointing, dwelling, trappy, pounding, and rolling. These defects can be related to conformation, injuries, or improper shoeing and trimming of the feet.

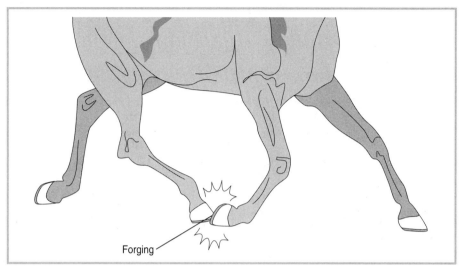

Figure 6-9   Forging occurs when the toe of the hind foot hits the sole of the forefoot.

Forging

Forging is striking the end of the branches of the hoof or the undersurface of the shoe of the forefoot with the toe of the hind foot (see Figure 6-9). This is the diagonal foot in pacers, and the lateral foot in trotters.

Interfering

Interfering is striking the supporting leg, usually at the fetlock, with the foot of the striding leg. Interference commonly occurs between the supporting front leg and a striding front leg or between a supporting hind leg and a striding hind leg (see Figure 6-10). Brushing is a slight interference. Striking is a severe interference resulting in an open wound.

Paddling

Paddling or **winging outward** is an outward deviation in the direction of the stride of the foreleg (see Figure 6-11). It is the result of a narrow or pigeon-toed standing position. Winging is exaggerated paddling and very noticeable in high-stepping horses. Paddling almost always causes interference.

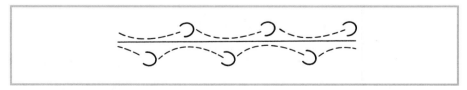

Figure 6-10   Interfering occurs when the foot of the striding leg strikes the supporting leg.

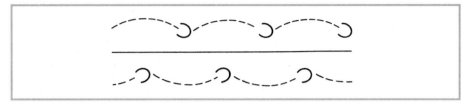

Figure 6-11 Paddling occurs when feet move forward in wide outward arcs. Base-narrow or pigeon-toed feet also cause paddling.

Winding

Winding is twisting the front leg around in front of the supporting leg as each stride is taken. Sometimes it is called threading, plaiting, or rope-walking. Wide-chested horses tend to walk in this manner. Winding increases the likelihood of interference and stumbling.

Scalping

Scalping occurs when the hind foot hits above or at the line of the hair (coronet) against the toe of a breaking-over (beginning the next stride) forefoot (see Figure 6-12).

Speedy-cutting

Speedy-cutting occurs when a trotter or pacer traveling at speed hits its hind leg above the scalping mark against the shoe of a breaking-over forefoot (see Figure 6-12). In trotters, legs on the same side are involved. In pacers, diagonal legs are involved.

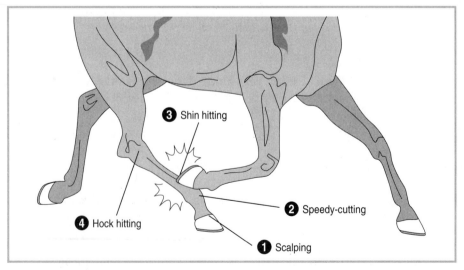

Figure 6-12 (1) Scalping, (2) speedy-cutting, (3) shin hitting, and (4) hock hitting

Several faults in conformation predispose a horse to scalping and speedy-cutting: short backs and long legs, leg weariness or hindlegs set too far under the body, short front and long back legs, and toes too long on the forefeet.

Cross-firing

Cross-firing is essentially the same as forging in a pacer in which the inside of the fore and hind foot strike in the air as the stride of the hind leg is about completed and the stride of the foreleg is just beginning (see Figure 6-13).

Pointing

Pointing is a stride in which extension is more pronounced than flexion. A horse with a pointed stride breaks or folds its knees very slightly and is low-gaited in front. Thoroughbreds at the trot are pointy-gaited. The term pointing is also used to indicate the standing position pose a horse frequently takes when afflicted with navicular bone disease or injury to the foot or leg: it stands on three legs and points with the fourth.

Dwelling

Dwelling is a perceptible pause in the flight of the foot as though the stride had been completed before the foot strikes the ground. It may occur either front or rear and is particularly common in heavy harness horses, heavy show ponies, and some saddlers.

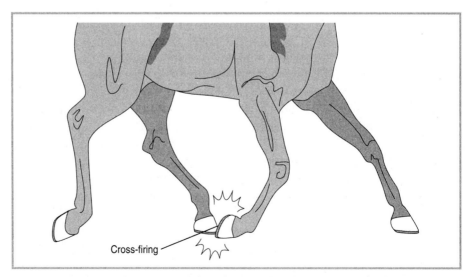

Figure 6-13   Cross-firing occurs when the hind foot hits the opposite forefoot.

Trappy

Trappy is a gait that is a short, quick, choppy stride. Horses with short and steep pasterns and straight shoulders tend to have a trappy gait.

Pounding

Pounding is heavy contact with the ground, usually accompanying a high, laboring stride. Faults in conformation that shift the horse's center of gravity forward tend to create pounding.

Rolling

Rolling describes excessive side-to-side shoulder motion. Horses wide between the forelegs and lacking muscle development in that area tend to roll their shoulders. The toe-narrow fault in conformation can also cause rolling.

CENTER OF GRAVITY

The **center of gravity**, where the mass of the horse is centered, is another important feature affecting the gait. Even though the center of gravity will vary with the horse's shape, it is most commonly located in the middle of the rib cage just caudal to the line separating the cranial and middle thirds of the body. Because the center of gravity is located more cranially, the forelimbs bear 60 to 65 percent of the body's weight. This puts increased stress on the forelimbs resulting in an increased incidence of lameness in those limbs. The horse that is taller over the croup than in the withers has an additional disadvantage because this shifts its center of gravity even further forward.

SUMMARY

The nervous, muscular, and skeletal systems work together to produce movement. The nervous system gathers information about the internal and external environment and provides the stimuli causing muscle contraction. As muscles contract and others relax they act on the joints and bones to produce movement. Muscular contraction requires oxygen and energy in the form of ATP. Muscular contraction produces waste products and heat, both of which must be removed from the muscles.

Muscles that produce the gaits in horses can be grouped according to their function—flexors, extensors, abductors, and adductors. All horses are capable of the four natural gaits: walk, trot, canter, and gallop. Other gaits require training and are said to be unnatural. Gaits can be described by the number of beats and the characteristics of the stride. Conformation can affect the gait as can the horse's center of gravity. How the horse moves its feet and/or legs while executing the gaits may appear as defects.

REVIEW

Success in any career requires knowledge. Test your knowledge of this chapter by answering these questions or solving these problems.

True or False

1. An adductor muscle decreases the angle of a joint.
2. Heat is a by-product of muscular contraction.
3. During a gallop, all four of a horse's feet are off of the ground at the same time.
4. A pigeon-toed horse will exhibit a trappy gait.
5. ATP is a neurotransmitter that excites muscle cells.

Short Answer

6. List four functional groups for muscles.
7. Contrast aerobic to anaerobic metabolism during muscular contraction.
8. List four natural gaits common to all horses.
9. What three factors of a horse's gait determine its speed?
10. Name three extensor and three flexor muscles on the hind leg.
11. Name the two phases of a stride.
12. Identify three types of muscle fibers.

Discussion

13. In relation to muscular contraction, why do working horses sweat?
14. Describe the sequence of events during muscular contraction.
15. How can conformation affect the movement or performance of a horse?
16. Describe six common defects in a horse's way of going.
17. Compare the walk, trot, canter, gallop, and rack.
18. Explain how the center of gravity affects the movement of a horse.

STUDENT ACTIVITIES

1. View videos of horses in motion showing different gaits. Show the video in slow motion and describe phases of the stride.
2. Attend a horse show or view a television broadcast of a horse show and learn to identify the gaits.
3. Using a drawing of the horse skeleton, draw in the muscles of the limbs and attach them to the proper location on the bones of the legs.
4. Research muscular contraction and make several drawings showing how the myofibril contracts. Specifically, show how ATP and calcium (Ca) are involved in contraction.

5. Learn about proprioceptors by closing your eyes or putting on a blindfold. Move your arms or legs to new positions and describe these without looking. Knowing where your arms and legs are in time and space without looking is proprioception.

6. Visit with a horse trainer and discuss how horses are trained to perform gaits.

7. Using numbered hoof prints for the left and right and front and hind legs, diagram the hooves on the ground through one complete cycle of a walk, trot, canter, and gallop.

8. Develop a rapid-fire game using the gaits and common defects matched to their descriptions.

ADDITIONAL RESOURCES

Books

American Youth Horse Council. 1993. *Horse industry handbook: a guide to equine care and management.* Lexington, KY: American Youth Horse Council, Inc.

Davidson, B., and Foster, C. 1994. *The complete book of the horse.* New York, NY: Barnes & Noble Books.

Dossenbach, M., and Dossenbach, H. D. 1994. *The noble horse.* New York, NY: Cresent Books (Random House).

Evans, J. W. 1989. *Horses: a guide to selection, care, and enjoyment.* 2nd ed. New York, NY: W. H. Freeman and Company.

Frandson, R. D., and Spurgeon, T. L. 1992. *Anatomy and physiology of farm animals.* 5th ed. Philadelphia, PA: Lea & Febiger.

Kainer, R. A., and McCracken, T. O. 1994. *The coloring atlas of horse anatomy.* Loveland, CO: Alpine Publications, Inc.

Videos

University of Kentucky. 1995. *Anatomy of lower leg.* VAS-0675. Agricultural Distribution Center.

University of Kentucky. 1995. *Biomechanics of movement.* VAS-0674. Agricultural Distribution Center.

University of Kentucky. 1995. *Muscle physiology.* VAS-0672. Agricultural Distribution Center.

Software

Compendia! on Horses. Diskettes or CD-ROM, Woodland, CA: Compendia! Inc.

1996 Annual Horse Chronicles. CD-ROM, Dover, NH: Orion Publishing, LLC (Internet address http://www.horse.chron.com).

Unsoundness

Unlike other farm animals, the horse is serviceable only when in motion. Any abnormal deviation in the structure or action of a horse can render it partly or completely useless. Any defect that affects serviceability, for example, lameness, blindness, faulty wind, and so on, is considered an unsoundness.

Those defects that detract from appearance but do not impair serviceability are considered blemishes, for example, scars, capped hocks, and elbows. Blemishes are looked down upon in gaited, parade, and some pleasure horses. They are more common in stock horses and tend to detract less from their value than from other types of horses.

An important part of selecting a horse is the ability to recognize common unsoundnesses and blemishes and faulty conformation that tends to predispose the animal toward unsoundness and blemishes.

O B J E C T I V E S

After completing this chapter, you should be able to:

- Distinguish between a blemish and an unsoundness
- Describe the common treatment for many of the problems that could develop into an unsoundness
- Name four common unsoundnesses associated with the head
- List five common unsoundnesses or blemishes that can be found on the body
- Describe two types of unsound lungs
- Name and describe ten unsoundnesses or blemishes of the front or hind leg
- Differentiate between a sprain and a fracture
- Name two types of sprains and two types of fractures
- Identify four conditions that predispose a horse to developing unsoundnesses
- Name and describe six stable vices affecting usefulness
- Describe how to methodically examine a horse for soundness

KEY TERMS

Acute	Curb	Poll
Allergy	Degree of finesse	Poll evil
Anti-inflammatory	Evaluation	Quidding
Atrophy	Fibrosis	Quittor
Bad mouth	Fistulous withers	Ringbone
Bang's	Flat foot	Road puffs
Bars	Flexion tests	Roaring
Base narrow toe-in	Founder	Scars
Base wide toe-out	Fracture	Scratches (grease heel)
Bench knees	Gingivitis	Sensitization
Biting	Gum disease	Shoe boil
Blemish	Hard at the heels	Shoe boil roll
Blindness	Heaving	Sidebones
Bog spavin	Hernia	Splints
Bone spavin (jack spavin)	High ringbone	Sprain
Boot	Hobble	Stall walking
Bowed tendons	Hock	Stifled
Buck-kneed	Hoof testers	Stomatitis
Calcification	Hydrotherapy	Straight shoulders
Calf-kneed	Implants	Stringhalt (stringiness,
Calks	Kicking	crampiness)
Camped out	Knocked-down hip	Sweeney
Camped under	Laminitis	Tail board
Capped hocks	Lateral cartilages	Tail rubbing
Chip fractures	Low ringbone	Thoroughpins
Club foot	Malocclusion	Thrush
Cocked ankles	Moon blindness	Ulcerate
Collar	Navicular disease	Umbilicus
Conformation	Ophthalmia	Unsoundness
Congenital	Osselets	Upright pasterns
Conservative treatment	Ossify	Veterinarian
Contracted feet	Over-at-the-knee	Vices
Contracted heel	Patella (upward fixation of)	Weaving
Corns	Pedal osteitis	Whistling
Cracks (quarter, toe, heel)	Plates	Windgalls
Cribbing	Plumb line	Wind sucking

BLEMISHES vs. UNSOUNDNESS

Basically an **unsoundness** is any condition that interferes or is apt to interfere with the function and performance of the horse. In horse show halter classes, horses with an unsoundness usually do not place. In performance classes, if the apparent unsoundness is not interfering with the horse's action, it is given little consideration. A **blemish** differs from an unsoundness in that it is unattractive, but does not and is not apt to interfere with the horse's performance.

Blemishes are usually an acquired physical problem that may not make the horse lame but may still interfere with the action of the horse. An unsoundness is usually caused by poor **conformation** and will tend to be a problem throughout the horse's lifetime. Often an unsoundness is also a blemish. Both are usually caused by stress and strain placed unevenly on the legs. The location and severity of the problem determines how the horse will be affected.

A blemish may or may not affect the level of performance. An unsoundness usually affects the performance of the horse, at least temporarily. How the horse is used must be considered when evaluating the importance of the problem. A pleasure horse that receives minimal stress and is ridden slightly will have a different **evaluation** than a horse that is at a high level of competition and is being shown and trained vigorously.

Treatment methods vary for these problems. The basic treatment for an injury usually consists of rest to stop further trauma, **hydrotherapy**—application of cold water to the affected area, usually hosing the leg for awhile—and medications to try to reduce swelling. Pain relievers may also be given if necessary. A **veterinarian** can inject medications into areas to reduce inflammation and swelling. Some problems also have surgical treatments. A veterinarian needs to be consulted to determine the amount and location of damage and the best treatment for the horse.

CAUSES OF UNSOUNDNESS

Horses may be lame due to some disease or affliction in the joints, tendons, ligaments, or muscles. Usually lameness from these causes cannot be seen and calls for a diagnosis by a veterinarian. Conversely, many unsoundnesses or indications of unsoundness can be seen. Many unsoundnesses and blemishes are due to excessive stress and strain beyond the endurance of the bone or muscle, injury to a bone or joint, inherited conditions, or nutritional deficiencies.

Xenophon (430[?] to 357 B.C.) was an Athenian soldier, writer, and disciple of Socrates. Besides many other writings, Xenophon wrote *On The Art of Horsemanship*. This is the earliest preserved book on the care and training of horses. He had a clear understanding of the nature of horses. Much of his advice is still good today. From his translated works, here is what Xenophon had to say about the conformation of a colt.

For judging an unbroken colt, the only criterion, obviously, is the body, for no clear signs of temper are to be detected in an animal that has not yet had a man on his back.

In examining his body, we say you must first look at his feet. For, just as a house is bound to be worthless if the foundations are unsound, however well the upper parts may look, so a war-horse will be quite useless, even though all his other points are good, if he has bad feet; for in that case he will be unable to use any of his good points.

When testing the feet first look to the hoofs. For it makes a great difference in the quality of the feet if they are thick rather than thin. Next you must not fail to notice whether the hoofs are high both in front and behind, or low. For high hoofs have the frog, as it is called, well off the ground; but flat hoofs tread with the strongest and weakest part of the foot simultaneously, like a bow-legged man. Moreover, Simon says that the ring, too, is a clear test of good feet: and he is right; for a hollow hoof rings like a cymbal in striking the ground.

Having begun here, we will proceed upwards by successive steps to the rest of the body.

The bones [of the pastern] above the hoofs and below the fetlocks should not be too upright, like a goat's: such legs give too hard a tread, jar the rider, and are more liable to inflammation. Nor yet should the bones be too low, else the fetlocks are likely to become bare and sore when the horse is ridden over clods or stones.

The bones of the shanks should be thick, since these are the pillars of the body; but not thick with veins nor

LOCATION OF COMMON BLEMISHES AND UNSOUNDNESSES

For discussion purposes, some of the common blemishes and unsoundnesses are grouped below according to their location on the body—the head, body, lungs, and limbs.

162

with flesh, else when the horse is ridden over hard ground, these parts are bound to become charged with blood and varicose; the legs will swell, and the skin will fall away, and when this gets loose the pin, too, is apt to give way and lame the horse.

If the colt's knees are supple when bending as he walks, you may guess that his legs will be supple when he is ridden too, for all horses acquire greater suppleness at the knee as time goes on. Supple knees are rightly approved, since they render the horse less likely to stumble and tire than stiff legs.

The arms below the shoulders, as in man, are stronger and better looking if they are thick.

A chest of some width is better formed both for appearance and for strength, and for carrying the legs well apart without crossing.

His neck should not hang downwards from the chest like a boar's, but stand straight up to the crest, like a cock's; but it should be flexible at the bend; and the head should be bony, with a small cheek. Thus the neck will protect the rider, and the eye see what lies before the feet. Besides, a horse of such a mould will have least power of

running away, be he never so high-spirited, for horses do not arch the neck and head, but stretch them out when they try to run away.

You should notice, too, whether both jaws are soft or hard, or only one; for horses with unequal jaws are generally unequally sensitive in the mouth.

A prominent eye looks more alert than one that is hollow, and, apart from that, it gives the horse a greater range of vision.

And wide open nostrils afford room for freer breathing than close ones, and at the same time make the horse look fiercer, for whenever a horse is angry with another or gets excited under his rider, he dilates his nostrils.

A fairly large crest and fairly small ears give the more characteristic shape to a horse's head.

High withers offer the rider a safer seat and a stronger grip on the shoulders.

Reading Xenophon's translation lets us know how much we rely on the knowledge of generations in the past. For the person interested in reading the entire translated text, it can be found on the Internet and downloaded.

Studying Figure 7-1 to become familiar with the location of some blemishes and unsoundnesses and reviewing the muscular and skeletal systems in Chapter 5 will aid understanding this chapter.

Unsoundness of the Head
Unsoundnesses around and relating to the head include blindness, **bad mouth**, **poll evil**, and **quidding**.

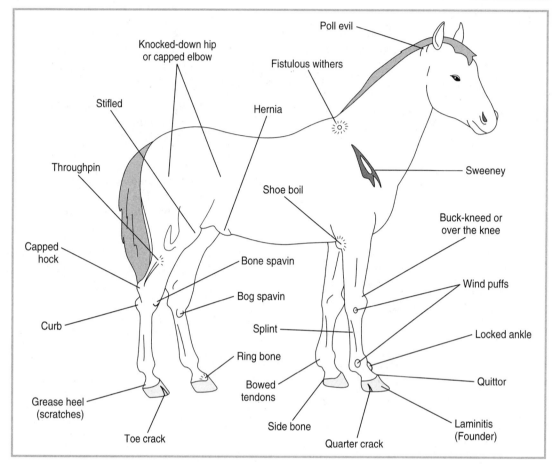

Figure 7-1 Locations of some unsoundnesses on the horse.

Blindness. **Blindness** seriously affects the usefulness of a horse. It is usually characterized by cloudiness of the cornea or complete change of color to white. Pale blue, watery eyes may indicate periodic ophthalmia (moon blindness). Watery eyes may appear in vitamin A deficiency. These conditions are not common in horses on pasture.

Moon Blindness. Periodic **ophthalmia** or **moon blindness** is an inflammation of the inner eye due in part to a vitamin B deficiency. It usually impairs vision and treatment is usually unsuccessful.

Bad Mouth. Bad mouth is a term used to describe various jaw or tooth misalignments. Bad mouth may be a **malocclusion** where the top and bottom teeth do not meet, or a monkey mouth (undershot jaw) where the lower jaw and tooth structure extend beyond the top teeth. Parrot mouth, or overshot

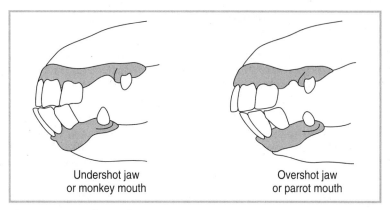

Figure 7-2   Bad mouth

jaw, is another example. In this case the top jaw and incisor teeth extend beyond the lower jaw. A bad mouth is considered an inherited unsoundness (see Figure 7-2).

Poll Evil. Poll evil is a fistula—a lesion or sore—on the **poll** that is difficult to heal. Poll evil (see Figure 7-1) is an acquired unsoundness resulting from a bruise or persistent irritation in the region of the poll. Its cause is *Brucella abortus*, the same organism that causes **Bang's** in cattle. Early symptoms are swelling and touchiness around the head and ears when the horse is being bridled. Severe inflammation, eruption, and bad scars may result if the wound is neglected.

Quidding. Quidding is seen when horses drop food from the mouth while in the process of chewing. This is usually caused by bad teeth or bad gums (**stomatitis** or **gingivitis**). It can also be caused by paralysis of the tongue. If there are sharp edges or points on the teeth, chewing will cause the horse pain when the points rub or cut the tongue and/or gums. Floating (filing) the teeth to remove the points and local treatment of sores will help. Soft palatable feed can be given to soothe sore gums.

When quidding is accompanied by pain while eating or when the bit is in the mouth, dental problems or the absence of teeth should be suspected. If quidding is accompanied by bad breath and weight loss, **gum disease** is a likely cause. Routine dental care and mouth washing will help. In severe cases, **anti-inflammatory** drugs and pain relievers may help the horse feel more comfortable.

Unsoundnesses and Blemishes of the Body
Fistulous withers, sweeney, knocked-down hip, scars, and **hernias** are considered unsoundnesses and blemishes of the body.

Fistulous Withers. This is an inflammation affecting the withers in much the same way as poll evil affects the poll. It may be present on one or both sides of the withers (see Figure 7-1). It should be treated early. Otherwise the disease can linger on, resulting in severe infection and occasionally a crestfallen condition of the neck immediately in front of the withers.

Sweeney. Sweeney (see Figure 7-1) applies to a wasting away of the shoulder muscle overlying the scapula of the horse. This is muscle **atrophy** of the shoulder caused by damage to a nerve in the shoulder. The damage is usually from direct trauma to the shoulder from a kick, running into a wall or solid object, or even running into another horse. It is characterized by the loss of muscle on either side of the spine of the scapula. The spine of the scapula is normally not seen but will become visible as the muscles atrophy. Depending upon the amount of nerve damage and the resulting muscle loss, there will be varying amounts of lameness. The gait of a horse with sweeney is usually characterized by swinging the leg out as it comes forward due to lack of support from the atrophied muscles. Nerve damage is almost always permanent.

Knocked-down Hip. This is a fracture of the external angle of the hip bone (ilium). It results in a lowering of the point of the hip that can be identified best by standing directly behind the horse. Hurrying through narrow doors, crowding in trailers, falling, and injury from other causes may be responsible. Usefulness is seldom impaired, but appearance is greatly affected.

Scars. Scars are marks left on the skin after the healing of a wound or sore. They may appear on any part of the body. A scar is often noticed because of the presence of white hairs. Working stock horses with scars are not discriminated against very much, but gaited and parade horses are seriously faulted for them.

Hernia. A hernia is generally the passage of a portion of the intestine through an opening in the abdominal muscle (see Figure 7-1). It may appear on any portion of the abdomen, but is more common near the **umbilicus**. Hernias are usually not serious enough to cause an unsoundness.

Unsoundness of the Lungs

Any permanent abnormality in the respiration process is a serious unsoundness. Two well-known conditions include **roaring** and **heaving**.

Roaring (whistling). A paralysis or partial paralysis of the nerves that control the muscles of the vocal cords may result in a roaring or **whistling** sound when air is inhaled into the lungs. The condition is seldom apparent when the horse is at rest, but it becomes obvious upon exertion. Roaring may be limited to one nostril and can be determined by plugging each nostril alternately.

Heaving. Heaving is caused by a loss of elasticity in the lungs resulting from a breakdown of the walls of a portion of the air cells. The condition is characterized by a visible extra contraction of the flank muscles during expiration. The expiration process can be seen, and often heard, to proceed normally to about two-thirds of completion, when it is stopped. The flank and lower rib muscles contract briefly, then expiration continues to completion. Dusty hay and/or atmosphere, severe exertion of horses out of condition, and respiratory infections are common causes of the condition.

Unsoundness of the Limbs

The **hock** is the most vulnerable, and the most important, joint of the body. All of the power of a pulling horse is generated in the hindquarters and transmitted to the **collar** by contact with the ground via the hocks. Working stock horses must bear most of the weight on the hind legs by keeping their hocks well under them if they are to attain maximum flexibility. **Degree of finesse** is determined with gaited and parade horses by how well they "move" off their hocks.

Structurally sound hocks should be reasonably deep from top to bottom; well supported by fairly large, flat, straight bone; characterized by clean-cut, well-defined ligaments, tendons, and veins; and free from induced unsoundness and blemishes.

There are many unsoundnesses and blemishes to the limbs. Some conditions are correctable; some are not. The discussion that follows does not identify and discuss every blemish or unsoundness of the limbs. For more detail and additional unsoundnesses the reader should follow up with some of the sources identified in the Additional Resources at the end of the chapter.

Bog Spavin. This is a serious discrimination. **Bog spavin** (see Figures 7-1 and 7-3) is a soft fluctuating enlargement located at the upper part of the hock due to a distention of the joint capsule. It is the result of horses trying to straighten the hock and trauma such as quick stops and turns or getting kicked by another horse.

Bone Spavin. A **bone** or **jack spavin** (see Figures 7-1 and 7-3) is a bony enlargement at the base and inside back border of the hock. It is a common unsoundness of light horses, especially those with sickle hocks or shallow hock joints from top to bottom surmounting fine, round bone. Such conformation should be seriously faulted in a working stock horse. In the early stages, lameness may be apparent only when the horse has remained standing for a while. Bone spavins, like ringbones, may fuse bones and render joints inarticulate.

Bowed Tendons. **Bowed tendons** (see Figures 7-1 and 7-3) are apparent by a thickening of the back surface of the leg immediately above the fetlock. One or more tendons and ligaments may be affected, but those commonly

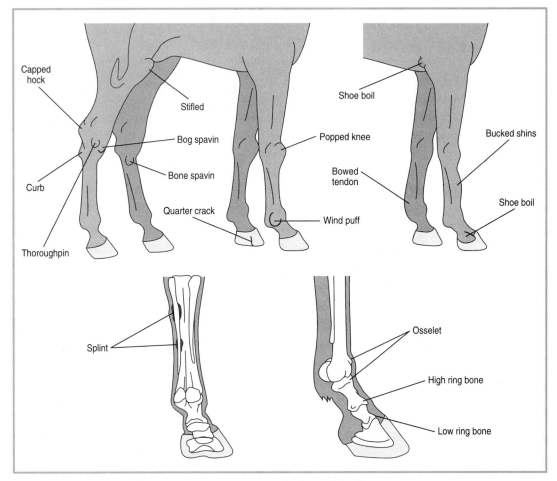

Figure 7-3 Locations of some unsoundnesses on the legs of the horse

involved are the superflexor tendon, deep flexor tendon, and suspensory ligament of one or both front legs. Predisposing causes are severe strain, wear and tear with age, and relatively small tendons attached to light, round bone. Bowed tendons usually cause severe unsoundness.

Buck-kneed. **Buck-kneed** (see Figures 7-1 and 7-4) is also called **over-at-the-knee**. Because the knee is in front of the **plumb line**, the leg is not straight. This can be the result of a shortening of the muscles on the front of the knee. It can be present in foals, but usually disappears by about three to four months. If it is **congenital** and a permanent condition, the forwardness will cause excessive strain on the leg.

Bucked Shins. Bucked shins (see Figure 7-3) occur on the front, top part of the cannon bone, below the knee. The forelimbs are affected more often than

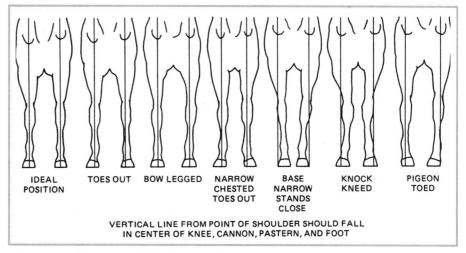

IDEAL POSITION   TOES OUT   BOW LEGGED   NARROW CHESTED TOES OUT   BASE NARROW STANDS CLOSE   KNOCK KNEED   PIGEON TOED

VERTICAL LINE FROM POINT OF SHOULDER SHOULD FALL IN CENTER OF KNEE, CANNON, PASTERN, AND FOOT

Figure 7-4   Front view of the chest and leg of the horse.

the hind limbs. Bucked shins are caused by trauma to the surface of the bone, possibly from stress on the tendons that run down the cannon bone or possibly from the forces distributed up the bone during fast work. This is usually the result of overwork and overtraining, especially in young racehorses. It is seen most commonly in racing Thoroughbreds, quarter horses, and Standardbreds.

Bucked shins can be mild to severe depending on the amount of stress applied to the bone. If severe, the bone may have a fracture. Because of this, the prognosis depends on the amount of injury. A swelling exists over the area that is affected and lameness may be present. In a mild case, rest and mild hand-walking are recommended. When the pain has decreased, a horse should be put on a controlled exercise program to get back into shape.

The greater the severity of the bucked shin, the longer the rest and controlled exercise program will need to be extended. A variety of surgical procedures are available to correct fractures, if necessary.

Calf-kneed.   **Calf-kneed** (Figure 7-5) is a deviation of the knee joint behind the plumb line (the opposite of buck-kneed) so the leg is not straight. This places great strain on the tendons and ligaments running down the back of the leg. Additionally, compression of the bones in the knee joint are increased leading to chip fractures. This is a serious problem. The horse is generally unable to tolerate heavy work.

Camped Out.   The **camped out** leg (see Figures 7-5, 7-6, and 7-7) is too far back and behind the plumb line. Usually the whole leg is involved and the plumb line is at or in front of the toe instead of behind the heel. This is often seen accompanied by upright pasterns and straight hocks in the hind limbs,

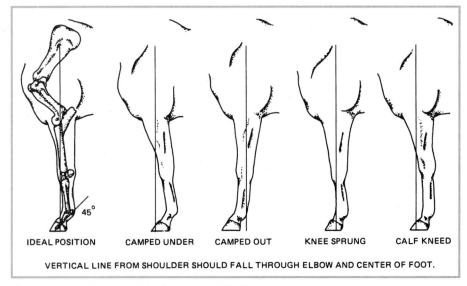

Figure 7-5   Side view of the front leg of the horse

which causes increased concussion on the navicular bone, pastern, fetlock joint, and hock. **Camped under** is the opposite condition.

Capped Elbow. Sometimes called a shoe boil (see Figure 7-3), this is a blemish at the point of the elbow. Capped elbow is usually caused by injury from the shoe when the front leg is folded under the body while the horse is lying down. Shoes with **calks** (heels) cause more damage than those with **plates**.

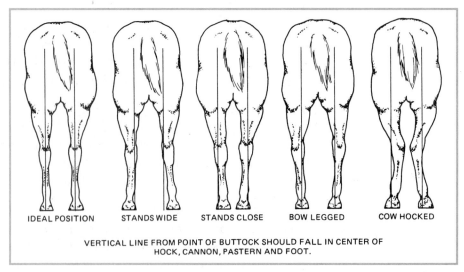

Figure 7-6   Rear view of the hind quarters and legs of the horse

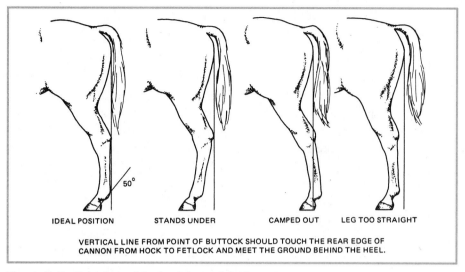

IDEAL POSITION   STANDS UNDER   CAMPED OUT   LEG TOO STRAIGHT

VERTICAL LINE FROM POINT OF BUTTOCK SHOULD TOUCH THE REAR EDGE OF
CANNON FROM HOCK TO FETLOCK AND MEET THE GROUND BEHIND THE HEEL.

Figure 7-7   Side view of the back legs of the horse

Capped Hock.   This is a thickening of the skin or large callus at the point of the hock. It is a common blemish. Many **capped hocks** (see Figures 7-1 and 7-3) result from bumping the hocks when being transported in short trailers or in trailers with unpadded tail gates.

Chip Fractures.   **Chip fractures** can occur in several different places but are most common at the knee and they are the most common problem that affects the knee. Chip fractures are small fractures that break off one of the bones in the knee. They are usually caused by high amounts of concussion and stress on the knee and are seen most frequently in racing horses: Thoroughbreds, Standardbreds, and quarter horses. They can be seen in any highly athletic horse, such as barrel racers, steeplechasers, hunters, and jumpers.

The knee is able to absorb a great deal of shock due to the many joint spaces and fluid within it. Normally, the horse is able to distribute shock by correctly lining up the bones of the knee. However, as the horse extends the knee and becomes fatigued while performing, the position of the bones may shift slightly. This decreases the knee's ability to disperse the forces. Additionally, as the horse fatigues, the flexor muscles on the back of the leg tire and allow for overextension of the knee. This compresses the front of the knee and creates high forces on the front of the bones.

The amount of damage that occurs is dependent on:

- Age of the horse (younger horses bones have not matured as much and are more likely to injure)

- Activity of the horse and its training level (the more strain put on the knee, the more likely it is to get a fracture)

- Conformation (horses with knees that do not line up properly at any time, will have greater strain on the knee all the time)
- Improper trimming and shoeing (an uneven foot or broken hoof-pastern axis will change the distribution of the forces across the knee)

Most chip fractures occur in the front of the bones at the radial carpal bone, intermediate carpal bone, and third carpal bone. Chip fractures are generally small. Names for other types of fractures indicate an increase in size, for example, corner fractures and slab fractures.

The horse with a chip fracture will show inflammation, swelling, pain, and lameness at the knee. The nature of the signs generally depends on the amount of trauma to the bones. If the fracture is new, the swelling tends to be diffuse across the knee, but over time, the swelling accumulates over the area of the chip. By feeling the knee, a veterinarian can usually find signs of tenderness. **Flexion tests** help to determine the extent and location of the fracture. X-rays and lameness tests are also used to diagnose the problem.

Two types of treatments are available—surgical or conservative. Treatment is usually dictated by the location and severity of the fracture and the amount of lameness. **Conservative treatment** is usually chosen for small fractures that do not appear to be troubling the horse much and are not affecting the joint. Treatment is stall confinement and hand-walking for a period of weeks or months. Anti-inflammatory drugs can be used to lessen the pain and decrease inflammation. Surgery is reserved for large fractures or fractures that have been displaced and are adversely affecting the horse and the joint.

The recovery of the horse is based on the amount of damage to the joint caused by the fracture. Mild cases generally have a very good prognosis. If the horse is not able to recover to its original athletic level, it may be suited for a less strenuous exercise program. A veterinarian will be able to provide advice.

Cocked Ankles. **Cocked ankles** (see Figure 7-1) may appear in front but are more common in hind legs. Severe strain or usage may result in inflammation or shortening of the tendons and a subsequent forward position of the ankle joints. Advanced cases impair movement and decrease usefulness.

Contracted Feet. **Contracted feet** (see Figure 7-8) are caused by continued improper shoeing, prolonged lameness, or excessive dryness. The heels lose their ability to contract and expand when the horse is in motion. Horses kept shod, those with long feet, and those with narrow heels are susceptible to the condition. Close trimming, going bare footed, or corrective shoeing usually produces sufficient cure to restore the horse to service.

Contracted Heel. This results when the back of the foot, or the heel region, becomes narrower than normal. **Contracted heel** is more common in the

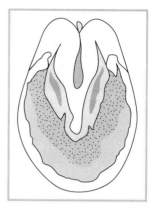

Figure 7-8   Contracted foot

front feet than the hind feet. Contracted heel is usually accompanied by a small frog (an elastic formation on the sole of the foot) and a sole that is concave or "dished." A small frog is an indication that the frog is not being compressed when the horse is walking.

As the frog shrinks in size with decreased use or over trimming, the heel contracts. If the heels become excessively contracted, lameness may result. Contracted heels are usually caused by improper shoeing/trimming and hoof growth. Lameness can also cause this condition if the horse is not applying weight to the foot. The treatment is via corrective shoeing. Depending on the severity of the problem, it may take a year or more to get the heels spread to the correct width for the foot.

Corns.   **Corns** appear as reddish spots in the horny sole, usually on the inside of the front feet, near the **bars**. Advanced cases may **ulcerate** and cause severe lameness. The causes are many, but bruises, improper shoeing, and contracted feet are the most common. Response to correct treatment and shoeing, is usually satisfactory.

Club Foot.   In this condition the foot axis is too straight and the hoof is too upright. **Club foot** is usually associated with a problem such as a contracted deep digital flexor tendon. It may be due to injury (one foot), improper nutrition (two or more feet), or possibly heredity. The upright foot causes the horse to be stiff and rough in its gait and may make it unrideable. Additionally, because the horse cannot flex and extend its foot correctly, it has a tendency to stumble. The nutritional condition may be correctable. The horse that inherits the problem should not be bred to prevent the further passage of the problem.

Curb.   **Curb** (see Figures 7-1 and 7-3) is an enlargement on the back of the leg, just below the hock. It is caused by trauma that causes the plantar ligament to become inflamed and then thickened. Curb is seen with faulty conformation (such as in sickle- and cow-hocked animals) and also with horses that slide too far, too fast in deep ground. Direct trauma, such as a kick from another animal or the horse kicking a hard object, may also stress the ligament and cause a curb. At first, a curb causes pain, then heat and swelling.

The horse should be rested until the swelling and pain have decreased. Anti-inflammatory drugs may be used. If the trauma does not affect the hock joint, there is often a good prognosis following treatment. The swelling will diminish and often just a blemish is left. If the horse has conformation problems that are causing the curb, it is unlikely to heal completely and permanent lameness may persist.

Flat Foot. This type of conformation lacks the natural concave curve to the sole. Instead, the sole is flat and predisposed to more contact with the ground. **Flat foot** increases the chances for sole bruises and resulting lameness. To avoid pain and pressure from hitting the ground with the sole, the horse may learn to place its heel down first. This is more common in the front feet than in the rear. Generally, this conformation is not seen in light horse breeds, but may be seen naturally in some draft breeds. Corrective shoeing may help alleviate some of the contact with the ground.

Founder or Laminitis. **Founder** or **laminitis** (see Figure 7-1) is an inflammation of the sensitive laminae that attach the hoof to the fleshy portion of the foot. Its cause is probably a **sensitization (allergy)**. When horses gain access to unlimited amounts of grain, founder often results. Other conditions conducive to founder are retained placenta after foaling and sometimes lush grass. All feet may be affected, but front feet usually suffer the most. Permanent damage usually can be reduced or eliminated with immediate attention by a veterinarian. Permanent damage results from dropping of the hoof sole and upturn of the toe walls when treatment is neglected.

Fractures. A **fracture** is a broken bone. These breaks range in degrees of seriousness. A fracture of any kind usually causes some degree of lameness depending on the bone that is fractured. In the past, fracture healing in horses often caused an altered function. With a greater demand for improved techniques, fracture repair has evolved. Some serious fractures of the long bones can be repaired with **implants** that can withstand the massive mechanical force applied to a bone.

Navicular Disease. **Navicular disease** is an inflammation of navicular bone and bursa. The condition causes lingering lameness and should be diagnosed and treated by a veterinarian. Often the exact course is difficult to determine. Hard work, upright pasterns, small feet, and trimming the heels too low may predispose a horse to navicular disease. Special shoeing, bar shoes, or pads may help. If not, the navicular nerve can be cut so the horse does not have any sensation in the foot. The horse may be dangerous to ride since it cannot feel the ground. Many horses, however, have had the nerve cut and remained useful for many years.

Osselets. These are soft swellings that occur on the front and sometimes sides of the fetlock joint. **Osselets** (see Figure 7-3) are due to injury to the joint capsule of the fetlock. This trauma affects the surface of the bone where the joint capsule attaches to it. The tearing of the joint capsule causes inflammation that stimulates the bone to heal by laying down more bone. During the inflammation stage, pain, heat, and swelling will be present. The horse is almost always lame and will have a shortened stride.

Osselets are often seen in racehorses, especially in young horses under a lot of strain from training. The earlier the injury is treated, the better the prognosis. Rest and hydrotherapy are important in the early stages to decrease pain and swelling. X-rays are often taken to check for complications, such as fractures. A veterinarian should be consulted for the best treatment. As long as the joint is not affected by the bony growth, the prognosis is often good. If the joint is involved, decreased joint movement, which leads to decreased performance, often results.

Pedal Osteitis. This condition is caused by chronic inflammation to the coffin bone, usually of the front feet. It is usually caused by persistent pounding of the feet, chronic sole bruising, or laminitis. **Pedal osteitis** is usually detected over the toe of the coffin bone and is caused by a decrease in the density of the bone in response to the trauma. Pedal osteitis is commonly associated with laminitis.

Clinically, the horse may be lame at all gaits depending on the progression of the bone demineralization. **Hoof testers** will pick up increased sensitivity, commonly over the toe. X-rays may show a roughening of the edge of the toe or wings of the coffin bone and an increased size of the channels of the blood vessels that run through the bone. Pedal osteitis can be hard to diagnose with X-rays. Clinical signs of lameness and location of the sensitivity are important clues.

Treatment usually consists of corrective shoeing to take pressure off the sole and toe. Pads may be used to help cushion the feet. Anti-inflammatory drugs can also help relieve some of the pain.

Pointing. The front legs bear about 60 percent of the weight of a horse. Healthy horses stand at rest with weight equally distributed on both front legs. Lameness in the foot or leg will cause "pointing." Pointing refers to a state of rest with one foot positioned about ten to twelve inches ahead of the other in an effort to reduce weight on the affected side. Weight is shifted habitually from one hind limb to the other by healthy horses during rest and does not indicate lameness (see Figure 7-9).

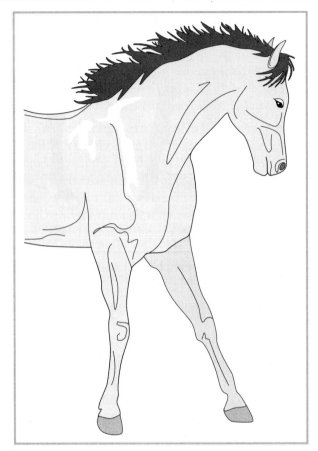

Figure 7-9 Pointing

Quarter Crack. **Quarter**, **toe**, or **heel cracks** (see Figures 7-1 and 7-3) can indicate poor owner management of the feet. These feet show cracks in either the toe, quarters, or heel of the hoof wall, or in any combination of these locations. Toe and quarter cracks are the most common. Usually this is associated with a hoof wall that is too long and has not been trimmed frequently enough. It can also develop with horses that are in rain and mud for long periods of time. The mud draws water out of the hoof wall and when the hoof dries it often cracks.

The cracks can be small and cause no problems. If cracks extend up the wall and into the sensitive lamina of the hoof wall, the horse is usually lame. Cracks often worsen if left untended since the weight of the horse, especially when moving, puts pressure on the hoof and the crack. For treatment, the hoof should be kept moist and treated as soon as cracks are detected to stop any permanent damage. If cracks are extensive, the veterinarian may need to remove the cracked area to promote new hoof growth.

Quittor. A **quittor** is a festering of the foot anywhere along the border of the coronet (see Figure 7-1). It may result from a calk wound, neglected corn, gravel, or nail puncture.

Ringbone. **Ringbone** (see Figures 7-1 and 7-3) is a bony enlargement on the pastern bones, front or rear. It occurs in two locations distinguished by the names high ringbone and low ringbone. They both occur around the pastern bone (the second phalanx). **High ringbone** occurs at the pastern joint. **Low ringbone** occurs at the pastern-coffin bone joint at about the level of the coronet band. Ringbone is caused by bony development around these joints due to tearing and damage of the ligaments and tendons at these bones.

The tearing damages the surface of the bone and stimulates production of new bone in these areas. Typically, this occurs in the forelimbs. It can occur on the front and sides of the joints, but rarely on the back of the joint. Damage to the tendons and ligaments can occur from overuse, because of the excess stress placed on the structures.

Conformation problems often worsen the stress placed on the tendons and ligaments. Especially stressful are **straight shoulders** and **upright pasterns**, **base narrow toe-in**, or **base wide toe-out** conformations. Direct trauma to this area can also lead to the development of ringbone. Ringbone usually occurs in the older horse unless it is due to trauma.

Ringbone typically causes heat, swelling, and pain. Lameness may not be seen in mild cases but is usually seen if the joint itself is also affected. It is hard to resolve the bony production of the joint because of the constant trauma and strain placed on the tendons and ligaments. Initially, treatment is rest and hydrotherapy to try to decrease the pain and swelling. Treatment then generally involves fusing the joint since these are low motion joints. The earlier treatment is initiated, the better the prognosis. If the joint is affected, the horse often progresses to degenerative joint disease and the prognosis is poor.

Scratches. **Scratches** or **grease heel** (see Figure 7-1) is a low-grade infection or scab in the skin follicles around the fetlock. It is caused by filthy stables and unsanitary conditions. Response to cleanliness and treatment is usually prompt and complete.

Shoe Boil. **Shoe boil** (see Figures 7-1 and 7-3) is seen mostly in horses that are stabled and lie down for extended periods of time. The horse lies with the foot tucked up against the elbow. This irritates the elbow from the trauma of being hit by the foot. The result is a swelling at the point of the elbow, which is also called capped elbow.

Lameness is uncommon and a shoe boil is usually just a blemish. The treatment is aimed at trying to reduce the swelling. Hydrotherapy and medicines to reduce the swelling are commonly used. **Fibrosis**, a thickening of the affected skin, may develop in chronic cases. This may leave a permanent blemish but usually does not affect the performance of the horse. If the boil occurs because of the shoe striking the elbow when lying down, a **shoe boil roll** or a **boot** can be applied to prevent the foot from injuring the elbow.

Sidebones.   This is a common unsoundness resulting from wear, injury, or abuse. On each side of the heel extending above the hoof are elastic cartilages just under the skin that serve as part of the shock-absorbing mechanism. They are commonly termed **lateral cartilages**. When they **ossify** (turn to bone) they are called **sidebones** (see Figures 7-1 and 7-3). In the process of ossification they may be firm but movable inward and outward by the fingers. The horse is then considered "**hard at the heels**." Sidebones are more common to the front outside lateral cartilage than to other locations.

Splints.   These are inflammations of the interosseous ligament that holds the splint bones to the cannon bone (see Figures 7-1 and 7-3). They are most common on the forelimbs and usually occur on the inside of the leg. **Splints** most commonly occur at the top of the splint bones, below the knee. They may also occur at the middle or end of the cannon bones. Splints are usually associated with conformation problems such as **bench knees** that place increased stress on the inside of the legs, trauma, or hard training. Trauma usually occurs due to slipping, being kicked, jumping, and playing hard.

   Splints are seen most in horses two to three years old. Older horses can "throw" splints from overwork. Damage to the interosseous ligament causes it to swell. The degree and size of the swelling is directly related to the area that is injured. Other signs include inflammation, pain, and swelling associated with the early splint. Lameness is often present at the trot or faster gait in the early stages, but it depends on the extent of the splint.

   A horse with splints should be rested, given hydrotherapy, and possibly other medications to reduce the size of the splint. The ligament heals by **calcification** of the injured area. Consequently, the result is a bony area after healing that becomes a blemish. The calcification in the healing process may also intrude on the tendons in the back of the leg. This occurs most commonly when the splint is at the end of the splint bone.

   If the pain seems intense or if the pain and swelling do not improve, the splint bone may have actually fractured and injured the tendons in the back of the leg. X-rays can be used to verify the problem.

Stifled.   **Stifled** (see Figure 7-3) refers to a displaced patella of the stifle joint. It sometimes cripples the horse permanently.

Stringhalt.   **Stringhalt**, also called **stringiness** or **crampiness**, is considered an unsoundness. It is an ill-defined disease of the nervous system characterized by sudden lifting or jerking upward of one or both of the hindlegs. Stringhalt is most obvious when the horse takes the first step or two (see Figure 7-10).

Figure 7-10 Stringhalt

Sprain. A **sprain** refers to any injury to a ligament. It usually occurs when a joint is carried through an abnormal range of motion such as in splints.

Thoroughpin. This is a soft fluctuating enlargement located in the hollows just above the hock. They can be pressed from side to side, hence the name. **Thoroughpins** (see Figure 7-3) are due to a distention of the synovial bursa, and considered a discrimination.

Thrush. **Thrush** is an inflammation of the fleshy frog of the foot. It is blackish in color, foul smelling, and associated with filthy stalls. It may cause lameness. Response to cleanliness and treatment is usually prompt and complete.

Upward Fixation of the Patella. This occurs when the patella is moved above its normal position and locks into place. It will prevent the horse from flexing its stifle and the stifle and hock will be extended. The horse will drag its leg since it cannot flex it and bring it back under the body. The young horse may outgrow this problem. In the older horse, the knee may be popped backed into place but this must be done carefully and correctly. Surgery can be done if this becomes a reoccurring problem.

Upward fixation of the patella is thought to be an inherited problem due to a straight hocked conformation. The hock and stifle are straighter than desired. Upward fixation of the patella may also be seen in the poorly-muscled horse. Conditioning may be enough to stop the problem in this case.

Wind Puffs. **Wind puffs**, **windgalls**, or **road puffs** (see Figure 7-3) are soft enlargements located at the ankle joints and due to enlargement of the synovial (lubricating) sacs.

STABLE VICES AFFECTING USEFULNESS

Although they are not really an unsoundness or blemish, **vices** affect the usefulness, desirability, and value of horses. Vices are habits acquired by some horses that are subjected to long periods of idleness. Hard work and freedom from close confinement are distinct preventives. Bad habits should be corrected or prevented early before they become confirmed.

Wind sucking, cribbing, weaving, and stall-walking horses are hard to keep in condition. And the latter two vices cause horses to be fatigued when they are needed.

Wind Sucking. A **wind sucking** horse identifies an object on which it can press its upper front teeth while pulling backward and sucking air into the stomach, usually accompanied by a prolonged grunting sound. The habit is practiced while eating, thus causing loss of food. Confirmed wind-suckers will identify an object in the pasture on which to suck wind, and will practice the habit when tied with bridle or halter as the opportunity is presented.

Cribbing (crib biting). **Cribbing** horses (see Figure 7-11) grasp an object (edge of a feed box or manger) between their teeth and apply pressure, gradually gnawing the object away if it is not metal. Wind sucking and crib biting are usually associated, although a horse may practice one without the other. Crib biting wears away the teeth to a point of decreased efficiency when grazing. Both habits may be partially prevented and sometimes stopped by a wide strap fitted sufficiently close about the throat to compress the larynx when pressure is borne on the front teeth. Normal swallowing is not impaired. The strap must be placed with care; it should be loose enough to prevent choking and tight enough to be effective.

Weaving. **Weaving** is a rhythmical shifting of the weight from one front foot to the other. It is not a common vice, but when carried to extremes it renders a horse almost useless. Its cause is obscure, but its occurrence is correlated with enforced idleness in confined quarters. Some horse owners condemn vertical bars that can be seen through, and others consider chain halter shanks that rattle when moved, as predisposing causes.

Figure 7-11   Cribbing

Stall Walking.   **Stall walking** is just what it sounds like. It is uncommon but reduces a horse's condition and induces fatigue.

Kicking.   Occasionally horses will learn to destroy partitions or doors in stalls by **kicking**. Some kick only at feeding time, thus giving vent to their impatience. They usually do not kick outside the stall. Padding the stall has been known to stop some kickers.

Biting.   Stallions often acquire the habit of nipping at the attendant for want of something to do. Gentle horses can be encouraged to nip when too much pressure is applied in grooming or during cinching the saddle girth. Many show horses use **biting** to defend themselves when agitated by pokes from well-wishers as they rest in their stalls on the show circuit. Removing the cause will usually correct the condition.

Tail Rubbing.   **Tail rubbing** starts by agitation from parasites and continues from habit. Parasite control and **tail boards** prevent it.

Halter Pulling.   Halter pulling develops when a horse becomes confident that it is stronger than the rigging that secures it. Young horses in training will not gain such confidence when secured by strong halter equipment tied to stationary objects. Bridles should not be used in tying young horses. The habit may be broken in early stages by a slip noose around the flank, with the rope shank passing between the forelegs, through the halter ring, and by being fastened securely. Pain experienced from hard backward pulling is usually given consideration before tightening a halter shank afterward. A second rather successful method is to pass the halter rope through a tie ring in the stall and fasten it to a **hobble** placed on a fore pastern.

EXAMINING HORSES FOR SOUNDNESS

An accurate diagnosis of a horse's soundness is never easy. Sometimes professional assistance is needed. Whenever possible, a potential buyer should take the horse on a trial basis for use under conditions to which it will be subjected under new ownership. Some guarantees of soundness are useful. Most horse owners can increase their competence in identifying unsoundness and blemishes by practice and by using a system of inspection (see Figure 7-12).

Natural Surroundings

Whenever possible, examine the horse in its stall under natural surroundings. Note the manner of tying—it may be a halter puller. If metal covers the manger or feed box, cribbing should be suspected. Look for signs of a strap around the throat latch. Note the arrangement of bedding. If the horse paws, bedding will

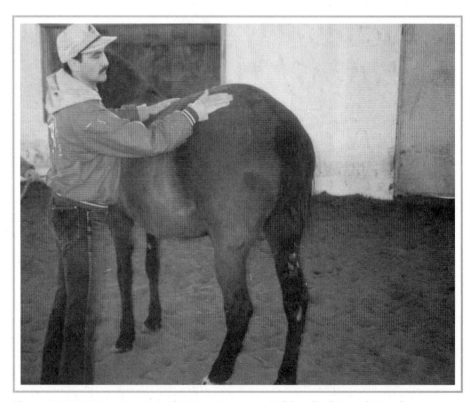

Figure 7-12 Horse owner and trainer Lawrence Valdez checking a horse for blemishes and unsoundnesses.

be piled up near its back feet. Slight lameness may be detected by movement of bedding caused from pointing. Signs of kicking may be noted. Move the horse around and observe signs of slight founder, stiffness, crampiness, and stable attitude.

Leading

Lead the horse from the stall and observe the eyes closely for normal dilation and color. Test eyesight further by leading it over obstacles, such as bales of hay, immediately after coming out of the stall into brighter light. Back the horse and observe hock action for string halt and crampiness. Stiff shoulders and/or stiff limbs are indicated by a stilted, sluggish stride.

In Motion

Examine the horse for lameness in motion. Lameness in a front limb is indicated by a nod of the head when weight is placed on the sound limb. The croup drops when weight is shifted from a lame hind limb to a sound one. Splint lameness usually gets worse with exercise, whereas spavin lameness may improve. The horse should be examined when cool, when warmed up, and when cooled off again, at both the walk and trot. Soundness of wind should be checked under conditions of hard work. Be alert for roaring and heaves or the appearance of a discharge from the nose. Cocked ankles may appear after sharp exercise, and weak fetlocks and knees may tremble.

Overall

Make a general examination with the horse at rest. It should not point or shift its weight from one forelimb to the other. Stand directly in front of the horse and observe the eyes for signs of cloudiness, position of the ears for alertness, and scars or indentations indicating diseased teeth. Pay particular attention to the knees, cannons, and hoof heads for irregularities. Move to the side at an oblique angle and note strength of back and coupling, signs of body scars, and shape and cleanness of hocks, cannons, fetlocks, and hoof heads.

Look for capped hocks, elbows, and leg set from a side view. Chin the horse at the withers for an estimation of height. Stand behind the horse and observe symmetry of hips, thighs, gaskins and hocks, and position of the feet. Move to the opposite side and the oblique angle previously described for final visual inspection before handling any part of the horse.

Determine age according to the instructions in Chapter 9.

The wall of a good hoof is composed of dense horn of uniform color without any signs of cracks in it or rings around it. The slant of the toe should be about 45 degrees and should correspond with that of the pastern. The heels should be deep and reasonably wide. Pick up each foot and look at the bearing surface. The frog should be full and elastic and help bear weight. The bars should be large and straight. The sole should be arched and should not appear flat as in dropped sole. Check for hard heels or sidebones, ringbone, corns, contracted feet, and thrush. If the horse is shod, check for wear on the shoe from contraction and expansion of healthy heels.

Examine the hocks for swellings, spavins, puffs, curbs, or other irregularities, by feeling when necessary.

A thorough examination combined with a week's trial will identify almost any unsoundness or blemish. Many horses serve faithfully for a lifetime without developing unsoundness, vices, or bad manners. Such service can come to horse owners only through patience, knowledge, and detailed attention to the needs of the animal.

SUMMARY

Some horses become unsound at an early age because of coarse, crooked legs, whereas others remain useful for years. Like cars, abusive treatment, excessive use, and poor care will render any horse unsound. Unsoundness interferes with the performance of the horse. A blemish is unsightly, but it is not apt to influence the performance of a horse. An evaluation of an unsoundness or blemish is influenced by the use the horse, for example pleasure versus competition. Many conditions leading to an unsoundness or blemish are preventable or treatable. The prognosis for treatment can also depend on the intended use of the horse. Unsoundnesses and blemishes can be found on any part of the body. For obvious reasons, those of the limbs are the most common.

Unsoundnesses and blemishes decrease the value and may alter the use of a horse. Stable vices may also decrease the value and use of a horse. Using a methodical inspection, unsoundnesses, blemishes, and stable vices can be spotted in an unknown horse.

REVIEW

Success in any career requires knowledge. Test your knowledge of this chapter by answering these questions or solving these problems.

True or False

1. A blemish will always interfere with the performance of a horse.

2. Poor conformation has little to do with unsoundness.

3. Unsoundness interferes with the function and performance of a horse.

4. Thrush is caused by unclean stables.

5. Quittor is the name for horses dropping feed out of their mouth while they are chewing.

Short Answer

6. Name six stable vices that can affect the use and value of a horse.

7. An unsoundness of the lungs or respiratory tract includes _____, a partial paralysis of the nerves to the muscles of the vocal cords, and _____, a loss of elasticity in the lungs.

8. Name two types of sprains and two types of fractures.

9. List four unsoundnesses associated with the head of the horse.

10. List five common unsoundnesses or blemishes that can be found on the body of a horse.

11. List four blemishes of the limbs.

12. Name six unsound conditions of the limbs.

13. Give two conditions caused by unclean stables.

Discussion

14. Describe the difference between a blemish and an unsoundness.

15. What is the common treatment for an injury that could develop into an unsoundness?

16. Identify four conditions that could predispose a horse to developing an unsoundness.

17. Briefly describe a system of inspection to check unknown horses for unsoundness, blemishes, and possible stable vices.

18. Describe the cause, diagnosis, and possible treatment of bucked shins, chip fractures, osselets, quarter crack, and ringbone.

STUDENT ACTIVITIES

1. Visit a veterinarian's office or invite a veterinarian to class. Ask the veterinarian to show and explain some X-rays of bone injuries.

2. Develop a report on new technologies used to treat leg fractures in horses.

3. Diagram the bones of the front or hind leg of a horse. Indicate on the diagram where an unsoundness can develop. Use the unsound conditions discussed in this chapter.

4. Visit a farrier or invite a farrier to class. Ask the farrier to describe how shoeing can correct some conformation problems and unsoundnesses.

5. View a video on conformation and/or lameness selected from the list in the Additional Resources section of this chapter.

6. Visit with a horse breeder and determine the most common types of blemishes and unsoundnesses.

ADDITIONAL RESOURCES

Books

Ensminger, M. E. 1990. *Horses and horsemanship.* 6th ed. Danville, IL: Interstate Publishers, Inc.

Evans, J. W. 1989. *Horses: a guide to selection, care, and enjoyment.* 2nd ed. New York, NY: W. H. Freeman and Company.

Frandson, R. D., and Spurgeon, T. L. 1992. *Anatomy and physiology of farm animals.* 5th ed. Philadelphia, PA: Lea & Febiger.

Fraser, C. M., ed. 1991. *The veterinary manual.* 7th ed. Rahway, NJ: Merck & Co.

Griffin, J. M., and Gore, T. 1989. *Horse owner's veterinary handbook.* New York, NY: Howell Book House.

Kainer, R. A., and McCracken, T. O. 1994. *The coloring atlas of horse anatomy.* Loveland, CO: Alpine Publications, Inc.

Simmons, H. H. 1963. *Horseman's veterinary guide.* Colorado Springs, CO: Western Horseman.

Stashak, T. S. 1996. *Horseowner's guide to lameness.* Media, PA: Williams & Wilkins.

Stoneridge, M. A. 1983. *Practical horseman's book of horsekeeping.* Garden City, NY: Doubleday & Company, Inc.

Videos

The following videos are available from Equine Research, Inc., Grand Prairie, TX:

    The Athletic Horse: Injuries, Problems, Solutions (no. 7V126)

    Conformation Evaluation (no. 7V01)

    Conformation Evaluation of the Trotter and Pacer (no. 7V138)

    Rooney's Guide to Lameness, parts I and II (nos. 7V135 and 7V136)

    Secrets of Conformation (no. 7V125)

University of Kentucky, 1995. *Anatomy of the Lower Leg* (VAS-0675), Lexington, KY: University of Kentucky.

CHAPTER 8

Selecting and Judging Horses

Selecting and judging horses requires combining and using knowledge and information about breeds, functional anatomy, age, height, weight, soundness, and movement. This is true for both selecting a horse for personal use and serving as a judge.

The phrase general appearance refers to and includes the horse's balance and symmetry of body parts, carriage of head and ears, and style. Each owner hopes these traits add up favorably. While appearance is mostly aesthetic, it is probably the largest single factor contributing to the value of a horse and the pleasure of being a horse owner.

O B J E C T I V E S

After completing this chapter, you should be able to:

- Describe ten factors to consider when selecting a horse to purchase
- Explain how expected use influences the selection of a horse
- Discuss why the age and sex of a horse are important considerations in selecting a horse
- Describe costs associated with owning a horse after the initial purchase
- Discuss why conformation is a more important consideration than breed when selecting a horse
- List five steps in judging a horse
- Name the views used and traits looked for in judging the conformation of a horse
- Describe ten qualities of a good judge
- Identify typical markings for the face and legs of horses
- List the terms used to describe common body colors of horses
- Discuss why the proper selection of a horse is so important

KEY TERMS

Ankle	Leg cues	Star
Bald face	Overo	Stock
Blaze	Pedigree	Stocking
Coronet	Performance record	Stocking plus
Dam	Puff	Strip
Distal spots	Registered	Stripe
Grade	Sires	Tobiano
Half-stocking	Snip	White spots

PARTS OF A HORSE

Figure 8-1 shows the parts of the horse and the terms used in referring to them. Familiarity with these terms will aid the reader's understanding.

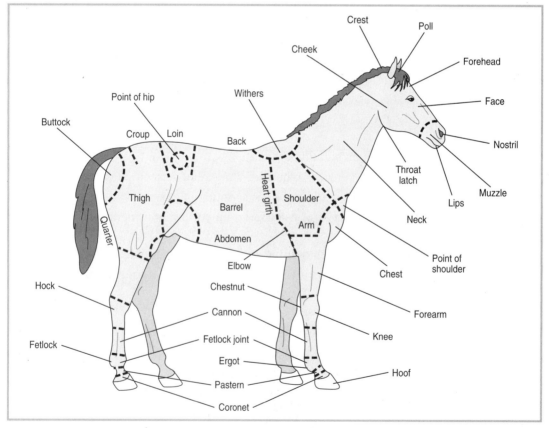

Figure 8–1 Parts of a horse

SELECTING A HORSE

Horses require time to care for them, facilities, knowledge, and money to pay for all the maintenance of the animal. Other alternatives to owning a horse include taking lessons, renting a horse at camps or parks, and leasing a horse and boarding it elsewhere.

If, after considering the realities of owning a horse, the decision is still to buy, then several factors must be considered. If the primary user is inexperienced, then disposition, soundness, and training become the most important factors. If the owner is investing in breeding **stock** or performance prospects, then the pedigree and performance records are crucial (see Figure 8-2).

Investment or Pleasure

Why is the horse being purchased? This is the first question to consider. If the horse is an investment, then the personal experience of the buyer may not be as critical as the knowledge and experience of the advisor. If the horse is a young race or show prospect, or a breeding animal, it will be managed by a professional horseman. Investing in horses is risky business. Although there

Figure 8-2 Pedigree record. *(Photo by Michael Dzaman)*

are some shining success stories, the odds of making enough money to pay the bills and get any return on the investment are very poor. A person should be cautious with the first investments, and be sure that the advisor, breeder, or trainer involved has respected credentials and is someone to be trusted.

If the horse is intended for personal recreation for yourself or the family, then the ability of the horse to cooperatively perform for all members of the family is essential. Owners eventually want to care for the horse themselves, so disposition, training, and soundness are important. Also, the recreation animal should be considered like any other form of recreation—money is spent for enjoyment and not expected to make money. The upkeep of the animal in money and time is ongoing and much different than buying a boat or a set of golf clubs. Selecting the right horse that owners can enjoy working with daily is the key to finding continued recreation from the horse.

Selection of a horse should also consider:

- Owner's experience with horses
- Expected use
- Soundness
- Grade or registered
- Breed
- Color
- Size of rider(s) and horse
- Age of rider(s) and horse
- Training
- Sex of horse
- Disposition and vices
- Facilities
- Price
- Conformation

Experience with Horses

Very experienced owners may purchase a young horse successfully if they have the skills and knowledge needed to train the horse. Inexperienced owners and young horses are a dangerous combination. The best horse for a novice owner is a mature animal that is well trained and accustomed to a variety of situations.

If the owner intends to pay someone else to board or train the horse, then the owner's expertise is not as critical. But this may lessen the owner's enjoyment of the horse. Considering who will ride the horse most is important. Just because an adult can make the horse obey, does not mean that a six-year-old child can enjoy the horse safely.

Expected Use

The type of horse purchased determines how easily it can perform the intended use. Any type of quiet horse will work for a trail and pleasure horse as long as it is physically capable of performing. A relaxed, mannerly horse that has a prompt, flat-footed walk will be best for trail riding.

If a horse is being purchased for show purposes, then the quality and type become more important. Western horses tend to be lower headed, quiet, and heavier muscled than English horses. Hunters have longer, flatter strides and move forward with more impulsion and a higher head carriage than western horses. English or saddle-type horses tend to be much higher headed, with their necks coming higher out of their withers. They move with more hock and knee elevation. Success in the show ring results directly from the horse's breed type and ability to perform. The type of horse selected should be based on the type of show situation.

Performance horses such as polo, dressage, reining, cutting, and roping horses need more specialized training and qualities. These skills make the horse higher priced because more training needs to be invested in them. Success in these activities depends on athletic ability and training rather than on characteristics of a specific breed of horse.

Breeding should be reserved for those horses of a quality that can improve the breed or type. If the primary purpose of the horse is to breed it, then the success of the ancestors in the horse's **pedigree** and the horse's own **performance record** are important. Purchasing quality breeding stock is expensive, and the outcome of breeding horses is unpredictable, so it is important to spend as much as possible to obtain truly superior mares.

Purchasing a breeding stallion should only be done for income purposes, and only about the top 5 percent of the horses should stand as sires.

Soundness

Horses must be sound enough to perform the expected activities. Horses that are lame may have permanent problems that will limit their performance or make using them inhumane. Horses with blemishes—scars or marks that do not interfere with their movement—should be less valuable as show horses, but blemishes should not be a consideration with breeding or pleasure horses.

Horses should also be sound in their breathing, vision, and reproductive capacity if they are purchased for breeding. A soundness examination should be done if much money is being spent or if a doubt exists. The more athletic the horse has to be, the more sound it must be. A race or competitive show or event horse must be very strong and sound. Pleasure trail horses and backyard horses must be sound enough to perform the expected activity, even if not perfectly. Soundness should always be measured in light of the expected performance of the horse.

Grade or Registered

Should a **grade** or **registered** horse be purchased? Grade, or nonregistered horses and ponies, can be successfully used as trail, pleasure, and performance horses (see Figure 8-3). But one can often buy a registered horse as cheaply as a grade horse and the resale value is much greater. If disposition and comfort of a recreational horse are the most important criteria, valuable grade horses are available. If the purpose is to produce foals, then only registered horses should be used. Many performance competitions such as dressage, reining, competitive trail riding, and combined training events do not require registered horses. Open and 4-H shows provide excellent areas for the nonregistered horses to compete. When the owner anticipates participating in breed shows, races, or other activities, owning a horse of that breed should be a priority.

Breed of Horse

The breed will dictate to some extent the activities and performance abilities of the horse. Usually saddle-type English horses are saddlebred, Morgan, Arabian, and saddle-type pintos. These horses lend themselves to the conformation and action to do well in English. Tennessee walking horses, Missouri fox trotters, Paso Finos, Peruvian Pasos, and racking horses do not trot. They are very comfortable for trail riding and showing in breed events, but they will not be competitive in English, Western, or Hunter classes requiring a walk, trot, and canter.

Figure 8-3 In this country most horses are used for recreational purposes so grade or nonregistered horses or ponies are as suitable as registered horses or ponies. *(Photo courtesy of Barbara Jensen)*

Most hunters are of breeds including Thoroughbred, quarter horse, European warm-blooded breeds, and ponies like Welsh and Connemara. Western event horses most often are the stock-type breeds like the Appaloosa, buckskin, paint, palomino, stock-type pinto, and quarter horse. Western ponies include the Pony of the Americas (POA), and Welsh.

Race horses are bred to trot, pace, or gallop. All harness-race horses either trot or pace and are Standardbreds. In the United States, horses that are ridden and gallop are most frequently Thoroughbreds.

All breeds of horses have calm, quiet horses as well as anxious, dangerous horses. Training and handling styles affect manners more than does the breed.

Color

The color of the horse has nothing to do with disposition, performance ability, or soundness. Color is, however, a significant determining factor in many people's purchase decision. Many breed registries such as buckskin, pinto, Appaloosa, palomino, Pony of the Americas, and Dominant Grey are based primarily on color. For individuals involved in breeding or using these breeds, color should be high on the priority list. Otherwise, the training, disposition, and soundness of the horse are more critical.

Size of Rider and Horse

The horse is capable of carrying a tremendous amount of weight. The only time the relative size of the rider and the horse is important is when showing. Then the suitability of horse to rider becomes an issue. Small children are better off on quiet-dispositioned large horses than on small ponies that are wild. Likewise, a small, quiet pony may be ideal for some; however the child will likely outgrow this mount.

A rider's leg ought to fit down the sides of the horse in order to give **leg cues** (signals to the horse), but not be so long that the leg from the knee down does not touch the ribs. Most adults buy horses over 58 inches at the top of the withers. As long as the mount is quiet enough for the child to work around and mount, the size of the animal should be considered secondary (see Figure 8-4).

Age of Rider and Horse

A good guideline is, the younger the rider the older the horse needed. This is a function of training, calmness, and experience that comes with an older horse rather than of age itself. Rarely will a horse under five years old be trained and quiet enough for a novice rider. Horses live to be twenty-five to thirty years of age, so the purchase of a six- to twelve-year-old is wise for amateurs and novices. Riders with more expertise and experience can buy,

Figure 8-4 A child with a horse.
(Photo by Michael Dzaman)

handle, and train yearlings or two-year-olds, but these young horses do not make predictable mounts for beginners.

Training

The willingness of a horse to respond to the handler's cues is a result of training. Horses that have "been around some" increase in value for the beginner. As more intricate maneuvers are desired for higher levels of competition, more training is needed. Sometimes, highly tuned horses are so responsive to the rider's cues that a novice confuses the horse and gets no response. A horse may be trained to the point that a person gets more response than is desirable. For example, the horse may give too fast a spin, too quick a start, too hard a stop, and hurt the rider. Adequate training for the intended use combined with an experienced disposition is important.

Sex

Mares and geldings are a better choice for riders with limited experience. Mares often look more refined and attractive. Still mares can have dramatic behavior changes when in estrus. Geldings are often quieter and more consistent. The only reason to own a stallion is either to breed mares or performance-test a potential breeding stallion. **Sires** in the horse industry should be of superior quality and have successful performance records. Only those able to improve the breed should be bred. Nonregistered males should be gelded within the first ten months of life to minimize stress on the horse and handlers.

Disposition and Vices

The manners of the horse may be changed with training and handling, but the natural disposition is genetic and/or acquired from the **dam**. Bad habits such as kicking, biting, wood chewing, and leaning on the handler can be corrected with firm, consistent, humane handling. Vices such as stall weaving, cribbing, digging, and being afraid of its own water bucket are likely part of the horse and not fixable. Horses that have been exposed to trailering, clipping, shoeing, and trail riding are usually quieter and have better manners. Less experienced owners should try to select horses with minimal vices. A good disposition should be near the top of the priority list.

Facilities

Housing for horses must be safe and adequate to contain the type of horse selected. Build or select housing that is suitable to the horse, rather than selecting an animal that can be housed conveniently (see Figure 8-5). If the proper facilities are not available, then boarding will need to be found. Facilities should not be a priority when selecting a horse. However, they should be decided on prior to purchasing a horse.

Price

The buyer determines the price that will be paid for a horse. The performance record, the breed type and conformation, the pedigree, and the degree of advertising will influence the price. Regardless of how much is spent to purchase the horse, monthly cost will be associated with keeping the horse.

Figure 8-5   Good horse facilities should be clean, well-ventilated, and safe.

Table 8-1 can be used to help calculate the cost of owing a horse for one year. Generally, horses do not increase in value with age; rather, they depreciate. The owner should not expect to get the full financial investment. The value of horses can be increased with training and subsequent race or show success.

New owners need to put priority on the criteria that are important to the expected use of the horse. They should not pay for flashy and showy if that is not the most important criteria for the use.

Most horseowners have their first horse less than three years. Either they gain interest and expertise and want to get a nicer horse, or they loose interest and get out of the horse business. The nicer and more appropriate the horse purchased, the better the resale value.

Table 8-1   Calculating the Annual Costs for Owning a Horse[1]

Operating Inputs	Units	Price/Unit	Quantity	Total Cost
Grain mix	cwt		19.8	
Grass hay	ton		2.8	
Salt and minerals	lbs		10.0	
Farrier	head		6.0	
Veterinary medicine[2]	head	$ 50.00	—	$ 50.00
Veterinary services	head	130.00	—	130.00
Utilities	dollar	80.00	—	80.00
Tack, misc. supplies	dollar	390.00	—	390.00
Bedding	head	70.00	—	70.00
Entry fees[3]	dollar		—	
Travel expenses	dollar		—	
Horse training	dollar		—	
Rider training	dollar		—	
Labor	dollar		—	
Interest	dollar		—	
Total cost of owning a horse for one year				

1. Assumes one mature light (1,100 pound) horse in a confined system.
2. Veterinary medicine, veterinary services, utilities, tack, and bedding are estimates. Use actual figures where available.
3. Use actual figures or best estimates for entry fees, travel expenses, horse training, rider training, labor, and interest for one year.

Conformation

The conformation or shape of the horse will dictate its athletic ability and ability to stay sound. Straight legs, especially through the knees and hocks, suggest that the horse will not break down as soon as a horse with crooked legs. Body conformation and the angle at which the neck ties into the shoulder determine whether the horse is capable of being a saddle-type English horse or is more suited to be a lower-headed Western-type horse. Short, strong-backed horses, horses with good angle to the shoulder, horses with long hips and strong hind quarters are desired. A bright, alert head and eye, a long neck, and a deep heart girth makes horses more athletic and consequently, attractive.

Some unsoundnesses in the feet, legs, and eyesight are serious and permanent. Horses with sight in only one eye are more easily frightened and are of less value. Horses with splay (turned-out) or pigeon-toed (turned-in) feet are more prone to unsoundness than are horses with straight legs. A horse's pasterns should be set at about a 45 to 52 degree angle with the ground and the toe at the same angle as the pastern. The steeper the pastern, the more concussion on the foot and the rougher the gait for the rider. The lower the angle, the more comfortable the ride, but the pastern will be weaker and more prone to tendon damage when worked hard. As with anything else, the importance of conformation depends on the intended use of the horse. Less than ideal conformation can be tolerated if the animal is sound and will not be shown in halter classes at shows.

Chapter 7 discusses many of the points to consider in evaluating conformation and soundness.

Conformation is different for the different breeds. For example, the quarter horse and the Thoroughbred are more muscled in the forearm, gaskin, and through the stifle region than the American saddle horse.

Muscling, especially through the rear quarters, is important in all breeds. Muscles in this area are what give horses their power. Viewed from behind, all horses should have as much width (muscle) through the center and lower part of the quarter as on top.

Figure 8-6 shows a horse with many faults. Too often this type is difficult to keep in good condition and it certainly lacks eye appeal.

ADOPTING WILD HORSES

Because it seems like a worthy cause, some people will consider adopting a wild horse. Seldom does this prove to be a good plan for first-time horse owners. At first glance it may seem like buying a piece of the old American West, but the natural "wild" instincts of these horses and burros are very strong. Only yearlings and occasionally two-year-olds should be considered because they are not as set in their ways as older horses. Adopting a younger wild horse means

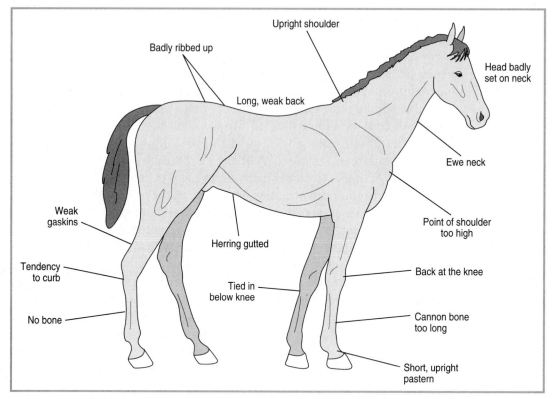

Figure 8-6 A horse with many faults

that the new owner will have to work with the horse or burro for one or two years before it can be ridden. Most of these horses have been underfed and their health management nonexistent before they were gathered up by the Bureau of Land Management. They have been kept in large groups of horses and have had very little exposure to humans. They are often underdeveloped for their age and take a lot of extra care, patience, discipline, and training to be useful.

Experienced owners could consider adopting horses if they have the expertise and energy to invest in a project horse. The horses are of very mixed breeding. Refined, quality horses are hard to find in the group available for adoption. Even when properly trained, the horses are seldom suitable for show purposes.

While this is commendable to want to adopt these gathered-up wild horses, the low adoption and transporting expenses should be balanced with the likelihood of a successful experience. For the money, purchasing domestically raised horses represents less risk, but it does not include the emotional benefit of adopting a wild horse or burro from the public rangelands of the American West.

JUDGING

To properly appraise or judge horses they should be viewed from at least three positions—front, side, and rear.

Stance

How a horse stands indicates how it will move. A long forearm contributes to a long stride. Sloping shoulders and pasterns are associated with a springy stride. If a horse stands straight, it is likely to move straight and true. If the legs are set properly, it is better able to move with collected action. The effect of conformation on a horse's movement was discussed in detail in Chapters 6 and 7.

Judging a horse requires close scrutiny, to unsoundness. The pasterns, cannon bones, knees, and especially the hocks should be examined for any swelling or protuberance that is out of the ordinary.

Nervous and continuous movement of the ears may mean impaired vision; protruding or bulging eyes, called pop eyes, usually indicate nearsightedness.

View from the Front

Figure 8-7 shows a front view of the fore limbs. A perpendicular line drawn downward from the point of the shoulder should fall upon the center of the knee, cannon, pastern, and foot. View A represents the correct alignment while views B through G represent common defects.

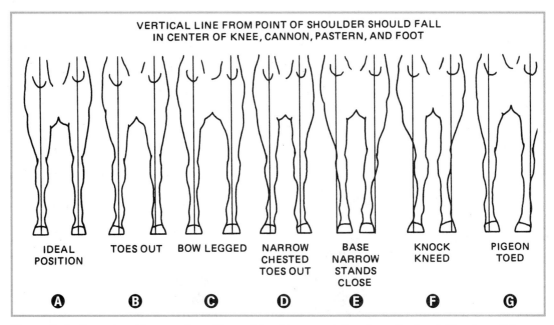

Figure 8-7 Correct and incorrect positions of front legs, as shown from the front. *(Courtesy of the Appaloosa Horse Club, Inc., ID)*

View from the Side

Figure 8-8 illustrates the different conformations of the hind legs as seen from the side. A perpendicular line drawn from the point of the buttock should just touch the upper rear point of the hock and fall barely behind the rear line of the cannon and fetlock.

Correct position of the leg as viewed from the side is most important in a horse. Figure 8-9 shows a side view of the fore limbs. A perpendicular line drawn downward from the center of the elbow point should fall upon the center of the knee and pastern, and back of the foot. A perpendicular line downward from the middle of the arm should fall upon the center of the foot. View A in Figure 8-9 represents the right conformation.

View from the Back

Figure 8-10 illustrates the conformation of the hind legs as viewed from the rear of the animal. A perpendicular line drawn downward from the point of the buttocks should fall in line with the center of the hock, cannon, pastern, and foot. View A in Figure 8-10 represents the correct conformation.

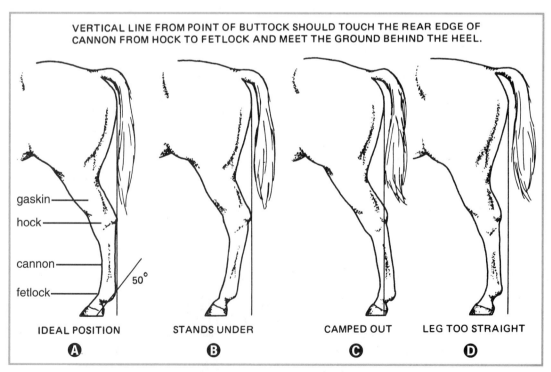

Figure 8-8 Correct and incorrect positions of the rear legs, as shown from the side. *(Courtesy of the Appaloosa Horse Club, Inc., ID)*

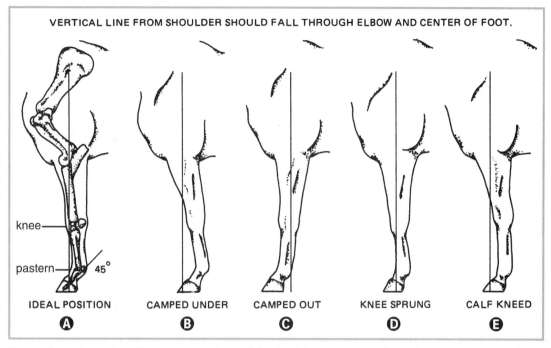

Figure 8-9 Correct and incorrect positions of the front legs, as shown from the side. *(Courtesy of the Appaloosa Horse Club, Inc., ID)*

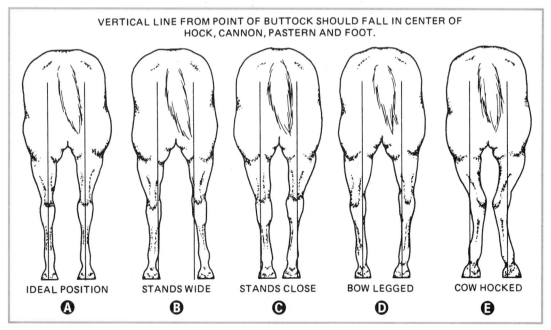

Figure 8-10 Correct and incorrect positions of the rear legs, as shown from the rear. *(Courtesy of the Appaloosa Horse Club, Inc., ID)*

Body Dimensions and Performance

Major contributions to a good-bodied horse include long, sloping shoulders; short, strong back; long underline; and long, rather level croup. These attributes increase the probability that the horse is or can become a good "athlete."

If the shoulders are long and sloping, they extend the stride in running, absorb shock, reduce stumbling, move the elbows away from the girth, and raise the head slightly. The shoulders should be surmounted by clean, high withers that extend well backward to afford maximum security of the saddle.

Short backs and long underlines move the fore and rear legs farther apart, tend to raise the croup and head, contribute to style and action, and increase height and length of stride. Also, short backs are stronger, reduce the length of coupling (hip bone to last rib), and are usually more muscular than others. Finally, well-sprung ribs that blend into hips and shoulders with minimum roughness tend to accompany short backs.

Long, rather level croups accommodate more muscling, increase style and balance, and are less often associated with crooked hind legs.

Since all of the power used in motion comes from the hindquarters, muscular development should be extensive, commensurate with breed requirements. Breeching, thighs, and gaskins should be especially muscular. Long, smooth muscles are preferred to those that are short and bunchy.

Leverage is gained with maximum length from hip to hock and minimum length of cannon. These dimensions are developed to a high degree in breeds that race. Smoothness, balance, and symmetry are a result of all parts blending together, being of proportionate size, and each contributing equally to the whole of a symmetrical individual. These, combined with refinement, alertness, and a proud carriage, contribute to style.

Leg Set

Proper leg set is essential to durability and good action. A leg should be properly positioned under each corner of the body, knees and hocks should not deviate inward or outward, and feet should point straight forward as viewed from the front, side, and back.

If a horse stands on crooked legs, it must move likewise. Crooked moving detracts from appearance, wastes energy, and predisposes a horse to unsoundnesses.

Pasterns should be medium in length, sloped at approximately 45 degrees, and flexible but strong. Hoofs should have the same angle as pasterns, and be deep and wide at the heels, moderate in size, dense of horn, and free of rings. White hoofs are softer (wear faster) than others. Slope of shoulders and pasterns and expansion of heels account for shock absorption when the horse is in motion.

Bone should be adequate in size, show definition of joints, and should appear flat viewed from the side, compared to a front view.

Bone spavins, bogs, thoroughpins, and weakness are common to sickle hocks. Jarring from short, straight pasterns and shoulders predisposes to side bones, stiffness, bogs, and lameness. Pigeon toes tend to wing, whereas splayed feet tend to swing inward in motion.

Effect of Quality on Wearability

Quality is indicated by refinement of head, bone, joints, and hair coat. It is reflected in thin skin, prominent veins and absence of coarseness, especially in the legs. Good circulation in the legs is important to durability. Coarse, "meaty" legs with reduced circulation tend to stock, **puff**, bog, and become unsound. A horse of quality is more attractive, and more appealing to the buyer.

Effect of Head and Neck on Flexibility

The length and shape of a horse's neck and the size of its head affect action. The neck should be long, slightly arched, and fine and clean at the throatlatch for maximum balance, style, and maneuverability. Fine throats enhance ease of breathing and allow maximum flexion of the chin without binding the jaws on the neck (see Figure 8-11). Short-necked, thick-throated horses "steer" hard and may be "head slingers" from jaw pressure when pulled up short. Size of

Figure 8-11 Past Grandma Rodeo Queen Myrtle Bean making her horse flex its neck for a sugar cube.

head should be in accord with breed requirements. Ears should not be over-sized and should be carried alertly. Eyes should be wide-spaced, large, and clear. Nostrils should be large but refined, and lips firm instead of pendulous.

Effect of Disposition on Usefulness

If riding is to be a joy and safety a requirement, good dispositions become a "must." They may be both "born" and "made." Some breeds are more docile than others, and wide differences exist among individuals within breeds. Any horse appropriately trained will have a satisfactory disposition for normal riding. Conversely, horses of excellent disposition can be spoiled by improper handling.

The ears and eyes of the horse show nervousness and resistance. Handling the feet can indicate the disposition of the horse.

Courage or "heart" is necessary for horses used for racing and sporting events. Intelligence or ability to learn is an asset in any horse. These can be identified in horses trained or in training and may be predicted in part by pedigree or family relationships.

A horse with the proper conformation and disposition is physically able to be an effective performer. To do so, it needs to be fed correctly and kept healthy.

What Does It Take To Be A Good Horse-Show Judge?

Being a good horse-show judge requires knowledge and skill. Here is a checklist:

- An exact knowledge of the rules of the division or breed

- A knowledge of correct conformation and movement

- A knowledge of standards of perfection for each breed or division in both halter and performance classes

- An organized mind to sort out top and bottom

- Good notetaking skills and a good method of remembering

- Physical stamina and good health

- Impartiality, ethics, and fairness

- Courtesy and good manners

- Control of the show ring: keeping track of entries, line-ups, work-outs, and time constraints

- Ability to live with a decision

Judges serve as their own conscience. The best show ring decision is the one a judge makes when he or she has the correct knowledge and is ethical and fair.

COLORS AND MARKINGS

To be a good judge, requires familiarity with the colors and markings of horses. Some of these are unique to a breed and others are purely descriptive. Colors and markings are also used for identification.

Body Colors

Body colors run the range of black to white. A good judge uses the proper color description when judging horses.

- Bay—Body color ranging from tan through red to reddish brown; mane and tail black; usually black legs

- Black—Body color true black without light areas; mane and tail black

- Blue Roan—More or less uniform mixture of white with black hairs on the body, but usually darker on head and lower legs; can have a few red hairs in mixture

- Brown—Body color brown or black with light areas at muzzle, eyes, flank, and inside legs; mane and tail black

- Buckskin—Body color yellowish or gold; mane and tail black; usually black on lower legs. Buckskins usually do not have dorsal stripes.

- Chestnut—Body color dark red or brownish-red; mane and tail usually dark red or brownish-red, but may be flaxen

- Dun—Body color yellowish or gold; mane and tail may be black, brown, red, yellow, white, or mixed; usually has dorsal stripe, zebra stripes on legs, and transverse stripes over withers

- Gray—Mixture of white with any colored hairs; often born solid-colored or almost solid-colored and get lighter with age, or more white hairs appear

- Grullo—Body color smoky or mouse-colored; not a mixture of black and white hairs, but each hair mouse-colored; mane and tail black; usually black on lower legs. Usually has dorsal stripe.

- Paint—The two most common paint color patterns are **tobiano** and **overo**. The tobiano horse will usually have head markings like a solid-colored horse; legs may be white, and body markings are often regular and distinct, being oval or round patterns. The overo horse will often have a **bald face**, at least one dark-colored leg, and body markings that are usually irregular, scattered, or splashy white. These markings do not cross the back between the withers and tail (see Figure 8-12).

- Palomino—Body color golden yellow; mane and tail white. Palominos do not have a dorsal stripe.

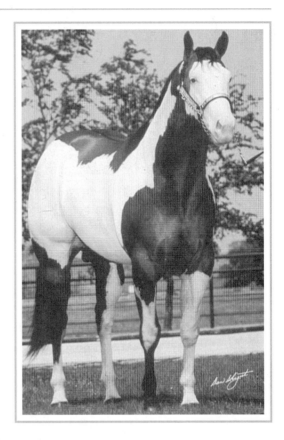

Figure 8-12 An American paint.
(Courtesy of Don Shugart)

- Red Dun—A form of dun with body color yellowish or beige; mane, tail and dorsal stripe are red

- Red Roan—More or less uniform mixture of white with red hairs on the body, but usually darker on head and lower legs; can have red, black, or flaxen mane and/or tail

- White—A true white horse is born white and remains white throughout its life. A white horse has snow-white hair, pink skin, and brown, hazel, or blue eyes.

Head Marks

Head markings include the **star**, **strip** or **stripe**, **snip**, **blaze**, and bald face. These are illustrated in Figure 8-13.

A star is a solid white mark on the forehead. The shape may range from oval to diamond to a narrow vertical, diagonal, or horizontal star. A strip or stripe is a white mark starting at eye level or below and ending on or above the upper lip. The size and shape of a stripe may vary widely and must be described in detail as to width, length, and its relationship (whether it is connected or unconnected) to a star.

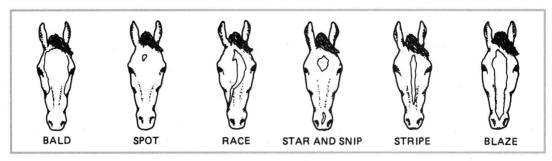

Figure 8-13 Head and face markings. *(Courtesy of the Appaloosa Horse Club, Inc., ID)*

A snip is a white or beige mark over the muzzle between the nostrils, while a blaze is a wide patch of white extending down the face and covering the full width of the nasal bones. A bald face is a wide white marking which extends beyond both eyes and nostrils.

Leg Marks

Descriptive words for leg markings include **coronet**, pastern, **ankle**, **half-stocking**, **stocking**, **stocking plus**, white on knee or hock, **white spots**, and **distal spots**.

A coronet is a white marking covering the coronet band. A pastern is a white marking from the coronet to the pastern, and an ankle a white marking from the coronet to the fetlock. A half-stocking is a white marking from the coronet to the middle of the cannon, a stocking a white marking from the coronet to the knee, and a stocking plus a white marking like the stocking, but one in which the white extends onto the knee or hock.

The designation white on knee or hock indicates a separate white mark on the knee or hock. White spots means white spots on the front of the coronet band or on the heel, while distal spots is the term used to indicate dark spots on a white coronet band.

Figure 8-14 shows the markings on the leg.

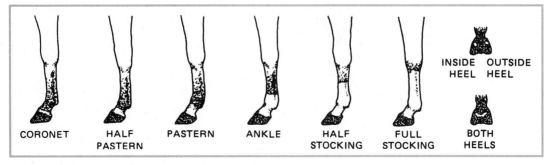

Figure 8-14 Leg markings. *(Courtesy of the Appaloosa Horse Club, Inc., ID)*

HORSE SCORECARD

Table 8-2 can be used as a scorecard when judging light horses. It helps the judge remember what to look for in each view and in specific areas.

QUALITIES OF A JUDGE

To be a successful judge either at a show on an individual basis or in a judging contest competing on a team, judges need these attributes (see Figure 8-15):

1. A desire to know thoroughly what is being judged

2. A clear knowledge of the ideal or standard type and the ability to recognize desirable and undesirable points of conformation

3. Quick and accurate powers of observation

4. Ability to form a mental image of many individual animals and to rank them by making comparisons

5. Reasoning power that takes into account practical considerations

6. Ability to reach a definite decision based on sound judgment

7. Extreme honesty and sincerity to avoid bias or prejudice

8. Decisions based on personal knowledge and judgment

Figure 8-15   Judging horses at halter at a horse show. *(Photo by Michael Dzaman)*

Table 8-2 Score Card for Judging or Selecting Light Horses

View or Item	What to Look For	Ideal Type	Points Assigned	Points Given
Front	Head	Well proportioned, refined, clean cut, with chiseled appearance; broad, full forehead with great width between eyes; jaw broad and strongly muscled; ears medium size, well carried and attractive.	15	
	Feminity or masculinity	Refinement and feminity in brood mares; boldness and masculinity in stallions.		
	Chest capacity	Deep wide chest		
	Set of front legs	Straight, true, and squarely set.		
Rear	Width of croup, and width through rear quarters	Wide and muscular over croup and through rear quarters.	15	
	Set of hind legs	Straight, true, and squarely set.		
Side	Style and beauty	High carriage of head, active ears, alert disposition, and beauty of conformation.	35	
	Balance and symmetry	All parts well developed and nicely blended together.		
	Neck	Fairly long; carried high; clean cut about throat latch.		
	Shoulders	Sloping at about 45 degrees.		
	Topline	Short, strong back and loin, with long, nicely turned and heavily muscled croup; high well-set tail; withers clearly defined and of same height as high point over croup.		
	Coupling	Short as denoted by last rib being close to hip.		
	Rear flank	Deep		
	Arm, forearm, and gaskin	Well-muscled		
	Legs, feet, and pasterns	Straight, true, and squarely set legs; pasterns sloping about 45 degrees.		
	Quality	Abundant, denoted by clean, flat bone, well-defined joints, tendons, refined head and ears; fine skin and hair.		
	Breed type	Enough characteristics of specific breed to meet breed specifications.		
Soundness	Soundness and freedom from any defects in conformation that may predispose to unsoundness	Sound and free from blemishes.	15	
Action	At walk	Easy, prompt, balanced; a long step, with each foot carried forward in a straight line; feet lifted off the ground.	20	
	At trot	Rapid, straight, and elastic with joints well flexed.		
	At canter	Slow and collected; readily executed on either lead.		
Total points			100	

9. Steady nerves and confidence in ability to make close independent decisions based entirely on the animals' merits

10. Ability to do the best work possible at the time and have no regrets about the results or accomplishments

11. Ability to evaluate and rank the individual animal according to its appearance on the day of judging, regardless of its rank at previous shows

12. Sound knowledge acquired through practice and experience to give effective reasons for decisions

13. A pleasant and even temperament without fraternizing with exhibitors or friends along the ringside

14. Firmness to stand by and defend placings without offending or in any way implying decisions are infallible.

SUMMARY

Horse ownership can be a very rewarding experience if the appropriate horse is selected. The criteria for selecting a horse include experience with horses, expected use, soundness, grade or registered, breed, color, size, age, training, sex, disposition and vices, facilities, price, and conformation. The importance of each of these factors will depend on individual needs. As experience is gained and interest grows (or changes), different types of horses will be needed. Sometimes expert help is needed when deciding what type of horse is right. Regardless of the horse selected, consistent and firm discipline and proper management are vital to maintain the animal. Horse ownership is a big responsibility. Continually gaining more knowledge can make horse ownership a more rewarding experience.

Judging horses requires extensive knowledge about horses to be able to make comparisons and reach a sound conclusion. Judges need to be able to describe horse colors, markings, and conformation. When judging conformation, emphasis is placed on the set of feet and legs. A good judge must recognize when a horse has a fault in the way it sets on its legs, since this determines how it will move. A crooked-legged horse cannot move true. Regardless of a horse's excellent head, neck, shoulder, top, and general balance and conformation, if it is crooked on its legs, it is not a top horse.

Besides a thorough knowledge of horses, good judges should possess personality traits that make them accurate and fair.

REVIEW

Success in any career requires knowledge. Test your knowledge of this chapter by answering these questions or solving these problems.

True or False

1. Buying a horse represents a sound investment that will increase in value.
2. Young horses are best for the first-time rider/owner.
3. Potential horse owners should consider only a registered horse.
4. Adopting a wild horse is the best risk for a first-time horse owner.
5. A horse with a bald face lacks hair on its nose.

Short Answer

6. List eight costs associated with owning a horse after the initial purchase.
7. Identify ten terms used to describe the body color of horses.
8. List ten factors to consider when selecting a horse.
9. Which is the best choice for riding: a mare, a gelding, or a stallion?
10. What is a good guideline for selecting the first horse for a young rider?

Discussion

11. When is breed an important consideration in selection?
12. Why is the proper selection of a horse so important?
13. When is the purchase of a horse considered an investment with the possibility of making money?
14. Why are age and sex important considerations when selecting a horse?
15. Describe five typical face markings found on horses.
16. Describe five typical leg markings.
17. Why is conformation a more important consideration than breed when selecting a horse for personal enjoyment?
18. Briefly describe how a horse should be judged.
19. What makes a person a good judge of horses?
20. Differentiate between selecting a horse for an investment and selecting a horse for personal recreation.

STUDENT ACTIVITIES

1. Collect color photographs of horses representing the following body colors: bay, blue roan, brown, buckskin, dun, grullo, tobiano, overo, red roan, palomino and white. The breed registries in Appendix Table A-16 may be helpful.

2. Select one of the breed registries in Appendix Table A-16. Write to the registry and ask them to send any guidelines they have for judging horses representative of their breed.

3. Attend a horse-judging event such as a horse show, county fair, 4-H, or FFA event. Take notes and photographs. Report on the event.

4. Using the Internet, newspaper want ads, or other sources, develop a table of horse prices. Include as much information about each horse as possible. For example, include the breed, age, sex, and any special training. Find this information on at least fifteen horses for sale.

6. Make a photocopy of Table 8-1. Using actual values from your area, complete the table and calculate the cost of owning a horse for one year.

ADDITIONAL RESOURCES

Books

American Youth Horse Council. 1993. *Horse industry handbook: a guide to equine care and management.* Lexington, KY: American Youth Horse Council, Inc.

Davidson, B., and Foster, C. 1994. *The complete book of the horse.* New York, NY: Barnes & Noble Books.

Dossenbach, M., and Dossenbach, H. D. 1994. *The noble horse.* New York, NY: Cresent Books (Random House).

English, J. E. 1995. *Complete guide for horse business success.* Grand Prairie, TX: Equine Research, Inc.

Silver, C. 1993. *The illustrated guide to horses of the world.* Stamford, CT: Longmeadow Press.

Wood, C. H., and Jackson, S. G. 1989. *Horse judging manual.* Lexington, KY: University of Kentucky, College of Agriculture, Cooperative Extension Service.

Software

Compendia! on Horses. Diskettes or CD-ROM, Woodland, CA: Compendia! Inc.

1996 Annual Horse Chronicles. CD-ROM, Dover, NH: Orion Publishing, LLC (Internet address http://www.horse.chron.com).

CHAPTER 9

Determining Age, Height, and Weight of Horses

Horses, like people, vary considerably in vigor and longevity. In general, they have passed their physical peak when they reach nine to ten years of age. At this age, the chance of an unsoundness being present has increased. Age and height are important considerations when selecting a horse for competition or for personal use.

OBJECTIVES

After completing this chapter, you should be able to:

- List the names used for different age groups of horses from birth to three years
- Discuss the importance of knowing the age of a horse
- Name four changes in teeth that are indicators of different ages
- Diagram a tooth showing the parts that change during the aging process
- Describe the changes horses' teeth exhibit during their lifetime
- List the temporary and permanent teeth of the horse and their approximate time of eruption
- List four abnormal tooth conditions
- Give four reasons why knowing the height and weight of a horse is important
- Tell how to determine the height of a horse in hands and inches
- Calculate the weight of a horse from the measurements of the heart girth and body length

KEY TERMS

Angle of incidence	Fillies	Neck
Baby teeth	Floating	Nippers
Birth date of foal	Foal	Parrot mouth
Bishoping	Galvayne's groove	Pincers
Canines	Girth	Premolar
Centers	Heart girth	Smooth mouth
Centrals	Incisors	Stallion
Colt	Infundibulum	Two-year-olds
Corners	Intermediates	Weanling
Cups	Mares	Wolf tooth
Deciduous	Milk teeth	Yearlings
Dental stars	Molars	
Dentition	Monkey mouth	

IMPORTANCE OF AGE

Young horses are referred to by their age. The young horse is called a **foal** until it is weaned. A male horse is referred to as a **colt** until it is three years old and then it is called a **stallion**. Young female horses are called **fillies** until they reach the age of three when they are referred to as **mares**. Sometimes the term **weanling** is used for horses that are six months to one year of age. Horses one to two years old may be called **yearlings** and horses two years of age may be called **two-year-olds**.

A horse's condition and training are more important than its age. Prime age for a horse is about seven to nine years, but this is not necessarily the ideal age. Horses frequently are active into their late twenties if they get proper care.

A buyer can often buy a top-quality older horse at the same price or less than he would pay for a younger horse of lesser quality. Although most older horses cannot perform as actively as they did when younger, they may have many years of useful service left.

Buyers should be ready to decide whether they prefer a younger horse or if an older one would do as well. This decision cannot be made until the buyer evaluates the individual horse. Finally, the age of the horse purchased depends on what the buyer can afford and what horses the buyer finds available.

Age is important too for competitive events. For racing or showing events, the foal's **birth date** is considered to be January 1, regardless of the actual month of birth during the year. So a foal born April 1, 1996, will be ten years old on January 1, 2006. Individuals who race or show try to have foals born as near to January 1 as possible. This gives the horse the advantage of more growth than those born later in the year.

USING TEETH TO DETERMINE AGE

Of course, the best way to determine the age of a horse is from good records. A record of the horse's birth is required by breed registries. When a record of age does not exist, the teeth furnish the best estimation of the age of a horse.

The art of determining the age of horses by inspection of the teeth is an old one. It can be used with a considerable degree of accuracy in determining the age of young horses. The probability of error increases as age advances and becomes a guess after the horse reaches ten to fourteen years of age. Stabled animals tend to appear younger than they are, whereas those grazing sandy areas, such as range horses, appear relatively older because of wear on the teeth.

Age determination is made by a study of the twelve front teeth, called **incisors**. The two central pairs both above and below are called **centrals** (**centers**), **pincers**, or **nippers**. The four teeth adjacent to these two pairs are called **intermediates**, and the outer four teeth are designated as **corners**.

Canine teeth or "tusks" may appear midway between the incisors and **molars** at four or five years of age in the case of geldings or stallions, but seldom appear in mares. Adult horses have 24 molar teeth (see Figure 9-1).

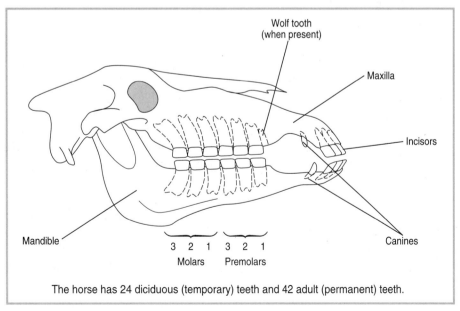

The horse has 24 diciduous (temporary) teeth and 42 adult (permanent) teeth.

Figure 9-1 The position of the teeth in the horse skull

Four key changes in the teeth can be used to estimate the age of horses:

- Occurrence of permanent teeth
- Disappearance of cups
- Angle of incidence
- Shape of the surface of the teeth

Occurrence of Permanent Teeth

Horses have two sets of teeth, one temporary and one permanent. Temporary teeth may also be called **baby** or **milk teeth**. Temporary incisors tend to erupt in pairs at eight days, eight weeks, and eight months of age.

A well-grown two-year-old may be mistaken for an older horse unless permanent teeth can be accurately identified. Permanent teeth are larger, longer, darker in color, and do not have the well-defined **neck** joining root and gum that temporary teeth do.

The four center permanent teeth appear (two above and two below) as the animal approaches three years of age, the intermediates at four, and the corners at five. This constitutes a full mouth.

Disappearance of Cups

Young permanent teeth have deep indentures in the center of their surfaces, referred to as **cups**. Cups are commonly used as reference points in age determination. Those in the upper teeth are deeper than the ones below, so they do not wear evenly with the surface or become smooth at equal periods of time. In general, the cups become smooth in the lower centers, intermediates, corners, upper centers, intermediates, and corners at six, seven, eight, nine, ten, and eleven years of age, respectively. A **smooth mouth** theoretically appears at eleven. A few horse owners ignore cups in the upper teeth and consider a nine-year-old horse smooth-mouthed. Although complete accuracy cannot be ensured from studying cups, this method is second in accuracy only to the appearance of permanent teeth in determining age.

As cups disappear, **dental stars** appear—first as narrow, yellow lines in front of the central enamel ring, then as dark circles near the center of the tooth in advanced age.

Angle of Incidence

The angle formed by the meeting of the upper and lower incisor teeth (profile view) affords an indication of age. This **angle of incidence** or "contact" changes from approximately 160 to 180 degrees in young horses, to less than a right angle as the incisors appear to slant forward and outward with aging. As the slant increases, the surfaces of the lower corner teeth do not wear clear to the back margin of the uppers so that a dovetail, notch, or hook is formed on the upper corners at seven years of age. It may disappear in a year or two, reappear around twelve to fifteen years, and disappear again. The condition varies considerably between individuals, but most horses have a well-developed notch at seven.

Shape of the Surface of the Teeth

The teeth change substantially in shape during wear and aging. The teeth appear broad and flat in young horses. They may be twice as wide (side-to-side) as they are deep (front-to-rear). This condition reverses itself in horses that reach or pass twenty years. From about eight to twelve years, the back (inside) surfaces become oval, then triangular at about fifteen years. Twenty-year-old teeth may be twice as deep from front to rear as they are wide (see Figure 9-2).

STRUCTURE OF THE TOOTH

Being able to estimate the age of a horse by its teeth also requires an understanding of the structure of the tooth and different stages of wear. Figure 9-2 shows the structure of a tooth as viewed longitudinally and in cross section. As the horse ages and the tooth wears, different portions of the tooth become visible.

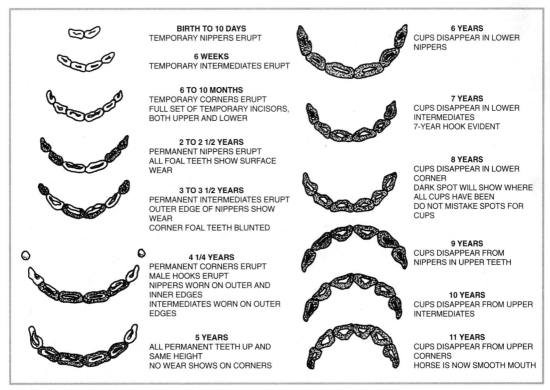

BIRTH TO 10 DAYS
TEMPORARY NIPPERS ERUPT

6 WEEKS
TEMPORARY INTERMEDIATES ERUPT

6 TO 10 MONTHS
TEMPORARY CORNERS ERUPT
FULL SET OF TEMPORARY INCISORS,
BOTH UPPER AND LOWER

2 TO 2 1/2 YEARS
PERMANENT NIPPERS ERUPT
ALL FOAL TEETH SHOW SURFACE
WEAR

3 TO 3 1/2 YEARS
PERMANENT INTERMEDIATES ERUPT
OUTER EDGE OF NIPPERS SHOW
WEAR
CORNER FOAL TEETH BLUNTED

4 1/4 YEARS
PERMANENT CORNERS ERUPT
MALE HOOKS ERUPT
NIPPERS WORN ON OUTER AND
INNER EDGES
INTERMEDIATES WORN ON OUTER
EDGES

5 YEARS
ALL PERMANENT TEETH UP AND
SAME HEIGHT
NO WEAR SHOWS ON CORNERS

6 YEARS
CUPS DISAPPEAR IN LOWER
NIPPERS

7 YEARS
CUPS DISAPPEAR IN LOWER
INTERMEDIATES
7-YEAR HOOK EVIDENT

8 YEARS
CUPS DISAPPEAR IN LOWER
CORNER
DARK SPOT WILL SHOW WHERE
ALL CUPS HAVE BEEN
DO NOT MISTAKE SPOTS FOR
CUPS

9 YEARS
CUPS DISAPPEAR FROM
NIPPERS IN UPPER TEETH

10 YEARS
CUPS DISAPPEAR FROM UPPER
INTERMEDIATES

11 YEARS
CUPS DISAPPEAR FROM UPPER
CORNERS
HORSE IS NOW SMOOTH MOUTH

Figure 9-2   The number of permanent incisors may be used to determine age to five years. After that, use the shape of the biting surface and the angle of the incisors. *(Courtesy of the Appaloosa Horse Club, Inc., ID)*

NUMBER AND TYPE OF TEETH

Table 9-1 summarizes the numbers, types, and appearance of teeth in horses.

Less than Two Weeks

Between birth and two weeks of age, the newborn has its first pair of incisors erupt. These are the central incisors. In profile, the mucous membrane of the gums will be a thin cover over the intermediate incisors. The temporary first, second, and third **premolar** should be present.

Four to Six Weeks

Between four and six weeks of age, eruption of the second set of incisors, the intermediates, occurs. The central incisors are in contact and coming into wear. The intermediates are not in contact, have no wear, and the cups are deep.

Six to Ten Months

The final set of temporary incisors, the corners, erupts between six and ten months of age. The permanent first premolar (or **wolf tooth**) erupts. Not all

Table 9-1	Horse Teeth and the Approximate Age at Eruption				
Type	**Tooth**	**Age**	**Type**	**Tooth**	**Age**
Temporary (Milk or Deciduous)	1st incisors (or centrals)	Birth or first week	Permanent	1st incisors (or centrals)	2.5 years
	2nd incisors (or intermediates)	4 to 6 weeks		2nd incisors (or intermediates)	3.5 years
	3rd incisors (or corners)	6 to 10 months		3rd incisors (or corners)	4.5 years
	Canine (or bridle)	Birth to first 2 weeks for all premolars		Canine (or bridle)	4 to 5 years
	1st premolar			1st premolar (or wolf tooth)	5 to 6 months
	2nd premolar			2nd premolar	2.5 years
	3rd premolar			3rd premolar	3 years
				4th premolar	4 years
				1st molar	9 to 12 months
				2nd molar	2 years
				3rd molar	3.5 to 4 years

horses have this wolf tooth. The dental surfaces of the centrals and intermediates start to show wear. The cup is shallower in the centrals than the intermediates because they have been in wear longer. The corners are not in contact with each other.

One Year

All the temporary incisors are visible from the front. In profile, the upper and lower corner incisors are not in contact. The dental surfaces of the centrals show considerable wear. The dental star is seen usually in the centrals and intermediates as a dark or yellowish-brown transverse line in the dentin on the labial (the surface of the tooth closest to the gums) side of the **infundibulum** (the funnel-shaped inside of the tooth). The first molar should be present by one year.

Two Years

The central and intermediate incisors are now quite free from the gum, especially the upper incisors. All pairs of the incisors should be in wear. The

How Old is Your Horse?

This poem helped old-timers remember how to use the teeth to tell the age of a horse.

To tell the age of any horse
Inspect the lower jaw of course.

Two middle nippers you'll behold
Before the colt is two weeks old.

Before six weeks two more will come;
Twelve months the corners cut the gum.

At two the middle nippers drop;
At three the second pair can't stop.

At four years old the side pair shows;
At five a full new mouth he grows.

Black spots will pass from view
At six years from the middle two.

The side two pairs at seven years,
And eight will find the corners clear.

The middle nipper, upper jaw,
At nine the black spots will withdraw.

At ten years old the sides are light;
Eleven finds the corners white.

As time goes on the horsemen know
The oval teeth three-sided grow.

They longer get, project before,
'Til twenty when we know no more!

—Anonymous

dental surface of the lower central incisors is smooth; the intermediates show decided wear; the corners show considerable wear. The dental star is clearly visible in the lower incisors. Eruption of the second molar occurs.

Two and One-Half Years

The first pair of permanent incisors erupt. The upper central incisors have not reached the level of the **deciduous** intermediates. The lower permanent central incisors have erupted through the gum but most of the labial surface of the lower incisors is covered by mucous membrane. On profile, the intermediate and corner incisors show distinct necks. The dental surface shows the intermediates worn smooth and the corners with noticeable wear. The second permanent premolar erupts as well.

Three Years

The first set of permanent incisors, the centrals, are now in wear. They are more solid in appearance, are larger and broader than the temporary teeth, and have vertical ridges and grooves. The dental table of the central incisors has a deep cup and the borders are sharp. Eruption of the third premolar takes place.

Three and One-Half Years

The second pair of permanent incisors, the intermediates, erupt. The central incisors are well in wear. The intermediates are nearing contact. In profile, the gap between the upper and lower intermediates is visible. The dental surfaces show wear on the centrals. The intermediates are sharp since they have not made contact yet. The temporary corners are nearly smooth.

Four Years

The first two sets of permanent incisors are now in wear. The jaws have acquired so much width that from the front, the corner temporary incisors are barely visible. **Canines** start to erupt. These may erupt as early as three and one half years or as late as five years. The dental surfaces of the central incisors show wear, but the cups are deep. The intermediates are in wear, but sharp. The fourth premolar erupts as well as the third molar.

Four and One-Half Years

The last pair of incisors, the corners, erupt. Head-on, the central and intermediate incisors are in contact. The permanent corners are visible, but they are not in contact. In profile, the upper and lower canines are erupting and are sharp. The dental surfaces have distinct cups in the centrals and intermediates, while the corners are sharp.

Five Years

The permanent **dentition** is complete. All the incisors are in wear. The canines have erupted completely. The dental surfaces of the centrals and intermediates are wide transversely and show wear, but their cups are readily visible and completely encircled by the central enamel.

Six Years

The dental surfaces of the lower centrals are usually smooth and the shape is more oval. The central enamel is not as wide as it was at five years, and it is closer to the surface of the tooth closest to the tongue surface. The intermediates have distinct cups, but otherwise resemble the centrals. The corners show wear. The canines have reached their full length and are in wear.

Seven Years

In profile, the dental surface of the lower corner incisor is narrower than that of the upper. This leads to a notch on the caudal corner of the upper incisor—the seven-year notch or hook. The dental surfaces of the lower central and intermediate incisors are smooth. Cups are no longer present. The lower corners retain their cups.

Eight Years

The lower dental surfaces are smooth and all cups are gone in lower corners. The dental star appears in the lower central incisors. The dental star appears first as a dark yellow or yellow-brown transverse line in the dentin on the cheek side of the infundibulum of the permanent central incisor. The central and intermediate incisors are oval.

Nine Years

The seven-year hook has usually disappeared by nine years. The distal end of **Galvayne's groove** may be visible at the margin of the gum on the upper corner incisors (see Figure 9-3). The central incisors are round, while their central enamel is triangular. Their dental star is more distinct and narrower and near the center of the dental surfaces. The intermediate incisors are becoming round, while the corners are oval.

Ten Years

The angle of the teeth is more oblique. The distal end of Galvayne's groove should be visible on the upper corner of the upper incisor. The dental surfaces of the lower central and intermediate incisors are round, while the corner incisors are oval to round. The central enamel is triangular in the central incisors and close to the lingual border. The dental star is more distinct and

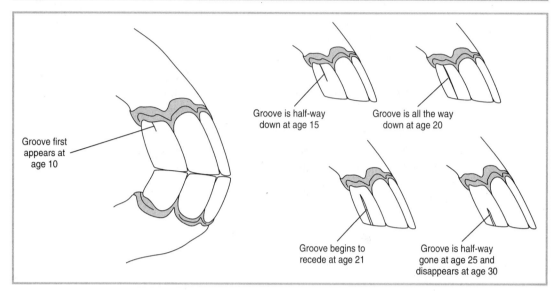

Groove first appears at age 10

Groove is half-way down at age 15

Groove is all the way down at age 20

Groove begins to recede at age 21

Groove is half-way gone at age 25 and disappears at age 30

Figure 9-3 Galvayne's groove

near the center of the teeth. Galvayne's groove appears at the gum margin of the upper corner incisor at about ten years of age, extends halfway down the tooth at fifteen years, and reaches the table margin at twenty. It then recedes and disappears at thirty years of age.

Eleven Years

The hook on the upper corner incisor returns (it may not appear until twelve years and it usually persists to fifteen years). The angle of the jaw increases in obliquity. The central enamel of each lower incisor forms a small ring close to the lingual border. The dental stars are narrower transversely and near the center of the dental table.

Twelve Years

The dental surfaces of all the lower incisors should be round. The central enamel is small and round, and it is disappearing from the centrals. The dental star is seen as a small yellow spot near the center of the dental surfaces.

Thirteen Years

The dental surfaces of the lower centrals may appear round or triangular. The central enamel in the lower incisors is small and round and, in many instances, disappearing. The dental stars are near the middle of the dental surfaces. The length of the teeth and the shape of the dental surfaces are the important markers for this age.

Fifteen Years

Galvayne's groove extends halfway down the labial (the surface of the tooth closest to the gums) surface of the upper corner incisors. The dental surfaces of the lower central incisors appear triangular. The intermediates are round to triangular and all lower incisors show a dark, distinct dental star.

Seventeen Years

The dental surface of each of the lower incisors is triangular. The dental stars are round and near the center of their respective teeth. Head on, the corner incisors are inclining slightly to the inside. In profile, the angle of the incisors is increasing.

Twenty Years

Galvayne's groove extends down the entire length of the labial (the surface of the tooth closest to the gums) surface of the upper corner incisors. The upper corner incisors deviate distinctly toward the median plane. Deviation of the intermediates is not as marked. The dental table of the lower incisors may be compressed transversely and may be worn almost to the gum (see Figure 9-4).

Abnormal Tooth Conditions

Several factors influence the wear and appearance of teeth, for example, bite (parrot mouth), cribbing, bishoping, and floating.

- **Parrot mouth** is a result of the upper and lower incisors not meeting because the lower jaw is too short. If it affects the molars also,

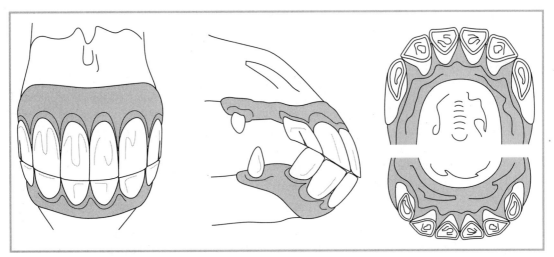

Figure 9-4   The teeth of an older horse

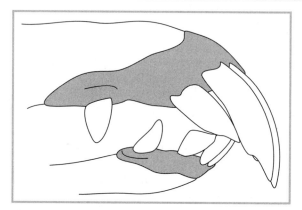

Figure 9-5 Parrot mouth

then sharp points and hooks may form during wear. This condition is rather common and may seriously interfere with grazing (see Figure 9-5).

- **Monkey mouth** is the opposite of parrot mouth and is seldom seen in horses.

- Cribbing is a habit common to stabled horses that damages incisors by chipping or breaking them.

- **Bishoping** is tampering with cups to make the horse appear younger than it is.

- **Floating** is filing high spots in molars to facilitate chewing. Molars should be checked regularly by veterinarians as the horse approaches mid-life and should be kept floated as needed (see Figure 9-6).

OTHER INDICATORS OF AGE

The features of older horses are a little like those of older people. The sides of the face become more depressed, the poll more prominent, and the hollows above the eyes deeper. The backbone becomes more prominent and starts to sag, and the joints appear more angular. Around the temples, eyes, nostrils, and elsewhere, white hair appears.

MEASURING HORSES

Typical measurements such as height, weight, and girth are influenced by age. These measurements are also affected by breed, type, sex, and nutrition.

Figure 9-6   Good horse owners check their horse's teeth on a regular basis.

Height

Height can influence price. Ponies are often cheaper because their use is limited. Horses are measured in hands. A hand is equal to four inches. With the horse on level ground, the point of measurement is the distance from the highest point of the withers to the ground. So, a horse measuring sixty inches is a fifteen-hand horse.

Height changes as a foal grows. Figure 9-7 shows how the height of Thoroughbred foals changes as they age.

Weight

Breed, type, and age determine the weight of a horse. Figure 9-8 graphically demonstrates how rapidly a horse increases in body weight as it matures. Knowing the weight of a horse is important for determining:

- Amount of feed needed
- Adequacy of a feeding program
- Potential health problems
- Optimal training and competing

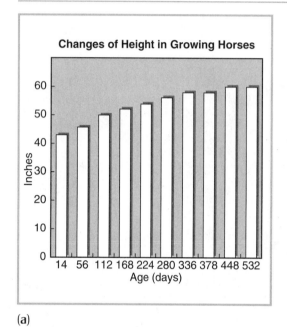

(a)   (b)

Figure 9-7 (a) Height chart. (b) Compare the height of this two-day-old Thoroughbred to her fourteen-year-old mother. *(Photo courtesy of Barbara Lee Jensen, After Hours Farms)*

- Maximal breeding efficiency
- Proper amount of medication

A University of Florida study found 88 percent of visual guesses on horse's weight resulted in underestimates. The best way is to use a truck scale: weigh the trailer and horse together, then weigh the trailer unloaded. Weight tapes give only rough estimates. For those who cannot use the truck scale method, researchers developed the following formula, using **heart girth** and body length.

Measure the heart girth just behind the elbow, taking the reading right after the horse exhales. Measure body length from the point of the shoulder to the point of the buttocks in a straight line. Avoid using a cloth measuring tape because it may stretch. A metal carpenter's tape is accurate but it is noisy and can spook the horse. A plastic coated tape works best. If one is not available, use cord or string that has no stretch and mark the spot with a pen. Then measure the cord with a carpenter's tape or yardstick.

Take the two measurements and multiply the heart girth in inches by itself, then multiply that by the body length in inches. Divide the total by 330 for the approximate weight in pounds.

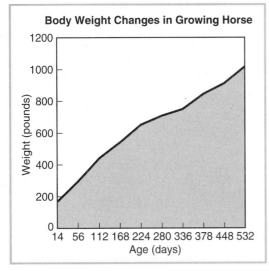

Body Weight Changes in Growing Horse

(a)

(b)

(c)

Figure 9-8 (a) Weight chart. (b) *Live From New York*, a two-month-old Thoroughbred filly. (c) *Live From New York* at age twelve. Note the difference in weight and muscling. *(Photo courtesy of Barbara Lee Jensen, After Hours Farms)*

For example, if the horse measures 75 inches around the heart girth and body length is 64 inches:

$$\frac{\text{heart girth} \times \text{heart girth} \times \text{body length}}{330} = \text{body weight}$$

$$\frac{75 \times 75 \times 64}{330} = 1{,}091 \text{ pounds}$$

For light horse foals from one to six weeks of age, a more accurate weight can be calculated using the following formula:

$$\frac{\text{heart girth in inches} - 25.1}{0.07} = \text{body weight}$$

Girth

A heart girth measurement is the circumference of the chest just behind the elbow. Heart girth gives some idea as to the space available for the heart and lungs.

Heart girth also can be used alone to estimate body weight. Some tapes are sold that give a direct reading of girth to weight. If this type of tape is not available, an ordinary measuring tape can be used and the girth converted to body weight using Table 9-2.

Table 9-2 Estimating a Horse's Weight from the Girth Measurement

Girth (inches)	Weight (pounds)[1]	Girth (inches)	Weight (pounds)[1]
32	100	66	860
40	200	68	930
45	275	70	1,000
50	375	72	1,070
55	500	74	1,140
60	650	76	1,210
62	720	78	1,290
64	790	80	1,370

1. For pregnant mares, multiply the value by 1.02, 1.06, 1.11 or 1.17 for their weight at 8, 9, 10, and 11 months of pregnancy, respectively.

SUMMARY

Age, height, and weight are important when considering a horse for competition or for personal use. For horses of a known age, the terms colt, filly, weanling, yearling, and two-year-old can be used until maturity. When accurate records are not available, age-related, specific changes in the teeth provide an accurate estimate of age.

The height of a horse relates to its use and value. Height can easily be measured. When scales are not available, the weight of a horse can be estimated from the girth measurement and a standardized table or from the girth and body length measurements and a standardized formula. Body weight is used for several management decisions.

REVIEW

Success in any career requires knowledge. Test your knowledge of this chapter by answering these questions or solving these problems.

True or False

1. A young female horse less than three years old is called a colt.

2. Horses have two sets of teeth.

3. The twelve front teeth of a horse are all called incisors.

4. Dental stars are the same as premolars.

5. A horse's girth can be used to estimate its weight.

Short Answer

6. List four reasons for needing to know the weight of a horse.

7. If the girth of a horse is 64 inches, about how much does it weigh?

8. If the girth of a horse is 70 inches and its body length is 62 inches, how much does it weigh?

9. If a horse is 64 inches from the highest point of its withers to the ground, how many hands tall is it?

10. What term is used to describe tampering with the tooth cups to make the horse appear younger?

11. What is the opposite of monkey mouth?

12. _____ is filing off the high spots in molars to improve chewing.

13. When do horses get their temporary and permanent centrals?

14. When do horses get their third permanent molar?

Discussion

15. Discuss four key changes in teeth that are used to estimate the age of a horse.

16. Define the following: foal, colt, stallion, mare, filly, weanling, and yearling.

17. Describe the importance of knowing the age of a horse.

18. What creates the dental star and when does it first appear?

19. Where is Galvayne's groove and when does it first appear?

20. Describe the changes in height and body weight as a foal grows.

STUDENT ACTIVITIES

1. Diagram the upper and lower jaw of a horse with a full complement of permanent teeth.

2. Compare the estimated body weight of a group of horses using the girth measurement method and the girth-body length formula.

3. Obtain an animal tooth and cut it in half longitudinally. Identify the parts of the tooth.

4. Sometimes the body weight of horses is given in kilograms and the height is given in centimeters. Develop a table of converted values for the following body weights given in kilograms: 80, 150, 220, 450, 500, and 550. Do the same for the following heights given in centimeters: 108, 126, 132, 139, 146, and 151. In this table convert the weight to pounds and the height to inches and hands.

5. Using different colors of modeling clay, make a model of a tooth that shows what happens as the surface of the tooth wears away.

6. Using the information in Table 9-2, create your own tape for measuring girth that gives a direct reading of a horse's weight.

ADDITIONAL RESOURCES

Books

Evans, J. W. 1989. *Horses: a guide to selection, care, and enjoyment.* 2nd ed. New York, NY: W. H. Freeman and Company.

Griffin, J. M., and Gore, T. 1989. *Horse owner's veterinary handbook.* New York, NY: Howell Book House.

Kainer, R. A., and McCracken, T. O. 1994. *The coloring atlas of horse anatomy.* Loveland, CO: Alpine Publications, Inc.

Lewis, L. D. 1996. *Feeding and care of the horse.* 2nd ed. Media, PA: Williams & Wilkins.

Price, S. D. 1993. *The whole horse catalog.* New York, NY: Fireside.

CHAPTER 10

Genetics

Genetics is a science that studies heredity and variation. Heredity is defined as the resemblance among individuals related by descent; variation is the occurrence of differences among individuals of the same species. Genetic material is contained on the chromosomes of all cells and the chromosomes are found in the nucleus of cells. Reproduction is the process of getting genetic material from the male in the form of sperm to the female egg.

Before reading this chapter, you should review Chapter 4. In Chapter 4, the function of the cell is discussed.

OBJECTIVES

After completing this chapter, you should be able to:

- Define a gene, an allele, and a chromosome
- Describe DNA
- Explain the difference between phenotypic and genotypic expression
- Give the correct number of chromosomes for horses and mules
- Discuss basic inheritance
- Distinguish between simple- and multiple-gene inheritance
- Distinguish among recessive dominance, codominance, and partial dominance
- Describe how DNA codes for proteins that make up the body and function in the body
- Compare qualitative traits to quantitative traits
- Discuss the relationship between genetics and environment
- Explain how genetics determines coat color
- Name five genetic diseases or abnormalities

KEY TERMS

Additive genes	Genome	Mule
Albino	Genotype	Nonadditive gene
Alleles	Heterozygous	Nucleotides
Chromosomes	Hinny	Phenotype
DNA (deoxyribonucleic	Homozygous	Qualitative traits
acid)	Jack	Quantitative traits
Dominant	Jennet	Recessive
Gametes	Messenger ribonucleic	
Genes	acid (mRNA)	

BASIC GENETICS

Genes are the basic unit of inheritance. They are carried on the chromosomes of all body cells. In the **gametes**—eggs or sperm—they pass inherited traits to the next generation. Different forms of the same gene—at the same location on the chromosome—are called alleles. Genes contain the "blueprint" or code that determines how an animal will look and interact with its environment. The number of chromosomes an animal possesses varies from species to species, but is consistent for a species—horses have 64 chromosomes.

Genes are made of **DNA (deoxyribonucleic acid)**. Resemblances and differences among related individuals are primarily due to genes. Genes cause the production of enzymes that control chemical reactions throughout the body, thus affecting body development and function. For normal body development and function, genes must occur in pairs. Genes are a part of the chromosomes that reside in the nucleus of all body cells. Chromosomes in the nucleus of a particular cell contain the same genetic information as the chromosomes in every cell of the body. So, the chromosomes in the cells of a horse's ear are the same as the chromosomes in its heart. The genes on the chromosomes, however, know their function in specific body tissues.

In the normal cell of a horse or any mammal, chromosomes occur in distinct pairs. Horses have 32 chromosome pairs for a total of 64 chromosomes.

Genomes

The complete set of "instructions" for making an organism is called its **genome**. The genome contains the master blueprint for all cellular structures and activities for the lifetime of the cell or organism. Found in every nucleus of the many trillions of cells in an animal, the genome consists of tightly coiled threads of deoxyribonucleic acid (DNA) and associated protein molecules, organized into distinct, physically separate microscopic structures called **chromosomes**.

The horse genome is organized into 64 chromosomes in 32 pairs. All genes are arranged linearly along the chromosomes. The nucleus of most horse cells contains two sets of chromosomes, one set from each of its parents. Each set has 31 single chromosomes, or autosomes, and an X or Y sex chromosome. A normal female will have a pair of X chromosomes; a male will have an X and Y pair. Chromosomes contain roughly equal parts of protein and DNA. DNA molecules are among the largest molecules now known. Chromosomes can be seen under a light microscope.

If unwound and tied together, the strands of DNA would stretch more than five feet long but would be only 50 trillionths of an inch wide. For every organism—from simple bacteria to remarkably complex horses—the components of these slender threads encode all the information necessary for building and maintaining life. Understanding how DNA performs this function requires some knowledge of its structure and organization.

Structure of DNA

In animals and humans, a DNA molecule consists of two strands that wrap around each other to resemble a twisted ladder whose sides, made of sugar and phosphate molecules, are connected by rungs of nitrogen-containing chemicals called bases. Each strand is a linear arrangement of repeating similar units called **nucleotides**, which are each composed of one sugar, one phosphate, and a nitrogenous base (see Figure 10-1).

Four different bases are present in DNA—adenine (A), thymine (T), cytosine (C), and guanine (G). Chromosomal DNA contains an average of 150 million bases. The particular order of the bases arranged along the sugar-phosphate backbone is called the DNA sequence. This sequence specifies the exact genetic instructions required to create a particular organism with its own unique traits.

The two DNA strands are held together by weak bonds between the bases on each strand, forming base pairs. Genome size is usually stated as the total number of base pairs. The horse genome contains over 3 billion base pairs.

Each time a cell divides into two daughter cells, its full genome is duplicated; for horses and other complex organisms, this duplication occurs in the nucleus. During cell division, the DNA molecule unwinds and the weak bonds between the base pairs break, allowing the strands to separate. Each strand directs the synthesis of a complementary new strand, with free nucleotides matching up with their complementary bases on each of the separated strands. Strict base-pairing rules are adhered to—adenine will pair only with thymine (an A-T pair) and cytosine only with guanine (a C-G pair). Each daughter cell receives one old and one new DNA strand. The cell's adherence to these base-pairing rules ensures that the new strand is an exact copy of the old one. This minimizes the incidence of errors (mutations) that may greatly affect the resulting organism or its offspring.

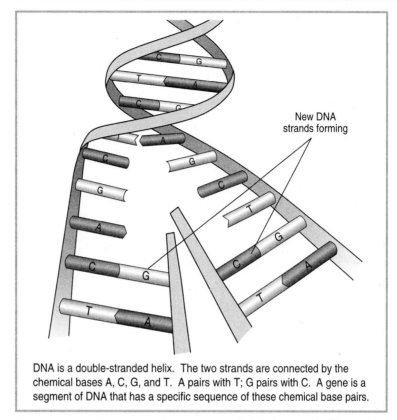

New DNA strands forming

DNA is a double-stranded helix. The two strands are connected by the chemical bases A, C, G, and T. A pairs with T; G pairs with C. A gene is a segment of DNA that has a specific sequence of these chemical base pairs.

Figure 10-1   The strands of DNA are similar to a twisted ladder with chromosomes contributed by each parent.

How the Code Works

Each DNA molecule contains many genes. A gene is a specific sequence of nucleotide bases, whose sequences carry the information required for constructing the proteins, that provide the structural components of cells and tissues as well as the enzymes for essential biochemical reactions.

Within the gene, each specific sequence of three DNA bases directs the cell's protein-synthesizing machinery to add specific amino acids. For example, the base sequence ATG codes for the amino acid methionine. Since 3 bases code for one amino acid, the protein coded by an average-sized gene (3,000 base pair) will contain 1,000 amino acids. The genetic code is thus a series of codons that specify which amino acids are required to make up specific proteins.

The protein-coding instructions from the genes are transmitted indirectly through **messenger ribonucleic acid (mRNA)**. This mRNA is moved from the nucleus to the cellular cytoplasm, where it serves as the template for protein synthesis. The cell's protein-synthesizing machinery then translates the code into a string of amino acids that will constitute a specific protein molecule.

Fundamentals of Inheritance

Chromosomes and gene numbers change during gamete (sex cell) formation (see Figure 10-2). Gametes are the eggs produced by sexually mature mares and the sperm cells produced by sexually mature stallions. During gamete formation in

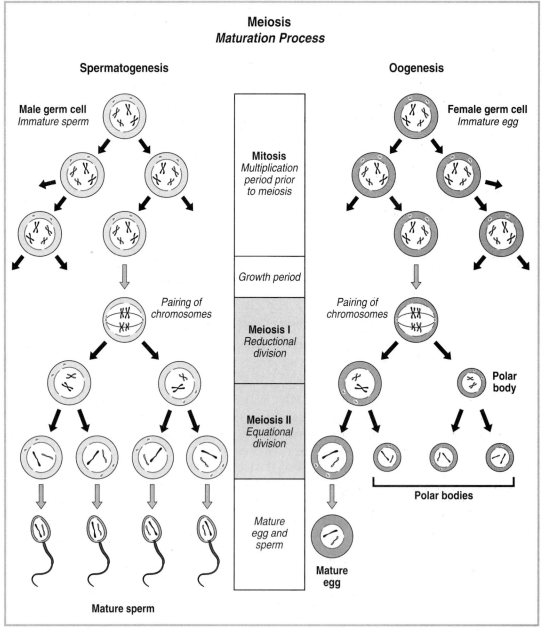

Figure 10–2 How the process of meiosis occurs in horses

the horse, the 32 chromosome pairs of a cell duplicate. Then, one of the four members associated with each of the duplicated chromosome pairs is randomly transferred to one of four forming gametes. The newly formed gamete now contains only one member of each original chromosome pair. This splitting of chromosome pairs causes a random transfer of each member into a forming gamete.

When an egg and sperm unite at fertilization, each carries only one member of each of its original chromosome pairs. The joining of a particular egg and sperm cell occurs at random. At fertilization, the chromosome number is restored to its original value. The new cell, the zygote that develops into a fetus, has one member of each chromosome pair from its sire and the other member from its dam. The resulting offspring will be genetically different from either parent because of the union of randomly matched gametes. Since horses have 64 chromosome pairs, the possible number of distinct assortments of genes in forming gametes is infinite. The possible number of genetically different horses can be much larger than the total number of horses being raised on the nation's farms.

How Sex is Determined

Chromosomes also determine the sex of a horse. The most common system of sex determination is the XY system of mammals. In this system, females carry XX chromosomes and males carry XY chromosomes. When females produce eggs, every egg possesses one X chromosome. When males produce sperm, half the sperm carry the X chromosome and half the Y chromosome. When the eggs and sperm unite, half the zygotes will be XX (female) and the other half will be XY (male). On the average, in a normal population, half of the offspring are males and half are females. Figure 10-3 illustrates how sex is determined with the XY system.

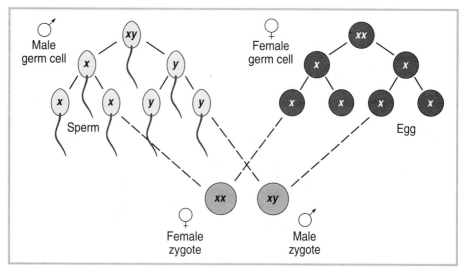

Figure 10-3   How sex is determined with the X and Y chromosomes

Chromosomes and Genes

The random transfer of chromosomes and their genes to forming gametes is called random segregation. Random segregation is the major cause of genetic differences among related individuals. These differences in genetic makeup are often referred to as genetic variation. Traits showing a large amount of genetic variation have a better chance of responding to selective breeding. If a large amount of genetic variation is present in a population, some animals will carry many favorable genes while others will have more undesirable genes for a given trait. If individuals with favorable genes can be identified and bred, there is an increased likelihood that their offspring will possess those favorable genetic traits. The specific genes that reside in the gene pairs that control a trait is called the animal's **genotype**.

Allelles.   Every cell contains a duplicate set of genes. Each set is derived from the single gene sets contributed at conception by both the mother and the father. The gene sets contain similar, but not necessarily identical, information. For example, both sets may contain a gene determining hair structure, but one set may contain the instructions for straight hair and the other for curly hair. The alternative forms of each gene are called **alleles**.

If both alleles are identical, the animal is said to be **homozygous** at that gene; if the alleles are dissimilar, then the animal is said to be **heterozygous** at that gene. Information about the homozygosity or heterozygosity for various genes can be inferred from information about parents and/or progeny and can be used for predicting the outcome of matings. For most of the alleles of horse coat colors, one cannot tell by looking at an animal whether it is homozygous at each coat color gene, so zygosity information will not be critical for purposes of identification. Sometimes, however, information about coat colors of parents may be used as an indication of incorrect parentage or erroneous identification, so some familiarity with genetic relationships may be useful.

Dominance.   Both sets of genes function simultaneously in the cell. Often when the gene pair is heterozygous, one allele may be visibly expressed but the other is not. The expressed allele in a heterozygous pair is known as the **dominant** allele, the unexpressed one as the **recessive** allele. The term dominant is given an allele only to describe its relationship to related alleles, and is not to be taken as an indication of any kind of physical or temperamental strength of the allele or the animal possessing it. Likewise, possession of a recessive allele does not connote weakness.

For simplicity in constructing models, geneticists symbolize genes by letters such as A, B, and so on. A dominant allele of a gene can be symbolized by a capital letter, for example R, and the recessive by an lower case letter—r.

For a given gene pair, the two genes can be alike or different. A homozygous gene pair has two identical genes while a heterozygous gene pair has different genes. Gene action can be grouped into two categories, nonadditive and additive.

- **Nonadditive Gene Action:** When **nonadditive gene** pairs control a trait, the members of the gene pairs will not be equally expressed.

- **Additive Gene Action: Additive genes** are those in which members of a gene pair have equal ability to be expressed. The expression of the gene pair is the sum of the individual effects of the genes in the pair.

Heredity vs Environment

All traits of horses are not controlled by just one gene pair. In fact, very few economically important traits are controlled by a single or even just a few gene pairs. Traits are controlled by possibly hundreds of gene pairs.

Consequently, traits are generally grouped into two categories, qualitative and quantitative.

Qualitative Traits.   **Qualitative traits** have four distinguishing characteristics:

1. Qualitative traits are controlled by a single or a few gene pairs.

2. Phenotypes (the trait characteristics we can see) of qualitative traits can be broken into distinct categories in which every member in that category looks the same.

3. The environment has little effect on the expression of the gene pair(s) controlling a qualitative trait. For example, coat color stays the same regardless of the environment.

4. The genotype of an individual for a qualitative trait can be determined (identifying the genes that occupy the gene pair(s)) with reasonable accuracy.

5. Qualitative traits show three types of gene action:
    - Dominance
    - Codominance
    - Partial dominance

An example of dominant inheritance is Combined Immune Deficiency. Blood type is an example of codominance, and coat color is an example of partial dominance.

Quantitative Traits. **Quantitative traits** are dissimilar in their attributes when compared to qualitative traits. Characteristics of quantitative traits include the following:

1. Quantitative traits are controlled by possibly hundreds or thousands of gene pairs located on several different chromosome pairs. Some gene pairs will contain additive genes while others can contain nonadditive genes. Most economically important traits are quantitative traits.

2. The environment does affect expression of the gene pairs controlling quantitative traits.

3. Phenotypes of quantitative traits cannot be classified into distinct categories since they will range from one extreme to another. It is impossible to accurately determine how many gene pairs are controlling a quantitative trait, so, an exact gene type can never be determined.

These factors make it difficult to identify individuals that have superior genotypes for quantitative traits.

Table 10-1 shows how much of some important traits in horses are due to genetics and how much are due to environment.

Table 10-1   Influence of Genetics and Environment on Some Traits in Horses

Trait	Due to Genetics (%)	Due to Environment (%)
Height at withers	45 to 50	50 to 55
Body weight	25 to 30	70 to 75
Body length	35 to 40	60 to 65
Heart girth circumference	20 to 25	75 to 80
Cannon bone circumference	20 to 25	75 to 80
Pulling power	20 to 30	70 to 80
Running speed	35 to 40	60 to 65
Walking speed	40 to 45	55 to 60
Trotting speed	35 to 45	55 to 65
Movement	40 to 50	50 to 60
Temperament	25 to 30	70 to 75
Reproductive traits	10 to 15	85 to 90

What do you get when you cross a male donkey or jack with a mare? You get a mule. So, what do you get when you cross a male Grant's zebra with a mare? You get a zorse, of course.

Horses and zebras have been crossed over the years, resulting in little more than pony-sized curiosities. Since zebras are hard to handle, they can only be bred to ponies. Zebras seem to defy domestication. But for someone with lots of patience, a Grant's zebra can become somewhat tame. This is accomplished by imprinting at birth and continuous handling.

By using a relatively tame Grant's zebra, semen can be collected in an artificial vagina when he is tricked into mounting a pony mare. This semen can then be used to artificially inseminate mares of any size—Arabians, quarter horses or even draft horses. This produces a larger zorse. Breeders of zorses hope that the zorses will combine the speed and savvy of zebras with the friendliness of horses. A zorse shows the coat color of the zebra and horses. For example a zorse show stripes like the zebra but may have the white stockings of the horse.

Like mules, zorses are infertile. The difference in the number of each species' chromosomes causes infertility. Horses have 64 chromosomes and zebras have 66 chromosomes. Unfortunately, zorses do not get species status because animals within a species must be able to interbreed.

Genotypic Expression. With all traits, the individual's **phenotype** is the sum of effects of the genotypic and the environmental effects (phenotype = genotype + environmental effects). Since qualitative traits are usually not affected by the environment, the phenotype of a qualitative trait is a good indicator of the genotype. Environmental effects do influence the phenotypic expression of a quantitative trait. An individual with an inferior genotype can rank higher phenotypically than individuals with superior genotypes because of favorable environmental effects. To reduce environmental effects, all animals must be treated the same. Usually an individual's phenotype, compared to an average for a similar group, is a good indicator or estimate of his genotype, or genotypic value.

Genetic evaluation programs often estimate the transmitting ability of an individual. The estimated transmitting ability is equal to one-half of an individual's estimated breeding value. The estimate of transmitting ability is the contribution that a stallion or mare is expected to make to the genotypic value of their offspring.

241

COAT COLOR

A system for classifying horse coat colors and markings is important in any program of identification of horses as individual animals. To have accurate and uniform application of the terminology for color classes, the system should stress recognition of basic, definable characteristics and should minimize the importance of subtleties that cannot be clearly defined. A scheme of coat-color classification based on recognition of the effects of the alleles of seven genes provides the tools necessary to define most of the common colors encountered in horses. For each of the seven genes of horse coat color to be considered, only two alternatives will be considered.

In any animal expressing the dominant allele of a gene, it cannot be determined by looking at the animal whether the second allele is a dominant or a recessive one. The presence of a recessive allele may be masked by a dominant allele, which leads to the expression "hidden recessive." Dominant alleles are never hidden by their related recessive alleles. Table 10-2 lists the genes and their action in the coat color of horses.

W and G Genes

The W (white) gene and the G (gray) gene represent alleles whose actions can obscure the actions of the other coat-color genes. If either allele W of the W gene or allele G of the G gene is present, the other coat-color genes cannot be determined by superficial examination.

A horse with the dominant allele W, will typically lack pigment in skin and hair at birth. The skin is pink, the eyes brown (sometimes blue), and the hair white. Such a horse is termed white. Sometimes such a horse is called **albino**. The W allele is only rarely encountered. All nonwhite horses are ww.

In horses, gray is controlled by the dominant allele G. A young horse with a G allele will be born any color but gray and will gradually become white or white with red or black flecks as it ages. Earliest indications of change to gray can be seen by careful scrutiny of the head of a young foal. Often the first evidence of the gray hairs will be seen around the eyes. In intermediate stages of the graying process, the horse will have a mixture of white and dark hairs.

In contrast to white (W) horses, gray horses are born pigmented, go through lightening stages, but always contain pigment in skin and eyes at all stages of coloration change.

A gray horse will be either GG or Gg. It is not possible to tell by looking at the horse whether it is homozygous for G. All nongray horses will be gg. For homozygous recessive colors, both alleles are written in the notation for color assignment, since a horse showing a color or pattern produced by recessives is by definition homozygous for the recessive alleles.

Table 10-2 Action of Horse Coat-Color Genes

Gene	Alleles[1]	Observed Coat Color
W	W w	WW: Lethal. Ww: Horse typically pigmented in skin, hair, and eyes and appears to be white. ww: Horse is fully pigmented.
G	G g	GG: Horse shows progressive silvering with age to white or flea-bitten, but is born any nongray color. Pigment is always present in skin and eyes at all stages of silvering. Gg: Same as GG. gg: Horse does not show progressive silvering with age.
E	E e	EE: Horse has ability to form black pigment in skin and hair. Black pigment in hair may be either in a points pattern or distributed overall. Ee: Same as EE. ee: Horse has black pigment in skin, but hair pigment appears red.
A	A a	AA: If horse has black hair (E), then that black hair is in points pattern. A has no effect on red (ee) pigment. Aa: Same as AA. aa: If horse has black hair (E), then that black hair is uniformly distributed over body and points. A has no effect on red (ee) pigment.
C	C Ccr[2]	CC: Horse is fully pigmented. CCcr: Red pigment is diluted to yellow, black pigment is unaffected. CcrCcr: Both red and black pigments are diluted to pale cream. Skin and eye color are also diluted.
D	D d	DD: Horse shows a diluted body color to pinkish-red, yellow-red, yellow or mouse gray and has dark points including dorsal stripe, shoulder strip, and leg barring. Dd: Same as DD. dd: Horse has undiluted coat color.
TO	TO to	TOTO: Horse is characterized by white spotting pattern known as tobiano. Legs are usually white. Toto: Same as TOTO. toto: No tobiano pattern present.

1. Different forms of the same gene.
2. In this table cr codes for a dilution factor.

Since gray is produced by a dominant gene, at least one parent of a gray horse must be gray. If a gray horse does not have a gray parent, then the purported parentage should be seriously considered incorrect.

Gene E

The first step for defining the coat color of a horse that is neither gray nor white is to decide if the animal has any black pigmented hairs. These hairs may be found in a distinctive pattern on the points (such as legs, mane, and tail), or black hair may be the only hair color (with the exception of white markings) over the entire body. If a horse has black hair in either of these patterns, then the animal possesses an allele of the E gene, which contains the instructions for placing black pigment in hair. The alternative allele to E is e. Allele e allows black pigment in the skin but not in the hair. The pigment conditioned by the e allele makes the hair appear red.

If an animal has no black-pigmented hair, it has the genetic formula ee. Basically, an ee animal will be some shade of red ranging from liver chestnut, to dark chestnut, to chestnut, or sorrel. Manes and tails may be lighter (flaxen), darker (not black), or the same color as the body. These pigment variations of red cannot yet be explained by simple genetic schemes. Shades of red are not consistently defined by breeds or regions of the country, so use of specific terms for the shades of red can be confusing.

Since the red animal is not gray and not white, its genetic formula is ww, gg, ee. When two red horses are bred (ww, gg, ee × ww, gg, ee), the offspring should also be red (ww, gg, ee). If the offspring has black pigment (E) or is gray (G) or white (W), the assumed parentage is incorrect.

Gene A

The gene that controls the distribution pattern of black hair is known as A. The allele A in combination with E will confine the black hair to the points to produce a bay. Various shades of bay, from dark bay or brown through mahogany bay, blood bay, copper bay, and light bay, exist. The genetics of these variations has not been defined. Any bay horse will include A and E in its genetic formula as well as ww and gg.

The alternative allele a does not restrict the distribution of black hair and thus, in the presence of the allele E of the E gene, a uniformly black horse is produced. In most breeds, the a allele is rare, so black horses are infrequently seen. Many black horses will sun-fade, especially around the muzzle and flanks; such animals may be called brown. The term brown can be used for several genetic combinations—various reds, bays and dark bays, as well as some blacks.

Neither A nor a affects either the pigment or its distribution in red (ee) horses. So, an examination of coat color cannot determine which alleles of the A gene a red horse has.

Gene C

An allele of the C gene, known as C, causes pigment dilution. Fully pigmented horses are CC. Heterozygous horses (CC) have red pigment diluted to yellow, but black pigment is not affected. A bay (E, A) becomes a buckskin by dilution of the red color body to yellow without affecting the black color of the mane and tail. The genetic formula for a buckskin is ww, gg, A, E, CC. A red horse (ee) becomes a palomino (see Figure 10-4) by dilution of the red pigment in the body to yellow with mane and tail being further diluted to flaxen. The genetic formula for a palomino is ww, gg, ee, CC.

A genetically black horse (E, aa) can carry the dilution allele without expressing it, since CC only affects red pigment.

In the homozygous condition, C completely dilutes any coat color to a very pale cream with pink skin and blue eyes. Such horses are often called cremello (also perlino or albino). Typically, such horses are the product of the mating of two dilute-colored animals such as palominos or buckskins. Cremello may be difficult to distinguish from white (see Figure 10-5).

Figure 10-4 Palomino-colored horse. *(Photo courtesy Palomino Horse Breeders of America, OK)*

Figure 10-5 An American creme draft horse, *Clar Ann Dick*, owned by Clarence Ziebell, Charles City, Iowa. *(Photo courtesy Elizbeth A. Ziebell, Sec, American Creme Draft Horse Association)*

Gene D

The D gene determines a second kind of dilution of coat color. Its effects can be confused with those of C, but several important differences in the effects of D and C on color exist. First, D dilutes both black and red pigment on the body, but does not dilute either pigment in the points. Red body color is diluted to a pinky-red, yellowish-red or yellow; black body color is diluted to a mouse-gray. Second, in addition to pigment dilution, a predominant characteristic of the allele D is the presence of a particular pattern that includes dark points, dorsal stripe, shoulder stripe, and leg barring. Third, homozygosity for D does not produce extreme dilution to cream as does C.

This pigment dilution pattern is called dun. In an otherwise red horse, the D allele produces a pinkish-red horse with darkened points known as a red dun or claybank dun (ww, gg, ee, CC, D). In an otherwise bay animal, the D allele produces a yellow or yellow-red animal with black points known as a buckskin dun (ww, gg, E, A, CC, D). An otherwise black animal with the D dilution allele is a mouse-gray color with black points known as a mouse dun or grulla (ww, gg, E, aa, CC, D).

The effect of D and C can be easily confused in A, E horses, so care must be taken in identification. An animal can have both the C and D dilutions, a situation that may be difficult to distinguish except by breeding tests. D is found in only a few breeds of horse, and probably in the United States would be seen only in stock horse breeds, as well as in some ponies.

Gene TO

Several different white spotting patterns exist in horses, but so far only that of the tobiano has been clearly shown to be conditioned by a single gene. Tobiano spotting, symbolized by TO, is a variable restricted pattern of white hair with underlying pink skin that can occur with any coat color. The pattern is present at birth and stable throughout life. Generally, white extends across the back in an apparent top-to-bottom distribution on the body. The white areas may merge to form an extensive white pattern of generally smooth outline. The legs are white, but the head is usually dark except for a facial marking pattern (see Figure 10-6). Tobianos can now be screened for their potential to be true breeding for the tobiano pattern.

Assignment of Coat Color by Genetic Formula

Defining the coat color of a horse is a stepwise process. The first step is to determine if either G or W is present. If yes, then the animal is gray or white and this is the end of the identification task.

If the horse is neither gray not white, then assignment of alleles of the other genes can be made to define the color. First, one must decide if the horse

Figure 10-6   Horse with Tobiano coat color. *(Photo courtesy American Paint Horse Association, TX)*

has E or not. If E, then it must be decided whether the horse has A or not. If the animal does not have E, then a decision about A cannot be made. If none of the colors is diluted and if no spotting pattern is present, these decisions about E and A will define the colors bay, black, and red.

If dilution of the basic colors to yellow, light red, mouse gray, or cream is present, then further definition can be made with addition of the alleles of C and D to the basic formula containing W, G, E, and A. In the absence of white spotting, these decisions will define the colors palomino, buckskin, cremello, red dun, buckskin dun, and mouse dun.

If a white spotting pattern that meets the definition of tobiano is present, TO can be assigned to the genetic formula.

The outcome of decisions about the genes W, G, E, A, C, D, and TO results in the assignment of alleles for each gene. Each assignment should be carefully reviewed to consider if the chosen alleles are likely to be found in the breed of horse being identified.

Some of the genetic formulas and the color definitions that can be assigned by this process are shown in Table 10-3.

Table 10-3 Genetic Formulas and Resulting Coat Colors

Genetic Formulas[1]	Color
W	White
G	Gray
E, A, CC, dd, gg, ww, toto	Bay
E, aa, CC, dd, gg, ww, toto	Black
ee, aa, CC, dd, gg, ww, toto	Red
E, A, CCcr, dd, gg, ww, toto	Buckskin
ee, CCcr, dd, gg, ww, toto	Palomino
CcrCcr	Cremello
E, A, CC, D, gg, ww, toto	Buckskin dun
E, aa, CC, D, gg, ww, toto	Mouse dun
ee, CC, D, gg, ww, toto	Red dun
E, A, CC, dd, gg, ww, TO	Bay tobiano
ee, CC, D, gg, ww, TO	Red dun tobiano

1. Refer to Table 10-2 for a description of the action of each gene.

GENETIC ABNORMALITIES

Defects in DNA can result in the failure to form essential proteins or in the formation of abnormal proteins. This may cause death or disease in the horse. These defects may be caused by abnormalities in a single gene, the cumulative effect of a group of abnormal genes, or some chromosomal abnormality. Table 10-4 lists and describes some of the genetic abnormalities or diseases in horses.

MULES

No discussion of genetics and horses would be complete without a mention of the mule (see Figure 10-7). In a way, mules are themselves a genetic abnormality—they have an uneven number of chromosomes. A **mule** is the offspring of a male donkey (**jack**) and a female horse (mare). A mule is much like the horse in size and body shape but has the shorter, thicker head, long ears, and the braying voice of the donkey. It also lacks, as does the donkey, the horse's calluses, or "chestnuts," on the hind legs. The reverse cross—between a stallion and a female donkey (called a **jennet** or jenny)—is a **hinny**, some-

Table 10-4	Some Genetic Diseases of Horses Caused by a Single or a Few Genes	
Genetic Disease	**Description**	**Comments**
Combined Immunodeficiency (CID)	Failure of immune system to form; animals die of infections.	Disease of Arabian and part-Arabian horses; transmitted as autosomal recessive; mutation of a single gene.
Hyperkalemic Periodic Paralysis (HyPP)	Defect in the movement of sodium and potassium in and out of muscle; animals intermittently have attacks of muscle weakness, tremors, and collapse.	Disease of quarter horses; transmitted as autosomal dominant; involves one gene.
Myotonic Dystrophy	Spasms occur in various muscles.	
Hemophilia A	Failure to produce blood-clotting factor; bleeding into joints, development of hematomas.	Disease of Thoroughbreds, quarter horses, Arabians and Standardbreds; transmitted as X-linked.
Hereditary Multiple Exostosis	Bony lumps develop on various bones throughout the body.	
Parrot Mouth	Lower jaw is shorter than upper jaw; incisor teeth improperly aligned.	
Lethal White Foal Syndrome	Failure to form certain types of nerves in the intestinal tract; foals die of colic several days after birth.	Affects some offspring produced by mating two overo paint horses; several genes involved.
Laryngeal Hemiplegia	Paralysis of the muscles that move the cartilages in the larynx; causes noise in the throat with exercise.	
Cerebellar ataxia	Degeneration of specific cells in the cerebellum of the brain; causes incoordination.	

(Continued on next page)

Table 10-4 Some Genetic Diseases of Horses Caused by a Single or a Few Genes *(concluded)*

Genetic Disease	Description	Comments
Gonadal Dysgenesis	Animals tend to be small and weak at birth; show disorders of the reproductive system; mares are sterile.	Presence of a single X chromosome in a female; caused by failure of the X chromosome to separate after duplication.
Hydrocephalus	Accumulation of fluid within compartments of the brain; results in crushing of brain.	
Neonatal Isoerythrolysis (NI)	Hemolytic disease of the newborn; foal's red blood cells destroyed by antibody in mare's colostrum; results in anemia and sometimes death.	Genetic makeup of animal predisposes to disease; underlying basis is incompatibility in blood type between the mare, the foal, and the foal's sire.
Umbilical Hernias	Opening in the body wall at the navel does not close normally; intestines may drop through opening.	
Inguinal Hernias	Openings through which testicles descend allow intestines to escape into the scrotum; may cause colic.	
Connective Tissue Disease	Skin is hyperelastic, easily stretched and injured.	
Epitheliogenesis Imperfecta	Skin fails to form over parts of the body or in the mouth.	
Cataracts	Cloudiness of the lens in the eye; result in blindness.	

times also called a jennet. A hinny is similar to the mule in appearance but is smaller and more horselike, with shorter ears and a longer head. It has the stripe or other color patterns of the donkey. Hinnies are more difficult to produce than mules. Although they may display normal sex drives, mules are infertile. Rarely, a female mule or hinny may come into heat and produce a foal. Horses have 64 chromosomes (32 pairs), donkeys, 62 (31 pairs). Mules and hinnies have 63 chromosomes.

Figure 10-7   The Clapier family's mule, *Rosey.*

SUMMARY

Genes are the basic unit of inheritance. They are composed of DNA and arranged along the length of the chromosomes. Foals receive one-half of their genetic material from the stallion and one-half from the mare. The genetic make-up of a horse is called its genotype. How the genetic information is expressed in terms of the physical appearance of the horse is called the phenotype. Most traits of a horse are controlled by many genes. The genetic make-up of a horse and the environment interact. Harmful genetic material passed to a foal may produce disease or death.

REVIEW

Success in any career requires knowledge. Test your knowledge of this chapter by answering these questions or solving these problems.

True or False

1. The complete set of instructions for making an organism is called its genome.
2. The possible number of distinct assortments of genes in forming gametes is not over 100.

3. Differences in genetic make-up are often referred to as genetic variation.

4. The environment has little effect on the expression of the gene pair(s) controlling a qualitative trait.

5. Alanine is one of the four bases in DNA.

6. Some genes actually dilute the coat color of horses.

Short Answer

7. How many pairs of chromosomes does a horse have?

8. In horses, which sex carries the XY chromosomes and which sex carries the XX chromosomes?

9. What is the difference between a homozygous gene pair and a heterozygous gene pair?

10. List the three types of gene action that qualitative traits show.

11. Name five economically important traits.

12. Every cell contains a duplicate set of genes. What is the alternative form of each gene called?

13. What do the W and G genes express in horse coat color?

14. List five genetic diseases or abnormalities.

15. How is a dominant and recessive allele of a gene indicated when making models?

Discussion

16. Briefly describe the structure of DNA.

17. What is a gamete?

18. Explain an animal's genotype.

19. Discuss the genetic control of coat color in horses.

20. Discuss the role of genetics and environment in the expression of economically important traits in the horse.

21. Compare qualitative to quantitative traits.

STUDENT ACTIVITIES

1. Using color photographs, videos, or real horses, try to describe the genetics of the coat color.

2. Obtain prepared microscope slides and view the chromosomes in the nucleus of some animal cells.

3. Draw or make a three-dimensional representation of DNA.

4. Using the Internet and/or other research methods, develop a report about any work that is being done to map the genome of horses.

5. To understand the randomness of the genetic process, have five people roll a pair of dice 12 times. Make a table that tracks how many times each person rolls two ones, two twos, two threes, two fours, two fives, and two sixes.

ADDITIONAL RESOURCES

Books

American Youth Horse Council. 1993. *Horse industry handbook: a guide to equine care and management.* Lexington, KY: American Youth Horse Council, Inc.

Asimov, I. 1962. *The genetic code.* New York, NY: New American Library.

Frandson, R. D., and Spurgeon, T. L. 1992. *Anatomy and physiology of farm animals.* 5th ed. Philadelphia, PA: Lea & Febiger.

Fraser, C. M., ed. 1991. *The veterinary manual.* 7th ed. Rahway, NJ: Merck & Co.

Hafez, E. S. E. 1993. *Reproduction in farm animals.* 6th ed. Philadelphia, PA: Lea & Febiger.

McKinnon, A. O., and Voss, J. L. 1993. *Equine reproduction.* Philadelphia, PA: Lea & Febiger.

CHAPTER 11

Reproduction and Breeding

Reproduction is the process of getting genetic material from the male to genetic material from the female through the union of sperm and egg cells. To produce offspring with regularity, maximize reproductive efficiency, and protect the future reproductive capabilities of the mare requires a sound, practical understanding of the mare's reproductive process and the development of breeding practices that coincide with her physiology.

OBJECTIVES

After completing this chapter, you should be able to:

- Discuss breeding periods
- List and discuss the major parts of the female reproduction tract
- List and discuss the major parts of the male reproduction tract
- Describe reproductive hormones during the estrus cycle
- Recognize fertility problems
- Explain gestation and parturition in horses
- Discuss and demonstrate methods of artificial insemination and heat detection
- Explain embryo transfer
- Describe the management of the mare and stallion prior to, during, and after the breeding season
- Describe the management of the mare, including care at parturition, nursing to weaning, and growing to maturity

KEY TERMS

Abortion	Follicle	Placenta
Allantois	Follicle stimulating	Polyestrous
Amniotic fluid	hormone (FSH)	Postpartum
Artificial insemination	Gestation	Progesterone
Bag up	Hand mating	Prolactin
Barren mare	Heat	Prostaglandins
Caslick	Hippomane	Relaxin
Colostrum	Involution	Scrotum
Condition score	Lactation	Semen
Corpus hemorrhagicum	Libido	Settle
Corpus luteum	Lochia	Short cycle
Defecation	Luteal phase	Silent heat
Diestrus	Maiden mare	Sprung
Dystocia	Metritis	Stillbirth
Ejaculation	Open mare	Vulva
Embryo transfer	Ovulation	Waxed teats
Estrogen	Parturition	Wet mare
Flushing	Pasture mating	Winking
Foal heat	Photoperiod	

PHYSIOLOGY OF REPRODUCTION

When it comes to breeding practices and the reproductive process, the mare's reproductive control mechanisms are quite efficient when left to function in the wild. Manipulation and confinement have reduced the efficiency of reproduction. Several factors contributing to poor reproductive performance include:

- Reproductive anatomy
- Long time period before an embryo can safely implant in the uterus
- Variable hormonal system synchronizing the whole process.

Reproductive organs of the stallion and mare as well as the endocrine system were introduced in Chapter 5. Figures 11-1 and 11-2 review the reproductive organs of the stallion and mare.

Stallion

The reproductive organs of the stallion consist of two testes, each suspended by a spermatic cord and external cremaster muscle; two epididymides; two

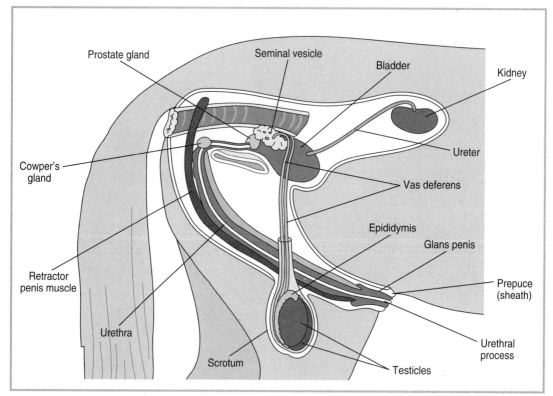

Figure 11-1 Stallion reproductive organs

deferent ducts; the penis and the associated muscles. Accessory sex glands are paired vesicular glands, one prostate gland and paired bulbourethral glands. The outside of the reproductive tract includes the **scrotum**, prepuce, and penis (see Figure 11-1).

The scrotum is an outpouching of the skin, divided into two scrotal sacs by a septum. The two sacs each contain one testis, located on either side of the penis. The testes should descend from the abdominal cavity through the inguinal canal into the scrotum between the last three weeks of gestation and the first two weeks after birth. If this does not happen before closure of the inguinal ring, the cryptorchid testis (usually the left one) stays in the abdominal cavity.

Mating Process. The process of **ejaculation** consists of three parts: erection, emission, and ejaculation.

Erection is stimulated by teasing the stallion. During erection, the penis lengthens and stiffens through engorgement with blood. Emission occurs in strong pulsatile contractions.

During emission, **semen**, which is spermatozoa and fluid from the cauda epididymis plus fluids from the accessory glands, arrive in the pelvic urethra.

During **ejaculation**, the semen is expelled through the urethra.

Reproduction in the stallion is also under hormonal control. The hormones directly involved include: FSH, LH, and testosterone. Actions of these hormones are described in Chapter 5 and Table 5-1.

Sperm Production. No spermatozoa production occurs until a stallion is well over one year old, and full reproductive capacity is not reached until the age of four. The stallion's reproductive capacity will then remain constant until he is about twenty years old. The tendency is to over-use a young stallion and under-use an old stallion.

Sperm output and sperm production is influenced by:

- Season
- Testicular size
- Age
- Frequency of ejaculation

Although stallions produce spermatozoa throughout the year, they are seasonal breeders. In the Northern Hemisphere, the three best months for testicular size, development, and function are May, June, and July, while from September through February the testes are regressed, especially in November and December. In December and January, the sperm count is 50 percent of that during June and July. Normal semen characteristics are given in Table 11-1.

If the majority of mares have to be bred early in the season, between February and June, an artificial lighting program may be useful. Starting in mid- to late December, the stallion should be exposed to sixteen hours of light and to eight hours of darkness per day; this should be continued until there are sixteen hours of natural daylight. However, the stallion has to be normally exposed to the decreasing daylight in the fall to eliminate a photorefractory condition that would prevent his being sensitive to increasing light. Artificial **photoperiods** do lead to early burnout and a decline in performance at the end of the breeding season.

Table 11-1 Quantitative Data on Stallion Semen and Spermatozoa		
Volume per Ejaculate in Milliliters	**Sperm per Cubic Millimeter**	**Total Sperm in Ejaculate**
25 to 150	60,000	6,000,000

Mare

The mare's reproductive anatomy is characterized by a simple uterus, unusual placenta arrangement, and inefficient cervical closure (see Figure 11-2). The uterus has a small uterine body with two long, narrow uterine horns. These structures are suspended within the abdominal cavity via ligaments and connective tissue to the abdominal wall.

The elongated shape of the uterus and uterine horns causes the uterus to drain inefficiently, predisposing this organ to infections. With each subsequent pregnancy, the uterus and supporting connective tissue decrease in tone, and the suspension of the uterus becomes lower and lower in the abdominal cavity. The closure of the cervix maintains pregnancy by retaining the embryo and its membranes within the uterus and preventing entry of bacteria. Maintenance of the cervical seal is vital for embryo survival by preventing infection. The closure of the cervix is controlled by hormonal levels and can be unstable. This has been cited as a possible cause for early embryonic death.

The placenta is attached to the endometrial lining of the uterus by innumerable tiny villi that project into the lining, forming a shallow one-cell-thick fusion through which the placenta transfers the embryo's blood, oxygen,

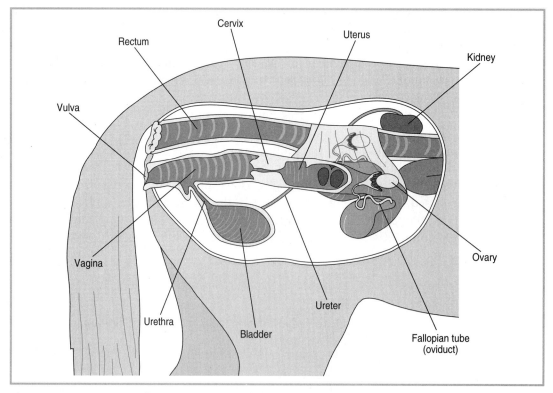

Figure 11-2 Mare reproductive tract

and nutritional needs. This type of placenta does not allow immunoglobulins to pass to the fetus. This attachment is responsible for maintaining embryonic life, and early shedding or detachment of the placenta drastically endangers the pregnancy.

Hormones of Reproduction. All reproductive functions in the mare are controlled by hormones produced in the glands of her endocrine system; hormonal balance controls all phases of reproductive tract stimulation and inhibition. When hormonal balance is not achieved, either because of a natural imbalance or a disturbance, a mare will have problems cycling, conceiving, maintaining pregnancy, delivering a foal, and providing an adequate milk supply. Table 5-1 describes the actions of the hormones involved in reproduction.

Hormonal Cycles. The mare has a strong **follicle** stimulating phase of her cycle, which is what causes her to come into **heat** and ovulate. The luteal phase occurs after she ovulates and is responsible for production of the **corpus luteum** and its production of **progesterone**.

If conception does not take place, **prostaglandins** are released that destroy the corpus luteum, reduce progesterone, and allow the mare to cycle again. If there is conception, no prostaglandins are released and the corpus luteum remains dominant. The high level of progesterone during the luteal phase maintains pregnancy by keeping **Follicle Stimulating Hormone (FSH)** and estrogen in check, thereby preventing the mare from coming back into estrus and disrupting the newly established pregnancy. Progesterone also relaxes the uterus to allow the embryo to implant and the new pregnancy to be established.

The mare is a seasonally **polyestrous** species, which means that she comes into estrus several times a year but does not cycle all year round. In the Northern Hemisphere, the mare begins cycling somewhat irregularly in January and February as the days get longer. She continues to have more regular cycles until the peak of her breeding season in June. In September or October, as the days shorten, the mare ceases to cycle regularly. By late November she stops cycling altogether and remains inactive through winter. The times of the year with irregular and subfertile cycles in February and March, and in September and October are called breeding transition months.

The current tendency of some breeders to mate mares earlier and earlier in hopes of producing foals earlier in the year is a reproductive problem for the mare. While this practice is perhaps economically advantageous when a larger and more developed foal is presented in the sale or show ring, it is incongruent with the mare's natural timing. The result is that reproductive efficiency is sacrificed for the chance to attain greater weanling and yearling sizes. An arbitrary January 1 birth date entices breeders to try to breed their mares before they are ready to accept and maintain a pregnancy or even to conceive.

When mares begin cycling in the spring, their estrus lasts six to eight days. The length of estrus progressively shortens until it is only three to four days in most mares at the peak of the season in May or June. These durations are thumb rules. Variation is more the rule.

During estrus, or heat, follicles develop in the ovaries. The follicles produce the hormone **estrogen** which causes the signs of sexual receptivity. Although several follicles develop simultaneously, usually only one follicle will emerge as the dominant follicle. **Ovulation**, the time at which a primary follicle is ready to shed an egg mature enough for fertilization from the ovary, occurs late in the estrus, no more than two days before the mare goes out of heat. The exact time of ovulation varies between individual mares. In fact, early in the season, a mare may exhibit signs of estrus and not ovulate at all, just as some mares may ovulate on schedule but not show outward signs of estrus (heat). The incidence of ovulation within the estrus period increases as daylight increases, peaking in late June.

After ovulation, the now eggless cavity in the follicle will fill up with a blood clot and is now called the **corpus hemorrhagicum**. This becomes the corpus luteum, which produces the hormone progesterone, which corresponds with the **diestrus**.

During estrus, the cervix relaxes and is soft and rose pink. The secretions in the vagina during estrus are clear and slimy and the vagina is red and vascular. In diestrus, the cervix will protrude into the vagina, will be pale pink and tightly constricted, and the secretions are scant, viscous and sticky.

At the time of ovulation, the follicle in large mares may be as large as 65 mm in diameter (ranging from 35 to 65 mm) when it enters the oviduct. By day twenty, the follicle may be detected by rectal palpation and also by ultrasound. Ultrasound can estimate follicular size and can also differentiate between a young corpus luteum and a soft follicle even though these structures feel similar during rectal palpation.

Usually only one follicle will ovulate. Occasionally two follicles will ovulate at the end of the estrus phase. This is undesirable because twin fetuses have a high risk of abortion and cause complications such as dystocia (long labor) and retained placenta.

If the fetus is aborted before day 45, the mare continues to show signs of being pregnant, due to special tissue secreting the hormone that maintains pregnancy, until day 120. Therefore, another ultrasound should be performed after day 45 (see Table 11-2).

Breeding Habits. Estrus cycles will start in a mare at puberty, which is usually between the age of fifteen to twenty-four months, but can be as early as one year of age and as late as four years of age. The mare usually goes into winter anestrus between November and February or March. The mare has a normal estrus cycle of twenty-one to twenty-two days. The first five to seven

Table 11-2 Breeding Characteristics of Mares	
Range in age when heat period begins	15 to 24 months
Recommended minimum age to breed	24 to 36 months
Duration of estrus (heat)	3 to 7 days
Best time to breed	Every other day beginning second day of heat
If not bred, estrus recurs in:	10 to 35 days

days, when the mare displays behavioral signs of sexual receptivity to the stallion, are called the estrus. When teased with a stallion, the mare will raise her tail, urinate, and the labia will open to expose the clitoris ("**winking**"), while she assumes a mating position.

During the second, or **luteal phase**, the behavioral pattern is that of sexual rejection of the stallion. This is called diestrus and lasts fourteen to fifteen days. During diestrus, the mare will switch her tail, pin her ears back, kick, and move away from the stallion when she is teased.

Fertilization.   The ovum leaves the ovary and enters the oviduct, where fertilization will take place. The ovum is viable for eight to twelve hours, while the spermatozoa coming up the oviduct can live for twenty-four to forty-eight hours (sometimes for several days) inside the mare's reproductive tract. The time involved for the spermatozoa to travel through the oviduct and reach the ovum is four to six hours. On the basis of these time constraints, breeding is recommended within one or two days prior to ovulation. After the egg is fertilized, it travels down the oviduct and enters the uterus in five to six days. Once in the uterine horn, the embryo is very mobile; it bounces around and may move from one uterine horn to the other. By day 16 to 18, the embryo settles in one part of the uterus, where it implants.

From ovulation to **parturition**, the average length of **gestation** is 335 days, plus or minus two to four weeks, depending on the season, nutritional status of the mare, and sex of the fetus.

Sterility

Abortion is the premature termination of pregnancy before 300 days of gestation, while termination after that time (when a foal may be born alive) is considered **stillbirth**. Abortion can be caused by both bacteria and viruses. If a mare aborts, a veterinarian should be consulted.

Checking the Mare for Breeding

Considerations for breeding include:

- Appearance
- Pedigree
- Hereditary disorders
- Disposition
- Conformation
- Performance

Factors that relate to the mare's reproductive potential are:

- Age
- Breed
- Status
- Past breeding records
- Previous athletic use
- General health

Age. From puberty to old age, a mare can conceive and carry a foal to term. The two- to three-year-old may have some abnormal cycling patterns. After multiple foals, mares may have anatomical changes in the **vulva** and vagina, predisposing for pneumovagina (air in the vagina) and urine pooling.

Breed. Breed organizations differ in policy on natural versus artificial insemination, semen transport, and preservation of semen, as well as on embryo transfer. Miniature breeds and the very large draft breeds show a greater tendency to reproductive failure.

Status. A **maiden mare** is one that has never been bred. A **barren** or **open mare** is one that was either not bred the previous season or did not conceive in the previous season. Unless she was not bred, this implies failure of conception or failure to maintain pregnancy. A **wet mare** has foaled during the current breeding season and is nursing the foal.

Past Breeding Records. Previous foaling data, such as gestation length, any complications, cycling patterns from previous years, previous reproductive surgery, previous uterine infection and treatment, as well as evidence of early embryonic death in previous seasons are all helpful information in reproductive evaluation.

Previous Athletic Use or Performance. After an athletic career, reproductive performance may be compromised because of injuries, diseases, or treatment with anabolic steroids.

As a general rule, most mares are seasonally polyestrous. This means that during a specific season of the year mares experience several reproductive cycles. In the Northern Hemisphere, as the days get longer, mares show behavioral estrus or heat beginning in February and extending through July. The same trends occur in mares in the Southern Hemisphere for the corresponding seasons. Mares kept on grass normally go into anestrus—no cycles—in the winter. Near the equator, where the amount of daylight remains fairly constant, the length of the estrus cycle shows little variation.

But as a whole population, mares can be classified into three major categories:

- **Defined breeding season.** Wild horses breed during the time of year that corresponds to the longest days of the year. So, the foals are born during the spring of the year when feed is apt to be the best.

- **Transitory breeding season.** Some domestic breeds and some individual mares show estrus cycles throughout the year. These mares breed, but matings in the winter months are not fertile since ovulation does not occur during the cycle. Actual ovulation occurs with estrus (heat) only during a breeding season defined by the increasing amount of light. Again foals are born during a limited foaling season.

- **Year-round breeding.** Some domestic breeds and some individual mares have estrus cycles accompanied by ovulation all year. Foals are born any time of year.

Of all the domestic animals, the reproductive cycle of the mare shows the greatest variation—no absolutes in biology.

General Health. Previous medical events such as chronic obstructive pulmonary disease (heaves) leading to coughing and difficulty breathing, cardiac disease, or pain from laminitis or tendinitis will all influence reproductive potential.

Pregnancy Diagnosis

Ultrasound has been used for pregnancy diagnosis since the early 1980s and is used in addition to rectal palpation. It may provide diagnosis of conditions that cannot be felt by rectal palpation.

The ultrasound probe is inserted into the rectum and moved across the reproductive tract. Ultrasonography is very useful in the study of the normal reproductive cycle, diseases of the ovaries and uterus, early detection of pregnancy, diagnosis of twins, diagnosis of embryonic death, and length of gestation.

In the nonpregnant mare, ultrasound study of the ovaries can distinguish between follicles, corpus hemorrhagica, corpus lutea, ovarian cysts, and tumors. In the uterus, cysts or an infection can be diagnosed.

With ultrasonography the diagnosis of pregnancy is possible as early as day 14 of gestation. The fetal heart beat can be detected as early as day 22 of gestation, and should be routinely looked for from day 25 on. Between day 60 and 70, it is possible to determine the sex of the foal.

CARE AND MANAGEMENT OF THE STALLION

Yearlings should not be depended upon for breeding. Two-year-olds may **settle** ten mares; three-year-olds, thirty; and mature stallions, fifty mares when hand mated. About half of this number can be pasture mated. A short breeding season will reduce the number, and sexual individuality of the stallion will greatly affect his siring ability.

The breeding stallion should be fed like a horse at hard work. An estimate is 1½ pounds of grain and 1 pound of hay per 100 pounds body weight. If he is worked under saddle, more feed will be required. Because of diverted interests, a ration high in palatability may be necessary for some stallions to get adequate intake. Grazing of good grass, even for short periods of time, is recommended.

Regular exercise usually results in increased sexual vigor (**libido**) and fertility.

For safety, fences should be strong and tall when stallions are grazed loose; and mares should not be in adjoining pastures unless extremely tall fences are used.

Methods of Mating

Two methods of mating are used when breeding horses: **pasture mating** and **hand mating**.

Pasture mating reduces labor, affords convenience to the owner, "catches" shy breeding mares, and creates an opportunity for a high settling percentage. It has the disadvantage of reducing the number of mares a stallion can serve and obscures breeding dates. Some risk to the stallion exists.

Stallions should be hand mated a few times as two-year-olds, then turned loose in a large pasture with a few older mares when they are to be used in a pasture breeding program. Even so, stallions are likely to carry some scars from their experience. For this reason, pasture mating is seldom used with breeds

	Number of Females to Mate in a Breeding Season	
Age of Sire	Hand Mating	Pasture Mating
Two-year-old stallion	10	5
Three-year-old stallion	30	15
Four-year-old stallion	35 to 40	20
Five-year-old stallion	40 to 75	20 to 25

Table 11-3 Mating Capacity of Sires

whose owners discriminate against blemishes. It is extensively practiced with stock horses in the range country.

A combination of hand mating followed by pasture mating will extend the number of mares bred and increase settling percentage (see Table 11-3).

Hand mating is practiced under a wide variety of conditions, ranging from rather casual selection of mares and sanitation conditions to those that are highly supervised with a veterinarian in attendance.

Stallions used with hand mating should be adept at teasing mares. This may be done at a teasing pole or over a stall door or any other sturdy fixture that does not injure the horses or attendants. A teasing stall and breeding stall are shown in Figures 11-3 and 11-4.

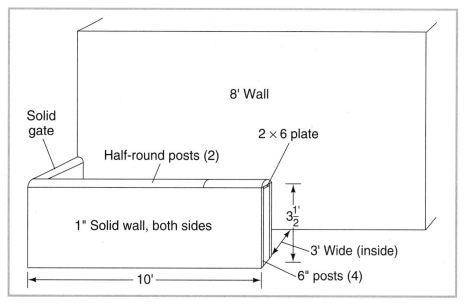

Figure 11-3 Teasing stall

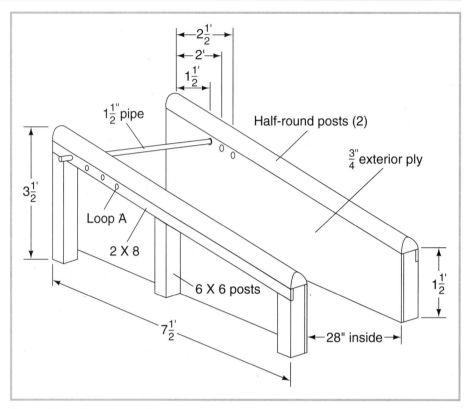

Figure 11-4 Breeding stall

CARE OF THE PREGNANT MARE

For the pregnant mare, aside from nutritional requirements, attention should be paid to regular dental and hoof care. Broodmares usually do not need shoes, but if they are shod, the shoes should be removed a few weeks before foaling to protect the foal at birth. A good exercise program is recommended (see Figure 11-5).

Parasite control to protect both mare and foal can be effected by deworming and helped along with general measures, such as sanitation and cleanliness. More details on parasite control are given in Chapter 15.

A vaccination program should include vaccinations against rhinopneumonitis (third, fifth, seventh, and ninth month of pregnancy), tetanus, equine encephalomyelitis, and influenza four weeks prior to foaling. In endemic areas, and under guidance from a veterinarian, additional vaccinations may be given for strangles, rabies, anthrax, and Potomac Horse Fever. In general, deworming and vaccination programs should be administered under the guidance from a veterinarian.

Figure 11-5 A pregnant mare. *(Photo courtesy of Barbara Lee Jensen, After Hours Farms)*

Since the placenta does not allow transfer of maternal antibodies to the fetus, the foal has to obtain antibodies against infection from the **colostrum**. The foal's intestine will absorb the antibodies only during the first twenty-four hours after birth, with the greatest absorption during the first few hours. If the mare loses colostrum in the last weeks of pregnancy when she starts to "**bag up**," this colostrum should be collected and saved for the foal.

Care at Foaling Time

Proper preparation of the mare for foaling is necessary if an owner is to realize the results of the time, effort, and money invested. The owner must prepare the mare, the environment, and the handlers for foaling.

During the final two months of pregnancy—the tenth and eleventh months of gestation—the mare's abdomen takes on the pendulous, enlarged characteristics of pregnancy. Mares that have had several foals tend to take on this appearance sooner. The mare develops a wider stride to compensate for the increased weight she is carrying, and her ribs may appear more "**sprung**." This is the time when the mammary glands begin to develop.

The pituitary hormone, **prolactin**, stimulates the udder to produce milk. If no mammary gland development is noted prior to foaling, the owner should be suspicious of hormonal inadequacies. This is a common occurrence if the mare is consuming fescue grass. An endophyte in some contaminated fescue seed heads blocks the action of prolactin. Normally, however, the udder slowly enlarges over the final two months of pregnancy. It becomes turgid and the teats fill out. Any leakage from the teats prior to foaling should be collected to avoid loss of any of the antibody-rich colostrum.

The mare's tail head, croup, and perineal area (between rectum and genitals) become relaxed several days prior to parturition. This is due to the hormone **relaxin**, which loosens the ligaments of the pelvis. The amount of relaxation varies with the age of the mare and the number of previous pregnancies. Relaxation can be very slight and difficult to observe in a maiden mare, and quite pronounced, even to the novice owner, in an older mare.

Signs of Impending Parturition

The following is a check list to help owners identify many of the signs present prior to foaling:

1. Large, pendulous abdomen, or sudden change in position of the foal. Change in gait; occurs one to two months before giving birth.

2. Udder enlargement. Change in its shape, texture, or temperature. A change in milk color from clear, or amber, to cloudy, or chalky white, means that delivery is very near. Calcium concentration in the mammary secretion increases immediately prior to foaling. If the milk calcium concentration is measured, by either the owner or a veterinarian, it can be used to predict the foal's arrival time.

3. **Waxed teats**. Drops of sticky, clear, or amber-colored fluid excreted prior to parturition become dried and hard, coating the ends of the teats, giving them a waxy appearance; occurs two weeks to just hours before foaling.

4. Relaxation of the tail head, croup, and perineal area; occurs one month to two weeks before foaling.

5. Enlarged abdominal milk veins; occurs two months to two weeks before foaling.

6. Loss of appetite. Does not occur in all mares, but if present, depressed appetite usually occurs during the last month of gestation.

7. Change in personality. May separate from the herd if pastured with other mares. May push hindquarters up against a wall. Usually, if present, this behavior will change in the last four to two weeks before foaling.

Mares vary widely in the degree and length of time that these signs are exhibited. The best predictor of foaling time is knowledge of the mare's gestation length and behavior during previous pregnancies.

Equipment Helpful for Foaling

Before foaling, the owner or manager should assemble the following equipment or supplies that will be needed during and/or after foaling:

- Four to five ounces of an iodine solution in a sterile jar
- Tail bandages or three-inch gauze bandages
- Roll of sterile cotton
- Package of gauze squares (3" or 4" square)
- Adhesive tape (one-inch wide)
- One pint povidone-iodine compound
- Six to eight clean towels
- Enema tube, soap, and lubricant
- Seamless pail
- Large animal thermometer

Labor and Parturition

The mare's labor is intense and rapid. Usually it is over within an hour. Because of this rate, an owner does not have time to develop a wait-and-see attitude. Because all of the foal's oxygen is obtained through the umbilical cord blood supply, a prolonged delivery can quickly endanger the foal. If the mare or the foal appears to be having difficulties, help must be summoned at once. Labor is commonly divided into three stages.

Stage 1. The time when the uterus begins to contract, the foal moves into position to be born, and the cervix relaxes is Stage 1. Stage 1 usually is imperceptible and can go unnoticed, even by the watchful observer. Signs of late Stage 1 labor, which may be observed in some mares, include restlessness, tail switching, pacing, and sweating over the neck, chest, and flanks. Because some of the symptoms are similar, Stage 1 may be confused with colic. The mare may urinate and defecate frequently, and carry her tail in an elevated position. The mare's respiration and heart rate may be increased, and her body temperature may be decreased. Stage 1 may be as short as several minutes or last longer than twenty-four hours. Recognizing this stage is important in order to prepare for the upcoming period of intense labor.

To prepare a mare in Stage 1 labor for foaling, the mare's perineal area should be cleaned thoroughly with a povidone-iodine solution, and the tail should be bandaged out of the way. The foaling stall should be cleaned and disinfected. If the mare has a **Caslick** in place (a suture closing her vulva to prevent infection), it should be removed now. Once the preparations have been finished, the owner should observe discreetly from a distance. Horses do not like to give birth in noisy surroundings, and the mare can shut-down the foaling process if she does not feel completely safe.

Stage 1 ends when the **allantois**, or fetal membranes, are pushed through the cervix by the advancing fetus and rupture, releasing **amniotic fluid**, (breaking water).

Stage 2. The time of intense labor contractions that push the foal through the birth canal is Stage 2. This stage usually lasts no more than thirty minutes. Little can be done to slow the labor or make corrections if problems arise at this point. The advancing fetus is enveloped in a white, filmy membrane known as the amniotic membrane. The **placenta**, a thicker pink membrane, is attached to the uterine wall and should not be visible at this time.

The foal normally is lying on its stomach, positioned upright in the birth canal with its forefeet slightly displaced, one before the other, and with the muzzle resting on top of the knees. The bottom of the feet should be facing toward the mare's hocks; the curving of the foal toward the hocks helps move the shoulders and hip through the mare's pelvis (see Figure 11-6).

The mare usually lies down as Stage 2 commences. If the mare is startled she may jump up and give birth standing. This results in quite a fall for the foal and premature rupture of the umbilical cord.

The foal's feet should appear within fifteen minutes of the appearance of fetal fluid and membranes. If both feet are not visible, a gentle hand should be inserted into the vulva to find the missing foot and guide it gently to the proper spot. This may prevent a sharp hoof from tearing the vaginal roof and perineal area.

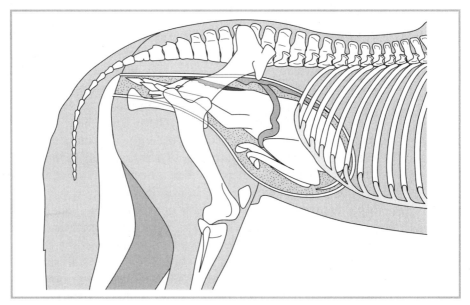

Figure 11-6 Normal presentation of foal.

Once the shoulders, the widest portion of the foal's body, are born, the rest of the foal usually follows shortly. Occasionally the hips may lock in the mare's pelvis; gentle, downward tension on the foal's front legs will help. This assistance should be very quiet and calm and timed with the mare's contractions.

If the mare has been pushing for forty-five minutes and no sign of a foal is seen, intervention is needed quickly if the foal is to survive. Should the mare require assistance, the attendant must first make sure that the foal's front legs and head are properly positioned. If they are, traction may be used to gently pull the front legs.

Stage 2 is complete when the foal has been born.

Dystocia. Many situations cause **dystocia**, or difficult birth, in the mare. A mare may become exhausted in the middle of labor and become unable to push the foal out, especially if she is older or in a debilitated condition. A foal will not fit through the birth canal unless it is in the proper position. Any deviation from the front-legs-first, head-facing-down-between-the-knees posture may result in dystocia. The presence of twins is a possibility when labor becomes extended. Usually, a leg or the head becomes turned back or tucked in such a way to be pointing against the direction of the foal's movement. The entire foal may also be completely backward, or breech, with the hindquarters presented first, or upside down, or both.

Signs of dystocia require immediate veterinary attention. Until the veterinarian arrives, the attendant may try to place the foal in the correct position by turning, locating, and extricating lodged limbs. If repositioning is attempted, care must be taken to ensure that the attendant's arm is sterile and well-lubricated. Inserting a length of rubber tubing into the mare's trachea and walking may prevent her from bearing down too hard on the foal or your arm. Whether the foal has been born or not, the placenta detaches from the uterine wall within an hour of the start of Stage 2. If the placenta detaches before the foal is born, the foal will lose its oxygen supply and die. Time is critical.

Foal. The newborn foal may have a blue tongue and bluish-white nasal mucous. As soon as it is born, the fetal membranes should be cleared away from the foal's head so that breathing can start. The umbilical cord should be allowed to break on its own. Once this has happened, the foal's navel should be immediately treated with an iodine solution to prevent entrance of pathogens through the opening. In some instances, the navel will bleed after the umbilical cord has broken. The cord should be tied shut with a length of sterile umbilical tape, gauze, or string.

Stage 3. Stage 3 is complete when the placenta and fetal membranes are expelled. Prior to being passed, the placenta hangs from the vulva. It should be tied up so that the mare does not tear it or become afraid of it. The membranes usually are passed five to forty-five minutes after the foal's birth. If they have not been passed in two hours, the membranes are considered retained, and a veterinarian should be called.

The placenta must be physiologically released from the uterus. Pulling may tear the placenta and leave small pieces in the uterus. These small pieces of placenta in the uterus may cause **metritis**, an inflammation of the uterus. Metritis is an infection that may result in laminitis or death. A mare with this condition may appear normal forty-eight hours after birth, then develop symptoms. Once serious symptoms develop, it may be too late to prevent permanent damage.

The placenta should be inspected after it has been passed to ensure that it is complete and that there are no tears or pieces missing. It is normal for a soft, dark-brown body of tissue called the **hippomane** to be floating among the membranes.

Immediately Postpartum

After the membranes have been properly expelled, the mare's uterus will undergo **involution**, during which the uterus will return to its nonpregnant size. Without infection or trauma, the uterus will involute within ten days. However, if a uterine infection is present, the process will be delayed. Older mares who have had many foals involute more slowly than younger mares. Involution may cause abdominal pain and some colic-like symptoms. Some mares become extremely agitated and the attendant should watch carefully to make sure the mare does not endanger the foal by rolling or getting up and down frequently.

The first twenty-four hours are the most critical for both the mare and the foal (see Figure 11-7). For this reason, they should be kept in a clean, quiet environment where they can be observed frequently. Many horse owners leave the pair stalled for the first day. After that, weather conditions and temperature should dictate whether the mare and foal are turned out or housed indoors. Damp or wet environments should be avoided the first few weeks. Some mares and foals will actually fare better and some foals show slightly greater resistance to adverse environmental conditions if they are kept outdoors.

Postpartum Mare Care

A mare needs special attention during the first week after delivery. The perineal region will be bruised and sore. It may hurt her to defecate. Bran, beet pulp, more salt, or other laxative-type feeds may loosen the stool and make **defecation** less painful. Because decreased appetite and water intake will result in decreased milk production, every effort should be made to keep the mare comfortable. The mare should have free-choice access to water and mineralized salt. During the first three months of lactation, the mare's energy

Figure 11-7 A newborn foal at the Northwest Equine Reproduction Laboratory, Moscow, Idaho.

requirements are double normal maintenance levels and she needs larger amounts of concentrates and high-quality hay.

A broodmare's stall is normally very dirty because she is eating and drinking larger amount of feeds. Because the foal is very susceptible to disease at this time, every effort should be made to keep the stall as fresh and clean as possible. This problem is easily prevented by housing outdoors.

Normal and Abnormal Postpartum Occurrences

The mare's uterus continues to involute through the first two weeks after delivery. During this time a dark-brown fluid may be seen on the vulva. This odorless discharge, called **lochia**, is normal. However, a foul-smelling discharge signals a uterine infection and requires medical attention. About the seventh or eighth day **postpartum** the mare's reproductive tract needs to be examined by a veterinarian if she is to be rebred.

Colic is relatively common during the first week postpartum. A more serious condition occurring immediately postpartum is internal bleeding caused by rupture of the middle uterine artery. This condition is occasionally seen in mares over fifteen years of age, and is usually fatal. The symptoms associated with a middle uterine artery rupture are colic, pawing, anxious-ness, and profuse sweating. In addition, the gums may look pale or white. A veterinarian should be called at once if this condition is suspected. Often a mare with a middle uterine artery rupture is unaware of her surroundings and of her foal. The foal should be removed from the stall to prevent injury.

Rebreeding

At six to twelve days postpartum, most mares will come into heat. This first estrus is called **foal heat**, and is part of the uterine involution process. Unless uterine involution is practically complete, conception is unlikely in a mare bred during the foal heat.

Any uterine or vaginal bruising, damage, or swelling slows down the involution process, and the uterus will be unable to support embryo life. The conception rate for mares bred during their foal heat is only 40 percent. These mares run a greater chance of developing uterine infections and scarring since the uterus is most susceptible to infection during the first thirty days after foaling. Semen is not sterile; thus every time a mare is bred, bacteria are introduced into her uterus. Excessive or improperly timed breeding attempts increase the chances of uterine infections.

The fertile egg is released from a follicle on the ovary during the mare's estrus. This usually occurs on approximately the seventh and again on the thirtieth day after delivery. A mare usually goes out of estrus within one day after ovulation, although there is variation between mares. Some remain in standing estrus for longer periods after ovulation, and some ovulate without showing any external signs of estrus. These mares are said to go through a **silent heat**. A common cause of silent heat in mares who have just foaled is their concern for their foals. Their maternal instinct is stronger than the instinct to display estrus. If the mare is allowed to see her foal at all times while she is being teased, she may relax and show some evidence of estrus. Older mares who have had many foals may not be completely involuted and ready to conceive, even on the second heat after foaling.

Prolactin, the hormone that stimulates milk production (**lactation**), may inhibit estrogen and the hormones necessary for ovarian activity. In some mares, no ovulation takes place while she is lactating. Lactational inhibition of ovarian activity is more common if the mare is on a substandard diet and in a negative energy balance. If this is the case, the mare's plane of nutrition needs to be improved, the foal weaned early, or hormonal treatments used under a veterinarian's supervision.

Transporting the Mare and Foal

Many consider the second heat cycle after delivery to be the most desirable time to breed. If the mare is to be transported to the stallion, she should arrive well before the onset of her second heat. Arriving at the breeding farm eighteen days postpartum should leave sufficient time for the mare to adapt to her new surroundings before estrus begins. Loading and unloading should become routine for the mare before the foal arrives. Many less-experienced owners prefer to have the mare foal at the breeding farm if the distance from home is great.

Body Condition of the Mare

A mare's body condition affects her reproductive efficiency and ability to reproduce. A mare should have the ideal body condition score of five before her second heat occurs.

Condition score is a method used to quantify the amount of subcutaneous fat cover. If a mare is not in acceptable condition, her diet must be changed to move her in the direction of a five score. Thin mares must gain weight since poor nutrition causes mares not to cycle normally. A score of four is termed moderately thin and is characterized by a negative crease down the back, with the vertebrae slightly protruding. A faint outline of the ribs is discernible, some fat can be felt about the tailhead, and neither the withers, shoulders, or neck are obviously thin.

A mare in ideal five, or moderate condition score, would have a level back with ribs that cannot be visually distinguished, but that can be easily felt. The fat around the tailhead is beginning to feel spongy, and the withers appear rounded over the backbone. The shoulders and neck blend smoothly into the body.

A condition score six is moderate to fleshy. A horse with a score of six has a slight crease down the back and spongy fat over the ribs. Fat around the tailhead feels soft and deposits of fat form along the sides of the withers as well as behind the shoulders and along the sides of the neck.

ARTIFICIAL INSEMINATION

Artificial insemination (AI) is the collection of semen from a stallion and the deposition of that semen into the mare, without conventional breeding. Advantages are the decreased risk of injury and infection. Additionally, because the semen can be evaluated, and divided in samples, many more mares can be bred to the same stallion.

Some horse breed registries, such as the Jockey Club, do not permit AI. Others, such as those for quarter horses, and Standardbreds will allow AI, but only with fresh semen collected on the premises of the mare. They will not allow transportation or freezing of the sperm. Before deciding on AI, the breed registry should be contacted.

The semen, once collected, should be immediately poured into a warmed graduated cylinder to be measured. In order to evaluate the percentage of progressively motile spermatozoa, an extender should be added to the raw semen before it is examined under the microscope. A sperm is called progressively motile only if it moves across the microscopic field rapidly.

When spermatozoa are to be stored or shipped, they have to be cooled. However, this cooling process stresses the cells and can injure them. To preserve the spermatozoa, they should be mixed with an extender, cooled slowly, and preferably kept at a low temperature for less than thirty-six hours.

Frozen sperm has both advantages and disadvantages. Aside from the fact that some breed registries impose restrictions on this practice, shipping liquid nitrogen containers anywhere in the world is cheaper than shipping a mare. A stallion's breeding season can continue while he is at shows and his semen can be preserved many years after his death. The disadvantages are lower pregnancy rates and possible inbreeding.

The practice of artificial insemination requires identification of the stallion with photographs, blood type, and possibly DNA-fingerprinting to avoid possible errors in identity and pedigrees. The breed registry should be contacted for regulations.

EMBRYO TRANSFER

An embryo can be nonsurgically removed from the uterus of one mare, transferred, and inserted into the uterus of another mare. Embryo transfer allows reproduction by older, less fertile mares, reproduction by two-year-old mares, and increased production of foals from genetically superior mares, but this method is very expensive and the yield is not high.

The basic embryo transfer procedure has changed little over the past decade. Success rates have improved due to improved quality control and precise timing of the different aspects of the procedure. Embryo transfer is more costly in horses than in cattle and other animal species due to the unavailability of medications to cause horses to produce many embryos to transfer per procedure on a routine basis.

Embryo transfer consists of three phases:

1. Synchronization of the donor and the recipient mare

2. Embryo flushing

3. Embryo transfer procedure

Synchronization of the Donor and the Recipient Mare

The donor is usually a valuable sport or subfertile mare, while the recipient is an inexpensive but healthy mare. Synchronization requires the use of hormones such as progesterone, prostaglandin, and HCG. To assure good synchronization, as the process is not perfect, two recipients are usually synchronized for each donating mare. The donor must be carefully palpated, cultured, and bred. The ovulation of the donor and the recipients must be timed to within twelve hours for best results, requiring twice-daily ultrasounds.

Embryo Flushing

Flushing is performed seven or eight days after insemination or breeding and involves lavaging (washing) the microscopic embryo out of the uterus using a special sterile solution. The embryo is then developmentally sized and graded: 1 being excellent, through 4 being poor quality.

Embryo Transfer Procedure

Embryo transfer is performed nonsurgically by loading the embryo into a uterine transfer catheter in a special nurturing solution and transferring the embryo into the uterus of the most synchronized recipient.

The odds of retrieving an embryo are very good (70 to 80 percent) from a healthy and reproductively sound donor mare. If the mare has a heavy show schedule or is subfertile, the success rate diminishes somewhat. Once recovered, embryos graded "good" or better will have a 60 to 70 percent chance of resulting in a pregnancy. The overall chances per cycle for a successful transfer with a young healthy donor are 50 to 60 percent and 30 to 40 percent with old problem donors. Breeding and flushing difficult mares on multiple cycles is the best way to ensure at least one pregnancy.

Recipient mares in no way contribute to the genetics of the foal. The characteristics of the foal are already programmed at conception with genetic material only from the donor mare and the stallion. The stallion is of great importance in the equation for success. Fresh sperm and the good timing of breeding is imperative.

ARTIFICIAL CONTROL OF BREEDING

Management of equine reproduction involves the use of photoperiod and hormones. The designated birth date of foals of many performance breeds in the Northern Hemisphere is January 1. This creates an economic pressure to start the breeding season in February so foals will be born in January. Since the mare is usually in winter anestrus, artificial lighting is used to induce follicular activity.

A mare requires approximately sixty days of artificial lighting before ovulation occurs. To induce ovulation in early February, the artificial lighting has to start in late November or early December. Mares must be confined to the area where the light (either one 200-watt incandescent or two 40-watt fluorescent bulbs per stall) is located. Also, they should be within seven to eight feet of the light source. Automatic timers can be used to provide artificial light, beginning at sunset. Sixteen hours of total daylight, natural plus artificial, and eight hours of darkness is the correct ratio.

Hormones can be used to make a mare "**short cycle**," or come in season early. Another use of hormones is to ensure that a mare will ovulate within twenty-four to forty-eight hours after being bred.

Special lighting techniques to simulate longer days and the hormone injections that are employed to bring mares into heat earlier in the winter require intensive management and often result in improperly timed breedings that are not successful. Mares becoming pregnant early in the year are more likely to lose their fetuses because of extreme fluctuations in hormone levels.

SUMMARY

Reproductive organs of the mare, under hormonal control, produce an egg (ovum) during the breeding season. Reproductive organs of the stallion, under hormonal control, produce sperm cells. Successful breeding during an estrus, or heat, period of the mare allows a sperm cell to unite with the egg. This brings together the genetic material of the mare and stallion to produce a new, unique horse. At first this new horse is a mass of cells called an embryo. The embryo implants into the uterus of the mare. As the embryo takes on the form of a new horse it is called a fetus. After about eleven months of developing in the uterus the mare goes through the process of labor and parturition, or birth. The new foal is cared for by the mare while the mare's reproductive tract goes through a process of involution in preparation for the next pregancy.

The reproductive process can be artificially controlled by lighting or hormones. Breeding can also be controlled by artificial insemination and embryo transfer.

REVIEW

Success in any career requires knowledge. Test your knowledge of this chapter by answering these questions or solving these problems.

True or False

1. Photoperiod is when a mare is receptive to photographers.
2. Dystocia means difficult birth.
3. The practice of artificial insemination is accepted by all breed registries.
4. Winking is a sign of estrus in mares.
5. Waxed teats are a sign of impending birth.
6. Artificial insemination is 100 percent successful.

Short Answer

7. List three parts of the mare's reproductive tract.
8. List three parts of the stallion's reproductive organs.
9. What are four signs of estrus or heat.
10. Name the four factors that influence sperm output and production.
11. How soon after parturition can a mare be bred?
12. What hormone stimulates milk production?

Discussion

13. What is the role of the hormone progesterone in the mare?
14. What are the advantages and disadvantages of both pasture and hand mating?
15. Discuss the meaning of polyestrus in the mare.
16. Explain the difference between abortion and stillbirth.
17. Describe Stage 1 and Stage 2 of parturition.
18. What is the meaning of Condition Scoring of the mare?

STUDENT ACTIVITIES

1. Invite a local veterinarian to visit your class to discuss reproductive problems in horses.
2. Visit a veterinarian's clinic or a horse farm when horses are being diagnosed for pregnancy.
3. Obtain prepared microscope slides of semen and compare the sperm cells of two different species.
4. Research and diagram the changes in a horse embryo through the fetal stages to the point of parturition.
5. Using Table A-16 in the Appendix, contact five breed registries and inquire about their policy for embryo transfer and artificial insemination.

ADDITIONAL RESOURCES

Books

American Youth Horse Council. 1993. *Horse industry handbook: a guide to equine care and management.* Lexington, KY: American Youth Horse Council, Inc.

Frandson, R. D., and Spurgeon, T. L. 1992. *Anatomy and physiology of farm animals.* 5th ed. Philadelphia, PA: Lea & Febiger.

Fraser, C. M., ed. 1991. *The veterinary manual*. 7th ed. Rahway, NJ: Merck & Co.

Hafez, E. S. E. 1993. *Reproduction in farm animals*. 6th ed. Philadelphia, PA: Lea & Febiger.

Kainer, R. A., and McCracken, T. O. 1994. *The coloring atlas of horse anatomy*. Loveland, CO: Alpine Publications, Inc.

Kellon, E. M. 1995. *Equine drugs and vaccines: a guide for owners and trainers*. Ossining, NY: Breakthrough Publications.

McKinnon, A. O., and Voss, J. L. 1993. *Equine reproduction*. Philadelphia, PA: Lea & Febiger.

Video

Squires, E. L., ed. 1996. *Semen Shipment and Embryo Transfer Seminar*. Lewisburg, TN: TWHBEA.

CHAPTER 12

Digestion and Nutrition

Digestion—the process that releases nutrients from feeds for use by the body—begins in the mouth where food is ground and mixed with saliva. Feed then moves to the stomach where the chemical breakdown starts releasing the nutrients for use by the horse.

The science of nutrition draws heavily on findings of chemistry, biochemistry, physics, microbiology, physiology, medicine, genetics, mathematics, endocrinology, cellular biology, and animal behavior. To the individual involved with horses, nutrition represents more than just feeding. Nutrition becomes the science of the interaction of a nutrient with some part of a living organism, including the composition of the feed, ingestion, liberation of energy, elimination of wastes, and all the syntheses for maintenance, growth, and reproduction.

OBJECTIVES

After completing this chapter, you should be able to:

- List six categories of nutrients
- Define terms associated with energy
- Describe each of the six energy nutrients
- List the sources of energy nutrients
- Describe the functions of the energy nutrients
- Describe the symptoms of energy nutrient deficiencies
- Explain the temperature and how it is important in horse management
- Describe the energy needs of horses for milk production, pregnancy, and work
- List the most important compound sugars
- Describe the digestion of fiber
- Describe the function of protein
- Describe digestible protein

- Explain the difference between essential and nonessential amino acids
- Identify at what stages of the horse's life protein requirements are the greatest
- Describe the functions of minerals in horse nutrition
- Describe the deficiency symptoms caused by the lack of minerals in the ration
- List the macrominerals needed by horses
- List the microminerals needed by horses
- List the vitamins that are essential in horse nutrition
- Describe the functions of vitamins
- Describe the deficiency symptoms caused by the lack of three vitamins in a ration
- List and discuss factors affecting the amount of water a horse will consume
- Describe the ways by which horses lose water from the body

K E Y   T E R M S

Absorption	Homeostasis	Net energy (NE)
Amino acids	Hyperparathyroidism	Nonruminant
As-fed basis	Hypochloremia	Nutrients
Concentrates	Impaction	Osteomalacia
Crude protein (CP)	Lactose	Rickets
Dehydration	Macrominerals	Roughage
Digestible energy (DE)	Metabolic alkalosis	Total digestible nutrients
Digestion	Metabolizable energy (ME)	(TDN)
Dry-matter basis	Microminerals	Water-soluble vitamins
Fat-soluble vitamins	National Research Council	
Fermentation	(NRC)	

EVOLUTION OF HORSE NUTRITION

Much credit has been attributed to oats and timothy hay for the nutrition of horses. But researchers have been unable to substantiate a need for either, when substitutions of other grains and hays were made. The Arabian horse, progenitor of most domestic breeds, reached its excellence in a country that produced no oats or timothy hay. As early as 1911, Trowbridge completed 365-day tests with hard-working mules that showed less weight loss and 28 percent less feed cost with corn compared to oats fed with mixed hay. Respiration counts showed no difference in heat tolerance. However, mules seemed to tire of corn over the year-long test more than oats.

Horses relish oats. This fact, combined with the knowledge that less care is needed to avoid digestive problems with oats than with corn because of the higher fiber content of oats, has always made oats popular.

Timothy hay and good oats fed together make a satisfactory ration for adult horses, but they are too low in protein, calcium, and vitamins for brood mares and growing horses.

Today, the subcommittee on Horse Nutrition of the Committee on Animal Nutrition of the **National Research Council (NRC)**, examines the literature and current practices in the nutrition and feeding of horses and publishes recommendations on horse nutrition. The latest NRC publication on horse nutrition was issued 1989. Many of the recommendations in this chapter and Chapter 13 are based on this NRC publication.

DIGESTIVE SYSTEM—ANATOMY AND FUNCTION

The anatomy of the horse's digestive system was discussed in Chapter 5. Figure 12-1 should serve as a review of the general anatomy.

The small intestine is a major site of **digestion** and **absorption** of many nutrients. Good parasite control is necessary for optimum function of the small intestine. Parasites not only reduce feed utilization, but can cause colic.

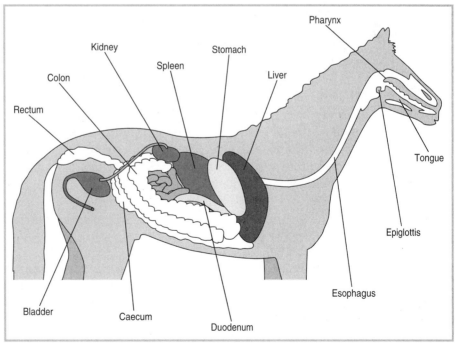

Figure 12-1 The horse digestive system

The large intestine consists of the cecum and colon. It has a large population of microorganisms (bacteria and protozoa) that digest the fiber in plant materials. If feed changes are made rapidly, the microorganisms do not have time to adapt.

To a degree, the cecum and colon serve the same purpose for the horse that the rumen does for the cow. However, the cecum's location toward the end of the digestive tract probably reduces its contribution to the horse's overall digestive efficiency. Feed passes through the gastrointestinal (GI) tract of the horse much faster than through the GI tract of ruminants. It is this faster rate of food passage that is largely responsible for lower digestion efficiency in horses than in ruminants.

NEEDS FOR FEED

Horses need the same feed ingredients as other livestock. These ingredients are carbohydrates, fats, protein, minerals, vitamins, and water. The first three of these can be converted to yield energy. Major sources of energy and protein are grains and roughages, including pasture.

Feeding horses is both an art and a science. There is considerable variation in individual horses' nutrient requirements, but a table of these requirements forms a useful basis for formulating rations. Feeding practices are discussed in Chapter 13.

All horses require **nutrients** for maintenance of body weight and to support digestive and metabolic functions. They need additional nutrients for growth, work, reproduction, and lactation.

Tables of nutrient requirements for horses are expressed in two ways: (1) daily nutrient requirements, and (2) nutrient concentration in the feed. This may be expressed on an **as-fed basis** or on a **dry-matter basis**.

Most horses receive their daily ration in two parts: **roughage** (hay or pasture) and **concentrates**. The concentrate portion contains grain and may include a protein supplement, minerals, and vitamins. It may also include bran, cane molasses, and/or dehydrated alfalfa.

NUTRIENT NEEDS

Feeds and feedstuffs contain the energy and nutrients essential for the growth, reproduction and health of horses. Deficiencies or excesses can reduce growth or lead to disease. Dietary requirements set the necessary levels for energy, protein, amino acids, lipids (fat), minerals and vitamins.

Table 12-1 indicates how the major nutrients are measured on a daily basis or in terms of the concentration in the feed.

Table 12-1 Requirements for the Major Nutrients Expressed and Their Units

Nutrient	Unit of Measure (amount/day)	Unit of Measure (concentration)
Digestible Energy	mcal./day	mcal./kg.
Crude Protein	kg./day	%
Calcium	grams/day	%
Phosphorus	grams/day	%
Sodium, Potassium	grams/day	%
Copper, Zinc, Iron, other trace minerals	mg./day	ppm or mg./kg.
Vitamins A, E, and D	I.U./day (international unit)	I.U./kg.
Thiamin, other B vitamins	mg./day	mg./kg.

Note: 1 kg. = 2.2 lbs.

Energy/Carbohydrates

Horses are **nonruminant** herbivores who use carbohydrates for their main energy supply. Carbohydrates eaten by the horse first pass through the stomach and intestine, where nonstructural carbohydrates—starch, maltose, and sucrose—are removed and enter the portal vein (see Figure 12-2). These carbohydrates are used for energy or are converted to other biochemicals needed for life. The liver stores carbohydates in the form of glycogen (see Figures 12-2 and 12-3).

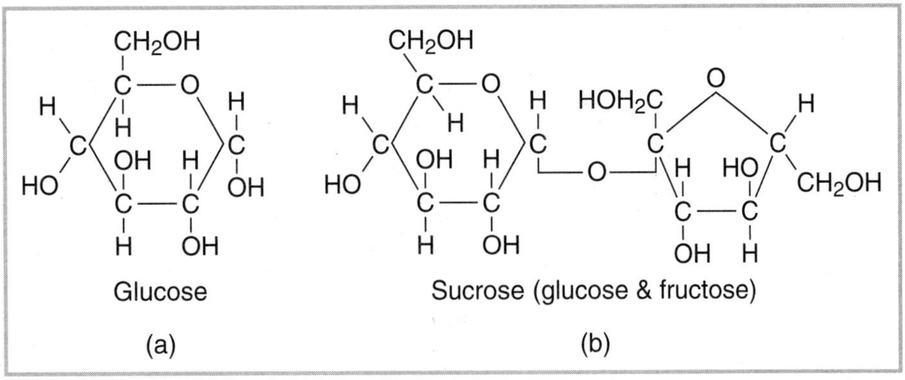

Figure 12-2 Diagram of two types of carbohydrates (**a**) a monosaccharide (1 sugar molecule) and (**b**) a disaccharide (two sugar molecules).

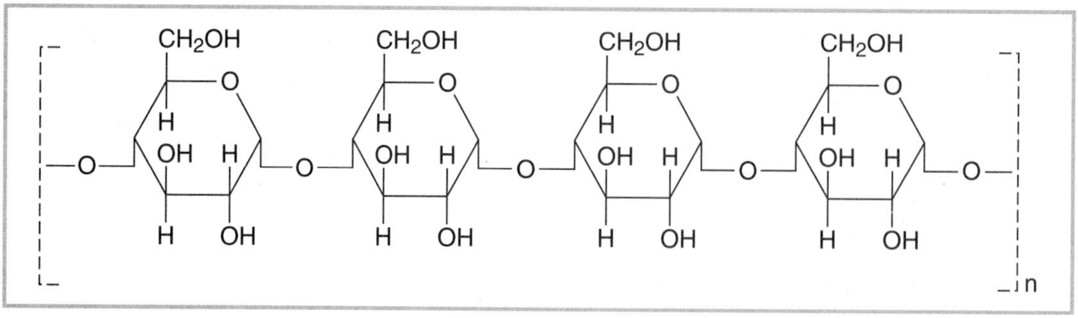

Figure 12-3 Diagram of starch

The remaining nonstructural carbohydrates, as well as the structural car-bohydrates—cellulose and hemicellulose—then reach the areas of the digestive tract that carry out **fermentation**. Fermentation results in the production of volatile fatty acids—acetic, propionic, isobutericbutyric, isovaleric, and valeric acid. These volatile fatty acids are easily absorbed and converted into energy.

As a side note, **lactose**—the sugar found in milk—is tolerable to horses up to three years of age. After that, the addition of milk or its by-products to the feed may disturb the gastrointestinal tract, causing diarrhea.

Fat/Lipid

Fat is a concentrated source of energy that can be readily utilized by the horse. Fat contains 2.25 times more energy per unit of weight than do carbohydrates or proteins. During exercise, especially strenuous activity such as galloping, body fat will be mobilized and converted to energy.

Fat in the diet seems to spare the glycogen storage; increasing the fat in the diet increases performance and maintains body condition. Horses accept fat addition to the diet as long as the fat is not rancid. Often performance horses are fed a percentage of their digestible energy concentration in the form of corn oil. The proportion of energy generated from fat and carbohydrate can be altered in exercising horses by dietary manipulation, but the ideal propor-tion is not yet known.

Energy Requirements

Energy can be measured as **digestible energy (DE)**, **metabolizable energy (ME)**, **net energy (NE)**, and as **total digestible nutrients (TDN)**. Most nutrient requirement tables for horses now use DE. Digestible energy (DE) is expressed in terms of calories, usually megacalories (million) or kilocalories (thousand). Energy requirements are needed for or altered by:

- Maintenance
- Reproduction

Humans count calories and humans avoid calories. Horses need calories in the form of digestible energy. Actually, both human and horses need calories, but neither needs calories in excess. The excess is stored as fat.

A calorie is a measure of the heat energy in food or feed. Digested food is actually burned in the body, so the number of calories in a food determines how much heat energy will be released when the food is burned in the body.

To directly measure calories in a feedstuff, a sample of feed is completely burned in a controlled environment. If the feedstuff contains one calorie, it has enough heat energy to raise the temperature of water exactly one degree centigrade from 14.5°C to 15.5°C.

For nutrition, one calorie is out of the realm of discussion. The energy in feedstuffs is expressed as thousands of calories or kilocalories (kcal.), or it is expressed as millions of calories or megacalories (mcal.). The fires of life burn the feeds at a slow controlled burn releasing the calories for energy to maintain life, reproduce, grow, and work.

- Gestation
- Lactation
- Growth
- Work
- Old age
- Stalling

Maintenance. The maintenance requirement of the horse is described as the energy that is needed to keep the animal from gaining or losing weight.

Reproduction. Energy is an important factor in the success of reproduction in mares. A mare that is in poor condition but gaining weight is twice as likely to conceive as a mare that is in poor condition while maintaining weight. Mares that are in good to fat condition have higher conception rates, whether they maintain or, even, lose body weight.

Gestation. During the last three months of pregnancy, the fetus experiences the greatest amount of growth. The ninth, tenth, and eleventh month of gestation requires an increase in DE.

Lactation. Milk production and composition varies between breeds, and even individuals. The National Research Council (NRC) for horses (1989) determined that the conversion of milk to DE was 792 kcal. of DE/kg. milk. This means lactating mares must be provided daily with more DE. Table 12-2 shows the increase in DE for mares of different weights and stages of lactation.

Growth. The tendency in the past has been to push the growth of foals to try to produce the biggest horse possible, as quickly as possible. This is not always in the best interest of the animal. Requirements for energy during growth are based on the age of the animal and the amount of weight being gained each day. The NRC recommendation for foals is—

$$DE(mcal./day) = maintenance + (4.81 + 1.17\,X - 0.023\,X2)\,(ADG)$$

Where X = age in months and ADG is the average daily weight gain.

Work. A working horse naturally needs more energy than a horse at rest. Many studies have been done to calculate the exact requirements of exercising horses. The huge variety in the types of work that horses do, as well as the ability of a given breed, or individual, to perform those tasks prevents us from having one formula to fit every animal. The NRC recommends that ponies and light horses (200 to 600 kg.) need an increase of 25 percent, 50 percent, or 100 percent for light, moderate, or heavy work, respectively. Draft horses should increase maintenance by 10 percent for every hour of field work.

Old Age. The older horse can be treated as a maintenance animal. However, the decrease in activity and use may lead to obesity and related health problems. It is wise to monitor the geriatric horse for weight gain and adjust the diet to include more forage.

Table 12-2   Increase of DE for Mares of Different Weights and Stages of Lactation

Formulas	Foaling to 3 Months
DE = maintenance + 0.04 BW × 0.792	200–299 kg. BW
DE = maintenance + 0.03 BW × 0.792	300–900 kg. BW
	3 Months to Weaning
DE = maintenance + 0.03 BW × 0.792	200–299 kg. BW
DE = maintenance + 0.02 BW × 0.792	300–900 kg. BW

Note: 1 kg. equals 2.2 pounds; BW is body weight

Stalling. A stalled horse has a lower requirement for energy, but is more likely to develop bad habits and vices. The stalled animal is better off with larger quantities of a lower energy roughage. This keeps the feed present for a greater portion of the day and gives the stalled horse an activity similar to normal behavior—grazing.

Underfeeding and Excess of Energy

Energy and protein are the major factors in evaluating a horse's ration. Underfeeding of either nutrient will cause a reduction in health and perform- ance. Overfeeding can result in excessive fat deposition. Overfeeding of protein can be wasteful and sometimes causes stress. A depressed appetite can be an indication of a protein deficiency and then cause an energy deficiency.

Protein

The protein requirement of the horse is related to the quality of the protein, the digestibility of the protein, and the requirements of the individual. Young animals undergo stages of rapid growth and require proteins to provide build- ing blocks for their bodies.

Proteins are composed of **amino acids**. These amino acids are used by the horse to build the proteins in its body. Various studies compared the growth of horses in different stages of life to the compositions of the proteins that the animals were eating. These studies indicated that the greatest growth was achieved when the horses were fed proteins high in lysine (an amino acid). Figure 12-4 shows diagrams of several amino acids. Mature horses do not need protein of as high a quality as younger animals.

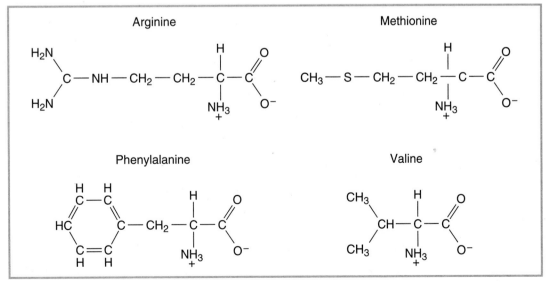

Figure 12-4 Diagram of several amino acids

Protein in the diet is expressed as **crude protein (CP)**. Digestible protein (DP) is a more accurate estimate of how much protein the animal is actually able to use. The DP of individual feeds has not been calculated for horses. Instead, DP values are estimated from CP values.

Deficiency.   Insufficient dietary protein decreases production of protein in foals. This results in smaller, less healthy foals, sometimes called "poor doers."

In an adult animal, a dietary deficiency of protein increases problems in areas of high protein turnover. The hair coat and hoof wall may be adversely affected and tissue wasting may occur.

Excess.   Excess protein has not been found to cause adverse effects on horses when fed in moderation. For adult horses, especially for hard-working horses, there is some debate about the value of excess protein.

Minerals

Minerals are important for energy transfer and as an integral part of vitamins, hormones, and amino acids. The horse obtains most of its necessary minerals from pasture, roughage, and grain. Depending on the amount required by the body, minerals in the diet are classified as **macrominerals** or **microminerals** (sometime called trace minerals). The seven macrominerals include:

- Calcium (Ca)
- Phosphorus (P)
- Potassium (K)
- Sodium (Na)
- Chloride (Cl)
- Magnesium (Mg)
- Sulfur (S)

Eight microminerals important in equine nutrition include:

- Copper (Cu)
- Iodine (I)
- Iron (Fe)
- Selenium (Se)
- Cobalt (Co)
- Manganese (Mn)
- Fluorine (F)
- Zinc (Zn)

Table 12-3 summarizes some of the mineral requirements and the signs of their deficiency in horses.

Table 12-3 Horses: Requirements, Functions, and Deficiency Signs of Some Minerals

Mineral (Requirement)	Some Functions	Some Deficiency Signs
Calcium (see first table)	Bone mineral; blood clotting; nerve, muscle, and gland function	Rickets, osteomalacia, NSH, osteoporosis; bones may be soft and easily deformed or broken
Phosphorus (see first table)	Bone mineral, part of many proteins involved in metabolism	Bone disease, decreased growth, reproductive problems, low blood phosphorus
Iron (50 mg./kg.)[1]	Part of hemoglobin and some enzymes, oxygen transport	Anemia: lack of stamina, poor growth
Copper (9 mg./kg.)	Iron absorption, hemoglobin synthesis, skin pigments, collagen metabolism	Anemia; hair pigment loss; bone disease: swollen joints, deformed thin bones
Magnesium (.1%)	Bone mineral, enzyme activator: energy metabolism	Nervousness, muscle tremors, ataxia, convulsions, mineralization of blood vessels, low serum magnesium
Sodium, potassium, and chloride	Tissue fluid pressure and acid–base balance, passage of nutrients and water into cells, nerve and muscle function	Craving for salt, hyperexcitability, decreased growth rate, loss of appetite
Zinc (36 mg./kg.)	Activator of many enzymes	Hair loss, scaly skin, poor wound healing; reproductive, behavioral, and skeletal abnormalities
Iodine (.1 mg./kg.)	Thyroid function	Goiter, poor growth, low body temperature, impaired development of hair and skin, foals weak at birth
Manganese (36 mg./kg.)	Synthesis of bone and cartilage components, cholesterol metabolism	Reproductive problems: delayed estrus, reduced fertility, spontaneous abortion, skeletal deformities in the newborn
Selenium (.2 mg./kg.)	Removal of peroxides from tissues, enzyme activation	White muscle disease, low serum selenium and serum glutathione peroxidase concentration

1. Units per kg. of air-dry feed

Calcium. Calcium is involved in **homeostasis**—the functions that maintain life—blood clotting mechanisms, and muscle contractions. Calcium also makes up 35 percent of the horse's bone structure.

Calcium deficiency in developing foals may lead to **rickets**, which shows up as poor mineralization of bone tissue, enlarged joints, and crooked long bones. Calcium excess does not seem to be detrimental, as long as the level of phosphorus is adequate.

Phosphorus. This mineral makes up 14 to 17 percent of the horse's skeleton. Phosphorus is required for many energy-transfer reactions, and for the synthesis of some lipids and proteins. Phosphorus requirements increase during late gestation and lactation. In the horse's diet, calcium and phosphorus are considered together in the calcium-phosphorus ratio.

Calcium-Phosphorus Ratio. Calcium and phosphorus are considered together because they work together and have an affect on one another's availability. The ratio and level of calcium and phosphorus must both be considered. Adequate vitamin D must be available for proper calcium and phosphorus use. Bone growth problems are a symptom of problems with the calcium, phosphorus, and vitamin D complex.

A ratio of less than 1:1 may be detrimental to calcium absorption. Even if the calcium intake is sufficient, the excessive phosphorus intake will cause malformations of the skeleton. On the other hand, very high ratios of calcium to phosphorus (as high as 6:4) in growing horses will not be detrimental as long as the phosphorus intake is adequate. Phosphorus deficiency will produce rickets in the developing horse and weakening of the bones (**osteomalacia**) in the mature horse, similar to what is found in deficiencies of calcium and vitamin D. Excessive phosphorus intake will lead to reduction of calcium absorption, chronic calcium deficiency, and nutritional secondary **hyperparathyroidism**. Hyperparathyroidism is an increase in the function of the parathyroids which lie next to the thyroid gland in the neck.

Table 12-4 gives the proper ratios of calcium to phosphorus for various classes of horses.

Table 12-4 Calcium to Phosphorus Ratios			
Status	**Minimum Ca:P**	**Maximum Ca:P**	**Optimum Ca:P**
Nursing Foal	1:1	1.5:1	1.2:1
Weanling	1:1	3:1	1.5:1
Yearling	1:1	3:1	2:1
Mature	1:1	5:1	2:1

Potassium. This mineral maintains the acid-base balance and osmotic pressure inside the cells. Forages and oilseed meals generally contain 1 to 2 percent potassium in the dry matter. Cereal grains (corn, oats, wheat) contain 0.3 to 0.4 percent potassium. Required potassium concentration in diet for growing foals is 1 percent. A mature horse requires 0.4 percent in the diet.

Since forages usually constitute a significant portion of the diet, the horse should get its potassium requirement from the diet. If only cereal grains are fed, potassium chloride and potassium carbonate can be used as supplements.

Potassium deficiency in foals causes loss of appetite and weight loss. These symptoms promptly reverse when potassium carbonate is given. Excess potassium in the diet is readily excreted, provided water intake is normal.

Sodium. Sodium maintains the acid-base balance outside the cells and regulates the osmosis of body fluids. Sodium is also involved in nerve and muscle function. Since the sodium concentration of natural feedstuffs for horses is often lower than 0.1 percent, salt is often added to concentrates at rates of 0.5 to 1.0 percent or fed free-choice as plain, iodized salt.

Chronic sodium deficiency in horses results in decreased elasticity of the skin, a tendency to lick sweat-covered tool handles, decreased appetite, and decreased water intake. Eventually, the horse will stop eating. If the deficiency is acute, the horse will have uncoordinated muscle contractions, impaired chewing, and an unsteady gait. In the blood, the sodium and chloride will be low and the potassium high. Horses are tolerant of high levels of salt in their diets as long as there is free access to fresh drinking water.

Chloride. In the diet, chloride normally accompanies sodium as NaCl or salt. This is an important extracellular anion (negative charge) involved in acid-base balance and osmotic regulation. Chloride is an essential component of bile, hydrochloric acid, and gastric secretions. The chloride requirement is assumed adequate when the sodium requirement is met.

Magnesium. More than half of the magnesium found in the body exists in the skeleton. Magnesium is an activator of many enzymes. Magnesium concentrations in common feedstuffs range from 0.1 to 0.3 percent.

Sulfur. Sulfur is a component of many biochemicals in the body including amino acids, biotin, thiamin, insulin, and chondroitin sulfate. Requirements for the horse have not been established and a deficiency has not been described.

Copper. This mineral is essential for several copper-dependent enzymes. Deficiency of copper may cause bone disease and bone malformation. A deficiency may also cause a dullness of the coat color. Horses are tolerant of excess copper.

Iodine.   Iodine is essential for the production of the thyroid hormones. These hormones regulate basal metabolism. If iodine has to be supplemented, iodized or trace mineralized salt containing 70 mg. of iodine per kg. can be used.

Iodine deficiency in pregnant horses, which may show no symptoms themselves, may lead to foals that are either stillborn, or born too weak to stand to suckle. These foals have an enlargement on the front side of the neck due to enlargement of the thyroid gland—a condition called goiter.

Excess of dietary iodine may result from using feedstuffs high in iodine, such as kelp (a seaweed) or from adding excessive supplemental iodine. This may lead to baldness in horses.

If a pregnant mare has had too much iodine in the feed, the foal will be born with an iodine-induced goiter. The milk of the mare will contain excess iodine. The foal needs to recover from the excess of iodine received before birth. An alternate source of milk, low in iodine, has to be found.

Before giving iodine supplementation to the pregnant mare or the foal with a goiter, the owner must determine whether the horse was fed too little or too much iodine.

Iron.   In the body of the horse, 60 percent of the iron is in the red blood cells and 20 percent is in the muscles. Common feedstuffs should meet the iron requirements.

Iron deficiency causes anemia since iron is not available for the formation of red blood cells. Young, milk-fed foals are most likely to have deficient iron. The body salvages iron from metabolism. Iron supplements have not proven effective in improving the oxygen-carrying capacity of red blood cells.

Excess iron is very toxic to young animals and death can result in a foal that has been given supplemental iron by mouth. Before death, the foal shows diarrhea, jaundice, dehydration, and coma.

Selenium.   This mineral is essential for detoxification of certain peroxides that are toxic to cell membranes. Selenium is closely connected with vitamin E in that the two work together to scavenge free radicals.

Selenium deficiency is linked to the status of vitamin E. Selenium/vitamin E deficiency causes white muscle disease that involves both the skeletal muscles and the heart muscle. The symptoms are weakness, difficulty in suckling and swallowing, troubled breathing, and heart dysfunction.

Excess selenium can be acute or chronic. The acute form (blind staggers), exhibits itself in blindness, head pressing, perspiration, colic, diarrhea, and lethargy. This acute form is seen if selenium is ingested from some toxic plants.

Chronic selenium toxicity results in hair loss about the mane and tail and cracking of the hooves around the coronary band.

Cobalt.   Cobalt is a part of the vitamin B_{12}. Microflora in the cecum and colon use dietary cobalt to make vitamin B_{12}. Specific dietary cobalt requirements have not been studied in the horse.

Manganese.   Manganese is necessary for carbohydrate and fat metabolism and for the synthesis of cartilage. Manganese requirements in horses are not established, but some recommendations are made based on information from other species.

Fluorine.   Fluorine is involved in bone and tooth development in other species. The dietary necessity is not established in horses. Excesses of fluorine in the diet can cause colored teeth, bone lesions, lameness, and unthriftiness.

Zinc.   Zinc is a component of many enzymes. Experimentally, a zinc deficiency can be produced in foals. Signs of the deficiency includes loss of appetite, reduced growth rate, and rough, scaly skin.

Common equine feedstuffs contain 15 to 40 mg. of zinc per kg. If a supplement is needed, zinc sulfate, zinc oxide, zinc chloride, zinc carbonate, and various zinc chelates can be used.

Vitamins

Vitamin requirements, like other requirements are affected by age, stage of production, gastrointestinal infections, and muscular activity. The type and quality of the diet and the extent of vitamin absorption determine the need for vitamin supplements. Bacteria in the gut produce vitamins while breaking down feedstuff. These vitamins are also available for absorption. Forages contain mostly fat- and water-soluble vitamins so that horses grazing high-quality pastures should not need vitamin supplementation. Vitamins are classified as fat-soluble or water-soluble.

The **fat-soluble vitamins** are vitamins A, D, E, and K. Conditions interfering with fat absorption also adversely influence absorption of these vitamins. Vitamin K is synthesized in the intestines. Vitamin D requires exposure to ultraviolet light. The other fat-soluble vitamins must be present in the diet. Table 12-5 summarizes some of the vitamins, their functions and signs of a deficiency.

Vitamin A.   Vitamin A is important for vision. Metabolites of vitamin A are found in visual pigments within the retina. This vitamin also plays a basic

	Table 12-5 Horses: Requirements, Functions, and Deficiency Signs of Some Vitamins	
Vitamins	**Some Functions**	**Some Deficiency Signs**
Vitamin A (2,000 I.U./kg.)[1]	Growth and development of bone and epithelial cells, vision	Night blindness, poor conception rate, abortion, loss of libido, testicular degeneration, convulsions, elevated cerebrospinal fluid pressure
Vitamin D (250 I.U./kg.)	Absorption of dietary calcium and phosphorus	Poor mineralization of bone, bone deformities
Vitamin E (15 I.U./kg.)	Antioxidant in tissues	Decreased serum tocopherol, increased red blood cell fragility; muscular dystrophy
Thiamin (3 mg./kg.)	Coenzyme in energy metabolism	Loss of appetite and weight; incoordination, muscular weakness, and twitching
Riboflavin 2 mg./kg.	Coenzyme in many enzyme systems	Conjunctivitis, lacrimation, aversion to bright light

1. Units per kg. of air dry feed

role in cell differentiation and, in the growing animal, in bone remodeling. Green plants and hays contain carotene, which the body normally converts to vitamin A.

Deficiency of vitamin A causes night blindness, excessive tearing of the eye, thickening of the horn layer of the skin and the cornea, lack of appetite, poor growth, respiratory infections, abscesses under the tongue, convulsive seizures, and progressive weakness.

Excess vitamin A, given over a long time, may cause fragile or thick bones, flaking skin, and tumors. In some animals, hair and skin are lost. They are severely depressed and lie down on their sides.

Vitamin D. Dietary vitamin D seems to be sufficiently present, especially if the horse is exposed to sunlight. Deficiency of vitamin D in the diet, while the horse is deprived of sunlight, seldom produces rickets, but does produce loss of appetite and slower growth. Supplementary vitamin D in the diet promotes absorption of calcium and phosphorus.

Excess vitamin D in the diet leads to calcification of blood vessels, the heart, and other soft tissues, and to bone abnormalities. Besides accidental addition of excess vitamin D to the diet, an excess may be caused by ingestion of certain toxic plants, such as day jasmine.

Vitamin E. An interrelationship exists between vitamin E and selenium. They both function as a part of a multicomponent antioxidant defense system.

Vitamin E activity in feedstuffs is reduced by moisture, mold growth, and grinding of the feedstuff during processing. Deficiencies of vitamin E and of selenium are difficult to distinguish separately. A deficiency of both nutrients in the foal will show pale areas in degenerating skeletal and cardiac muscles, as well as swelling of the tongue. If the deficiency is not corrected, pulmonary congestion occurs.

Claims that vitamin E is beneficial in restoring fertility in horses have not been substantiated by research. No symptoms of excess vitamin E in the horse are known.

Vitamin K. Vitamin K plays an important role in blood clotting. Deficiency of vitamin K results in decreased production of thrombin, which in turn interferes with the formation of fibrin clots. This leads to excessive bleeding with blood that will not clot. Excess intake of vitamin K does not seem to cause problems in the horse.

Water-Soluble Vitamins. **Water-soluble vitamins** include thiamin, riboflavin, niacin, pantothenic acid, biotin, folacin, ascorbic acid (vitamin C), choline, and vitamin B_{12}. Some of the water-soluble vitamins are often grouped as the B vitamins. This includes thiamin, riboflavin, niacin, pantothenic acid, biotin, and folacin. The water-soluble vitamins are available in feedstuffs or synthesized by microorganisms in the intestine. Only a couple of the water-soluble vitamins have a required level in the diet.

A need for thiamine in the diet has been demonstrated. A deficiency of riboflavin contributes to periodic ophthalmia (moon blindness). Common sources of B vitamins are green plants, dried legumes, and soybean meal.

Less is known about vitamins than minerals, but supplementation is easy and inexpensive.

Horses with access to good pasture, if only for a brief time, and those receiving good quality hay, especially if it is half legume, will probably need no vitamin supplementation (see Figure 12-5). Deficiencies are more likely to appear with horses confined for long periods of time on poor quality roughage.

Establishments with horses confined for long periods should consider an economical commercial source of vitamins as insurance against deficiencies when the roughages are not of top quality. On the other hand, "stuffing" an animal with vitamins many times beyond the known requirement increases expenses and contributes nothing to its health.

Figure 12-5 Green forages are a good source of vitamins such as vitamin K, thiamine, and riboflavin. *(Photo courtesy of Kentucky Horse Park)*

WATER

A source of fresh, clean drinking water is essential for horses at all times. Daily consumption may average ten to twelve gallons, with much higher amounts consumed at hard work and/or in hot weather. When water is not available free-choice, idle animals should be taken to it at least twice daily at regular intervals. **Impaction** is a common and rather serious problem resulting from infrequent drinking. Hot horses should be cooled out or permitted small amounts of water before drinking their fill. Those at work should be watered frequently whenever possible, as it refreshes the animal and reduces heat exhaustion (see Figure 12-6).

Dehydration and Electrolyte Balance

Dehydration from sweating results in the loss of both water and electrolytes—sodium, chloride, and some potassium. During extended workouts in

Figure 12-6 Horses need plenty of clean, fresh water.

hot dry weather the losses can be very significant. In heavy sweating, the loss of chloride can result in **hypochloremia** and **metabolic alkalosis**.

An adequate water supply, a balanced diet, and free-choice mineralized salt should provide all the necessary fluid and electrolytes for racing or extended work.

REQUIREMENTS AND ALLOWANCES

The subcommittee on Horse Nutrition of the NRC's Committee on Animal Nutrition has made recommendations for the nutrient allowance and requirement of horses, based on all the available information about their nutritional needs and by making inferences from other species. Tables 12-6, 12-7, and 12-8 are adapted from the NRC information. Table 12-6 gives the minimum daily requirements for digestible energy, crude protein, calcium, phosphorus, and vitamin A for different conditions and weight. Table 12-7 presents the same minimum daily requirements for growing horses with different mature weights, with or without training. Finally, Table 12-8 lists the mineral and vitamin recommendations for mineral and vitamins for maintenance, growth, pregnancy and work.

The next task is to put all of this information about digestion and nutrition into action and feed horses. The feeding of horses is covered in Chapter 13.

Table 12-6 Minimum Daily Nutrient Requirements for Mature Horses[1]

Condition	Mature Body Weight (lbs.)	Digestible Energy	Crude Protein		Calcium		Phosphorus		Vit. A
		mcal. daily	lbs. per day	% of diet	grams per day	% of diet	grams per day	% of diet	I.U.'s per day
Mature horse at rest (maintenance)	440	7.4	0.65	8	8	.25	6	.20	6.0
	880	13.4	1.18	8	16	.25	11	.20	12.0
	1,100	16.4	1.45	8	20	.25	14	.20	15.0
	1,980	24.1	2.13	8	36	.25	25	.20	27.0
Mature horse at moderate work[2]	440	11.1	0.98	10	14	.30	10	.25	9.0
	880	20.1	1.77	10	25	.30	17	.25	18.0
	1,100	24.6	2.17	10	30	.30	21	.25	22.0
	1,980	36.2	3.20	10	44	.30	32	.25	40.0
Mares, last 30 days of pregnancy	440	8.9	0.86	11	17	.50	13	.40	12.0
	880	16.1	1.56	11	31	.50	13	.40	24.0
	1,100	19.7	1.91	11	37	.50	23	.40	30.0
	1,980	29.0	2.81	11	55	.50	28	.40	54.0
Mares, peak of lactation[3]	440	13.7	1.52	13	27	.50	18	.35	12.0
	880	22.9	2.52	13	45	.50	29	.35	24.0
	1,100	28.3	3.15	13	56	.50	36	.35	30.0
	1,980	45.5	5.67	13	101	.50	65	.35	54.0

1. From Nutrient Requirements of Horses, 1989, National Research Council.
2. Examples are horses used in ranch work, roping, cutting, barrel racing, jumping, etc.
3. Lactation level is assumed to be 3% of body weight/day.

Table 12-7 Minimum Daily Nutrient Requirements for Growing Horses[1]

Current Age (Months)	Current Body Weight	Expected Daily Gain	D.E. mcal. per day	Crude lbs. per day	Protein (%)	Calcium		Phosphorus		Vit. A 1,000 I.U.'s per day
						Grams per day	(%)	Grams per day	(%)	
Growing horses[2] 440 lb. mature weight										
4	165	0.88	7.3	0.81	16	16	.70	9	.50	3
12	308	0.44	8.7	0.86	14	12	.55	7	.40	6
18NT[3]	374	0.22	8.3	0.83	13	10	.45	6	.35	8
24T[4]	407	0.11	11.4	1.07	12	13	.45	7	.35	8
Growing horses 880 lb. mature weight										
4	319	1.87	13.5	1.49	16	33	.70	18	.50	7
12	583	0.88	15.6	1.55	14	23	.55	13	.40	12
18NT	726	0.55	15.9	1.58	13	21	.45	12	.35	15
24NT	803	0.33	21.5	2.02	12	27	.45	15	.35	16
Growing horses 1,100 lb. mature weight										
4	385	2.00	14.4	1.59	16	34	.70	19	.50	8
12	1,100	1.98	31.2	3.10	14	49	.55	27	.40	22
18NT	1,463	1.54	33.6	3.34	13	49	.45	27	.35	30
24T	1,672	0.99	42.2	3.96	12	61	.45	34	.35	34

1. From Nutrient Requirements of Horses, 1989, National Research Council.
2. Moderate rate of gain.
3. Long yearling (18 months) not in training.
4. Two-year-old (24 months) in training.

Nutrient	Adequate Levels			Maximum Tolerance Levels
	Maintenance	Growth and Broodmares	Working	
Minerals				
Sodium %	.10	.10	.30	3.0%
Chloride %	.3	.40	.40	5.0%
Magnesium %	.10	.1	.15	.5
Sulfur %	.15	.15	.15	1.25%
Iron ppm[2]	40	50	40	1,000
Zinc ppm	40	40	40	500
Manganese ppm	40	40	40	1,000
Copper ppm	10	10	10	800
Iodine ppm	.1	.1	.1	5.0
Cobalt ppm	.1	.1	.1	10
Selenium	.1	.1	.1	2.0
Fluorene ppm	—	—	—	50
Vitamins				
Vitamin A I.U./lb.	910	1,667	910	7,273
Vitamin D I.U./lb.	135	135	365	1,000
Vitamin E I.U./lb.	25	37	37	450
Thiamine ppm	3	3	5	3,000
Riboflavin ppm	2	2	2	—

Table 12-8   Minerals and Vitamins for Horse Rations[1]

1. NRC, Nutrient Requirements of Horses, 1989
2. Parts per million (ppm) = mg./kg. = mg./2.2 lb.

SUMMARY

The purpose of digestion and nutrition is to supply the horse with the proper amounts and kinds of nutrients for maintenance, growth, reproduction, lactation, and work. These nutrients include carbohydrates, fats, protein, minerals, and vitamins. Water is also an important part of proper nutrition. Carbohydrates, fats, and protein provide energy, which is measured in calories. Protein in the diet also provides amino acids, which become the building blocks for protein in the body of the horse. Depending on the quantity in the body, minerals are classified as macrominerals or microminerals. Minerals become a part of the skeleton, function in energy production, and become a part of enzymes. Vitamins are either fat-soluble or water-soluble. Forages contain the vitamins and some vitamins are produced by microoganisms in the digestive tract. Vitamins serve in biochemical reactions that influnce how the other nutrients are used in the body. Deficient, out-of-balance or excess nutrients can reduce production or lead to health problems.

REVIEW

Success in any career requires knowledge. Test your knowledge of this chapter by answering these questions or solving these problems.

True or False

1. Carbohydrates are made up of amino acids.

2. Macrominerals suppy most of the energy in the diet.

3. Vitamin A is fat-soluble vitamin.

4. Fats contain more energy than proteins or carbohydrates.

5. A lactating mare needs more energy than a pregnant mare in the early stages of gestation.

Short Answer

6. Name four fat-soluble vitamins and six water-soluble vitamins.

7. List five macrominerals and five microminerals.

8. Name five conditions that alter the energy requirements of a horse.

9. Give two terms used to express energy in a feed and two terms used to express protein in a feed.

10. What other element is always associated with sodium in the diet?

Discussion

11. Compare the energy needs of a mature mare that is being maintained to that of a mare that is lactating and a mare working out each day.

12. What is a calorie?

13. Describe two functions of protein in the diet.

14. Explain the importance of the calcium-phosphorus ratio.

15. Describe the symptoms of three vitamin deficiencies.

STUDENT ACTIVITIES

1. Obtain five samples of high protein feeds common to your area. Take these samples to a laboratory for protein analysis.

2. Fecal material is a reality of any livestock production operation. Research the differences or similarities in the composition of horse, sheep, beef, and dairy fecal waste.

3. Compare the nutrition information contained on a cereal box for humans to the information contained on a feed tag for horses. What did you learn about human nutrition?

4. Obtain feed labels from horse, dairy, pig, and chicken feed. Compare the contents of each and compare the price.

5. Collect samples of horse feed and develop a display in small bottles or plastic bags. Label the type of feed and its protein, energy, and mineral content. Use a feed composition table to find this information.

ADDITIONAL RESOURCES

Books

American Youth Horse Council. 1993. *Horse industry handbook: a guide to equine care and management.* Lexington, KY: American Youth Horse Council, Inc.

Asimov, I. 1954. *The chemicals of life.* New York, NY: New American Library.

Ensminger, M. E., Oldfield, J. E., and Hieneman, W. W. 1990. *Feeds and nutrition.* Danville, IL: Interstate Publishers, Inc.

Frandson, R. D., and Spurgeon, T. L. 1992. *Anatomy and physiology of farm animals.* 5th ed. Philadelphia, PA: Lea & Febiger.

Lewis, L. D. 1996. *Feeding and care of the horse.* 2nd ed. Media, PA: Williams & Wilkins.

Subcommittee on Horse Nutrition, National Research Council. 1989. *Nutrient requirements of horses.* 5th ed. Washington, D.C.: National Academy Press.

Feeds and Feeding Horses

The greatest expense of owning a horse is the feed. This can be lessened by keeping a healthy horse, feeding a balanced ration according to need, and purchasing feeds that meet the needs of the animal.

OBJECTIVES

After completing this chapter, you should be able to:

- Identify and describe sources of hay

- List and describe sources of concentrates

- Describe how to feed roughages and concentrates

- Name and describe sources of proteins

- Explain how horses are fed according to their activity level

- Make feeding recommendations or management suggestions

- Describe some typical rations for horses at different stages and activity levels

- Calculate the nutrient level of a mixed feed using a feed composition table

- Discuss how to feed minerals

- Identify sources of minerals

K E Y   T E R M S

Azoturia	Forage	Pasturing
Bloom	Grass founder	Ration
Bolt	Heating	Rotational grazing
Bulk	Heaves	Roughage
Concentrates	Laxative	Silage
Crude protein (CP)	Legume	Stocking rate
Digestible energy (DE)	Monday morning disease	Supplement
Dry matter	Palatable	Trace mineralized salt

DETERMINING WHAT TO FEED

In balancing rations, the goals of a horse owner are to:

- Furnish horses with a daily supply of nutrients in the correct amounts

- Supply palatable, easily obtained feedstuffs

- Provide feedstuffs economical for the conditions

By nature, horses consume **forage**. Under natural conditions, they spend several hours a day grazing. Basing rations on adequate amounts of good-quality roughage minimizes digestive disturbances such as colic. Supplementing hay or pasture with the correct amount of the right concentrates will meet all requirements for energy, protein, minerals, and vitamins.

Since individual horses vary considerably in their nutrient requirements, feeding horses is both an art and a science. But tables such as Tables 12-6, 12-7, and 12-8 in Chapter 12 form a useful basis for formulating rations. Horse owners need to be able to read and understand these tables.

All horses require nutrients to maintain body weight and to support digestive and metabolic functions. In some cases they need additional nutrients for growth, work, reproduction, or lactation.

Tables of nutrient requirements for horses are expressed in two ways:

1. Daily nutrient requirements.

2. Nutrient concentration in the feed. This may be expressed on an as-fed basis or on a **dry-matter** basis.

Most horses receive their daily **ration** in two parts: **roughage** (hay, silage, and/or pasture) and **concentrates**. The concentrate portion contains grain and may include a protein **supplement**, minerals, and vitamins. It may also include bran, cane molasses, and/or dehydrated alfalfa.

The horse owner must decide:

- How much and what kind of roughage to feed
- The correct concentrate mixture and the amount of it needed to supply the nutrients not present in adequate amounts in the roughage

FEEDS FOR HORSES

Feeds for horses are discussed in four groups:

1. Roughages (hays, silage, pasture)
2. Concentrates (grains)
3. Protein supplements
4. Minerals

Most grains and hays contain 88 to 90 percent **dry matter**. If a horse receives insufficient dry matter, it may become bored and chew on its stall or eat its bedding. If the feed has too much **bulk** (excessive amounts of fiber or water), the horse might not be able to eat enough to satisfy all its nutritional requirements—carbohydrates, protein, minerals, and vitamins.

Table 13-1 provides the composition of some common feeds for horses. Dry matter, **digestible energy (DE)**, **crude protein (CP)**, calcium (Ca), phosphorus (P), and vitamin A content is listed for hays, concentrates and protein supplements, and mineral supplements.

Roughages

Adequate amounts of roughage in the ration decrease the risk of colic and laminitis. Roughage also helps maintain the correct calcium-to-phosphorus ratio, because roughages—especially legume hays—are high in calcium and because grain is low in calcium. Rations should always contain more calcium than phosphorus. Calcium to phosphorus ratios between 1.1 to 1 and 2 to 1 are within an acceptable range. Even higher calcium levels can be tolerated. However, when phosphorus levels are higher than calcium levels, severe skeletal abnormalities may result.

Adequate hay in the ration of horses kept in stalls also is beneficial because they eat it over a longer time span than they do grain. This helps prevent vices such as wood chewing.

A good rule of thumb is to feed at least one pound of hay per day for every 100 pounds body weight of the horse. A 1,000-pound horse would be fed about ten pounds of hay per day. Mature, idle horses in good condition, fed excellent hay in increased quantities (about two pounds per 100 pounds of body weight) may do well without grain added to their ration. Growing or working horses, mares during late pregnancy, and mares during lactation need grain and other concentrates in addition to the roughage.

Table 13-1 Composition of Some Common Horse Feedstuffs[1]						
Feedstuffs	**Dry[2] Matter %**	**D.E. mcal./lb.**	**C.P. lb./lb.**	**Ca g./lb.**	**P g./lb.**	**Vitamin A 1,000 I.U./lb.**
Hays						
Alfalfa: Early bloom	90.5	1.02	0.180	5.81	0.86	23.00
Full bloom	90.9	0.89	0.155	4.90	0.99	10.74
Red Clover	88.4	0.89	0.132	5.53	0.99	9.88
Orchard grass: Early bloom	89.1	0.88	0.114	1.09	1.36	6.08
Late bloom	90.6	0.78	0.076	1.09	1.22	3.29
Bromegrass: Mid-bloom	87.6	0.85	1.260	1.13	1.13	2.45
Timothy: Early bloom	89.1	0.83	0.096	2.04	1.13	8.51
Late bloom	88.3	0.72	0.069	1.54	0.59	7.23
Fescue: Full bloom	91.9	0.86	0.1181	0.81	1.32	8.73
Mixed: 30% legume	89.0	0.93	0.1332	0.66	1.10	11.72
Concentrates/Protein Supplements						
Barley	88.6	1.49	0.117	0.23	1.54	0.37
Corn	88.0	1.54	0.091	0.23	1.27	0.98
Oats	89.2	1.30	0.118	0.36	1.54	0.02
Wheats, red	88.4	1.55	0.114	0.14	1.77	–
Wheat bran	89.1	1.33	0.154	0.59	5.13	0.48
Soybean meal	89.1	1.43	0.445	1.59	2.86	–
Linseed meal	90.2	1.25	0.346	1.77	3.63	–
Molasses (blackstrap)	74.3	1.18	0.043	3.36	0.36	–
Vegetable oil	99.8	4.08	–	–	–	–
Mineral Supplements						
Limestone, CaCO$_3$	100	–	–	178.67	0.18	–
Oystershell	99	–	–	170.64	0.31	–
Bone meal, steamed	97	–	–	135.12	56.58	–
Rock phosphate, defl.	100	–	–	145.15	81.65	–
Dicalcium phosphate	97	–	–	96.81	83.73	–
Sodium triphosphate	96	–	–	–	108.86	–

1. Derived from *Nutrient Requirements of Horses*, 5th ed. NRC. 1989.
2. All nutrients are expressed on an as-fed basis.

Hays

The most important consideration in selecting a dry roughage is that it be free of dust and mold. Otherwise, **heaves** and colic may result. Early-cut hays, properly cured, are preferred. They can be identified by color, head development on grass hays, leaf-to-stem ratio, and size of stems in legumes. Bales should be broken to check for dust and moldy odor.

Confined idle adult horses will eat about fifteen to twenty pounds of good-quality mixed hay daily when no grain is fed. Feeding just what the horse will eat takes some experience and observation. Table 13-2 provides guidelines for feeding hay to a mare in a dry lot or stall.

Legume Hays. **Legume** hays are higher in protein and minerals and are more palatable than grass hays. They make excellent horse feed and should be included in the rations of young growing animals, breeding animals, and many adult working horses.

- **Alfalfa.** When properly cured, alfalfa is the best of the legumes from a nutrient standpoint. Its high protein, calcium, and vitamin content make it especially useful in balancing rations for brood mares and young growing horses. Some halter show people make extensive use of top quality alfalfa in show rations, especially with horses that are finicky about eating.

Although western work horses in irrigated districts in the past consumed much alfalfa free-choice, some horsemen consider it "**heating**" or having too **laxative** an effect and feed it only as half of the roughage.

- **Clovers.** Many varieties of clover are used alone or in combination with grass hays for horses. Red clover is similar to alfalfa and can be substituted for it, usually with slightly less beneficial results. It is lower in protein and usually has a higher ratio of stems to leaves than alfalfa. Properly-cured Alsike clover is a good hay for horses.

Table 13-2 Average Amounts of Good-Quality Hay to Feed to a Mare in Dry Lot or Stall	
Body Weight (lbs.)	**Daily Amount of Hay (lbs.)**
800	12 to 16
900	14 to 18
1,000	15 to 20
1,100	17 to 22
1,200	18 to 24
1,300	20 to 26

- **Lespedeza.** When cut early, lespedeza makes an excellent hay. It is higher in protein than red clover. When it is cut late, many leaves are lost from shattering and the stems become wiry and low in digestibility. The calcium content is about half that of alfalfa.

Grass Hays. Grass hays yield less per acre and are lower in protein, calcium, and vitamins, but they are less likely to be moldy and dusty than legume hays. They are usually cut too late to yield quality hay and often are priced higher than their feeding value justifies.

Grass hays often are grown and harvested in mixtures with legumes, which produces an excellent combination suitable for almost any kind of horse-feeding program.

- **Timothy.** No other hay has attained the lasting popularity of timothy. Its wide range of climatic adaptability, ease of curing, bright color, and freedom from dust and mold make it the horse owner's favorite. Since it is low in protein, it is a better feed for mature work horses than for stallions, mares, or young growing stock. If it is fed as the only roughage, it should be supplemented with protein or fed with a high-protein grain such as oats instead of corn. Special effort need not be made to obtain timothy, as it can be satisfactorily substituted for in all horse rations. Mature, late-cut timothy is a poor feed for any class of livestock.

- **Prairie grass.** Some horse owners substitute prairie hay satisfactorily for timothy. However, it is lower in protein, less bright in color, and usually less palatable than timothy.

- **Bromegrass.** Bromegrass makes good horse hay. It is palatable when harvested in the bloom stage.

- **Orchard grass.** Orchard grass is much like bromegrass but not quite as satisfactory.

- **Cereal grasses.** Cereals make good hays when cut early. They should be cut in the soft to stiff dough stage. They are seldom cut early enough. Oats, barley, wheat, and rye hays are preferred, in that order. Extensive use is made of these in the Pacific Coast region.

- **Fescue.** Fescue hay infected with endophyte fungus causes reproductive problems in mares if fed during late pregnancy. It is also low in energy and horses don't like it very much. If harvested before it gets too mature, however, it usually works for mature geldings or open mares providing they have adequate supplementation.

Silage

Various types of **silage** can be used to replace half of the hay ration. It must be of good quality and free of decay or mold and should be chopped fine. Good corn silage is preferred, but grain sorghum and grass silage can be used. It should be worked slowly into the rations of mature idle horses, growing horses, brood mares, and stallions. It is too bulky for hard-working animals and foals. Legume haylage can replace silage with equal or superior results. The cost for haylage for most horse owners is too high unless they combine it with a cattle feeding program.

Pastures

Grass is the natural feed for horses (see Figure 13-1). No onefeed stuff is as complete in nutrients as green pasture grown on fertile soil, and few feeds are fed in a more healthful environment. Grass reduces cost, provides succulence in the ration, and furnishes minerals and vitamins that are sometimes lacking in other feeds. However, hardworking horses need supplemental energy feeds because of the high water content of grass. Dry grass is usually low in protein and vitamins, and heavy stocking rates pose a parasite problem.

Pasturing can reduce stable vices caused by boredom or mineral deficiencies. Pasture rotation reduces the problem of parasites. **Rotational grazing** will also reduce patch grazing; placing minerals away from the water source encourages more evenly distributed grazing (see Figure 13-2). A horse requires two to five acres of pasture for maintenance. Table 13-3 can be used to determine the **stocking rate** for horses on pastures.

Animal units in Table 13-3 describe a pasture in terms of the needs of a horse. For example, a ten-acre pasture may have a carrying capacity of three animal units. This means it could provide feed for three 1,100 pound horses in a maintenance condition, but it could provide feed for less than two lactating

Figure 13-1　Horse on pasture at River Grove Farm, Hailey, Idaho.

Figure 13-2 Pasturing and rotational grazing are good feeding practices.

Table 13-3 Pasture Stocking Rate for 1,100–Pound Mare	
Condition	**Animal Units**[1]
Maintenance	1.0
Light work (2 hours/day)	1.4
Medium work (2 hours/day)	1.8
Last 90 days of pregnancy	1.1
Peak lactation	1.8

1. Animal units describe the carrying capacity of a pasture.

mares (3/1.8). Lactating mares represent 1.8 animal units because of their increased nutritional needs.

In the spring, horses should be pastured for only a limited time each day, since **grass founder** can occur on lush pasture.

Concentrates

Concentrates are high-energy feeds. Grains are concentrates used with hay to regulate energy intake and ensure that nutrients are sufficient to meet the work, growth, or reproductive performance required of the animal. Medium-sized, hard-working horses may need as much as twelve pounds or more of grain and an equal amount of hay daily to maintain body weight, whereas idle adult horses may get fat on grass alone (see Figure 13-3).

Figure 13-3   Horses love oats!

Horses like grain. Some even **bolt** it to the point of choking. Most grains are improved by grinding or rolling, but none should be ground fine. Frequent feedings in small amounts are preferred, with at least a half-hour's rest for tired horses before grain is fed. Continued heavy grain feeding during a day off can cause a serious disease called **Azoturia** ("**Monday morning" disease**). In general, grain rations should be cut in half and hay increased on days that working horses are idled. Substituting one or more grains for others needs to be a gradual process. Grains for horses include oats, corn, grain sorghum, barley, wheat, wheat bran, and cane molasses. Table 13-4 provides some guidelines for feeding grain with hay or pasture.

Table 13-4   Amount of Hay and Grain to be Fed to a 1,100-Pound Mare During Late Gestation

Status	Alfalfa		Grazing or Grass Hay	
	Hay	Grain Mix (10% CP)[1]	Hay	Grain Mix (14% CP)
	Pounds	Pounds	Pounds	Pounds
Open or early to mid pregnancy	17 to 22	—	17 to 22	—
Pregnant (9 months to term)	14 to 15	5 to 6	14 to 15	6 to 7
Early lactation (0 to 3 months)	14 to 15	10 to 12	14 to 15	12 to 14
Late lactation (>3 months)	14 to 15	6 to 7	14 to 15	7 to 8

1. CP = Crude protein

Oats. Oats are the grain of choice for most horseowners and horses. The bulky nature of oats permits horseowners maximum liberty in their use with minimum danger of digestive disorders. Even the most picky horses find oats to their liking. Oats are higher in protein than most grains, which makes them useful with low-protein grass hays. However, half-legume hay ensures a more complete ration when oats are fed as the only grain. Some disadvantages are expense on a digestible energy basis and variability in quality. Federal grades are No. 1, No. 2, No. 3, No. 4, and Sample. Grades 1 and 2 are the best buy.

Although oats are an excellent horse feed, when cost and/or convenience dictate, most rations can be formulated satisfactorily without them.

Corn. Corn is a good feed and is used extensively in the Midwest. About 15 percent less corn would equal a given weight of oats in energy value. For this reason, corn is especially useful for improving the condition of thin horses and maintaining condition of those at hard work. It is often a good buy on an energy basis, even exceeding hay on occasion.

Because of its high energy content and low fiber, corn must be fed with more care than oats to avoid colic. Corn and oats, in equal parts, make an excellent grain ration. Corn can supply all of the grain when fed according to the work that horses are performing and when large amounts are not given at one time.

Some people call corn a "heating" feed in warm weather. This theory is not easy to explain because "heat" produced by digestion is greater for fibrous feeds, such as hays and oats, than for corn. Probably a major reason is that horses eating corn tend to stay fatter than others, especially if they are not regularly exercised.

Grain Sorghum. Grain sorghum can be substituted for corn in most rations. It varies in protein content from 6 to 12 percent, it has little vitamin A, and some varieties are unpalatable. Grain sorghum is best when used in a grain mixture. In some areas, grain sorghum is often a better buy than corn. It should be cracked or rolled.

Barley. Barley is a very satisfactory feed when ground and fed as described for corn. Fifteen percent wheat bran or 25 percent oats fed with barley almost eliminates the risk of colic.

Wheat. Wheat is seldom fed to horses except in the Pacific Northwest. It can be fed as a part of the grain ration—about one-third—when fed with a bulky feed. Wheat should be rolled or coarsely ground. Wheat tends to be doughy when moist and produces palatability problems.

Wheat Bran. Wheat bran is highly **palatable**, slightly laxative, and very bulky. Horse owners have long preferred "bran mashes" for animals stressed by extreme fatigue, foaling, or sickness. Bran is reasonably high in protein,

Nutrition is complex. For the novice, remembering everything can be difficult. Specific information can be learned from books, tables, and people. General "rules of thumb" are helpful and easy to remember.

Here are some "rules of thumb" about the nutrient content of feedstuffs for horses.

- **Rule #1 for Energy**: Total digestible nutrients (TDN), or calories, for hay is about 50 percent; for grains it is about 75 percent.

- **Rule #2 for Protein**: The protein content for alfalfa and clovers (legumes) is about 14 to 16 percent; for grasses it is 7 to 10 percent.

- **Rule #3 for Minerals**: Legumes are rich in calcium; grasses are fair. All forages are low in phosphorus. Grains are high in phosphorus, but low in calcium.

- **Rule #4 for Vitamins**: Forages are high in vitamins; grains are low.

Four rules are easier to remember than entire feed tables. Of course, when accurate information is needed, feed tables and laboratory tests should be used.

high in phosphorus, and, like other grains, low in calcium. Because of its high cost on an energy basis, it is generally used at levels of 5 to 15 percent of the ration.

Cane Molasses. The addition of 5 to 10 percent molasses reduces dust and increases palatability of a ration. Greater amounts will have too great a laxative effect. It is very low in protein and usually expensive on an energy basis. Dried molasses is often added to the grain ration to increase consumption.

Protein Supplements

The horse's need for protein is relatively low and easy to meet with practical rations. With the exception of milking mares, most 600 to 1,200-pound horses need from three-quarters to one pound of digestible protein (DP) daily. If the roughage is half-legume hay fed in adequate amounts, the protein need will be met. However, supplementing rations of young growing horses is insurance against a deficiency and stimulates appetite. The hair coat of horses being fitted for show will "**bloom**" to a higher degree when about one pound of an oil meal is supplied daily. Large amounts will cause a laxative effect (see Figure 13-4).

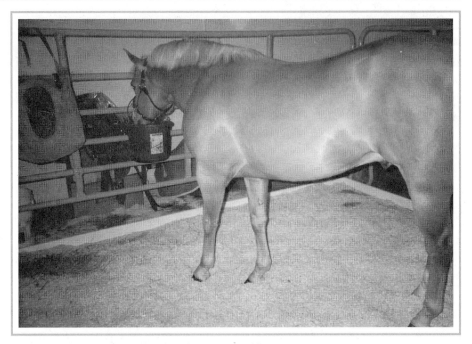

Figure 13-4 Good nutrition produces a good hair coat.

Protein supplementation is needed when poor-quality late-cut grass hays are fed. Some of the protein supplements used for horses are linseed meal, soybean meal, and cottonseed meal.

Linseed Meal (30 to 32 percent protein). "Old process" or "expeller-type" linseed meal was considered by horse owners to be effective in blooming the hair coat. It contained a fatty acid (linoleic) that may be deficient in standard horse rations. "New process" or "solvent" processing removes this fatty acid from linseed meal. Since linseed meal is not well balanced in amino acids, use of "solvent" process is hard to justify.

Fitters of show horses who use legume hay may find linseed meal has too laxative an effect in their programs.

Soybean Meal (42 to 50 percent protein). Soybean meal is a preferred supplement for horses. It is higher in protein, has a better balance of amino acids, and is cheaper in the Midwest than other supplements.

Cottonseed Meal (40 to 45 percent protein). Cottonseed meal is used extensively for horses in the Southwest. It seldom costs less than soybean meal in the Midwest and is not as palatable, so the extent of its use is limited.

Commercial Protein Supplements. These vary in composition, protein level, and price. They often contain needed minerals and vitamins and are convenient for those who do not wish to formulate their own horse rations.

Some may be expensive. Commercial supplements are usually formulated for a specific feeding program. They should be fed according to directions.

Other Protein Supplements.   Alfalfa meal, corn gluten meal, meat meals, and others can be used with horses as protein supplements.

Minerals

Trace-mineralized salt contains no calcium, and phosphorus and dicalcium phosphorus are not a source of selenium, manganese, or other trace minerals. A horse has a natural craving for salt, but has neither a particular appetite for nor a natural instinct to seek out sources of calcium or phosphorus. Therefore, the way to supplement horses with calcium or phosphorus is to mix trace-mineralized salt with limestone or dicalcium phosphate. Limestone and dicalcium phosphate are rich but unpalatable sources.

Commercial Feeds

Commercially prepared feeds may actually provide nutrients such as trace minerals, vitamins, and protein supplements in a less expensive form than the individual horse owner can provide.

But a word of warning. Aside from providing adequate nutrition, no nutrient or supplement will (a) make the hoof grow faster and stronger; (b) cure a curb, spavin, ringbone, etc.; (c) increase conception in mares or libido in stallions; (d) increase intelligence; (e) prevent colic; or (f) cure heaves, sleeping sickness, and equine infectious anemia (EIA). In short, horse owners should not be fooled into buying magic from a bottle or a can.

EXAMPLE RATIONS

Based on the information in this chapter, the following example rations can be modified to fit the needs of individual horses and horse owners. Tables 13-5 through 13-8 provide sample rations for a foal, a weanling, a two-year-old, a preganant or lactating mare, and an adult horse. Information for each ration provides the amount of each ingredient needed to make a half ton or a ton of feed. The amounts can be adjusted mathematically to make smaller or larger amounts of feed.

Table 13-5 provides an example ration for a creep-feeding foal. This ration should not be used after weaning because it is too high in protein and calcium unless it is fed with a nonlegume hay (see Figure 13-5).

Table 13-6 is an example of a ration for a weanling horse. This ration is lower in protein and calcium but higher in energy than the ration in Table 13-5.

The example ration shown in Table 13-7 is lower still in protein and calcium than either Table 13-5 or 13-6. If a mare is obese in late pregancy she does not need grain and can be maintained on a good-quality hay.

Table 13-5 Foal Creep Ration		
Ingredients	Pounds to Make ½ ton	Pounds to Make 1 ton
Oats, crimped or crushed	440	880
Corn, coarsely cracked	220	440
Soybean, meal 44 percent	240	480
Molasses, liquid	70	140
Dicalcium phosphate	15	30
Limestone	10	20
Salt, trace mineral	5	10
Vitamins A, E, D to supply 4,000 I.U./lb.	11	21
Total pounds	1,001	2,002

Crude protein in the diet is 18 percent.
Calcium in the diet is 0.88 percent.
Phosphorus in the diet is 0.60 percent.
Feed this grain ration free-choice with good legume hay to foals from two weeks old to weaning or to early-weaned foals from three to eight months old.

Figure 13-5 Care must be taken when planning the care and feeding of foals.

Table 13-6 Weanling Horse Ration

Ingredients	Pounds to Make ½ ton	Pounds to Make 1 ton
Oats, crimped or crushed	440	880
Corn, coarsely cracked	270	540
Soybean, meal 44 percent	190	380
Molasses, liquid	75	150
Dicalcium phosphate	10	20
Limestone	5	10
Salt, trace mineral	5	10
Vitamins A, E, D to supply 4000 I.U./lb.	11	21
Total pounds	1,001	2,002

Crude protein in the diet is 16.31 percent.
Calcium in the diet is 0.75 percent.
Phosphorus in the diet is 0.55 percent.
Feed this grain ration to weanlings. Add good legume or at least half-legume hay at the rate of 1 to 1½ pounds of grain per 100 pounds of body weight. Feed hay free-choice.

Table 13-7 Yearling, Two-year-old, Late Pregnancy, and Lactating Mare Ration

Ingredients	Pounds to Make ½ ton	Pounds to Make 1 ton
Oats, crimped or crushed	440	880
Corn, coarsely cracked	340	680
Soybean, meal 44 percent	130	260
Molasses, liquid	70	140
Dicalcium phosphate	5	10
Limestone	10	20
Salt, trace mineral	5	10
Vitamins A, E, D to supply 4000 I.U./lb.	11	21
Total pounds	1,001	2,002

Crude protein in the diet is 14.3 percent.
Calcium in the diet is 0.61 percent.
Phosphorus in the diet is 0.43 percent.
Feed this ration at the beginning of the yearling year with good legume or at least half-legume hay or good pasture. Regulate intake to control the desired degree of condition. Four to eight pounds daily should be adequate.

As growing horses approach 18 months of age, nonlegume hay is adequate with enough grain to maintain condition.

Table 13-8 Adult Horse, Early Pregnancy, and Late Two-year-old Ration		
Ingredients	Pounds to Make ½ ton	Pounds to Make 1 ton
Oats, crimped or crushed	500	1,000
Corn, coarsely cracked	390	780
Soybean, meal 44 percent	30	60
Molasses, liquid	65	130
Dicalcium phosphate	3	6
Limestone	7	14
Salt, trace mineral	5	10
Vitamins A, E, D to supply 4000 I.U./lb.	11	21
Total pounds	1,001	2,002

Crude protein in the diet is 11.0 percent.
Calcium in the diet is 0.43 percent.
Phosphorus in the diet is 0.36 percent.
This ration is designed for adult and 2-year-old idle and working horses and for mares until the last three months of pregnancy. It may be fed with either legume or nonlegume hay, but nonlegume hay will result in fewer digestive upsets with hardworking horses eating large amounts of grain.

The ration example in Table 13-8 is too low in protein, calcium, and phosphorus for weanlings and lactating mares. It is marginal for mares in late pregnancy.

CALCULATING NUTRIENTS

Formulating an adequate ration for a horse is simple if these steps are followed:

1. Know what the horse requires.

2. Know what kind of feed will fill those requirements economically.

3. Know what feeds are palatable.

4. Know how much of a given feed the horse can eat.

5. Know how to calculate the amount of a nutrient in a feed.

The most common feeding problem confronting horse people is figuring what percentage of a given nutrient is in a mixed ration. Referring to tables will show how much protein, digestible energy, or calcium is in corn or oats, but will not be specific for a mixed feed of unequal parts of corn, oats and soybean meal. In order to figure the nutrient content of a mixed grain ration, simply multiply the pounds of each of the feedstuffs in the mixture (corn, oats, soybean meal, etc.) by the percentage of the nutrient (digestible energy, protein, calcium, etc.) that each feed contains. Total the amounts obtained and divide by the number of pounds of feed in the mixture. This procedure provides a weighted average.

An Example

For example, what is the protein content of a feed that contains 500 pounds of oats, 400 pounds of corn, and 30 pounds of soybean meal? To find the protein content of a mixed feed:

1. Find the protein content of each of the feedstuffs in Table 13-1.

2. Multiply this value by the number of pounds of that feedstuff in the mixture.

3. Next find the total pounds of protein in the feed mixture.

4. Finally, divide the total amount of protein in the feed mixture by the total weight of the feed mixture and convert this to a percentage.

The following example shows how.

Feedstuff	Protein in Feedstuff		Pounds Feedstuff in Mix		Protein in Mix
Oats	0.118 lb./lb.	×	500	=	59 lb.
Corn	0.091 lb./lb.	×	400	=	36 lb.
Soybean meal	0.445 lb./lb.	×	30	=	13 lb.
Total			930		108 lb.

$$\frac{108 \text{ lb. of protein in mix}}{930 \text{ lb. of feed mix}} \times 100 = 11.6 \text{ percent protein in the mixture}$$

This process will work with the digestible energy, calcium, and phosphorus of the feed mix.

The common error is to add up the protein content of the corn, oats, and soybean meal and divide by three. But if corn and oats constitute 90 percent of the mixture, they naturally have a greater effect on the average composition than soybean meal, which makes up only 10 percent of the mixture.

FEEDING MANAGEMENT RECOMMENDATIONS

Since feeding horses is as much an art as it is a science, the following guidelines will help horse owners successfully feed their horses.

1. Feed only quality feeds.

2. Feed balanced rations.

3. Feed half the weight of the ration as quality hay.

4. Feed higher protein and mineral rations to growing horses and lactating mares.

5. Feed legume hay to young, growing horses, lactating mares, and out-of-condition horses.

6. Use nonlegume hays for adult horses.

7. Regulate hay-to-grain ratio to control condition in adult horses.

8. Feed salt separately, free-choice.

9. Feed calcium and phosphorus free-choice.

10. Keep teeth functional. Horses five years old and older should be checked annually by a veterinarian to see if their teeth need floating (filing).

11. See that stabled horses get exercise—they will eat better, digest food better, and be less prone to colic.

12. Feed according to the individuality of horse. Some horses are hard keepers and need more feed per unit of body weight.

13. Feed by weight, not volume. A gallon of two different grains may vary in nutrient content.

14. Minimize the use of finely ground feedstuffs in a prepared ration. If a ration is ground fine, horses will be reluctant to eat it and the chances of colic will increase.

15. Offer plenty of good water, no colder than 45 degrees Fahrenheit. Free-choice water is best. Horses should be watered at least twice daily.

16. Change feeds gradually. When changing from a low-density (low-grain), high-fiber ration to one of increased density, change gradually over a period of a week or more.

17. Start horses on feed slowly. Horses on pasture should be started on dry feed gradually. Start this on pasture if practical and gradually increase the feed to the desired amount in a week to ten days.

18. Do not feed grain to tired or hot horses until they have cooled and rested, preferably for one or two hours. Instead, feed hay while they rest in their blankets or out of drafts.

19. Feed before work. Hungry horses should finish eating at least an hour before hard work.

20. Feed all confined horses at least twice daily. If horses are working hard and consuming a lot of grain, three times is mandatory.

21. When feeding hay, give half the hay allowance at night, when horses have more time to eat and digest it.

SUMMARY

Feeding horses is part art and part science. High-quality roughages form the basis for feeding horses. The nutritional needs of the horse change depending on its condition, activity level, age, gestational stage, or lactation. Additional nutritional needs can be met by feeding concentrates and protein supplements. The ration of a horse may also need mineral supplementation to cover its calcium, phoshorus, and other mineral requirments. Some vitamins may be added from a premix. Horse owners can make their own rations or buy commercially prepared feeds.

REVIEW

Success in any career requires knowledge. Test your knowledge of this chapter by answering these questions or solving these problems.

True or False

1. Founder in horses can occur on lush pasture.

2. Legume hays are low in protein.

3. Oats are the horse's favorite grain.

4. Feed changes for horses should occur slowly.

5. Commercial feeds can cure all ills of the horse.

Short Answer

6. How many acres of pasture does a horse require for maintenance?

7. List the four groups of feeds for horses.

8. List five types of hay for horses.

9. Name five horse feeding/management recommendations.

10. If a horse is working hard, how many times a day should it be fed?

Discussion

11. What is the role of roughage in the horse's diet?

12. How does the horse owner supplement horse rations with minerals?

13. Explain the five guidelines or steps for formulating a ration for a horse.

14. What is an advantage of stable feeding? Of pasture feeding?

15. Using Table 13-1, calculate the digestible energy, protein, and calcium content of a feed that contains 600 lb. of oats, 200 lb. of corn and 30 lb. of soybean meal.

16. Define the term "dry matter."

STUDENT ACTIVITIES

1. Identify in a diagram the external and internal structures associated with the digestive system of the horse. (Use Chapter 5 as a guide.)

2. Obtain samples of protein feeds, for example soybean meal, cottonseed meal, and commercial protein supplements. Use published composition tables and compare each feed. Observe differences in the smell, texture, etc. Compare the costs.

3. Use a computer program to balance the diet of a racehorse or a pregnant mare. Figure the cost of the diet.

4. Contact suppliers of horse equipment and obtain information on feeders and compare the different types.

5. Collect fresh samples of some typical horse feedstuffs and determine their dry matter content. Weigh the samples at collection time. Dry (do not cook) these samples in an oven and weigh them again. Use the data generated to calculate the dry matter.

ADDITIONAL RESOURCES

Books

American Youth Horse Council. 1993. *Horse industry handbook: a guide to equine care and management.* Lexington, KY: American Youth Horse Council, Inc.

Blakely, J. 1981. *Horses and horse sense: the practical science of horse husbandry.* Reston, VI: Reston Publishing Company, Inc.

Ensminger, M. E. 1990. *Horses and horsemanship.* 6th ed. Danville, IL: Interstate Publishers, Inc.

Evans, J. W. 1989. *Horses: a guide to selection, care, and enjoyment.* 2nd ed. New York, NY: W. H. Freeman and Company.

Feedstuffs, 1996 Reference Issue, Carol Stream, IL.

Lewis, L. D. 1996. *Feeding and care of the horse.* 2nd ed. Media, PA: Williams & Wilkins.

Price, S. D. 1993. *The whole horse catalog.* New York, NY: Fireside.

Prince, E. R., and Collier, G. M. 1986. *Basic horse care.* New York, NY: Doubleday & Company, Inc.

Self, M. C. 1963. *The horseman's encyclopecia.* New York, NY: A.S. Barnes and Company.

Stoneridge, M. A. 1983. *Practical horseman's book of horsekeeping.* Garden City, NY: Doubleday & Company, Inc.

Subcommittee on Horse Nutrition, National Research Council. 1989. *Nurient requirements of horses.* 5th ed. Washington, D.C.: National Academy Press.

University of Missouri-Columbia Extension Division. (n.d.) *Missouri horse care and guide book.* Columbia, MO: Cooperative Extension Service, University of Missouri and Lincoln University.

CHAPTER *14*

Health Management

All horseowners want their horses to be healthy, to look good, and to be physically fit. Most have a sense that these three things are connected. But how? How can an owner or trainer tell whether or not a horse is healthy just by looking at it? What does a healthy horse look like? And, once the state of a horse's health is determined, what can be done to improve and maintain it? These are good questions. The problem is that almost every horse owner would answer them differently.

O B J E C T I V E S

After completing this chapter, you should be able to:

- Define terms associated with disease conditions
- Discuss disease resistance and immunity
- Define terms associated with the severity of a disease or condition
- Describe immunization
- Discuss how a vaccination program relates to immunity
- List signs of disease
- Discuss common diseases caused by viruses
- Discuss common diseases caused by bacteria
- List five noninfectious diseases
- Describe the signs of good health in a horse
- List the objectives for first aid for horses
- List four digestive diseases
- List five respiratory diseases
- Discuss laminitis and colic
- Relate body condition to health

K E Y   T E R M S

Active immunity	Disease	Passive immunity
Agglutination	Immunoglobulins	Plasma
Antibiotics	Infectious	Titer
Antibodies	Lesions	Toxins
Antigen	Mutagens	Vaccines
Capillary refill	Noninfectious	

SIGNS OF HEALTH

For the horse owner or anyone working with horses, the first step in health management, is learning to recognize a healthy horse. Disease can then be recognized and treated early.

Healthy horses show good body condition. Other ways to quickly assess the health of a horse include evaluating its general appearance and behavior, examining specific parts of the body (such as hoofs and eyes), observing its manure and urine, and measuring vital signs like heart rate, respiratory rate, and temperature.

Normal Body Condition

The body condition, or degree of fat cover, of horses is a good indicator of their general health. A scoring system that assesses fat cover has been designed to gauge reproductive efficiency in mares. An evaluation—or **condition score**—based on this system can also serve as a guide to judging the health and fitness of all horses. This system is presented in Table 14-1.

Animals with a score between 4 and 6 can be considered healthy. Scores lower than 4 or higher than 6 on this scale indicate the likelihood of metabolic and other health problems.

Other Signs of Good Health

Observing horses at horse events like shows, races, and other competitions is another good way to establish a standard for what is normal.

A bright, actively interested horse can be recognized at a glance (see Figure 14-1). It will be alert, inquisitive, and attentive. It will not have the dull, lethargic look that can indicate overtraining, overuse, or ill health.

When in pastures, lots, and paddocks, horses normally will try to stay in a group, so one off by itself may be hurt or ill. Normal healthy horses also chew evenly with both sides of their mouth and show predictable enthusiasm for eating.

Table 14-1 Horse Condition Scoring System

Score	Condition Description
1	**Poor.** Animal extremely emaciated. Spinous processes, ribs, tailhead, and point of hip and point of buttocks project prominently; bone structure of withers, shoulders, and neck easily noticeable; no fatty tissue can be felt.
2	**Very thin.** Animal emaciated. Slight fat covering over the base of spinous processes; transverse processes of the lumbar vertebrae feel rounded; spinous processes, ribs, tailhead, and point of hip and point of buttocks prominent; withers, shoulders, and neck structures faintly discernible.
3	**Thin.** Fat built up about halfway on the spinous processes; transverse processes cannot be felt; slight fat cover over the ribs; spinous processes and ribs easily discernible; tailhead prominent, but individual vertebrae cannot be identified visually; point of buttocks appear rounded but easily discernible; point of hip not distinguishable; withers, shoulders, and neck accentuated.
4	**Moderately thin.** Slight ridge along back; faint outline of ribs discernible; tailhead prominence depends on conformation, but fat can be felt around it; point of hip not discernible; withers, shoulders, and neck not obviously thin.
5	**Moderate.** Back is flat (no crease or ridge); ribs not visually distinguishable but easily felt; fat around tailhead beginning to feel spongy; withers appear rounded over spinous processes; shoulders and neck blend smoothly into body.
6	**Moderate to fleshy.** May be slight crease down back; fat over ribs spongy; fat around tailhead soft; fat beginning to be deposited along the side of withers, behind shoulders, and along the sides of neck.
7	**Fleshy.** May have crease down back; individual ribs can be felt, but there is noticeable fat between ribs; fat around tailhead soft; fat deposited along withers, behind shoulders, and along neck.
8	**Fat.** Crease down back; difficult to feel ribs; fat around tailhead very soft; area along withers filled with fat; area behind shoulder filled with fat; noticeable thickening of neck; fat deposited along inner thighs.
9	**Extremely fat.** Obvious crease down back; patchy fat appearing over ribs; bulging fat around tailhead, along withers, behind shoulders, and along neck; fat along inner thighs may cause them to rub together; flank filled with fat.

Figure 14-1   A healthy horse is a pleasure to watch. *(Photo courtesy of Michael Dzaman)*

Hair Coat.   A shiny, glossy hair coat is one of the best indicators of a healthy horse. Hair coat is reflective of good nutrition and health and certainly can be improved by regular grooming.

Hoof Growth.   Normal, healthy horses have healthy hoof wall tissue. The wall should grow at a rate of ¼ to ½ inch per month. The hoof should be smooth and uncracked and it should form a straight line with the front of the pastern when viewed from the side.

Eyes.   The eyes should be bright, fully open, and clear, without discharge or a glazed, dull appearance.

Hydration.   The water balance of a horse is vital to its health. A skin fold test can be done by pinching a fold of skin on the neck, pulling it out, and recording the number of seconds the skin takes to return to its normal position. One-half to one second is normal; longer means the horse is dehydrated.

Manure/Urine.   Horses normally have firm manure balls that are not loose and watery and do not show undigested grains and other feed stuffs. Urine is normally wheat-straw colored and not cloudy or dark red in color.

Mucous Membranes.   The membranes of the horse's gums and lips should be a healthy pink color. Pale, white, yellow, or deep purple colors are all cause for concern.

Table 14-2 Signs of Health in the Horse	
Sign	Normal
Temperature	99.5° F to 101.5° F
Heart rate	32 to 48 beats per minute
Respiratory rate	8 to 16 breaths per minute
Mucous membranes	Pink color
Capillary refill time	1 to 2 seconds

Capillary Refill. The circulation of a horse can be assessed by gently pressing the thumb against the gums of the horse and counting the number of seconds it takes for the color to return to the area once the thumb is removed. One to two seconds is normal.

Heart Rate. The normal heart rate of an adult, resting horse is thirty-two to forty-eight beats per minute. This will vary with the age of the horse, ambient temperature and humidity, exercise, and excitement levels.

Respiratory Rate. The normal respiratory rate of an adult, resting horse is eight to sixteen breaths per minute. Exercise, ambient temperature, humidity, fever, distress, pain, and anxiety will increase the respiratory rate.

Temperature. The normal body temperature of a horse is 99.5° F to 101.5° F. High environmental temperature, exercise, or dehydration can increase this by two to three degrees (see Table 14-2).

Body Weight

Regardless of body type and breed, the nutritional requirements, medical dosages, and management practices vital to maintaining healthy horses are based on body weight.

This is one area where observation, no matter how close, is not sufficient. The only way to get an accurate, reliable weight for a horse is to weigh it. It is important to realize, however, that a horse's weight can vary by as much as sixty pounds, depending on how recently it has eaten, drunk, urinated, and defecated.

Overweight horses are more common than malnourished ones, but the health of the horse is compromised in either case. Health management begins with proper feeding to meet the horse's nutritional needs and keep it in top physical condition.

MANAGING HEALTH THROUGH PROPER NUTRITION

The nutritional requirements of horses depend on the level of their exercise, their reproductive state, their age and growth rate, and their mature body weight. These must be determined in order to feed horses properly. Determinations must be based on the actual use and condition of each horse, not on good intentions or wishful thinking on the part of the owner. Each horse is different and needs to be fed according to its individual temperament, metabolic rate, and genetic makeup (see Figure 14-2).

Feeding Classification

Nutritional requirements and proper feeding of horses were covered in Chapters 12 and 13. The basic feeding classifications to ensure good health include maintenance, pregnancy, lactation, growth, and work.

Maintenance. The horse in this feeding classification is mature, maintaining its body weight, and is not pregnant, lactating, breeding, or being exercised. The nutritional requirements are very low relative to other classifications.

Pregnancy. The nutritional requirements during the first eight months of pregnancy are the same as for a mare being maintained. However, during the

Figure 14-2 Sisu, a Finnish word meaning perseverance, was rescued from a negligent owner who did not feed and care for her properly. She had lost a tremendous amount of weight and muscle tone, and her health suffered. The patches on her back are the result of rainrot, which is slowly clearing up. Sisu is now being well cared for but it will be many months before she is back to normal weight and good health. This could have been avoided if she was simply provided the basic necessities of food, water, and shelter. *(Photo courtesy of Cathy Esperti)*

ninth, tenth, and eleventh months of pregnancy, the mare's requirements increase 11 percent, 13 percent, and 20 percent, respectively.

Lactation. During the first three months after foaling, mares can produce milk equivalent to 3 percent of their body weight every day and 2 percent per day during months four to six. Feeding requirements are about 70 percent above maintenance during the first three months of lactation and 50 percent during months four to six.

Growth. Growing foals require feeds of higher quality than those for maintenance. Growth rate and age of the foal determine the requirements. Horses are considered to be growing for up to thirty months, and in the slower-maturing breeds the period is longer. Optimum growth rates for various breeds have not been well defined, but overfeeding can cause developmental orthopedic diseases, and underfeeding can cause permanent stunting (see Figure 14-3).

Work. Horses being exercised or worked require more nutrients, especially energy, than horses being maintained. As the intensity or duration of the work increases from light to moderate to intense, the requirements for energy increase 25 percent, 50 percent, and 100 percent above maintenance, respectively.

The Essential Nutrients

Horses need the same basic nutrients that humans do: water, energy, protein, minerals, and vitamins. In horses, as in humans, too much or too little of any of these essentials can lead to serious health problems.

Figure 14-3 Growing foals require plenty of free exercise and feeds of higher quality than those for maintenance.

Water. Horses need a good supply of clean water daily. Water is essential for all body functions, including temperature regulation and feed digestion. The amount of water needed depends on the level of exercise, the ambient temperature, the quality of the feeds in the ration, and the proportion of the diet that is forage. A minimum of one gallon of water per 100 pounds of body weight per day should be provided.

Energy. This is the nutrient that primarily determines the weight and condition of the horse. Energy is derived primarily from carbohydrates, fats, and any protein excesses. Energy is needed for body functions, including maintenance, temperature regulation, digestion, and work.

All energy that is provided beyond what the body needs will go toward the formation of fat. A deficiency of energy will reduce the body condition score (fat) before the biological needs will be sacrificed. Prolonged deficiencies will result in unthriftiness and starvation. Excesses of energy can cause obesity, which can, in turn, cause many metabolic diseases such as laminitis (founder), osteochondrosis, epiphysitis, tying-up syndrome, and colic.

Protein. The body needs proteins for muscle and bone growth, milk production, fetal growth, and normal metabolism. The protein requirements of many horses, especially those in the maintenance, early pregnancy, and exercise classifications, can be met with good quality hay or pasture forage.

Minerals. The development of bone and many essential reactions within the body require adequate levels of calcium, phosphorous, sodium, chloride, selenium, and other minerals. Many of these minerals can be adequately supplied with good quality mixed hays or a mineral supplement in the concentrate mixture. Trace mineralized salt with selenium should always be available.

Vitamins. Both the fat-soluble vitamins (A, D, E, and K) and the water-soluble vitamins (B complex and possibly C) are required as coenzymes throughout the body for normal metabolism. Because forages are rich sources of vitamins, very few horses grazing good pastures or fed predominantly good hay are likely to need supplementation. High stress situations may make vitamin B complex supplementation useful.

HORSE HEALTH PROGRAM

Keeping a horse healthy requires diligence to details, but prevention is always better than treatment.

General Program

Following are a few minimum guidelines essential for normal horse care. Horses vary a great deal in their metabolism and must be managed as individuals in regard to maintaining health. For more specific guidance and when questions arise or problems develop, help should be sought from a veterinarian, extension agent, farrier, feed store operator, or some other equine professional.

- Shelter from wind and weather with trees, a shed, or barn is adequate in most climates. Much of the tradition about horse housing is for the comfort of the owner, not the horse.

- To reduce chance of injury, a safe environment that is free of hazards such as nails, barbed wire, broken fences, glass windows, and unsecured pesticides, should be provided.

- Adequate clean water should be provided at least three times a day; free access to water is best.

- A routine schedule of feeding and exercise should be maintained. Sudden changes in feeds, feeding schedule, or work/activity can cause lameness, colic, and muscle problems. Regular exercise, either free choice or regulated, is important in maintaining athletic horses.

- Horses are natural nibblers. They can be fed once a day but will be more efficient (digest more of their feed) if fed two or three times daily.

- Horses should be fed at least 1.5 to 2.5 percent of their body weight per day in hay or pasture. Hay (forage) helps prevent intestinal problems and abnormal behavior (vices) caused by lack of fiber and boredom.

- Commercial concentrate feed mixtures should be used if necessary to supply the nutrients needed. The concentrate should be selected to complement the hay or pasture composition. Feed should be fed on the basis of weight not volume.

- Hays and feeds should be free of dust and mold.

- When necessary, feeds should be changed gradually over a ten- to fourteen-day period.

- Floating (filing) horses' molars to decrease the sharp points that interfere with normal chewing is often needed. Regular dental checkups and floating will prevent mouth problems.

- Horses should be dewormed regularly, and parasite load should be assessed with occasional fecal floatation tests and treated accordingly.

- All horses should be on a regular vaccination schedule that includes tetanus at least.

- Regular hoof care is important. Feet may need to be trimmed if hoof growth exceeds wear, but it is not usually necessary to keep shoes on horses that are not in training or being ridden or driven on rough terrains.

IMMUNITY

The **immune system** forms the body's defense against a foreign substance, whether microorganisms (bacteria, fungi, viruses, protozoa, and parasites), a potentially toxic material (foreign protein, carbohydrate, or nucleic acid), or an abnormal cell (one invaded by a virus or that has become malignant). It attacks the foreign substance and maintains a memory of the invader so that a second exposure will provoke a greater, faster response.

Immunity refers to the ability of an animal that has recovered from a disease to remain well after a second exposure to the same disease. Immunology is the branch of medicine concerned with the body's response to foreign substances and abnormal cells; immunization is the ability to create a response to fight an illness without exposing the body to that illness.

Under normal circumstances the immune system responds to foreign organisms by producing **antibodies** and stimulating specialized cells that destroy the organisms or neutralize their toxic products. The immune system monitors the cells of the body constantly to ensure that they are not abnormal. Cells infected with viruses or cells from another animal (even of the same species) have protein markers on their outer membranes that signal the immune system to destroy them. The immune system also has the ability to recognize and eliminate malignant or abnormal cells within the body. These mutant, or cancer, cells may occur spontaneously or they may be induced by certain viruses or chemicals (**mutagens**).

Cells and Tissues in the Immune Response

The immune system has two general responses: it activates cells to destroy a harmful cell with cell-to-cell interaction, or it activates other cells to produce large protein molecules called **antibodies** that bind to bacteria, yeast, some viruses, and even **toxins** (poison) and make them harmless. In many cases, both responses occur. To activate the immune cells, large cells called macrophages eat and partially digest the invading material and place pieces of it, called the **antigen**, on its surface (see Figure 14-4). This attracts and activates T cells from the thymus.

Since the T cell and the antigen must fit together in order to bind, there are many different T cells to match the many antigens. If the T cells are what are called the helper type, they will attract B cells, formed in the bone marrow, whose receptors must also match the antigen. The combination of antigen prepared by the macrophage and T helper cell will activate the B cell specific for that antigen, which multiplies to form a clone. The clone then begins to make antibodies. Other clonal cells form the immune memory for this antigen by remaining indefinitely in an alert state, ready to multiply again should the antigen appear in the future.

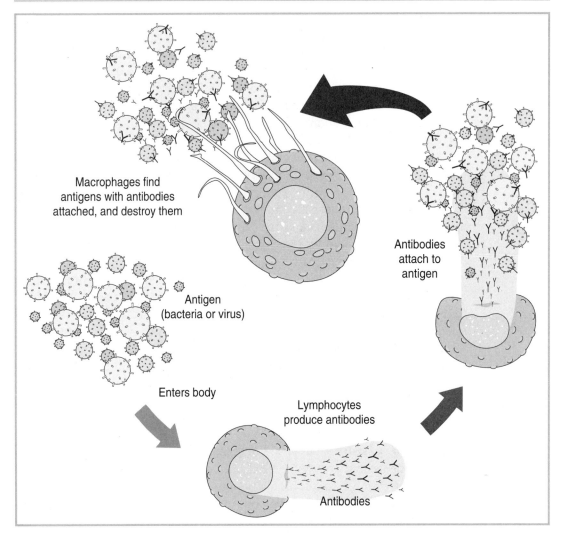

Macrophages find
antigens with antibodies
attached, and destroy them

Antigen
(bacteria or virus)

Enters body

Antibodies
attach to
antigen

Lymphocytes
produce antibodies

Antibodies

Figure 14-4 The formation of antibodies—the immune response

Antibodies control not only bacteria but also viruses, fungi, yeast, parasites, protozoa, and many toxic chemicals. In addition to T helper cells, there are T killer cells that can recognize a body cell that has been invaded by a virus and kill that cell so the virus cannot multiply.

Macrophages found throughout the tissues of the body and in the form of monocytes make up about 3 percent of white blood cells. Lymphocytes, two-thirds of which are T cells, and one-third B cells, constitute 30 to 40 percent of white cells of the blood. T cells are also found in the thymus gland and in lymph nodes. B cells make up the cells of the outer portion of lymph nodes.

Antigens

An antigen is a substance that, when introduced into an organism, induces an immune response consisting of the production of a circulating antibody. This type of immunity is known as humoral immunity. Protein molecules are potent antigens. Within a few days after injection, an antigen summons large amounts of the antibody capable of interacting with it. The interaction of an antigen with its specific antibody does not involve the entire antigen but only small areas on its surface.

Antibodies

The molecules responsible for recognizing antigens on foreign molecules on cell surfaces are called antibodies. Antibodies are members of a related group of gamma globulin molecules known as **immunoglobulins** (Ig). A typical immunoglobulin is made of four protein chains joined together in two pairs.

Five classes of immunoglobulins exist, based on structural differences. These differences are identified by the Greek letters gamma, mu, alpha, delta, and epsilon, and the immunoglobulins that contain them are called IgG, IgM, IgA, IgD, and IgE, respectively. Each class has different biological and structural properties and is distributed throughout the body.

IgG, the most abundant immunoglobulin, occurs primarily in blood serum, as well as throughout the internal body fluids. Produced in response to bacteria, viruses, and fungi that have gained access to the body, IgG is a major line of defense against such organisms.

Antibodies in Defense

The simplest and most prevalent means by which the immune system defends the body against bacteria and viruses is by the combination of a specific antibody with the antigens located on the surface of invading organisms. An aggregate of cells, called an agglutination, is formed by antibodies bound by one of their two combining sites to one cell, and to another cell by their other site. These aggregates are then engulfed and digested by the body's wandering scavenger cells, the macrophages. Antibodies also bind to toxic molecules (toxins) given off by microorganisms, forming large, insoluble aggregates (precipitates) that are also removed by macrophages. Antibodies also cover up the attachment sites of viruses preventing their ability to infect cells. Precipitin and **agglutination** reactions are used as diagnostic tools for identifying and quantifying the antibodies of infectious organisms in blood samples and other body fluids.

The Newborn Foal

Because protective antibodies are too large to pass through the mare's thick placenta and into the foal's bloodstream, antibodies are not provided to the foal by the mare during gestation. Immunity for the foal is available only through the antibodies in her first milk, the colostrum. The colostrum, however, is available for only forty-eight hours, after which it is replaced by normal milk, which has no immunity value. Ingestion of colostrum is the most critical factor influencing the foal's survival, growth, and future health.

The foal's ability to absorb antibodies from the colostrum through the lining of the duodenum of the intestine is based on the presence of special absorption cells in the small intestine. These special cells decrease rapidly over the first twenty-four hours of the foal's life. After forty-eight hours they are replaced with normal duodenal lining; a foal can only absorb antibodies during its first two days and absorbs most of them within its first twelve hours of life. The foal must receive colostrum during the critical first twelve hours of life to receive the antibodies it will need for immunity in the following three to four months.

The process by which antibodies are passed from mare to foal is called **passive transfer**. Occasionally a foal exhibits a failure of passive transfer; even though the antibodies are readily available in the colostrum, the foal is unable to absorb and use them. An immunoglobulin, IgG, test can be performed to identify failure of passive transfer of blood. An immunoglobulin level of less than 600 to 800 mg./dl. (milligrams per deciliter) indicates a problem. Blood IgG levels below 400 mg./dl. indicates a failure of passive transfer, and a veterinarian should be consulted.

Antibodies against certain diseases will be present in the mare's colostrum if she is vaccinated with booster injections thirty days before parturition. The mare should be vaccinated against all common diseases, because the colostrum can contain antibodies against only those diseases to which she has been exposed. Tetanus is one disease to which horses are extremely susceptible. If the mare has not been given a tetanus toxoid thirty days prior to parturition, then the foal should be given 1,500 I.U. of tetanus antitoxin at birth in order to protect it from the tetanus pathogen, which is ever present in equine feces.

No antibiotics or probiotics, such as special vitamin formulations, should be administered the first day of life in an effort to bolster the foal's vigor. These tend only to increase illness and diarrhea in the newborn foal.

Occasionally a mare does not produce milk in adequate quantities, produces poor-quality colostrum, leaks her colostrum, or dies. A source of frozen colostrum taken from other mares will provide the foal with the immune protection it needs. Milking six to eight ounces of colostrum from all mares after their foals have suckled will provide a reserve that can be frozen. This resource should be on hand even if no problems are expected. When used, this colostrum should be slowly heated, not microwaved, to body temperature. Microwaving destroys the protein molecules that make the antibodies protective. The best colostrum sources will be from mares on the same farm as the foal in order to provide the most specific immunity. Nurse mares can be used as a source of milk.

Plasma transfusions can be used to raise the IgG levels with foals that are more than twenty-four hours old. Generally one liter of plasma is administered over a thirty- to sixty-minute period, depending on the foal's vigor. The best plasma source is from a horse living on the same farm as the foal. Commercial plasma sources are available, but the antibodies in this plasma may not be specific enough to give the foal adequate protection against the particular disease organisms in its immediate environment. Typically, a plasma transfusion of one liter raises the IgG level only 200 mg./dl. More than one transfusion may be needed for foals with very low IgG levels.

Premature Foals

Foals that arrive prematurely (before 320 days of gestation) will be especially susceptible to risks from the environment and disease. Since their bodies are physiologically underdeveloped, they are unprepared, even with sufficient colostrum, to provide adequate resistance to disease. The lungs of premature foals are especially underdeveloped. They tend to have low body temperatures, are especially susceptible to colds, infections, and hypothermia, and require a lot of intensive care if they are to survive.

VACCINATIONS

Horses establish an immunity to a specific disease first by being exposed and then by developing their own antibodies to fight off that specific disease. The foal is born with no immunity (antibodies of its own) and the protection it first receives must come in the form of colostrum from the dam. This is termed **passive immunity**—when an animal receives antibodies that were produced by another animal. **Active immunity** is when an animal is challenged and stimulated to produce its own antibodies. This challenge is usually a disease that is present and introduces a foreign protein or antigen into the animal. If the horse has a high level of antibodies in its blood, the antibody **titer** (level) is said to be high for that specific antibody or disease (see Figure 14-5).

Vaccinations are given to do one of two things:

1. Give the animal antibodies that were produced by another animal (passive immunity)

2. Challenge the horse with just enough antigen that it will build its own antibodies (active immunity)

In both cases, the object is to protect the vaccinated animal against a specific disease.

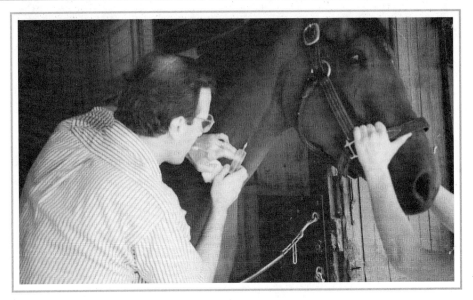

Figure 14-5 A horse receiving a vaccination

Vaccines contain either inactivated killed organisms or modified live organisms. The ideal vaccine:

- Prevents clinical signs of the disease
- Stimulates the immune response
- Produces durable immunity with a single dose
- Is safe with no side-effects
- Is incapable of producing the disease
- Is stable during movement and storage
- Is economical

The efficacy of any vaccine can be influenced by the type of vaccine, the site of action, the normal antigenic variation, and the age of the horse at the time of vaccination. Various stresses also reduce the ability of an animal to respond to a vaccination.

Table 14-3 gives examples of vaccination programs and schedules for some common equine diseases.

CAUSES OF DISEASE

Disease is any condition of a horse that impairs normal physiological functions. Disease increases costs, reduces performance, and can limit growth in the young horse.

Table 14-3 A Suggested Vaccination Program		
Disease	**Initial Vaccine Series**	**Booster**
Tetanus	2 injections; 1 month apart	Annually or at time of injury
Influenza	2 injections; 2 months apart	Annually or every 3 to 4 months for horses at risk
Rabies	2 injections; 3 and 12 months of age	Annually
Eastern, Western Venezuelan Encephalomyelitis	3 injections; 3, 4, and 6 months of age	Annually, before onset of insect season
Rhinopneumontis (Equine Herpesvirus Type I and II)	2 injections; 3 and 6 months of age	Mares in fifth, seventh, and ninth months of pregnancy; every 3 to 6 months for horses in areas of risk
Botulism	3 injections; 1 month apart	Annually for brood mares; 1 month prior to foaling
Equine Viral Arteritis	1 injection	Mares and stallions; 3 weeks prior to breeding; inadvisable in late gestation
Strangles	2 to 3 injections; 1 month apart	Annually or prior to exposure
Potomac Horse Fever	2 injections; 1 month apart	Annual booster in May/June in high-risk areas
Anthrax	2 injections; 1 month apart	Annually to horses at risk of exposure

Source: University of Kentucky, Ag Sat Corporation, *Art and Science of Equine Production.*

Two broad categories of disease affect horses: **infectious** and **noninfectious** diseases. Infectious diseases are caused by pathogenic organisms present in the environment or carried by other animals. In contrast, noninfectious diseases are caused by environmental problems, nutritional deficiencies, or genetic defects. Noninfectious diseases are not contagious and usually cannot be cured by medications. Noninfectious diseases are often a management problem.

Infectious Diseases

Infectious diseases are broadly categorized as parasitic, bacterial, and viral. Both internal and external parasites can affect horses (see Figure 14-6). Parasitic diseases are discussed in Chapter 15.

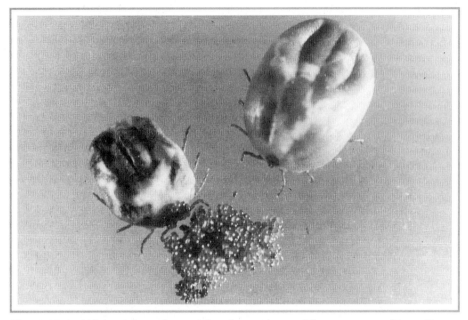

Figure 14-6 Winter ticks and eggs taken from a horse. *(Photo courtesy of Iowa State University)*

Bacterial diseases are often internal infections and require treatment with antibiotics. Bacterial diseases can also be external, resulting in erosion of the skin and ulceration.

Viral diseases are impossible to distinguish from bacterial diseases without special laboratory tests. They are difficult to diagnose and there are no specific medications available to cure viral infections. Immunization can protect horses from some viral diseases, but vaccines do not exist for all the viruses that cause disease in horses.

Noninfectious Diseases

Noninfectious diseases can be broadly categorized as environmental, nutritional, or genetic. Many nutritional diseases are caused by the lack of or excess of a nutrient. Many of these are described in Chapter 12. Environmental diseases include natural or man-made toxins in the environment. Genetic diseases are covered in Table 10-4 in Chapter 10.

DISEASES OF HORSES

For the purpose of discussion, following are some of the more common and/or serious diseases of horses. They are categorized as respiratory diseases, digestive diseases, and other common diseases of horses.

Respiratory Diseases

The principal use for most horses depends on their athletic ability, which requires physical soundness. Any disease that affects the respiratory system may potentially interfere with the horse's soundness of wind and thus its overall athletic ability.

A number of disease conditions may interfere with respiratory health. Respiratory disease can affect horses of any age, and chronic problems may seriously reduce the usefulness of horses at maturity.

Respiratory diseases may be due to bacterial or viral agents, anatomical problems, allergic responses, or a combination of these.

Sinusitis. In this condition, the sinuses of the horse's head become inflamed. Causes include infectious agents, structural problems, and tumors.

Sinusitis is rarely a contagious disease unless it is caused by a specific virus or bacterium. The person who cares for the horse may notice a discharge from one or both nostrils. This may be quite thick and tinged with blood, and frequently has a strong, unpleasant odor. Occasionally, the side of the horse's face may appear swollen.

Veterinarians diagnose sinusitis by physical examination, cultures of the discharges, X-ray examination, and sometimes by surgery accompanied by biopsy or the removal of a small piece of the involved tissue.

Prevention includes periodic examinations for abnormalities, an active immunization program to prevent infectious diseases, and periodic dental exams to prevent tooth disease from affecting the sinuses.

Palate Elongation. In this structural abnormality, the soft portion of the roof of the mouth extends too far back into the upper throat or pharynx where it may interfere with breathing during strenuous exercise. Veterinarians diagnose the condition by examining the horse's throat with an endoscope, a flexible instrument with lights that allows them to look into body cavities or spaces.

Horses cannot breathe through their mouths due to the physical design of the pharynx. Elongation of the soft palate reduces normal airflow to the point that a horse becomes unable to tolerate hard exercise.

This is a developmental problem of the horse's anatomy. Treatment consists of surgically removing the excess portion of the palate.

Bleeders. EIPH or Exercise Induced Pulmonary Hemorrhage (bleeding) is a serious condition in equine athletes. Racehorses are most frequently affected. The more strenuous the exercise, the more frequently bleeding occurs.

The hemorrhage occurs in minute vessels in the lungs. Although seldom fatal, bleeding can interfere with the horse's breathing and result in what appears to be choking or difficulty obtaining air.

Since blood does not always appear at the nostrils, owners may not be aware that bleeding is occurring. In some horses, blood flows from the nostrils when they lower their heads.

Heaves. Allergic equine respiratory disease is primarily seen clinically as a condition called heaves by horse owners or pulmonary emphysema by veterinarians. It resembles asthma or emphysema in humans. Horses suffering from heaves exhibit reduced tolerance to exercise, a frequent soft cough, a distinct push with the abdominal muscles when air is being expelled, and a crackling or squeaking sound over the lung fields that can be heard with a stethoscope.

The condition frequently follows a bout of respiratory disease accompanied by severe coughing. It sometimes appears suddenly with severe respiratory distress in bronchial asthma-like attacks. The condition usually is progressive. Frequently it is associated with feeding of roughage that contains a high content of dust, pollen, or mold spores.

Aside from avoiding feeds that cause the condition, preventive steps include a vigorous vaccination program against infectious respiratory disease. Prompt treatment and adequate rest are required until the horse recovers. Treatment is a combination of several procedures. Nonallergenic, dust-free feeds and as much green pasture turnout as possible is needed. This may be combined with antihistamines, bronchial dilators, and/or regularly decreasing doses of corticosteroids. Sometimes atropine or atropine-like drugs help in acute cases. Some degree of follow-up nursing care will be needed for the remaining life of the horse.

Rhinopneumonitis. Equine herpesvirus types 1 and 4 (EHV-1 and EHV-4) cause rhinopneumonitis. In susceptible horses, especially foals, it is an acute upper respiratory infection with severe nasal discharge, particularly in foals. Direct contact spreads the virus between horses.

After exposure, susceptible horses may develop a temperature of 102° F to 107° F for up to a week. Other signs include depression, loss of appetite, a watery nasal discharge, and a mild cough. Immunized horses usually develop a milder infection or even an infection without any signs. Occasionally, the virus may enter the central nervous system and cause mild to severe incoordination that can progress to total paralysis.

Pregnant mares without immunity to EHV may abort. Abortions usually occur from the eighth through the eleventh month of pregnancy. Occasionally, weak foals are born but die shortly after birth. Mares abort without warning and breed back without difficulty. Although some specific signs may occur in aborted foals, laboratory diagnosis is best.

Prevention by immunity from either natural exposure or vaccination is relatively short lived. Repeated exposure or repeated immunizations enhance protection.

Vaccination programs vary, but generally include vaccinating with either a modified live virus vaccine at two-to-three month intervals year round or using the killed product at the fifth, seventh, and ninth month of pregnancy. No effective treatment is known at present. Vaccinating foals at 90 days of age and repeating at 120 days of age reduces the clinical signs in foals.

Horse farms must isolate pregnant mares and foals from contact with temporary stock. Show and racehorses returning to the farm or mares visiting for breeding are sources of infection. No other animals or humans are known to be affected by this virus.

Influenza. Influenza is an acute, highly contagious disease that causes a high fever and persistent cough. Flu is caused by at least two distinct myxoviruses that are widely spread throughout the horse population. Exposure occurs at shows, sales, races, trail rides, and other events where horses come together from different areas.

Following exposure to nasal discharges containing influenza virus, the susceptible horse develops a temperature of up to 107° F within three to five days. The fever may persist for up to three days. A hard persistent cough develops early and persists for up to two to three weeks.

Secondary bacterial infections sometimes develop as a complication. Muscle soreness and stiffness are occasionally seen.

Good nursing care is the best treatment. This includes providing a soft, palatable, dust-free diet and fresh, clean water, while preventing drafts. Bandaging of legs and blanketing may help.

Horses should not return to work of any kind for ten to fourteen days after complete cessation of all flu symptoms. If the horse returns to work too soon, recurring bouts or other complications usually result.

Use of antibiotics or fever-controlling drugs are seldom needed and should only be given following a veterinarian's advice. Serious complications occur when a horse's symptoms are masked by medication, since the horse returns to work too soon.

A vaccine containing two inactivated viruses provides some prevention. An initial vaccination followed by a booster in four to six weeks is recommended. Followup boosters are given at three- to six-month intervals. The more likely the horses are to be exposed, the more frequent the boosters are recommended.

Pinkeye. Viral arteritis, or pinkeye, is a separate viral disease that can cause respiratory symptoms along with swelling of the legs and abortion in pregnant mares. This usually is a sporadic disease spread by contact with infective nasal discharges.

Few horses die unless complications occur. Young or very old horses are most severely affected. Up to 80 percent of infected pregnant mares may abort. Severity apparently varies greatly between outbreaks.

Symptoms occur one to eight days following exposure. Swelling and redness appear around the eyes with flowing of tears and squinting. Horses are dull and go off their feed. Eyelids, legs, and the underside of the body become swollen. Some yellowing or jaundice may be noticed. Pregnant mares abort during or shortly after the fever occurs.

Complications can include fatal lung disease due to accumulation of fluid. This is especially dangerous if the horse has already had a lung disease.

Veterinarians diagnose viral arteritis by clinical signs and differentiating it from several other viral diseases.

Treatment in general consists of very careful nursing care. The specific treatment depends on the individual case. Vaccines are available at this time. Prevention is best accomplished through good sanitation and isolation.

Strangles. This is a highly contagious abscess-producing infection caused by a specific streptococcus bacterium, *Streptococcus equi*. Pus from ruptured abscesses can contaminate the environment including mangers, fences, and water tanks. This material remains infective to other horses for months and is a major means of prolonging an outbreak of the disease on a farm.

Young horses are most frequently affected, but susceptible horses of any age may develop the disease following exposure. Stress created by hauling, weaning, weather changes, hard work, and poor nutrition all weaken a horse's defense. Outbreaks tend to be prolonged and may follow introduction of an apparently healthy carrier.

Only horses, mules, ponies and related equidae (zebras, and others) are susceptible. Other domestic animals and humans do not develop the disease.

Three to six days following exposure, infected horses stop eating. They may extend their heads and drool. Swallowing is painful. A temperature of 106° F is not unusual.

Swelling between the jaws and near the base of the ear may occur. These swellings enlarge, become soft, burst, and drain a creamy, blood-tinged pus within five to seven days.

Once the abscesses drain, the temperature frequently returns to normal or near normal. During the period when the temperature is high, the horse appears very depressed and may lose a considerable amount of weight.

The entire course of the disease in an uncomplicated case may take four to six weeks. Complicated cases may be prolonged for months, with failure of lesions to heal, severe weight loss, and extensive abscess formation throughout the animal's body.

If new horses are added regularly to a herd, the disease can reoccur. The infection rate is high, with over 80 percent of exposed horses showing signs of the disease. Death losses usually are low if no complications occur.

Death may occur either from abscesses rupturing within internal organs, or choking due to large abscesses blocking the horse's ability to breathe or breaking into the horse's airway, resulting in suffocation.

After the apparent recovery of an affected horse, considerable rehabilitation time is necessary to avoid a flare-up of infections in the throat (guttural pouches) and development of internal abscesses, swelling of the legs, head, or abdomen, and joint infections, especially with foals and yearlings.

Veterinarians diagnose strangles by its clinical appearance plus culture and identification of the organism *Streptococcus equi*. Other streptococci can cause a similar but milder disease that resembles strangles.

Commercially available vaccines prevent strangles. Two or three injections at one month intervals are given deep in heavy muscles. Boosters are given annually. Vaccination of horses during an outbreak or while recovering from the disease can be disastrous. Only healthy, uninfected, unexposed horses should be vaccinated.

Treatment is best determined by a veterinarian. Hot packs on forming abscesses, cleaning of abscesses, and providing easily eaten feed and fresh, clean water are essential. Isolating infected animals in one spot and leaving them there until completely healed prevents widespread contamination from the contents of abscesses.

Streptococcus equi is highly sensitive to penicillin, but vigorous and prolonged treatment is necessary. Recommended doses of penicillin should be administered daily, and for several days after the horse's temperature returns to normal. This usually requires treatment for a minimum of ten to fourteen days.

Discontinuing treatment too soon, reducing dosage, skipping days of treatment, or using inappropriate drugs or dosages are the most frequent causes of treatment failure.

Pleuropneumonia.

This disease condition usually is the aftereffect of an earlier respiratory problem and can be caused by a number of different organisms. Inflammation develops in the tissues that line the chest cavity and surround the lungs. This space fills with fluid, debris, and infective bacteria.

Pleuropneumonia results from incomplete or improper treatment of a previous lung disease. In most cases, corticosteroids or other anti-inflammatory drugs were used. Occasionally, inadequate dosages of antibiotics or sulfa drugs were given or treatment was not carried out for a long enough period of time.

Pleuropneumonia is usually considered a contagious disease. The condition is painful, and the horse is reluctant to move. Pressure over the chest area causes discomfort, and some horses act as though they have colic or abdominal pain.

The veterinarian diagnoses the disease by listening to the horse's chest with a stethoscope for abnormal lung sounds, by observing the horse's painful attitude, by noting sounds of increased density on percussion, and finally, by performing a chest tap—drawing fluid for examination.

Avoiding respiratory disease in horses is the best prevention. This includes providing a clean, healthy environment, a complete immunization and internal parasite control program, and avoiding undue stress. If respiratory disease occurs, prompt, appropriate treatment is necessary for the indicated length of time.

Treatment may consist of drainage by tubes surgically implanted in the chest, specific antibiotic treatment, and good nursing care.

Abscess Pneumonia. This is a serious disease of foals. The bacterium *Rhodococcus equi* causes heavily encapsulated abscesses to develop in the lung where they displace lung tissue. In severe cases, extensive destruction of lung tissue occurs. The bacterium seems to be very irritating to tissue and causes severe tissue reaction.

Symptoms seldom are evident until the disease has progressed to a critical stage. In the early stages, only a dry persistent cough is present.

As the disease progresses, the foal develops severe breathing difficulty. Any exercise worsens the condition. Rectal temperatures generally are in the 102° F to 104° F range.

Death is usually due to asphyxiation as lung tissue is destroyed. Foals that recover may show no after-effects as adults, despite the severity of symptoms during the disease's course.

Veterinarians diagnose the disease in foals by the symptoms, cultures of the trachea (windpipe), and age of the foal (usually four to eight weeks when the disease is first noticed).

Abscess pneumonia appears to be increasing in frequency, and several colleges of veterinary medicine are studying the disease.

At present no effective commercial vaccine is available for use in horses. Prevention is best accomplished by superior management, especially internal parasite control, dust control, avoiding overcrowding, increased use of pasture, and minimizing enclosed housing.

Treatment with antibiotics should be vigorous and prolonged. Foals that respond must be treated for extensive periods of time—twice daily for up to six weeks is not an uncommon treatment schedule.

Hot, dry, dusty conditions tend to result in an increase in this condition. The disease is not felt to be highly contagious, as large numbers of foals in a group usually are not affected.

When respiratory disease complications occur, problems can become more serious or even fatal. These complications occur due to a number of causes. For example, some drugs cause horses to appear better, but they suppress the horses' own defenses. The infection spreads and the condition worsens. Then secondary bacterial infections follow. Cultures and identification are necessary to treat these conditions specifically.

Adequate time for complete recuperation is absolutely essential. Green grass, fresh air, and being outside are all beneficial to recovery.

Digestive Diseases

Digestive diseases are a frequent problem of horses; insurance companies cite colic as the most frequent cause of death in horses insured against loss. The following is a discussion of some of the more frequently seen causes of digestive disturbance.

Problems in the mouth—the beginning of the digestive tract—can result in improper chewing of food which can then hamper swallowing and digestion. Malformed mouth parts on newborn foals should always be evaluated. Some abnormalities may be repaired surgically, while others may be impossible to correct. Because many of these conditions are considered heritable, use of these animals as breeding stock should be discouraged.

Wry muzzle, cleft palate, and overshot or undershot jaws are all conditions affecting horses. Their importance depends on their severity. These problems not only affect eating, but can seriously interfere with the use of bits for control when the horse becomes old enough to train for riding or driving.

Since dental disease may occur at any age, horses should have their mouths examined twice yearly for dental abnormalities. Although caries, or cavities, are uncommon, abnormal or uneven wear is frequently observed and needs correction.

Young horses lose their baby teeth from two to five years of age. At times, these are not shed normally, and it becomes necessary to assist in their removal. When sharp points develop along edges of the grinding or jaw teeth, they irritate the inside of the lips and edges of the tongue. These sharp points are filed down with specially designed files called "floats."

Horse owners can recognize possible dental problems when horses begin chewing abnormally, twist their heads sideways, drop excessive feed from their mouths, or refuse to eat hard grains or pellets.

Foreign bodies in the mouth cause similar problems. Grass awns, pieces of wood or metal, corncobs, and other items may lodge in the mouth and interfere with eating.

Proper feeding practices, such as using only clean feed boxes and avoiding hay with foxtail or similar type awns (bristles) in it, can prevent foreign bodies in the mouth.

Choke. Choke is a condition in horses that occurs when feed becomes lodged in the esophagus. While choke seldom is life threatening, it is uncomfortable to the horse. Many horses become excited and lunge about trying to dislodge the material causing the choke. In the process, they may injure the handler or themselves.

Choke usually occurs when horses attempt to eat too fast, or are fed very finely ground or very dry feed. Grass clippings from lawns also can cause choke.

A horse with choke should be placed in a stall and allowed free movement of its head. Veterinarians usually attempt to remove a choke with a naso-gastric tube and lavage, or flushing with water. Sedation may be necessary.

Owners should not attempt to dislodge a choke themselves as injury to the esophagus or lungs may occur.

Since horses that choke are prone to do it repeatedly, such horses should be denied access to the type of feed or circumstances that may cause choking to occur.

Use of large, flat-bottomed feed troughs or placing large rocks (softball size) in the grain box will slow down gluttonous eaters. Extremely dry or finely ground grain should not be fed. Adequate eating space also helps.

Colic. **Colic** is a broad term that describes a horse showing abdominal pain. This can be caused by a number of conditions.

When colic occurs, it is important to determine the exact cause if possible. Successful treatment often depends on a correct diagnosis.

A distended stomach, acute inflammation of the small intestine, parasites that cause a decrease in blood flow to the intestine, dry food impaction, or gas distention of the large bowel are all types of colic; the degree of severity as well as treatment required varies.

Any colic, no matter how mild, is an emergency. The potential for the condition to worsen is too great to risk delay in treatment.

Owners first notice that a horse has colic when it stops eating and drinking. The horse may curl its upper lip, paw at the ground, and turn its head toward either side. More severe pain causes colicky horses to sweat, to get up and down, and to attempt to roll. The horse with colic indicates it is in severe discomfort.

Rapid breathing, profuse sweating, violent activity, and a cold, clammy feeling may indicate the horse has gone into shock and is in need of immediate professional attention.

Veterinarians attempt to diagnose the specific type and cause of colic. They use medication to control pain and the horse's response to help evaluate the severity of the condition. Reducing the pressure in the stomach is important; oral medication may be needed to lubricate a mass or prevent further gas distention.

Administering oral medication is dangerous and should be done only with great care. Any foreign material, including medicine, that accidentally enters the lungs can cause pneumonia.

In some cases, surgery is the treatment of choice. This means moving a very hurting, sick horse to a veterinary hospital that has surgical facilities. The decision needs to be made as soon as possible and necessary supportive treatment must be provided until the horse arrives at the hospital.

The most frequent causes of colic are internal parasites and sudden drastic changes in the feeding schedule, either in the amount or kind of feed. Autopsies of many horses that die from colic reveal related lesions due to internal parasites. Diarrhea, especially in young horses, can result in colic due to telescoping of the bowel.

Any severe digestive upset has the potential to result in colic symptoms. Prevention of colic includes:

1. An ongoing parasite control program, especially for young animals

2. Maintaining a regular feeding schedule using only quality feeds

3. Avoiding sudden dietary changes in kinds or amounts of feeds

4. Providing salt and clean, fresh water free choice at all times

Treatment of this emergency condition is best left to professional veterinary care.

Potomac Fever. Acute infectious diarrhea syndrome, or Potomac fever, is a severe diarrhea condition of horses. Most affected animals are adults that may have recently been under stress. About 30 percent of horses with Potomac fever die.

Potomac fever seems to be caused by the bacterium, *Ehrlichia risticii*. It can occur sporadically, with only one horse on a farm being affected. An arthropod carrier is suspected but no particular species has been identified.

The majority of cases have occurred in Maryland, Virginia, and southeastern Pennsylvania, but similar cases have been reported in other areas of the country.

Infected horses become depressed, stop eating, and develop a profuse watery diarrhea. Some horses will have a fever of up to 105° F before the diarrhea starts. With continued diarrhea, the affected horse becomes weaker and develops signs of shock. The disease does not seem to be contagious and does not affect humans.

Treatment to replace fluids and control the diarrhea must begin as soon as possible. Large volumes of intravenous fluids and antidiarrheals are necessary. Time required for almost constant treatment becomes extensive and fairly expensive. Antibiotics are effective against *Ehrlichia risticii*.

Other Diarrheas. Other causes of acute diarrheal disease in horses include colitis, salmonellosis, and other diarrheal syndromes. Most of these are related to or follow stress, such as hauling, respiratory disease, or surgery. In some instances, such as salmonellosis, the diarrhea is contagious between horses and may infect humans.

Most of these diarrheas respond to vigorous treatment, although laminitis or founder frequently occur following a severe diarrheal episode. On occasion, a horse will recover and be left with chronically soft stools.

Specific treatments or preventions are not available at this time. A vaccine is available for salmonellosis.

Foal Heat Diarrhea. Most newborn foals develop diarrhea at seven to twelve days of age. At about the same time, the mare comes into what is called foal heat.

Affected foals usually show no problems due to the diarrhea, but occasionally they become ill or the diarrhea persists. Some may even develop serious intestinal problems and colic.

The cause is felt to be related to larvae of the intestinal threadworm, *Strongyloides westeri*. The immature larvae locate in the mare's udder and the foal becomes infected by nursing. Within eight to ten days, the parasites are established within the foal's intestinal tract where they irritate the gut wall, causing diarrhea.

Deworming pregnant mares during the last thirty days of pregnancy prevents the *S. westeri* larvae.

Affected foals may respond to intestinal protectants, appropriate deworming agents, and fluids, if needed. Oral antibiotics seldom are of much value.

A secondary problem is the scalding of the foal's rear quarters, with resultant burning, irritation, and hair loss. To prevent this, horse owners must clean the foal's rear parts and place some protective ointment on the area. Applying Vaseline or zinc oxide ointment to the foal's tail helps, as it becomes a self-made applicator.

Seemingly harmless diarrheal conditions can rapidly become critical. Horse owners need to keep a close watch on any animals with diarrhea. Proper treatment and aftercare are essential to minimize resultant problems.

Laminitis. **Laminitis** is defined as an inflammation of the lamina of the inner hoof wall. Laminitis and a process called founder are often thought to be the same problem. Laminitis is due to metabolic changes that affect the lamina. Founder is the mechanical displacement of the coffin bone within the foot. Founder is associated with laminitis but it is possible for a horse to have laminitis without founder.

Laminitis is difficult to prevent because there are many different causes, some of which may act together to cause this lameness. One cause of laminitis is colic, especially with grain overload. Eating too much grain results in a high production of lactic acid in the horse's intestinal tract. The lactic acid damages the gut wall and allows bacteria to enter the blood. This results in endotoxemia (the presence of toxins in the blood), which affects the lamina by decreasing the blood flow to the lamina. Colic can also cause laminitis by direct damage to the intestinal wall, such as with a torsion (twist of the intestine). The wall will die in that area and, again, allow bacteria to get into the blood.

Another cause is a **dystocia** or retained placenta in the mare. Stress, exhaustion, or infections that stress the horse for extended periods can also cause laminitis.

Leg problems are another major cause of laminitis. With excessive, repetitive stress and concussion put on the leg, the blood flow to the lamina can be decreased. This damages the lamina. Lameness causes a horse to shift its weight onto the good leg. This excess stress can also lead to laminitis.

The first signs of laminitis are often subtle and can be easily missed. However, the condition develops rapidly, and if not caught immediately, the horse can quickly become quite lame. Initial signs include restlessness or agitation—the horse will pace around the stall and shift its weight back and forth between its feet. Within the first day, the horse is often reluctant to turn and its gait will be stiff. In the first two days, the horse often assumes the classic laminitis stance, to shift weight onto its heels and off its toes, which hurt. Digital pulses will often be moderate to bounding by this time and there is usually a depression at the apex of the coronary band. By this stage, it is essential to get the horse under a veterinarian's care.

The next signs that develop include a bulging of the sole downward toward the ground so that it is not concave anymore. The coffin bone in the foot rotates (founder), which may be visible on a radiograph taken by the veterinarian. Within four to five days, a separation of the hoof wall from the skin at the level of the coronary band is possible. The coffin bone may have rotated enough by this time to have perforated the sole. This is a serious situation (see Figure 14-7).

As soon as signs of laminitis are noted in a horse, a veterinarian should be called. The horse is usually treated with drugs to decrease the inflammation. Special pads are put on the sole to support it. Early in the course of laminitis, special shoes can be put on the feet to elevate the heel and take pressure off the deep flexor tendon that is responsible for the rotation of the coffin bone.

Hot- and cold-water soaks are often used to increase circulation to the lamina and cool the feet. This also helps make the horse more comfortable. The stall should be deeply bedded to help cushion the feet. If a medical reason for the laminitis exists, for example, colic, treatment of that problem needs to be initiated.

A horse that has suffered only a mild case of laminitis may recover without complication and may be able to return to its normal level of exercise. The veterinarian will have to work with the owner or trainer to determine the amount of damage and monitor the horse's recovery.

Chronic laminitis can be detected by changes in the hoof wall. Regular disturbances of the blood flow to the lamina will result in changed growth rate of the hoof wall. The heel will grow faster than the toe and the growth rings

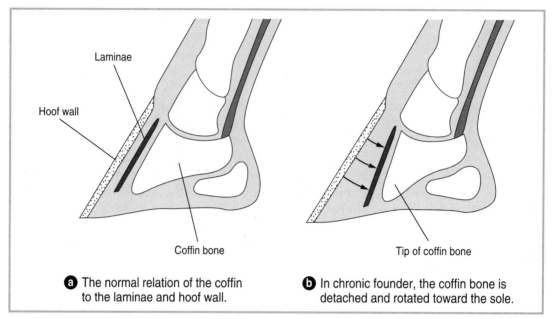

Laminae

Hoof wall

Coffin bone

Tip of coffin bone

ⓐ The normal relation of the coffin to the laminae and hoof wall.

ⓑ In chronic founder, the coffin bone is detached and rotated toward the sole.

Figure 14-7 A rotated coffin bone from laminitis

will be farther apart at the heel. These changes should be looked for when purchasing a horse.

Blister Beetles

Blister beetles, of the insect family Meloidae, defend themselves with a toxic secretion, cantharidin, that causes severe irritation to the skin and mucous membranes of warm-blooded animals. If whole or crushed parts of blister beetles are ingested by a grazing animal, the cantharidin can cause irritation and hemorrhages in the stomach. The amount of cantharidin produced varies from male to female and among species of blister beetles (see Figure 14-8).

Among domestic grazing animals, horses are most susceptible to this toxin. Only a few beetles, eaten with hay, can cause severe illness or even death to a horse. Affected horses exhibit signs of colic, frequently void small amounts of blood-tinged urine, and at times have muscle tremors. If blister beetle poisoning is suspected, a local veterinarian should be consulted immediately so treatment can be started.

Blister beetles commonly feed on alfalfa and the flowers of a number of plants that frequently grow in hay fields. The problem occurs when the hay harvesting process crushes or grinds up blister beetles and toxic parts remain in hay that is fed to horses.

Some steps can be taken to reduce the possibility of incorporating blister beetles in hay. First-cutting hay seldom contains blister beetles if it is cut in early to mid-June, before the adult beetles are present in alfalfa. Blister beetle

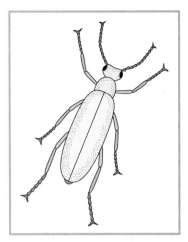

Figure 14-8 A blister beetle

poisonings have increased since the advent of swather-conditioning equipment that runs hay between rollers or crimpers. Hay conditioning equipment will kill many beetles as they pass through the rollers, contaminating several feet of windrow with crushed beetle parts. However, separate cutting followed by windrowing allows the beetles to find their way out of windrows while the hay is drying and prior to baling.

Other Common Diseases

Clostridial diseases as a group are caused by a family of bacteria that grow in the gut or tissue and produce gas and very powerful toxins that affect the nervous system.

Tetanus. Tetanus, a common disease of horses sometimes called lockjaw, is caused by a neurotoxin produced by *Clostridium tetani.* This nerve-tissue poison causes spasms and rigidity of the skeletal muscles. Affected horses cannot eat and have difficulty drinking. Over half of affected horses die due to suffocation, starvation, or dehydration.

Due to the large number of tetanus bacteria in the horse's digestive tract, people working around horses should consult their physicians concerning tetanus immunization for themselves.

Infected horses acquire the problem through puncture wounds or other deep wounds. Within ten to fourteen days following injury, horses become increasingly nervous, then stiff or rigid and as a result, have difficulty moving. The more rapid and severe the onset of symptoms, the less chance of recovery.

Persistent treatment and much nursing care are needed. Affected horses need to be protected from light and sound that can stimulate nervousness. Horses are placed in darkened stalls and their ears plugged with cotton to reduce stimuli from sound.

Veterinarians usually administer tetanus antitoxin, antibiotics, and sedatives repeatedly for several weeks. One-third to one-half of affected horses may recover if diagnosed early and treated vigorously. Prevention is twofold. Unvaccinated animals should receive tetanus antitoxin within twenty-four hours following injury or surgery. This provides temporary protection for ten to fourteen days. If healing is not complete at that time, tetanus antitoxin should be repeated at two-week intervals until healing is complete.

Vaccination with tetanus toxoid provides a very stable immunity. All horses should be vaccinated against tetanus and receive boosters on an annual basis and following an injury.

Botulism. Botulism is caused by *Clostridium botulinum* and can occur in adults as "forage poisoning" or in foals as the "Shaker Foal Syndrome."

With forage poisoning, adult horses become weak and stagger, have difficulty swallowing, and may go down and be partially or completely paralyzed. Silage, incompletely cured hay, or forage with spoiled areas are usually the cause. Prevention is best accomplished by careful selection of hay, silage, or other harvested forage for horses.

The Shaker Foal Syndrome appears as a problem in young foals nursing mares that are being fed high energy, high protein diets. Experimental vaccination of pregnant mares prior to foaling has prevented the condition, and a vaccine may become available commercially.

Care in nutritional management of mares nursing foals may control or prevent the condition from developing in foals. Affected foals become uncoordinated, develop jerky movements, and eventually become paralyzed and die. Once symptoms are apparent, treatment has little effect.

Clostridial Myositis. Infections of muscle masses by one or more of several clostridial bacteria families can occur. These organisms usually enter through wounds, needle injections, or other muscle injury.

The bacteria grow rapidly, form gas pockets, create severe pain, and cause shock from the toxins produced. Hand pressure causes both crackling sounds and sensation due to the gas formed under the skin. Often, cattle are or have been present when a problem occurs.

Veterinarians treat this condition by promptly establishing drainage and using adequate dosages of appropriate antibiotics. Despite vigorous treatment, some cases fail to respond.

Appropriate injection techniques, avoiding the use of irritating drugs, and not injecting excessive volumes at one site are all important in preventing clostridial myositis. Prevention of injuries and prompt attention to any wound are helpful in preventing this disease. No approved vaccines are available.

Sleeping Sickness. This disease is also called equine encephalomyelitis. Three forms of this disease are caused by viruses that affect the nervous system. Wild animals and birds act as reservoirs. Mosquitoes are the principal means of virus transmission between victims. Horses at pasture are more susceptible than stabled horses. The viruses also can affect humans. In the Eastern and Western forms of the disease, horses are last hosts and the virus does not spread from them. Venezuelan encephalomyelitis, however, spreads between horses and from horses to humans.

Infected horses initially develop a fever, act as though they have problems seeing, wander aimlessly, stagger, grind their teeth, and have a drooping lip. The disease may progress until paralysis occurs. Horses with mild cases recover slowly over several weeks.

From 25 percent to 50 percent of horses infected with the Western form may die, over 90 percent of those infected with the Eastern form die, and 75 percent die with the Venezuelan virus.

No specific treatment is available, but veterinarians can provide supportive care. Mosquito control is an important preventive measure, as is annual vaccination. The Venezuelan form has not been a problem in the United States since 1971, but could enter this country from Latin America.

Highly effective vaccines are commercially available and should be administered annually, before the mosquito season. They may be combined with other vaccines.

Swamp Fever. Equine Infectious Anemia (EIA), or swamp fever, is a viral disease. In the acute form, it causes severe red blood cell destruction resulting in anemia. Recovered animals are carriers. The virus is spread by bloodsucking insects and repeated use of needles or instruments without adequate sterilization between patients.

The disease causes severe anemia, fever, weakness, weight loss, edema, and sometimes death. Inapparent infections show few, if any, symptoms. Horses without visible signs of infections that receive regular hard physical work or some other stress frequently begin to show clinical signs of the disease.

Clinical diagnosis of EIA is by a positive antibody level test. Titers (levels) causing a positive test occur two to four weeks after exposure to the initial disease.

No effective treatment is available. Prevention is best accomplished by maintaining horses that test negative for the antibody. Fly control and use of disposable needles among horses are also important aspects of control. Any horse with a positive antibody test should be maintained away from uninfected horses, especially during the insect season.

Rabies. Rabies (Hydrophobia) is a universally fatal viral disease of the central nervous system. The virus is transmitted in saliva and infects humans, as well as other mammals. Wild animals, especially raccoons and skunks, appear to be important reservoirs of the disease.

When horses come in contact with rabid wild animals, their curiosity often results in their being bitten on the muzzle. Symptoms of rabies usually occur within two weeks following the bite.

A sudden change in behavior is the first indication of rabies. Drooling may or may not occur. After one to three days, horses may suddenly become vicious, attempting to bite without a reason. Some roll extensively, as if they had colic. The size and strength of horses makes them dangerous and potentially unmanageable. Self-mutilization is not uncommon.

Treatment is not considered effective, feasible, or safe for the humans involved. An animal suspected of having rabies should be confined for two weeks. If the horse is then destroyed, care should be taken not to damage the brain. A veterinarian will remove the head and prepare it for submission to a laboratory for examination at once to confirm the diagnosis of rabies.

Horses may be protected against rabies by vaccination with an approved product properly administered by a licensed veterinarian.

Vesicular Stomatitis.   Vesicular stomatitis occurs in the United States from time to time; some strains of the virus that causes it are foreign. Vesicular stomatitis gets its name from the appearance of blisters (vesicles) and raw ulcers in the mouth (stoma) of infected horses, swine, cattle, and humans. It is characterized by blisters on the tongue, teats, soles of the feet, and the coronary band.

Insects and the transporting of animals are probably responsible for the spread of the disease. The incubation period is about two to eight days, but may be longer.

No specific treatment is available. Animals should be protected from a secondary infection in the **lesions**. Prevention involves restricting the movement of animals during outbreaks, disinfecting trailers and stalls, and controlling insects. A vaccine is available.

POISONOUS PLANTS

Poisonous plants in the form of wildflowers or weeds may be found in both hays and pasture. Plants poisonous to horses can also include cultivated plants that were never intended for horse feed and trees near where horses are kept. Some of these common poisonous plants and their effect on the horse are listed in Table 14-4.

Poisonous plants are not harmful until a horse eats them. The best way to protect your horse from eating poisonous plants is good management.

- Provide sufficient high quality forage.

- Manage pasture to prevent overgrazing and to control weeds.

- Be knowledgeable and particular about the plants growing on your property.

- Take time to inspect the hay you feed for weeds.

Table 14-4 Common Poisonous Plants				
Common Name	**Scientific Name**	**Location[1]**	**Toxicity[2]**	**Signs**
Wildflowers and Weeds				
Blue flax	*Linum* spp.	Throughout North America	+++	Rapid labored breathing, frothing at mouth
Bracken fern	*Pteridium aquilinum*	Forested areas	+	Loss of flesh, lack of coordination, depression, paralysis
Castor-oil plant	*Ricinus communis*	Tropical areas	+++	No appetite, constipation and diarrhea, hard breathing, sweating
Death camas	*Zigadenus* spp.	North America	++	Stiff leggedness, hypersensitivity, weakness, convulsions
Fiddleneck	*Amsinckia intermedia*	Pacific Coast	++	Photosensitization, weight loss, anemia, jaundice
Foxglove	*Digitalis purpurea*	Western U.S.	+++	Heart irregularities, diarrhea, labored rapid breathing
Jimsonweed	*Datura stramonium*	North America	++	Colic, diarrhea, dilation of pupils, excitability and depression
Larkspur	*Delphinium* spp.	West, Midwest U.S.	+++	Hypersensitivity, trembling, collapse, convulsions
Locoweed	*Astragalus* spp.	West, Southwest North America	+	Strange behavior, incoordination, odd head carriage, weight loss

(continued on next page)

	Table 14-4	Common Poisonous Plants *(continued)*		
Common Name	**Scientific Name**	**Location**[1]	**Toxicity**[2]	**Signs**
Wildflowers and Weeds *(continued)*				
Milkweed	*Asclepias* spp.	North America	+++	Lack of coordination, depression, shallow breathing, unsteadiness, coma
Monkshood	*Aconitum* spp.	West, Midwest U.S.	+++	Hypersensitivity, trembling, collapse, convulsions
Nightshade	*Solanum* spp.	North America	++	Trembling, incoordination, diarrhea
Poison hemlock	*Conium maculatum*	North America	+++	Trembling, incoordination, salivation, colic, shallow breathing, coma
Pokeweed	*Phytolacca americana*	Eastern, Southern U.S.	++	Diarrhea
Ragwort/ groundsel/ hound's tongue	*Senecio* spp.	North America	++	Weakness, liver failure, yellow mucous membranes, incoordination
Sagebrush	*Artemisia* spp.	Western North America	++	Excitabiity, falling, front leg incoordination
Saint John's wort/Klamath weed	*Hypericum* spp.	Western U.S. plains areas	++	Photosensitization
Water hemlock	*Cicuta* spp.	North America	++++	Rapid respiration and heart rate, violent spasms, coma

(continued on next page)

Common Name	Scientific Name	Location[1]	Toxicity[2]	Signs
Table 14-4 Common Poisonous Plants *(continued)*				
Wildflowers and Weeds *(concluded)*				
White snakeroot/ Jimmy-weed	*Eupatorium* spp.	Eastern, Southern U.S. forested areas	+++	Sweating, stumbling, spread-legged stance, congestive heart failure
Yellow star thistle/Russian knapweed	*Centauria* spp.	Western U.S., Canada	+	Inability to swallow food, tongue lolling, smile expression
Cultivated Plants				
Alsike clover	*Tribolium hybridum*	Eastern, North central North America	+	Depression, colic, diarrhea, photosensitization
Avocado	*Persea americana*	Southern U.S.	++	Colic, diarrhea, noninfectious mastitis in lactating mares
Azalea/laurel/ rhododendron	*Rhododendron* spp.	North America	++	Diarrhea, colic, excessive salivation, depression, incoordiantion, stupor
Johnsongrass/ Sudan grass	*Sorghum* spp.	North America	+	Frequent urination/defecation, tremors, gasping, convulsions
Oleander	*Nerium oleander*	Southern U.S.	++++	Sweating, bloody diarrhea, colic, difficult breathing, arrhythmia
Yellow oleander	*Thevetia peruviana*	Southern U.S.	++++	Sweating, bloody diarrhea, colic, difficult breathing, arrhythmia, tetany

(continued on next page)

	Table 14-4 Common Poisonous Plants *(concluded)*			
Common Name	**Scientific Name**	**Location[1]**	**Toxicity[2]**	**Signs**
	Cultivated Plants *(concluded)*			
Yew	*Taxus*	U.S.	++++	Nervousness, difficult breathing, incoordination, convulsions
	Trees			
Black locust	*Robinia pseudoacacia*	Central, Southern U.S., Canada	++	Diarrhea or constipation, stupor, laminitis, appetite loss
Black walnut	*Juglans nigra*	Northern, Central U.S.	+++	Increased temperature, laminitis, swelling in legs, heart and respiratory rates increased
Chokecherry/ wild black cherry	*Prunus* spp.	Southern, Northeast, Northwest U.S.	+++	Convulsions, frequent urination/defecation, gasping, tremors
Elderberry	*Sambucus* spp.	Forested areas of U.S.	+++	Gasping, tremors, frequent urination/ defecation, convulsions
Horse chestnut/ buckeye	*Aesculus* spp.	Southern, Eastern U.S.	++	Incoordination, muscle tremors
Red maple	*Acer rubrum*	Eastern U.S.	+++	Mucuous membranes dark, depression, colic, urine red or brown

1. Location refers to areas where the plant occurs naturally. Many of these plants are cultivated outside the location indicated on this table.
2. The number of plus signs (+) indicate the relative toxicity of the plant; four plus signs (++++) indicate the most toxic while one plus sign (+) is the least toxic.

Adapted from J. Moore, "Poisonous Plants: A Survival Guide," *Equus*, June, 1995, pages 28–37.

Some plants are not poisonous but they are mechanically injurious. These plants can cause discomfort or pain to a horse. Plants that can cause mechanical injury to a horse include such plants as sand burrs, thistles, foxtail, cactus, goat head, stinging nettle, and cockle burrs. Most of these plants cause sores in the mouth causing the horse to slobber, and have difficulty eating. Some may cause skin irritations or eye injury.

FIRST AID FOR HORSES

Equine first aid is the emergency care and treatment given to an injured or ill horse before treatment can be administered by a veterinarian or until the horse can be transported to a facility where help is available. The objectives include:

1. Intervening with a life-threatening situation

2. Recognizing serious or potentially serious, life-threatening conditions such as hemorrhage (bleeding), fracture, dehydration, and shock

3. Use measures to minimize further damage and prevent complications or after-effects such as:
   Extension of bone damage
   Damage to blood vessels or nerves
   Damage to soft tissue
   Secondary laminitis

Preliminary information—temperature, heart rate, respiratory rate, color of mucous membranes, and **capillary refill** time—are vital statistics that will help a veterinarian evaluate an emergency situation over the phone. Also,

---

First-Aid Kit

- Water-soluble antibacterial ointment and/or spray (e.g. Betadine)
- Bandage material:
  1 roll nonsterile cotton
  Sheet cottons or quilts for leg wraps
  Brown gauze or polo wraps
  Elastikon
  Tefla pads
  Kling rolls
  4 in. × 4 in. sponges
  Vetwrap™ or other self-adhesive wrap
- Poultice
- Rubbing alcohol
- Soap
- Epsom salts
- Thermometer with string and clip
- Scissors
- Clean towels
- Hoof pick and knife
- Shoe pullers
- Fly repellant
- Flashlight
- Prescribed, in-date medications
- Large syringe

Modern equine medical centers provide services once reserved for humans. High-speed treadmills, aquatreds, recovery pools, completely equipped equine ambulances, and in-house laboratories are frequently available at these facilities.

High-speed treadmills are used for gait analysis, stress testing, respiratory examination, and physical therapy. The speed of the treadmill ranges from one to thirty-three miles an hour and the treadmill can be elevated to a slope of six degrees to increase the work level desired.

An aquatred is a water-filled treadmill sunk into the ground. It assists in the conditioning and recovery of orthopedic, laminitis, and soft-tissue injury cases. The aquatred offers an adequate level of exercise while protecting the skeletal system and providing soft-tissue hydrotherapy.

The eight-foot deep recovery pool is heated and equipped with jacuzzi jets to provide a warm, relaxing, safe environment for a horse recovering from anesthesia. The pool is frequently used for postsurgical care of horses with fragile orthopedic problems. Jacuzzi jets provide soft-tissue therapy and also serve to relax the recovering horse.

The recovery pool is also used for physical therapy. A high-powered ski jet propels water from the front of the pool to provide a force for the horse to swim against. It is ideal for use in soft-tissue therapy in laminitis or recumbent (lying down) horses. The warm water and jacuzzi also relieve sore muscles and other problems associated with chronic recumbence.

Completely equipped equine ambulances are another service provided by modern clinics. The ambulance is equipped with a winch and transport mat capable of moving any size recumbent patient from up to 100 feet away into the ambulance quickly and safely. A complete triage or medical assessment facility, the ambulance and accompanying team are prepared to aid fracture patients, combat shock, deal with cardiovascular emergencies, provide respiratory support, and anesthetize a patient safely for the entire trip to the medical facility if necessary. Complete anesthetic monitoring is possible on the ambulance as are the ability to provide intravenous fluids, emergency drugs, hemostasis, and triage fracture stabilization.

Modern equine medical centers may also be equipped with an in-house hematology, chemistry, and microbiology laboratory that provides many results within minutes of obtaining a sample. These results can be essential to rapid treatment of emergency cases.

Based on the value of some horses, no cost is too great for full recovery, which now can often be achieved due to the services of a modern equine medical center.

a description of what happened to create the emergency can help with initial diagnosis and/or treatment. A well-equipped first-aid kit will help the horse owner treat many minor ailments and cope with emergencies while waiting for the arrival of a veterinarian.

SUMMARY

Horse owners need to be able to recognize a healthy horse by observing the signs of good health. The immune system is essential to the health of the horse. Antigens induce antibodies that destroy harmful substances in the body. The immune system also remembers disease-causing invaders and is prepared for the next invasion. Vaccinations provide immunity to many problem diseases. Before the foal's immune system is developed, it receives its first immunity in the colostrum from the mare.

Many diseases can affect horses. These diseases can be grouped as infectious and noninfectious diseases. The infectious diseases can be caused by bacteria, viruses, or parasites. Noninfectious diseases are often due to a management problem. If owners are aware of the signs of these diseases they can identify them in the early stages. Through good management and knowing which diseases are the most serious threat to horses, horse owners can minimize the spread of contagious diseases and effect prompt treatment of sick animals.

REVIEW

Success in any career requires knowledge. Test your knowledge of this chapter by answering these questions or solving these problems.

True or False

1. A horse's weight can change by as much as fifty pounds depending on recent eating, drinking, or elimination.

2. All energy provided above maintenance for the horse, will go towards the formation of fat.

3. Immunity refers to the ability of an individual who has recovered from a disease to remain well after a second exposure to the same disease.

4. Antibodies are provided to the newborn foal during gestation.

5. EIPH is a serious problem in pregnant mares.

6. Horses can breathe through their mouths like humans.

7. Colic in a horse is always an emergency.

Short Answer

8. List ten signs of good health in a horse.

9. What are the normal heart rate, respiratory rate, and temperature of the horse?

10. Name eight points of a good horse health program.

11. What term describes the process by which the mare passes antibodies to the foal?

12. List the two reasons vaccines are given.

13. Name the virus that is the cause of rhinopneumonitis in foals.

14. What disease is produced by the neurotoxin *Clostridium tetani* and affects humans as well as horses?

15. List four diseases of a horse caused by bacteria and four caused by viruses.

Discussion

16. List at least four pieces of information given that will help a veterinarian evaluate an emergency situation over the phone.

17. Explain the two general responses of the immune system and define the terms *antigen* and *antibody*.

18. Discuss the differences between infectious and noninfectious diseases and give examples of each.

19. Describe the symptoms of and treatment for strangles.

20. Describe the general symptoms of colic and four steps to prevent its occurance.

21. What is foal heat diarrhea?

22. Explain the difference between laminitis and founder and how they are related.

23. What is the best way to prevent a horse from eating poisonous plants?

24. Why is first aid important for horses and horse owners?

STUDENT ACTIVITIES

1. Based on the information in this chapter, develop a checklist of things to observe when checking horses for signs of disease. Make the checklist complete enough that it could be given to a new employee.

2. From a biological supply house or the biology department at a college or high school, obtain prepared microscope slides of bacteria. Observe the different shapes—rods, spheres or spirals.

3. Research and report the role of bacteria in human health and commerce. Some bacteria cause disease while other bacteria have a positive role in human welfare.

4. Using Table 14-4, identify poisonous plants in your area. Make a collection of these plants.

5. Visit with a veterinarian to discuss types of antibiotics and their effectiveness.

6. Put together a first aid kit for a horse and develop a brief first aid manual to go in the kit.

7. Ask a veterinarian to describe a typical vaccination schedule for horses in your area. Make a table showing the annual cost to vaccinate a horse.

8. Using the Internet, research new vaccines being developed and find out what progress is being made on other horse diseases. Report your findings.

9. Working with a professional, take the vital signs of a healthy horse.

10. Construct a table that compares the vital signs of a horse to those of an adult pig, cow, and sheep.

ADDITIONAL RESOURCES

Books

American Youth Horse Council. 1993. *Horse industry handbook: a guide to equine care and management.* Lexington, KY: American Youth Horse Council, Inc.

Frandson, R. D., and Spurgeon, T. L. 1992. *Anatomy and physiology of farm animals.* 5th ed. Philadelphia, PA: Lea & Febiger.

Fraser, C. M., ed. 1991. *The veterinary manual.* 7th ed. Rahway, NJ: Merck & Co.

Hill, C. 1994. *Horsekeeping on a small acerage.* Pownal, VT: Storey Communications, Inc.

Lewis, L. D. 1996. *Feeding and care of the horse.* 2nd ed. Media, PA: Williams & Wilkins.

Price, S. D. 1993. *The whole horse catalog.* New York, NY: Fireside.

Prince, E. R., and Collier, G. M. 1986. *Basic horse care.* New York, NY: Doubleday & Company, Inc.

Simmons, H. H. 1963. *Horseman's veterinary guide.* Colorado Springs, CO: Western Horseman.

Stoneridge, M. A. 1983. *Practical horseman's book of horsekeeping.* Garden City, NY: Doubleday & Company, Inc.

University of Missouri-Columbia Extension Division. (n.d.) *Missouri horse care and guide book.* Columbia, MO: Cooperative Extension Service, University of Missouri and Lincoln University.

CHAPTER 15

Parasite Control

Internal parasites are those that are found on the inside of the horse. They include single-celled animals (protozoa), roundworms (nematodes), and flatworms (flukes and tapeworms). These parasites usually are found in the gastrointestinal tract, but may be located in other internal organs. External parasites—lice, ticks, and mites—are found on the skin, in the ears, and at other places on the body.

Signs associated with both internal and external parasites depend on the type and number of parasites present and can range from no apparent effects to general unthriftiness, weakness, debilitation, and ultimately death of the host.

Every horse is infected by one or more of these parasites. So, horses should be on a parasite prevention and control program. A general knowledge and understanding of the nature of these parasites and their development is essential before necessary prevention and control measures can be effectively applied.

O B J E C T I V E S

After completing this chapter, you should be able to:

- Describe the life cycle of a typical internal parasite with an intermediate host
- Describe the symptoms of a parasite-infected horse
- List management techniques that help prevent parasite infections
- Give the scientific names for five common parasites
- Name the general categories of chemicals used to treat horses with parasites
- Describe the life cycle and damage caused by strongyles and ascarids
- Identify flying insects that are carriers of disease
- List the parts of the digestive system that internal parasites may affect
- Distinguish the different effects of mites, ticks, chiggers, and lice
- Discuss why young horses are more severely affected by parasites than older horses

- Discuss the six sanitation and management practices used for reducing or controlling parasites.

- Explain how a horse is checked for parasites

- List outward appearances of parasites on horses

KEY TERMS

Abdominal worm	Flukes	Protozoans
Aneurysm	Hosts	Roundworms
Anthelmintics	Intermediate host	Stomach hair worm
Ascarids	Ivermectin	Stomach worms
Babesia	Larvae	Strongyles
Benzimidazole	Lice	Summer sores
Bots	Lungworm	Tapeworms
Chiggers	Mange	Threadworm
Dermatitis	Midges	Ticks
Ectoparasites	Mites	Unthriftiness
Equine encephalomyelitis	Nematode	
Eye worm	Pinworms	

INTERNAL PARASITES

More than 150 types of internal parasites are known to infect horses. The most important ones are strongyles, ascarids, tapeworms and bots. The digestive tract, or stomach and intestines, is the most commonly affected area, although larvae migrate through all tissues of the horse's body. **Larvae**, the first stage of the parasite, must go through several stages to become adults; each stage appearing somewhat different (see Figure 15-1).

Life cycles of **strongyles** and **ascarids** are similar. They are classified into a large group of parasites known as **roundworms**. **Bots** are the larvae of an insect, the bot fly.

Internal parasites are widespread. Unless control measures are practiced, they are likely to increase and cause severe injury or death of the horse. Injury or harm inflicted on the horse is related to:

1. The kind of parasites

2. The number involved

3. The time over which the parasites are acquired

Strongyles are the most injurious. Ascarids, bots, and tapeworms are generally less harmful. A few parasites may be tolerated by the horse without

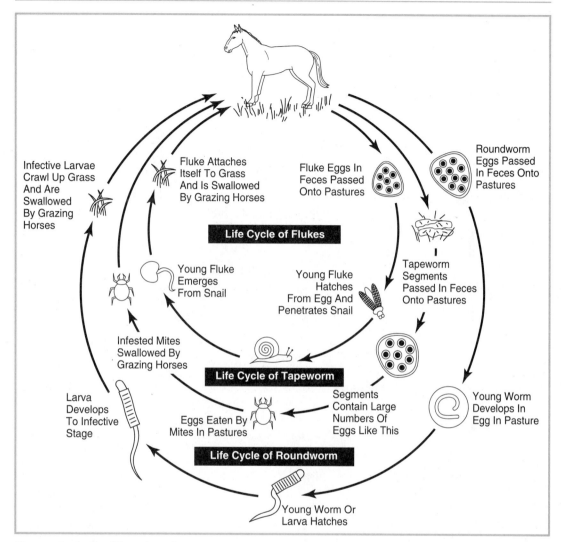

Infective Larvae Crawl Up Grass And Are Swallowed By Grazing Horses

Fluke Attaches Itself To Grass And Is Swallowed By Grazing Horses

Fluke Eggs In Feces Passed Onto Pastures

Roundworm Eggs Passed In Feces Onto Pastures

Life Cycle of Flukes

Young Fluke Emerges From Snail

Young Fluke Hatches From Egg And Penetrates Snail

Tapeworm Segments Passed In Feces Onto Pastures

Infested Mites Swallowed By Grazing Horses

Life Cycle of Tapeworm

Larva Develops To Infective Stage

Eggs Eaten By Mites In Pastures

Segments Contain Large Numbers Of Eggs Like This

Young Worm Develops In Egg In Pasture

Life Cycle of Roundworm

Young Worm Or Larva Hatches

Figure 15-1 The life cycles of some common internal parasites

apparent signs of ill effect, but larger numbers are quite likely to be harmful. Acquiring a large number within a few days may overwhelm and kill a horse. Getting the same number over a period of weeks or months is generally much less harmful.

Horses affected the most by parasites are young sucklings or weanlings and yearlings. In general, ascarid problems are restricted to young horses. This is because in most cases, resistance or immunity is built up by the time a horse is two or three years old. Strongyles and bots affect horses of all ages. Even so, the young are much more severely affected than older horses. Table 15-1 gives a brief outline of some common internal parasites.

Table 15-1 Common Internal Parasites of the Horse			
Parasite	**Where Found**	**Damage**	**Signs**
Habronema adult *Habronema* larvae (stomach worm)	Stomach Injured skin	Causes tumors of wall Granulomatous ulcers	Gastritis, digestive disorders, Summer sores, often healing spontaneously after first frost.
Gasterophilus (Bots)	Stomach Gums	Inflammation, perforation of stomach wall, gums	Digestive upsets and bowel irritation.
Parascaris (large white worm)	Small intestine	Irritate intestinal wall, possible obstruction	Flatulence, diarrhea, rough hair coat, "Hay belly" more common in young horses.
Strongyloides (threadworm)	Small intestine	Erosion of intestinal mucosa, enteritis	Anorexia, loss of weight, diarrhea, anemia, common cause of trouble in suckling foals.
Anoplocephala (tapeworm)	Small intestine	Ulceration of ileocecal valve, enteritis	Unthriftiness.
Strongylus (Bloodworm)	Large intestine and colon	Adults suck blood, cause ulcers on mucosa. Larvae cause enlargement and aneurysms of anterior mesenteric artery	Anemia, unthriftiness, colic, anorexia, malaise, soft feces with a foul odor. In large infections legs and abdomen swell.
Triodontophorus Poteriostomum Trichonema and others (Small strongyles)	Large intestine and colon	Irritate intestinal wall causing thickening and nodules with larvae in them feeding on blood	Anemia, anorexia, dark or black manure, soft feces with a foul odor. In large infections, legs and abdomen swell.
Oxyuris (pinworm)	Large intestine	Adults feed on gut contents. Larvae feed on mucosa.	Restlessness, irregular feeding with consequent loss of condition, dull hair coat, tail rubbing.

Life Cycles and General Characteristics

To survive and propagate themselves, well-adapted parasites live in harmony with their **hosts**, the animals from which the parasite obtains food. If a parasite were always to kill its host, it would be responsible for its own death, since by definition parasites are organisms that live in or on another organism of a different species for the purpose of obtaining food.

Parasites include protozoa, nematodes (roundworms), cestodes (tapeworms), trematodes (flukes), and acanthocephalans (spiny-headed worms). Most of the parasites affecting horses are nematodes, or roundworms.

Protozoans. **Protozoans** are single-celled animals that occur in the bloodstream and intestinal tract of horses. These organisms multiply by dividing and may be transmitted from horse to horse by an arthropod vector (carrier) or simply by being ingested in food or water as a result of fecal contamination (see Figure 15-2).

Nematodes. Roundworms are by far the most serious and economically important of the worms that occur in horses. These, as their name implies, are elongated, cylindrical worms ranging in size from two millimeters to thirty-five centimeters in length. Although the large worms cause significant problems, the small worms are far more important from both an economic and health point of view.

Some roundworm parasites damage the host by sucking blood, others cause damage by migrating through body tissues such as the lungs, and still others can cause severe colic in horses by forming a mass of worms in the intestine that interferes with intestinal motility and to some extent, absorption of nutrients.

Most equine **nematode** parasites have a direct life cycle. This type of parasite does not require any other organism except the definitive, or final, host to complete its life cycle.

Figure 15-2 Protozoans are one-celled animals that can cause disease. *(Courtesy of David Tyler, The University of Georgia)*

Typically, females that live in the digestive tract lay eggs, which are passed to the outside with the horse's feces. The eggs hatch in two to three days, depending on temperature and humidity, into small wormlike organisms called first-stage larvae (L_1). First-stage larvae develop and molt to second-stage larvae (L_2), which mold to third-stage larvae (L_3). The L_3 stage is infective to the final host. They migrate up blades of grass and the horse ingests them when grazing. These preparasitic stages are much the same for most of the strongyle parasites of the horse.

When the horse ingests the third-stage larvae, they develop into fourth-stage larvae, which may wander extensively through the body of the horse before becoming adults in the intestinal tract—large strongyles—or they may develop into adults in the gut with no migration through other organs—small strongyles.

Some nematode parasites require a second host in order to complete their life cycles. This second host is an invertebrate and is called the **intermediate host**. Typically, the intermediate host eats the eggs or first-stage larvae, which then develop in the intermediate host instead of on the ground. The definitive host becomes infected when the intermediate host (fly, tick, etc.) injects the infective stage of the parasite while it is taking a blood meal. Sometimes the definitive host gets the infective stage by eating the infected intermediate host.

Nematodes have a complete digestive system. They have a mouth through which they suck blood or intestinal juices and they excrete their waste through an anus (see Figure 15-3).

Cestodes. All cestodes—**tapeworms**—that occur in horses use pasture **mites** as intermediate hosts. The final host becomes infested by ingesting the mite containing the infective cysticercoid while grazing. Mites become infected by ingesting tapeworm eggs deposited on the pasture with the host's feces.

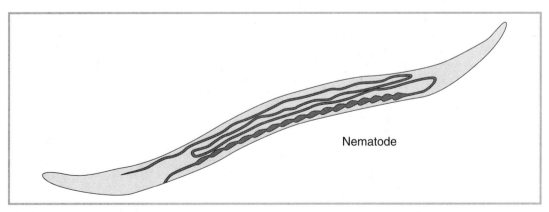

Nematode

Figure 15-3 A nematode

The tapeworm that occurs in horses, is a large worm consisting of a head, which attaches to the intestine of the horse, and a long ribbon-like body with similar segments called proglottids. Unlike nematodes, in which the sexes are separate (males and females mate to produce the next generation), tapeworms contain both sexes within the same worm.

Tapeworms absorb nutrients through their skin, having no mouth or anus (see Figure 15-4).

Trematodes. **Flukes** also require an intermediate host, most often a snail. Although flukes do occur in horses, they are of minor significance and will not be discussed here.

Large Strongyles

The group of nematodes called the large strongyles are the most damaging of all the parasites that occur in horses. Adult worms range in size from approximately twelve and one-half millimeters up to about thirty-one millimeters in length. They live in the large intestine and cecum where they feed by eating plugs of the mucosal lining.

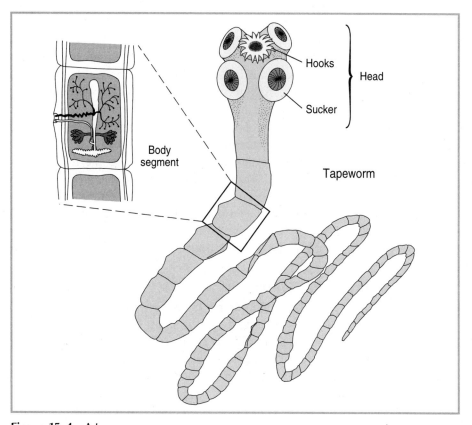

Figure 15-4 A tapeworm

Far more damaging than the adult are the larvae that migrate through internal organs of the host. Some prefer to live in one of the large arteries supplying the small intestine of the horse. These larval strongyles damage the artery's lining causing it to react and become very thickened, producing an **aneurysm**.

Often, blood clots form and are carried by the bloodstream to smaller vessels where they can block the blood supply to a part of the intestine. Where other vessels supply this part of the intestine, no real damage is done. But if no other blood supply exists, this part of the intestine dies and unless corrected surgically, the condition can be life threatening. Sometimes these blood clots find their way back to the arteries that supply the hind legs and can cause rear limb lameness.

The parasite causing these problems is called *Strongylus vulgaris*. Other large strongyles (*S. edentatus* and *S. equinus*) migrate through different organs, notably the liver and pancreas, and inflict damage in their own particular way.

Small Strongyles

As their name implies, small strongyles are much smaller than the large strongyles, usually about thirteen millimeters in length although some are smaller. These nematodes are present in much larger numbers than the large strongyles. There may be hundreds of large strongyles in a horse, but usually there are thousands of small ones. While the small strongyles do not cause the damage or present the danger that large strongyles do, they can cause colic due to decreased intestinal motility, in addition to producing **unthriftiness**, diarrhea, rough haircoat, and other signs associated with heavy parasitic infections. These species of parasites usually are clumped into one or two genera (*Triodontophorus* or *Trichonema*).

As with the large strongyles, horses become infested by these parasites through ingesting the infective third-stage larvae while grazing. But unlike the large strongyles, these parasites require little time to reach maturity, start producing eggs, and further contaminate pastures. Consequently, they quickly build up large numbers of larvae to reinfest horses and assure their propagation. Like the large strongyles, small strongyles inhabit the large intestine and the cecum.

Hair-like Worms

Smallest of the nematode parasites occurring in horses is the **stomach hair worm**, *Trichostrongylus axei*. It is about four or five millimeters in length and very thin and hair-like. As with the large and small strongyles, the life cycle is direct. Eggs are passed in the feces, hatch, and develop into third-stage infective larvae in four or five days. Horses become infested by eating the larvae on the grass.

This parasite occurs in the stomach and the small intestine and damages the lining of these organs, sometimes causing bleeding into the gut. This is associated with dark, fetid diarrhea and with heavy infections can cause a rapid loss of condition.

The equine intestinal **threadworm**, *Strongyloides westeri*, is somewhat unique. These small hair-like worms are eight to nine millimeters long and only the adult female is parasitic.

To add to the uniqueness of this parasite, the adult males and females can exist outside the host in a free-living state. When conditions become unfavorable for existence on the outside, the females produce eggs that hatch into third-stage infective larvae and either are eaten by the horse or penetrate the skin. If they penetrate the skin, the larvae migrate to the lungs, penetrate the alveoli, and after reaching the trachea, are coughed up and swallowed where they continue to develop to adulthood in the gut.

Some of these migrating larvae do not develop, but remain dormant in muscle tissue of mares until they foal. Then they migrate into the mammary gland and infest nursing foals via the colostrum. *S. westeri* has been thought to contribute to foal heat diarrhea, which occurs twelve to thirteen days after birth.

Tapeworm

The most common tapeworm in horses is in the genus *Anoplocephala*. This cestode, or flatworm, *Anoplocephala magna*, is called the "large horse tapeworm" and occurs most often in the small intestine. It also is found in the stomach and sometimes in the cecum.

This is a fairly robust tapeworm about twenty-five centimeters long, with very short segments. It retains its position in the host by attaching to the small intestine lining by four suckers located on the head (scolex). Like all tapeworms, both sexes are contained in each individual segment (proglottid).

This worm has an indirect life cycle: the eggs are passed in the host's feces and are eaten by pasture mites, which are the intermediate hosts. After the horse eats the oribatid mite containing the infective cysticercoid (larva), the larva develops into an adult in the small intestine in six to ten weeks. Typically, these worms don't live very long in their host.

With light infestations, horses show no signs. Heavy infestations cause horses to suffer colic and diarrhea, and possibly go off feed. They often are depressed, may become dehydrated, and spend a lot of time lying down. When many worms are present, sometimes few feces are passed because the worms cause an intestinal obstruction. Heavy infestations of this parasite can produce complications that result in death.

Horse tapeworm infection is diagnosed by finding the eggs in the feces, but many horses do not pass eggs—especially with heavy infestations, so tapeworms may not be diagnosed when present.

Anoplocephaliasis in the horse is usually a disease of yearlings at pasture.

Lungworms

Although most common in donkeys, horses also harbor worms (*Dictyocaulus arnfieldi*) that live in their lungs (**lungworm**). This is another nematode, or roundworm, parasite. The females are about sixty millimeters long and the males can be a little over half that. The adults live in the lungs of horses where they mate; the eggs produced by the females are coughed up, swallowed and passed with the feces.

Dictyocaulus has a direct life cycle. Horses become infected by ingesting the L_3, or third-stage larva. In horses, the adults may never produce eggs, while in donkeys, they may start producing eggs in three to four months. Although lungworms cause very little clinical problem in donkeys, in horses they may cause coughing, an increased respiratory rate, and some nasal discharge. Because eggs often are not produced in horses, diagnosis becomes difficult since veterinarians cannot find eggs in the feces and must rely on history—including whether donkeys are grazing with horses—and clinical signs to diagnose the disease.

Stomach Worms

The Habronemas, which consist of *Habronema muscae, H. majus,* and *H. megastoma* (*Draschia megastoma*), are the equine **stomach worms** that cause two rather distinct diseases in horses called gastric and cutaneous habronemiasis. The Habronemas have indirect life cycles, with house and stable flies serving as intermediate hosts.

Habronema eggs, which pass with the horse's feces, are eaten by fly maggots and mature with the fly as it becomes an adult. Infective larvae are deposited around the horse's lips and nostrils where the flies feed, thereby gaining entrance into the horse's mouth. Horses may also become infected by ingesting infected adult flies that have become entrapped in food or water. The larvae are freed in the horse's stomach and develop into adults in about two months. In the stomach of horses, these parasites produce fibrous tumors, or numerous nodules, which, if close together, form a tumor.

Another type of disease caused by this parasite is cutaneous habronemiasis or "**summer sores**." This also is caused by species of *Habronema*, but is due to the larvae that the intermediate host deposits in existing wounds in the skin. (Some parasitologists think that the larvae can penetrate healthy skin.) Cutaneous habronemiasis occurs during the summer, and is most common on areas where horses cannot switch flies, such as the inside of the legs, over the withers, the penile sheath, and fetlocks.

These lesions are brownish-red, angry-looking sores that may ooze serum tinged with blood. They seem to itch very badly, and often disappear when cold weather sets in—only to reappear when the weather warms up again. The appearance of cutaneous skin lesions in the summertime when flies are numerous would suggest summer sores.

The gastric form of habronemiasis is more difficult to diagnose since few eggs are passed and, because the larvae don't float very well, they are sometimes missed during routine fecal flotation examination. Adult females are about twenty-five millimeters long. Males are somewhat smaller—usually about two-thirds to three-quarters the size of the female.

Ascarids

Parascaris equorum is the horse ascarid. This is a very large, robust roundworm. Females grow up to thirty-five centimeters long, and the males are somewhat smaller. The life cycle is direct, but instead of ingesting the larvae on the pasture, foals ingest the infective eggs that contain larvae. Because it takes about two weeks for the eggs to become infective, a foal could ingest freshly passed feces (a common habit) and not become infected.

After infective eggs are ingested, they hatch and penetrate the wall of the intestine, migrate in the bloodstream to the lungs where they may cause some respiratory problems, and are then coughed up, swallowed, and mature in the small intestine.

The adults start producing eggs about twelve weeks after the foal becomes infected. Because of a developing immunity, foals often shed the infection at about seven months of age.

Clinical signs of ascarid infection in foals include a dry hair coat, potbelly, and abdominal discomfort (sometimes these foals kick their flanks). They will be undersized for their age and breed, and very often they have dry stools covered with mucus, although diarrhea sometimes is present. It should be noted that ascarid eggs are very resistant and can survive for years in the soil.

Because some **anthelmintics** render these parasites unable to move, impaction due to a large mass of immobile worms sometimes can occur following deworming. A veterinarian can suggest an appropriate anthelmintic to use with a heavy ascarid infection.

Pinworms

Horses, like people and unlike dogs, can have **pinworms**. Two kinds of pinworms occur in horses. A rather large one, the females of which can be up to sixty-three millimeters long, is *Oxyuris equi*. The minute horse pinworm, *Probstmayria* sp., is only about two millimeters long and is of little consequence.

The life cycle of *Oxyuris* species is direct, and like horse ascarids, the egg is the infective stage. It is infective for three to five days after being laid, and is ingested by the horse with food or water. The parasite matures in the mucosa of the cecum, colon, and rectum, and starts producing eggs in 120 to 150 days.

Because the females migrate out of the anus to lay eggs and then return to the colon, this disease causes an intense itching around the anus of horses. Owners will see horses rubbing their hind quarters, often resulting in all the hair being rubbed off over the tailhead.

Often these horses become restless, go off feed, and lose condition. Sometimes young mares may appear to be in heat. Although adult female pinworms occasionally can be seen around the horse's anus, diagnosis is by finding the eggs, usually with a Scotch Tape swab.

Research Achievements in Equine Health

Research is necessary for improving and maintaining the health of horses. Scientists and researchers at the University of Kentucky, Lexington, have played a major role in improving equine health.

In 1993, the Office International des Epizooties, the animal equivalent of the World Health Organization, designated the Maxwell H. Gluck Equine Research Center at the University of Kentucky for three significant equine viral diseases: equine rhinopneumonitis, equine influenza, and equine viral arteritis.

Leaders from the Thoroughbred business, veterinary, and research communities convened an historic meeting at the Gluck Equine Research Center in November of 1993 to initiate formulation of an action plan to follow in case of future outbreaks of infectious disease at racetracks (and possibly other public equine facilities). In only four months, the plan was developed and implemented under the guidance of personnel from the Gluck Center.

Thousands of mares throughout the United States aborted due to herpesvirus infection prior to the Department of Veterinary Science's development of the first reliable vaccine against this disease. Subsequent research spawned an even safer vaccine and offered improved management strategies for preventing the spread of herpesvirus infection in mares.

Researchers also developed a vaccine against strangles (*Streptococcus equi*), which causes severe illness in a wide range of breeds and ages of horses.

Other research highlights at the University of Kentucky include:

- Development of ELISA tests (enzyme-linked immunosorbent assays) for drug detection.

Babesiasis

Equine piroplasmosis, or equine babesiasis, is a protozoan disease occurring in horses, mules, and donkeys in the southeastern part of the United States, particularly Florida and Georgia.

Two species of **Babesia**—*B. caballi* and *B. equi*—are known. These are small, protozoan parasites that occur in red blood cells. The life cycle is indirect. The tropical horse tick (*Dermacentor nitens*) serves as intermediate host. The brown dog tick (*Rhipicephulus sanguineus*) may be able to serve as an intermediate host for *B. equi*.

Equine babesiosis causes horses to have a fever and to become listless and depressed. They may go off feed, and may develop central nervous system disturbances causing rear leg weakness or even paralysis. The limbs may become swollen—stocked-up.

- Discovery of the mare's response to extended light in controlling her reproductive cycle changing forever the struggle to get mares to foal earlier in the year.

- Development of the first multivalent vaccine against equine influenza.

- Identification of Tyzzer's disease, a highly fatal liver disease in young foals caused by *Bacillus piliformis*.

- Release of a vaccine against equine viral arteritis. When the Thoroughbred breeding industry in Kentucky faced a major crisis in 1984 because of an epidemic of equine viral arteritis, a vaccine developed in the Department of Veterinary Science in the 1960s saved the industry from considerable economic loss.

- Development of a blood test to detect evidence of contagious equine metritis (CEM) infection in mares—a disease introduced to the United States in 1977 that threatened the economic well-being of the entire Thoroughbred industry because of international sanctions on the movement of horses.

- Development of a vaccine and vaccination protocol resulting in the prevention of the shaker foal syndrome, cooperation with private practitioners, state agencies, and the federal government.

- Development of the first diagnostic test for equine *protozoal myeloencephalitis* (EPM) in living horses using spinal fluid or serum.

- Validation of the effectiveness of many of the current equine anthelmintics in current use.

And the research goes on. Each year scientists and researchers confront new challenges in the form of health problems in the equine industry.

This disease usually lasts eight to ten days and can cause death, although most horses recover and return to normal.

Onchocerciasis

Onchocerca species are nematodes that occur as adults in connective tissue of horses, mules, and donkeys. They are fairly common parasites. About three-quarters of horses surveyed in the Midwest were infected with *Onchocerca cervicalis*. Adult females are quite long, up to thirty centimeters in length, but the males are small, six to seven millimeters long. The females of *Onchocerca cervicalis* occur in the ligamentum nuchae of horses and mules.

Onchocerca reticulata occurs in the flexor tendons and suspensory ligaments. This nematode also requires an intermediate host and uses **midges** (*Culicoides* sp.) as an arthropod vector. Biting midges pick up the microfilaria in the skin of horses; these develop to an infective stage in the midge in about three weeks. When the midge takes a blood meal from a horse, the infective stage is injected—thus, completing the cycle.

In addition to the **dermatitis** this organism can cause, it sometimes causes eye problems. *O. reticulata* causes occasional lameness. Species of *Onchocerca* do not cause fistulous withers or poll evil as formerly believed. Because the new parasiticide, **ivermectin**, kills *Onchocerca* microfilaria very quickly, horses sometimes mount an immune response to these dead microfilaria—resulting in tissue edema. This resolves itself spontaneously in about seven days.

Eye Worm

The equine **eye worm**, *Thelazia lacrymalis*, is about nineteen millimeters long and lives in the tear duct and conjunctival sac of the horse's eye. The female worms produce living larvae; they don't lay eggs. These L_1 larvae wander into the eye secretions and are picked up by face flies that serve as the intermediate host. In the fly, the larvae develop into the infective stage and can be transferred to another host when the face fly feeds on eye secretions.

Although most eye worm infections go undetected, heavy infections cause mild eye irritation and can, on rare occasions, result in blindness, probably due to secondary bacterial infection.

Diagnosis is made by observing adult worms in the eye. Treatment is best achieved by removing the adults from the conjunctival sac under ophthalmic anesthesia and tranquilization. Decreasing the prevalence of eye worms is best achieved by controlling face flies.

Sometimes the **abdominal worm**, *Setaria equina*, develops in the eye and causes damage. Normally, these nematode parasites, which use mosquitoes as intermediate hosts, occur in the abdominal cavity and are of little or no consequence.

EXTERNAL PARASITES

External parasites of horses include the ticks, mites, chiggers, and lice. These cause irritation and may carry disease.

Ticks

Three kinds of ticks occur commonly on horses. Each has a preferred location on the horse. Some are more common in specific parts of the country.

The winter tick, *Dermacentor albipictus*, has become widely distributed because horses now are commonly transported from one part of the country to another. Although this tick occurs primarily on the horse, it is found on other farm animals, such as cattle, sheep, and goats, so these animals can also be involved in its spread.

Ticks differ one from another in that some use only one host as they develop from larvae, to nymphs, and adults, while other ticks use more than one host. The winter tick is one of the ticks that uses only one host. The entire life cycle takes place on the horse (see Figure 15-5).

Tick infestations, like those of lice, are more common in the winter than in the warm seasons. Large numbers of winter ticks can cause horses to become weak, lose their appetite and become thin, and, because of the blood loss, sometimes develop an anemia that makes them more susceptible to other diseases. Ticks can cause death, especially in foals.

The Pacific coast tick, *Dermacentor occidentalis*, is found chiefly in coastal areas of the West. Unlike the winter tick, this tick drops off the host to lay its eggs, and the larvae and nymphs feed on small mammals before becoming adults and parasitizing horses.

The Pacific coast tick can transmit Rocky Mountain spotted fever, Colorado tick fever, and other diseases affecting rabbits and cattle. It also can produce a condition called tick paralysis that can affect humans, dogs, and

Figure 15-5 Winter tick

calves. Consequently, horses should be inspected for ticks after trail rides or cross-country pleasure rides in areas where these ticks occur.

The ear tick, *Otobius megnini*, like *Dermacentor albipictus*, is a one-host tick. It is common on horses but is also found in the ears of cattle, sheep, dogs, cats, and occasionally, people. These ticks, however, do not occur on the horse in their adult stage: only the larvae and nymphs are found in the horse's ears. Adults have nonfunctional mouth parts, but may survive for two years on the ground.

Ear ticks on horses cause irritation as evidenced by excessive head tossing or rubbing of the ears. Horses with drooping ears that shake their heads a lot may have ear ticks. Ear ticks also predispose the animal to secondary bacterial infection of the middle and inner ear and can, consequently, cause serious problems. This tick, unlike *Dermacentor*, does not itself transmit any diseases. Several topical preparations are available for treating ear ticks.

Mange Mites

Mites are **ectoparasites** that are closely related to ticks and cause a skin condition called **mange**. The entire life cycle of mange mites occurs on the horse. Mating occurs on the skin or in burrows the mites make in the skin. The eggs hatch on the host after about four days and are mature, egg-laying adults twelve to fourteen days later.

Sarcoptic mange (*Sarcoptes*) causes lesions usually found on the neck, shoulders, head, chest, and flanks of horses. These mites burrow under the skin and cause severe irritation and itching. In trying to relieve the source of itching, horses will bite and rub the affected area until the hair is lost and large, scabby areas often result (see Figure 15-6).

Chorioptic mange (*Chorioptes*) produces lesions like Sarcoptic mange, but since the mites occur more commonly on the lower extremities, it often is called foot mange. Horses affected with these mites will paw, lick, and bite at their lower legs in an attempt to relieve the itching.

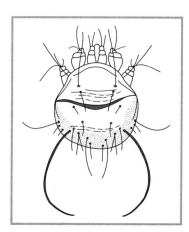

Figure 15-6   Sarcoptic mange mite

Psoroptic mange occurs primarily on the poll or the tail. This mange mite (*Psoroptes*) also causes intense itching, with hair loss and scabs if the horse traumatizes itself extensively.

Mange mites can live off the host for a short time and can be transferred from one host to another on combs, blankets, and so forth. In the past, mange has been extremely difficult to control, but with the new ivermectins, mange should become a less serious problem.

Chiggers and Lice

Chiggers affect horses in much the same way they affect people. **Chiggers** are the larval stage of harvest mites (*Trombicula*) and affect horses' feet and muzzles as they walk and feed on infested pastures.

Lice can be a very serious problem in horses. There are two kinds of **lice**: biting lice (*Damalinia*), which feed on skin and hair, and sucking lice (*Haematopinus*), which pierce the skin and suck blood and tissue fluids (see Figure 15-7).

Mites and lice are very host specific. They will not pass from horses to cows, sheep, goats, dogs, or other animals.

Infestation with both sucking and biting lice can be debilitating to horses. Biting lice cause skin irritation and itching and horses will rub, bite, and kick at themselves in an attempt to relieve the source of irritation. This results in a rough coat with loss of hair; if serious enough, secondary bacterial infection can cause major skin lesions. In addition, heavy louse infections can produce serious unthriftiness and weight loss.

Sucking lice and a heavy infestation of biting lice can remove enough blood to cause a horse to become seriously anemic, in addition to producing irritation and debilitation because of itching.

Louse infestations usually are more severe in late winter and early spring. Frequent grooming and applications of topical pesticides are helpful in louse control.

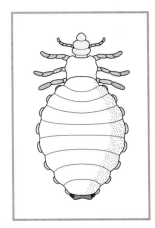

Figure 15-7   Horse sucking louse

FLYING INSECTS

Although not permanently associated with their host as worms and mites are, flies, mosquitoes, gnats, and other flying insects are important not only because of the worry and loss of condition they cause but also because some are carriers of disease.

Fly control depends to a great extent on sanitation, good grooming, and common sense. Flies breed in manure and sometimes spilled grain, especially if it's wet. Removing spilled grain and manure from stalls on a regular basis, and changing bedding as it becomes soiled with feces and urine, will aid in fly control. Some flies serve as carriers of parasitic worms and viral diseases.

Mosquitoes transmit **equine encephalomyelitis**. Black flies and "no-see-ums" very often cause intense itching and attendant lesions in horses' ears, although they will bite other thin-skinned areas of the horse as well.

Bots

Bots are fly larvae that are parasites in the stomach of horses. *Gastrophilus intestinalis*, the common horse bot, and *G. nasalis*, the throat botfly, are the two common botflies found in this country.

Adult flies look somewhat like bees and are not seen often. These adult flies lay eggs on the hair of the legs or around the chin and throat of horses.

G. intestinalis lays its eggs on the forelegs and shoulders of horses. The eggs hatch when the horse licks itself—so the larvae quickly gain entrance to the horse's mouth. *G. nasalis* lays its eggs around the chin and throat where they hatch spontaneously (the horse doesn't need to lick the eggs for them to hatch). *G. nasalis* eggs hatch and burrow under the horse's skin into the mouth.

Both species remain for about a month in the lining of the tongue and cheeks where they may cause severe ulcers around the teeth and cause horses to go off feed. After about a month, the larvae are found in the stomach where they produce a condition called gastric myiasis. Although in small numbers, bots cause virtually no clinical signs, in heavy infections there may be almost no part of the horse's stomach wall that does not have a bot attached (see Figure 15-8).

These botfly larvae are fairly large, about two centimeters long, and have large oral hooklets that they attach to the stomach wall. Sometimes they completely penetrate the stomach wall causing peritonitis and subsequent severe problems.

Adult flies seriously annoy horses when depositing their eggs on the legs and chin. Washing the legs and chins of horses with warm water containing an organophosphate insecticide every week during botfly season aids in control.

Warble flies, which cause cattle grub, can affect horses, but seldom are a problem except in cow ponies used to work range cattle.

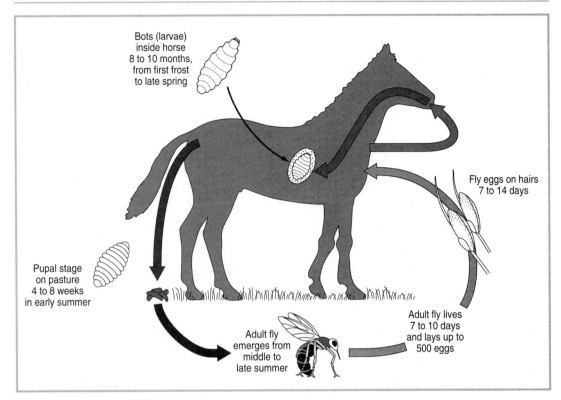

Figure 15-8 The botfly life cycle

PREVENTION AND CONTROL

Sanitation and good management practices should be used to control parasite infections. Foals are born free of internal parasites; their buildup of internal parasite infections is related to the degree of contact, either direct or indirect, with older animals carrying the infections. All of the worm parasites discussed here use feces or manure to spread infections by contamination of feed and water supplies or the environment.

Transfer stages of these worm parasites do not actively seek the host to complete the infection process. Instead, they rely on chance to be picked up and swallowed. Thus, only a very small percentage actually complete this step in the life cycle. To compensate for this, large numbers of eggs are produced by the female worms to start the transfer process.

Sanitation and sound management practices aid in controlling or minimizing the spread of the infections. These practices assist the natural destructive forces such as sunlight and drying during transfer stages. Susceptible animals also should be allowed only limited contact with contaminated pastures, paddocks, or stables. A checklist of sanitation and management practices effective in reducing numbers of parasites and flying insects includes the following:

1. Proper manure disposal

   - Stable manure: Compost before spreading on pasture, or spread on cropland and other ungrazed areas.

   - Small corrals or paddocks: Pick up all manure and compost or dispose of as above.

2. Pasture management

   - Practice frequent mowing and chain harrowing.

   - Avoid overstocking.

   - Rotate grazing as much as practical.

   - Graze young animals separate from older horses.

   - Follow horses with cattle or sheep before returning pastures to horses.

3. Feed

   - Provide mangers, racks, or bunks for hay and grain.

   - Do not feed off the ground.

4. Water

   - Provide a clean water supply.

   - Avoid water sources contaminated with feces.

5. Removal of bot eggs

   - Clip egg-bearing hairs or sponge affected areas with warm water.

6. Regular deworming of horses should be practiced under the supervision of a veterinarian who is familiar with local land and weather changes.

Anthelmintic Drugs

In addition, horses often need to be treated with specific drugs, commonly referred to as anthelmintics, to obtain effective control of parasites (see Table 15-2). These drugs remove the parasites from the intestinal tract. The treated animal is relieved of the immediate damage or injury caused by parasites, but probably more important, removal of parasites breaks the cycle. This serves to reduce contamination of the environment with transfer stages, limiting the spread of the infections and protecting animals from reinfection.

The best method of strongyle control currently recommended is to administer a small quantity of a deworming compound to the individual horse's ration daily. This compound is very effective in those management systems in which horses are fed individually.

Table 15–2 Antiparasitic Compounds for Internal Parasites

Class	Generic Name	Methods[1]	Percent Effectiveness				
			Bots	Strongyles	Pinworms	Ascarids
Avermectins	Ivermectin	P, T	95–100	95–100	95–100	90–100
Benzimidazoles	Fenbendadzone (FBZ)	T, F, P	0	95–100	95–100	90–100
	Mebendazole (MBZ)	T, F, P	0	65–95	95–100	95–100
	Oxfendazole (OFZ)	T, F	0	95–100	95–100	90–100
	Oxibendazole (OBZ)	T, F, P	0	95–100	95–100	90–100
	Thiabendazole (TBZ)	T, F	0	90–100	90–100	10–75
	MBZ + TCF	T, F, P	95–100	65–95	95–100	65–95
	OFZ + TCF	P	95–100	95–100	95–100	95–100
	TBZ + TCF	T, F	95–100	90–100	90–100	95–100
Organophospates	Trichlorfon (TCF)	T, P	95–100	0	90–100	95–100
Phenylguanidines	Febantel (FBT)	T, F, P	0	95–100	95–100	95–100
	FBT + TCF	P	95–100	95–100	95–100	95–100
Pyrimidines	Pyrantel-pamoate (PRT)	T, F, P	0	65–100	95–100	90–100
	Pyrantel-tartrate[2]	F			60–70	

1. F = Feed, P = Paste, T = Stomach tube.
2. Infective larvae are prevented from entering the tissues.

The best routine deworming program for any particular management system should be designed around individual client/horse requirements and may differ depending on the number of horses on a pasture, amount of time spent at pasture compared with time spent in a stable, availability of alternative pastures (for rotational grazing), age of the horses, and other conditions. The local veterinarian should guide the horse owner regarding the optimal deworming strategy.

If daily deworming cannot be undertaken, regular use of ivermectin is recommended (every six to ten weeks depending on the local circumstances). Ivermectin has had a profound effect on parasite burdens in horses. It is so effective that the old-fashioned deworming strategies that used stomach-tubing or promoted the "rotation" of a number of weaker deworming compounds are no longer recommended. Traditionally, dangerous and toxic compounds were used to eliminate bots. Fortunately, ivermectin is as effective against bots as it is against strongyles and ascarids. Although tapeworm problems are uncommon, once-yearly deworming against tapeworms is recommended for horses that are dewormed with ivermectin on a regular basis. This is especially true if the horse has access to permanent pasture grazing. The best method of deworming against tapeworms is to use a high dose of pyrantel pamoate.

The veterinarian's services can include a microscopic examination of fecal samples for an indication of the kinds and relative numbers of worm parasites in the animals. This, along with other information such as numbers and ages of animals and type and amount of pasture, provide the veterinarian with a rational basis for the selection of drugs and frequency of treatments for the particular situation.

With primary emphasis on strongyle control, some operations may require only one or two treatments per year, whereas others with factors or circumstances favoring heavy infections may take as many as six treatments per year to maintain effective control. All horses on a farm should be included in the control program. New stock or temporary boarders should be treated and quarantined for a week or so before they are placed on pasture or otherwise allowed to mingle with other horses.

Drug Resistance

Many anthelmintics introduced over the last several years are **benzimidazole** analogs to which nematode parasites are starting to develop a resistance.

Resistance develops when a wormer does not kill all of the target worm population; some survive contaminate pastures. Over time, pastures become contaminated with a high proportion of larvae that, when eaten, will develop into adults able to tolerate the doses of anthelmintic normally administered.

Before concluding that lack of response is due to benzimidazole resistance, other reasons should be considered to explain why horses have eggs in their feces after worming. Among them might be a low plane of nutrition, rapid reinfection, wrong choices of anthelmintic, an inappropriate dose, or faulty administration.

If anthelmintic resistance is a problem, two processes can be used. One is to use a given class of anthelmintic—for instance, a benzimidazole—for a year, and then use a different compound, such as pyrantel, an organic phosphate, or ivermectin, for a year. Simply changing from one benzimidazole to another does not constitute changing anthelmintics. The second option is to change the classes of anthelmintics each time horses are wormed.

A prime objective of any strongyle control program should be to keep pasture contamination of larvae to a minimum. The worming protocol needed to accomplish this objective will depend on worm burden, stocking rate, and climatic conditions, and so will vary from farm to farm. The horse owner with fifty acres to support two horses will have far fewer problems than one who is trying to keep two horses on two acres.

SUMMARY

Parasites are grouped as internal and external. Many parasites can affect horses, but internal parasites create the greatest health problems. The person working with horses must learn to recognize the signs of parasitic infections early. Also, many good management practices will prevent or lessen the chance of severe infections. Based on local geography and weather conditions, a veterinarian can help develop a control program that also combines the use of chemicals to kill internal parasites.

Parasite infections are difficult to identify and hard to eliminate. Success in controlling them must be a determined and sustained effort. A continuing battle must be waged against internal parasites, the most common danger to the health and well-being of horses.

REVIEW

Success in any career requires knowledge. Test your knowledge of this chapter by answering these questions or solving these problems.

True or False

1. Keeping pasture contamination to a minimum will help in parasite control.

2. Bot flies transmit equine encephalomyelitis.

3. All parasites require an intermediate host.

4. Lungworms cause more problems in horses than they do in donkeys.

5. Horses and people can have pinworms.

Short Answer

6. Name six sanitation and management practices for reducing parasites.

7. List three hosts of parasites.

8. Name five symptoms of any parasite infestation.

9. List five internal and five external parasites.

10. Name three ways the veterinarian can help in a deworming program.

11. List the major internal parts of the digestive system that any type of worm infestation may affect in the horse.

Discussion

12. What is the purpose of an anthelmintic?

13. Explain the life cycle or stages of a parasite.

14. Explain the difference in how ticks, lice, chiggers, and mites affect horses.

15. Why are young horses affected more by parasites than older horses are?

16. How are horses checked for parasites?

STUDENT ACTIVITIES

1. Based on the information in this chapter, develop a checklist of things to observe when examining horses for signs of parasites. Make the checklist complete enough that it could be given to a new employee. Using a word processor, put this checklist in a table format.

2. Visit or contact a veterinary clinic. Find out how to submit a fecal sample to be checked for parasitic infection.

3. Using the generic names for the antiparasitic drugs in Table 15-2, identify the brand name or trade name used to sell these drugs. This can be done by visiting a livestock supply store, veterinary clinic, or reading a livestock or equine supply catalog.

4. Invite a veterinarian or sales representative to discuss the types of anthelmintics sold. Also, find out what types of restrictions are placed on the sale of these medicines.

5. Research and report the role of bacteria in human health and commerce. Some bacteria cause disease while other bacteria have a beneficial role in human welfare.

6. Develop a program for regular deworming based on your location and weather conditions. Using a word processor, put this deworming program into a table format.

7. Learn how to check a horse for ticks. Write a paper describing this process.

8. Compare the types of parasites found in cattle, pigs, and sheep to those found in horses. Describe any similarities or differences. Put this information into a table format.

ADDITIONAL RESOURCES

Books

American Youth Horse Council. 1993. *Horse industry handbook: a guide to equine care and management.* Lexington, KY: American Youth Horse Council, Inc.

Blakely, J. 1981. *Horses and horse sense: the practical science of horse husbandry.* Reston, VI: Reston Publishing Company, Inc.

Davidson, B., and Foster, C. 1994. *The complete book of the horse.* New York, NY: Barnes & Noble Books.

Evans, J. W. 1989. *Horses: a guide to selection, care, and enjoyment.* 2nd ed. New York, NY: W. H. Freeman and Company.

Frandson, R. D., and Spurgeon, T. L. 1992. *Anatomy and physiology of farm animals.* 5th ed. Philadelphia, PA: Lea & Febiger.

Fraser, C. M., ed. 1991. *The veterinary manual.* 7th ed. Rahway, NJ: Merck & Co.

Griffin, J. M., and Gore, T. 1989. *Horse owner's veterinary handbook.* New York, NY: Howell Book House.

Hill, C. 1994. *Horsekeeping on a small acerage.* Pownal, VT: Storey Communications, Inc.

Kellon, E. M. 1995. *Equine drugs and vaccines: a guide for owners and trainers.* Ossining, NY: Breakthrough Publications.

Lewis, L. D. 1996. *Feeding and care of the horse.* 2nd ed. Media, PA: Williams & Wilkins.

Price, S. D. 1993. *The whole horse catalog.* New York, NY: Fireside.

Prince, E. R., and Collier, G. M. 1986. *Basic horse care.* New York, NY: Doubleday & Company, Inc.

Simmons, H. H. 1963. *Horseman's veterinary guide.* Colorado Springs, CO: Western Horseman.

University of Missouri-Columbia Extension Division. (n.d.) *Missouri horse care and guide book.* Columbia, MO: Cooperative Extension Service, University of Missouri and Lincoln University.

U.S. Department of Agriculture. 1923. *Special report on diseases of the horse.* Washington, D.C.: U.S. Government Printing Office.

CHAPTER 16

Common Management Practices

Successful ownership and enjoyment of horses depends on a solid knowledge of horses and good management practices. Management is knowledge in action. Many of the management practices for horses have already been discussed in other chapters. This chapter covers management topics not discussed elsewhere: dealing with stress in horses, identifying horses, neonatal, weaning, and castration decisions, stall maintenance, fly control, pasture maintenance, and wound management.

OBJECTIVES

After completing this chapter, you should be able to:

- Recognize stress in horses
- Describe methods of marking and identifying horses
- Discuss procedures for the neonatal foal
- List methods of fly control
- Describe common fly problems and the habits of several species
- Explain the best management practices for pastures and forages
- Describe wound types and their proper management
- Discuss the management considerations when weaning a foal
- Explain the importance of records to the management of horses

K E Y   T E R M S

Auditory	Hierarchy	Pupae
Bedding	Mastitis	Roan
Chestnuts	Meconium	Soil test
Cowlicks	Microchip	Stress
Dun	Neonatal	Tattoo
Freeze brand	Olfactory	Wound
Grullo	Paddocks	

RECOGNIZING STRESS IN HORSES

Stress is a demand for adaptation. Some stress is necessary, but each horse has its own tolerance level and when that level is exceeded failure results. Knowing the signs of stress, monitoring the signs of stress, and reducing stress are all signs of good management.

Anyone working around horses needs to learn to recognize the signs of stress. A horse experiencing stress may appear frightened or nervous, it may be pacing or running, or it may develop a vice such as cribbing or stall weaving. Abnormal sweating may signal physical or psychological stress. Muscle tone can provide some clues. If the horse is tense, sweating, and the muscles are contracted, it may be tying-up. If the muscles are flaccid and extremely relaxed and the horse is depressed, the central nervous system may be damaged. If any of these signs are observed, a closer inspection is needed. Intervention may be necessary.

Permanent Records

To recognize changes in the horse's condition, normal values must be known. They will be different for each horse so each horse needs its own permanent record. Horse owners must keep a file on every horse including the following information:

- Permanent identification, birth date, and registrations
- Reproductive history, breeding dates, and foaling dates
- Weight and condition scores
- Normal temperature (T), pulse (P), and respiration (R) or TPR
- Deworming dates and products used
- Vaccination dates, diseases, and products used
- Illness dates, diagnoses, and treatments
- Injury dates and treatments
- Surgery dates and outcomes
- Allergy causes

This record can be a hand-written form or it can be computerized. Figure 16-1 illustrates a computerized form.

Horse Name:			
Gender:		Date Foaled:	
Breed:		Color:	
Registration No.:		ID/Tatoo:	
Markings:			
Owner Name:		Phone:	
Address:			
City:		State:	Zip:
Insurance Carrier:		Policy No.:	

Medical Record

Date	Problem	Vital Signs	Treatment

Maintenance Record

Vaccinations	Date	Date Due
Eastern Equine Encephalitis		
Western Equine Encephalitis		
Rhinopneumonitis		
Tetanus		
Rabies		
Potomac Horse Fever		
Lyme Disease		
Influenza		
Wormer		
Farrier		

Figure 16-1 Example of a computerized horse record form

Stress can be grouped into four different categories for horses:

1. Behavioral or psychological
2. Mechanical
3. Metabolic
4. Immunological

Behavior

Managing horses in a low stress environment requires understanding how their senses perceive the world and a few principles of their behavior. For more information on behavior refer to Chapter 19.

Sensory Perception. Horses do not see the world as humans do. Horses have both binocular and monocular vision. Monocular vision allows them to see 220 degrees around them when their head is down to graze. Binocular vision allows them to focus on objects in front of them.

Horses hear much better than people do, but olfaction is even more acute—smell is their strongest sense.

Horses are also very sensitive to touch. Their untrained natural response is to move into pressure.

Communication. Horses communicate with each other through visual signals. Recognizing these signals can help owners understand their horses. Anger is demonstrated by laying the ears back, pursing the lips, and swishing the tail. Interested horses cock their ears forward and have a relaxed body. A fearful horse may put its ears forward or to the side; its body is tense and its tail clamped or stiff. Relaxed horses have relaxed ears and one hind leg cocked. They may chew or lick their lips.

These behaviors can be easily recognized and may alert the owner to certain stereotypes of horse behavior. For example, some horses are sullen and difficult most of the time while others are actually treacherous. Bad tempered, resentful horses may bite, strike, or kick at any time.

Social Behavior. The social behavior of horses is controlled by the herd instinct. Horses seek out and enjoy the company of other horses (see Figure 16-2). Social order is important and there is an established dominance **hierarchy** in any herd of horses. Dominance is the ability to control access to resources. The dominance hierarchy requires that each horse recognize the other horse and determine through some initial aggressive acts (biting or kicking) and submissive acts (running away) which horse is dominant and which is subordinate. After the initial conflicts establish the hierarchy, just the signs of anger from the dominant animal will be enough to warn subordinates. Pecking order can change if a horse is removed from a group for several weeks or if a mare is in estrus or going to have a foal.

Figure 16-2 Horses are gregarious animals that prefer to be in the company of other horses where they feel secure. This is important to know when working with young and/or anxious horses.

Initial contact across safe wooden fences can alleviate some social stress and introducing horses gradually can help avoid injuries associated with fighting. Providing extra feeding stations and dividing the feed up so horses all get adequate portions is another way to avoid conflict.

Mechanical Stress

Structural injury can be detected by lameness, local inflammation, swelling, heat, and/or pain. Checking for injuries should be part of the daily routine.

Nutrition and Metabolic Stress

The digestive system of the horse is designed to handle frequent small meals. They are continuous grazers by nature and usually do best when kept at pasture. If this is not possible, good-quality hay fed in frequent meals is the next best thing to pasture.

Nutritional programs are designed by evaluating each horse's weight and condition. Horses must be fed individually in most cases. The energy requirements of "easy keepers" and "hard keepers" can differ by up to 30 percent. About 80 percent of the horse's feed goes to meet its energy requirement. The hay used should contain the most nutrients per dollar.

To reduce stress, horses require that a certain proportion of the diet be roughage. Vitamin and mineral requirements must also be met but not exceeded for the stage and condition of the horse. Nutritional requirements and feeding guidelines are given in Chapters 12 and 13.

Three metabolic problems in horses are closely associated with nutrition:

- Colic
- Laminitis
- Tying-up

Horse owners need to learn to recognize these conditions. They can be serious health concerns for a horse.

Colic. Colic is an acute abdominal crisis or stomach ache. In its most serious form it can cause death and often does. Abdominal pain can lead to shock, which causes dramatic changes in fluid balance. The signs of colic include pawing, looking at the belly, stretching out, rolling, not eating, violent movement, kicking and biting at the flank, and depression. Every sign will not be seen in every case. If colic is suspected, the horse's temperature, pulse, and respiration should be taken and recorded and a veterinarian called. If the heart rate and/or respiration is elevated an emergency may exist. The numerous causes of colic include sudden changes in diet, parasites, twisting of the intestines, gastric ulcers, lack of a quality water source, cribbing, compaction, and gas colic. Colic is discussed more completely in Chapter 14.

Laminitis. Laminitis is any condition that leads to separation of the sensitive and insensitive lamina of the hoof. The early signs include a rapid heart, pounding digital pulse, depression, elevated temperature, and circulatory impairment. As the condition progresses over time the horse will become lame. The horse will extend its legs out in front and rock back over its hocks and will be reluctant to move or pick up its front feet. A veterinarian needs to be called at the earliest signs of laminitis. Laminitis is covered more completely in Chapter 14.

Tying-up. Tying-up, or exercise-related muscle problems are metabolic conditions related to nutrition and exercise. The signs include altered gait, rapid breathing, stiffness in the hindquarters, rigid back, trembling muscles, sweating, reluctance to move, collapse, muscle damage, brown urine, kidney failure, and laminitis. Some of the contributing factors include excess soluble carbohydrates in the diet, stress, metabolic and hormonal disturbances, electrolyte imbalances, and selenium and vitamin E deficiencies.

Tying-up can occur shortly after exercise begins or after the horse has been worked hard, such as on an endurance ride. If the horse is exhausted and over heated, rest it immediately in the shade, cool its extremities by bathing them with cool water, and give the horse an electrolyte solution. Commercial electrolyte solutions are available through a veterinarian or feed store.

If the horse ties-up shortly after the start of exercise, it should not be moved. Instead the horse should be dried off, covered on the hindquarters with a blanket, and a veterinarian called.

Once a horse has tied-up the condition will likely recur. Altering the diet may possibly help correct the problem.

Immunological Stress

Stress caused by disease and/or parasites can range from superficial discomfort to death. A good vaccination program is the best defense against infectious diseases. A veterinarian will help form a vaccination schedule based on disease prevalence, the use of the horse, the season, and the effectiveness of the vaccine. Some of the vaccines available include protection against tetanus, influenza, rhinopneumonitis, Eastern, Western, and Venezuelan encephalomyelitis, strangles, Potomac horse fever, rabies, leptospirosis, and clostridium. Diseases and vaccination programs are discussed more completely in Chapter 14.

An effective deworming program must include good management practices as well as regular use of antiparasitic drugs. Parasite infections and control are discussed in Chapter 15. Some important guidelines include:

- Treat all horses at the same time.
- Rotate clean horses to clean pastures.
- Design feed and water facilities to prevent fecal contamination.
- Remove manure frequently from stalls and paddocks.
- Clip and harrow pastures regularly.
- Consult with a veterinarian on selection and use of antiparasitic drugs.
- Monitor the effectiveness of the parasite-control program by checking egg counts in the feces.

MARKING OR IDENTIFYING HORSES

In today's competitive world of equine sports, proper identification takes a high priority. Thorough and effective identification ensures that the horse being bought, sold, raced, or bred is indeed the horse claimed. Some of the circumstances in which positive identification is important include:

- Health and disease control
- Theft prevention, documentation, and recovery
- Slaughter
- Breeding
- Recovery of animals lost or killed in natural disasters
- Fraud prevention

The Jockey Club was the first organization in the United States to set up an accurate identification system for horses. In the early 1900s, the Thorough-bred racing industry was having problems with "ringers" running under assumed names. A ringer is a falsely identified horse entered in a race below its class, giving it an almost certain chance to win.

Today there are many methods used to identify a horse, including body markings, tattooing, freeze branding, blood typing, and **microchip** implantation.

Body Markings

Obviously, coat colors such as bay, blue **roan**, **dun**, **grullo**, and palomino can be used to identify horses (see Figure 16-3).

Distinctive markings or patterns such as a star or blaze on the head and stockings or distal spots on the legs provide more detail (see Figure 16-4). The correct terms to use in identifying coat colors and facial and leg markings are listed and explained in Chapter 8.

Unique body markings used for identification include chestnuts, cowlicks, and dimples. Body markings are recorded in a record as a picture or drawings.

- **Chestnuts** or "night eyes" are horny, irregular growths on the inside of the horse's legs. On the front legs, they are just above the knee. On the rear legs, they are toward the back of the hock. Chestnuts are like human fingerprints because no two are alike, and they do not change in size or shape throughout the horse's adult life.

Figure 16-3 Palomino quarter horse. *(Courtesy of Palomino Horse Breeders of America, OK)*

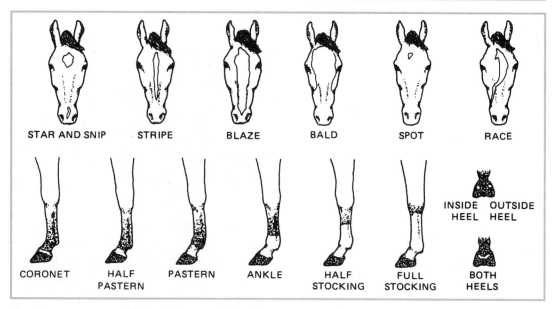

Figure 16-4 Face and leg markings

- **Cowlicks** are permanent hair whorls that cannot be brushed or clipped out. They are located mainly on the forehead and neck.

- Dimples are permanent indentations in the muscle under the skin. They are usually located at the point of the shoulder or in the neck muscles.

- Others markings can include white or black patches on the body and scars. Firing marks on the legs are also useful for identifying horses.

Tattooing

A **tattoo** consists of a letter corresponding to the year the horse was born and a number matching the registration number of the horse. The tattoo may be placed in several areas, but the upper lip is the most common site (see Figure 16-5). The actual tattoo instrument consists of a chrome-plated brass block that contains a needle pattern with a varying number of needles, depending on the particular number or letter. The needle pattern was developed over several years until a specific pattern was obtained that could not easily be altered.

Before the tattoo is applied, the horse is carefully examined for color, markings, cowlicks, chestnuts, and other easily identifiable traits. Once the identity of the horse is assured, the mucous membrane on the upper lip is exposed using a lip clamp. The area is cleaned with alcohol, and the proper digits are placed in the tattoo gun. The gun is then dipped in an antiseptic and applied to the lip. Finally, ink is rubbed into the perforations.

Figure 16-5 Checking a lip tattoo.

Lip tattooing was perfected by the Thoroughbred Racing Protective Bureau (TRPB). The Jockey Club uses this method of identification to guarantee the identity of every racing horse at a track that is a member of the Thoroughbred Racing Association (TRA).

Freeze Branding

A **freeze brand** uses an unalterable system of angular symbols developed by Dr. Keith Farrell, a veterinary medical officer with the U.S. Department of Agriculture. As with tattooing, the first symbol represents the year the horse was born followed by the registration number. The brand is most commonly applied to an area approximately two inches by seven inches midway on the neck, underneath the mane.

The identity of the horse is double-checked before the brand is applied. Copper stamps or marking rods are cooled in liquid nitrogen or dry ice. An area under the mane is shaved and washed with 98 percent alcohol, which aids in conducting the intense cold. The copper stamp is applied to the animal's skin for ten to twenty seconds. An indention is left in the skin immediately after the brand is applied. Some swelling may occur in the first few days. However, after two months, a distinct and permanent mark remains. The intense cold kills the pigment-producing cells, called melanocytes, leaving an area of pigment-free skin. On dark-colored animals, the hair grows back white, and on white animals, an area with no hair results.

Freeze branding has many advantages over fire branding. A freeze brand produces minimal changes in the hide, is more distinct and legible, does not produce an open wound, and is relatively painless.

Freeze branding is used by the Arabian Registry to identify purebred Arabians. An "A" is placed in the first position of the system of marks to indicate "Arabian."

Blood Typing

Although markings, tattooing, and freeze branding are effective in differentiating individual horses, blood typing has been developed over recent years and is an equally effective alternative. Serologists test for the sixteen most common blood antigens and serum proteins. The combinations seem limitless, as there are some 125 billion possible blood types in horses. Blood typing is used by the Jockey Club, the American Quarter Horse Association, the Arabian Horse Registry, and others.

Microchip Implantation

To discourage thieves from stealing horses and to aid in recovery of stolen horses, a new method of identifying horses has emerged over the past few years. The implantation of a microchip the size of a grain of rice containing the horse's registration number or identification number can assist in identifying a horse if by chance it turns up stolen.

The veterinarian uses a specially designed needle and syringe to implant the microchip. A local anesthetic is administered about midway down the horse's neck just below the crest. The chip is then inserted into the ligament in the neck, using the custom syringe. The chip is actually lodged about an inch underneath the skin's surface. It is equipped with a nonmigratory tip to ensure that it stays in place.

The microchip identification cannot be altered. The chip is read by using a hand-held scanner, similar to those used in grocery stores. Many slaughterhouses and brand inspectors have scanners for identification in several states.

NEONATAL PROCEDURES

The first important **neonatal** task is treatment of the foal's navel with a tincture of iodine solution. The umbilical cord should be allowed to sever naturally, rather than being cut or tied off. The navel stump should be saturated with tincture of iodine immediately after tearing. Treating the navel stump will prevent joint infection or navel ill. This is a bacterial infection usually resulting from poor navel treatment—umbilical infection is its source. It develops in foals less than thirty days of age and is considered a medical emergency. Aggressive therapy is needed to prevent permanent cartilage or bone damage. Symptoms of joint ill include sudden lameness with or without systemic illness, swollen and painful joints, and stiffness. More than one joint may be involved, such as stiffness through the back or neck pain.

The naval stump is an ideal growth environment and entrance for potentially life-threatening bacteria into the system of the newborn. Simple iodine treatment minimizes the risks of navel ill and other infections quickly and easily. The treatment should be repeated daily until the navel stump is completely dry. If the navel does not close properly, or begins to leak urine, additional treatment and veterinary attention may be needed.

The second task to be attended to is a foal enema to prevent impacted meconium and to stimulate intestinal peristalsis. The **meconium** is a soft, dark greenish-brown substance consisting of digested amniotic fluid, glandular secretions, mucous, bile, and epithelial cells that accumulates in the digestive tract during development. Natural elimination of the meconium should occur within three hours after birth, but retention and constipation may occur anywhere from six to twenty-four hours after birth and after prior fecal passages. The meconium needs to be passed for normal intestinal activity to begin. Foals often have trouble expelling the meconium and have a tendency toward painful constipation if it is not passed in a timely fashion soon after birth.

There are two types of meconium retention: in the large colon (high), and in the rectum (low). The signs of meconium retention and constipation are similar: restlessness, tail switching, attempts to defecate, elevated tail, straining, colic pain, rolling, getting up and down often, and lying upside down with knees and forelegs extended toward the head.

An enema is used to treat and prevent meconium retention. The foal's rectum and anus are easily damaged. The enema should be administered with one pint of warm, soapy water through a soft, narrow, pliable rubber tube inserted no more than four to five inches into the rectum. Mixtures of warm water and mineral oil or commercially obtained human enemas can also be used.

Exercise often helps foals pass their meconium. Weather permitting, a foal with signs of impaction should be turned out.

Exercise

The timing for a mare and foal to be first turned out depends on the climate, time of year of foaling, and weather conditions. Turn-out on the first day after birth is fine if the temperature is mild and the ground is not icy or muddy. The pair should not be allowed to become fatigued or overly stressed by temperature extremes or by overheating as a result of exercise or excitement. This can be minimized by frequent turn-out of the mare before she foals. Often, only the mare is led to the turn-out area and the foal is simply guided behind her. If the foal is led to the turn-out area, a rump rope and halter should be used properly, without jerking or pulling, so as not to damage the foal.

Figure 16-6 Mares and foals can be turned out together as long as there is adequate space and appropriate precautions are taken.

The mare should be released first into the turn-out area since she will usually want to run and kick for a moment or two. This prevents the potential disaster of the foal being accidentally in her way. The pair should be turned out alone for the first week or so, and then allowed to join any other mares and foals. Observation is required for the following ten days, as problems could still occur (see Figure 16-6). The mare's udder should be closely watched and the foal's temperature monitored daily. A subnormal, out-of-range temperature indicates an infection and requires immediate attention.

WEANING AND CASTRATION

After being separated from the mare, the foal usually experiences more contact with human handlers who require certain standards of behavior. Therefore, the foal should be taught to accept basic handling and discipline before weaning. Haltering, brushing, and leading the foal while it is still on the side of the mare will make training easier.

Because weaning can be very stressful, the foal should be in good health before being separated from its dam. The only exception would include weaning to facilitate medical treatment of the foal or mare, as recommended by a veterinarian.

Elective surgery to correct conditions such as umbilical hernias and angular limb deformities should be performed well ahead of weaning. Not only will early treatment aid in more complete correction of some conditions, the stressful period surrounding the weaning process can be avoided.

Time of Weaning

The best age for weaning foals depends on the health of the mare and foal, the temperament and vices of the mare, the environment in which the foal will be weaned, the maturity of the foal at a given age, and the management level on a given farm. If necessary, foals can be weaned as early as a few days after birth, but the usual age for weaning is between four and six months. A newborn foal relies on the mare for nutrition, protection, and security. A foal weaned at an extremely young age requires intense nutritional and behavioral management. By four months of age, the foal should be eating feed and less dependent on its dam for protection and emotional support. Weaning before this age may increase weaning stress, especially if environmental conditions are harsh, the foal is not eating grain, or the foal is heavily dependent on the mare.

Little nutritional or social support is gained by waiting until six months of age to wean. In fact, later weaning may promote some unwanted behavior in foals. A breeder may want to separate a mare that has an adverse disposition or vices from her foal as early as possible. Some behavior patterns can be learned from the mare and, with early separation, the dam's behavior will have less influence on the foal's behavior.

Weaning Methods

The management level of the breeding farm, the condition and temperament of the mare and foal, the facilities available, and the number of foals to be weaned during a given period all affect the method with which foals are weaned. Foals weaned together and those consuming creep feed prior to weaning experience less weaning stress.

Weaning methods range from an abrupt separation in which the foal and mare are separated immediately from all contact, to gradual separation in which the foal and mare are allowed visual, **auditory** (sound), and olfactory (smell) contact before complete removal. Complete abrupt separation usually involves moving the mare to another turn-out area, or moving the foal into a confinement completely separated from any contact with the mare. Advocates of abrupt weaning suggest that mare and foal injury is lessened when contact is completely prohibited.

However, some research indicates that foals weaned by complete, abrupt separation exhibit more behavioral problems associated with weaning stress than foals weaned by a more gradual separation. Gradual methods usually involve placing the mare and foal next to one another in enclosures that allow for visual contact for a period of several days to weeks. Fences or stall partitions used in gradual systems must restrict suckling (see Figure 16-7). Again, weaning in pairs and preconditioning the foal to solid feed before weaning will reduce weaning stress.

Figure 16-7 In order to lessen the possibility of behavior problems, some research indicates that gradual separation is best. Advocates of abrupt weaning believe that there is less chance of injury to the mare and foal when any contact is completely prohibited.

One of the best ways to lessen weaning stress is to maintain familiar surroundings. This can be done by leaving foals of like size and age together. When other foals are not available, an older, nonlactating, well dispositioned mare or gelding may be used for companionship. Some farms have successfully used goats for the same purpose. Use of other livestock species or mature horses as weaning companions may be especially beneficial when it is necessary to wean single foals that are very young or unusually nervous. The foal appears to experience far less stress when other elements of the environment are the same and when companionship is available. This limits weight loss, decreases the incidence of disease, and makes the transition to self-sufficiency less traumatic.

Regardless of the method used, facility construction and design must emphasize safety. Any protrusions such as feed troughs can cause injury to nervous foals. Also, any area that is wider than a foal's hoof has the potential for trapping the leg of a foal if it should strike or rear next to a stall wall.

Mare Care During Weaning

A mare usually calms down more quickly than a foal, although the time required for her to resume normal behavior may vary from a few hours to a few days. Just as foals should be weaned in pairs or small groups, newly separated dams may need to be maintained in pairs or small groups. Unless aggressive behavior between mares is evident, such grouping may aid in more rapid calming following separation from their foals.

If the mare still has significant milk production, grain intake should be decreased and exercise increased. If the udder becomes very tight, a small amount of milk may periodically be extracted by hand, but this practice is discouraged unless necessary. If the udder is still tight four days after weaning and the mare's temperature rises significantly or if other indications warrant it, the milk should be checked for the presence of **mastitis** (infection) and appropriate therapy started. A veterinarian's assistance is recommended.

Castration

A castrated horse is called a gelding. Geldings are easier to care for and not as prone to injury as stallions, they are easier to haul, and several geldings can be kept safely in a paddock.

Horses can be castrated at any age, but most horses are castrated between birth and two years of age. Colts with poor conformation or poor pedigrees are usually castrated as soon as the testicles descend into the scrotum. Testicles usually descend at birth or certainly by the tenth month after birth. If a horse has good conformation and a good pedigree he is generally kept in tack until he fails to meet certain performance criteria.

Timing of castration depends on weather conditions and development of the individual animal. Spring, before fly season and hot weather, is the best time for castration. Most often castration is performed by an experienced veterinarian.

When one or both testes fail to descend into the scrotum, the animal is referred to as a cryptorchid or ridgeling. In horses, colts are considered cryptorchid if the testes are not descended by fifteen months. These animals are difficult to castrate and since some tissue may continue to produce testosterone, they may retain their stallion attitude.

BEDDING

One of the more fundamental and less glamorous aspects of horse ownership is stall maintenance. A significant part of this is the choice of **bedding**. Good bedding protects the horse's feet from thrush. It encourages the horse to lie down to rest and cushions its feet and legs from the hard stall floor. The horse needs to have some material under it that will also soak up or drain off the urine and the moisture from the manure. Ordinary stall floors are unable to do these things.

The best bedding material should be absorbent, dust-free, readily available, easily disposed of, unpalatable, and affordable. One of the first things to

Heatstroke

Heatstroke is not as common today as it was when horses were used to power machines before the use of gas-powered engines. But heatstroke still does occur in horses during the summer months in the heat of the day. It can be caused by overexertion on a hot, humid day, by confinement to a poorly ventilated stall on such a day, by transportation in hot vans, and on exercisers that are not shaded from the sun. Horses that have been idle and are not conditioned to the work or the climate are the most susceptible. A lack of adequate drinking water can predispose a horse to heatstroke.

Signs of heatstroke can include a collapse, a staring expression, vomiting, and diarrhea. The inside of the horse's mouth may be bright red and the rectal temperature reach as high as 109.5° F.

Veterinary assistance is required immediately in cases of heatstroke and emergency first-aid measures started. First-aid consists of placing ice packs on the horse's head between the ears and cold cloth packs along its spinal column. If ice packs are not available, cloth sacks saturated with cold water should be directed over the head and down the spinal column. The veterinarian will provide other cooling methods and supportive treatment.

look for in a bedding is its absorbency. The more absorbent the material, the less of it required, and the less frequently it must be replaced.

Cost is often the major consideration in choosing a bedding. Since many crop wastes can be made into bedding, owners should look around and see what is plentiful in the area. Even though a specific bedding material may not be the most dust-free or the most absorbent, the fact that it is plentiful and cheap may be enough to justify overlooking the disadvantages, and it may be the most practical choice.

Of the common kinds of bedding, the most popular is straw. Straw makes an attractive bed, and many people are willing to put up with its disadvantages just because they like to see their horses knee-deep in a nice, shiny yellow bed. Straw is the bedding of choice for foaling stalls on many breeding farms, since there is the potential for finer bedding materials, such as shavings, to readily stick to the newborn's body and airways. However, straw is bad for horses that like to eat their bedding. It is also highly combustible.

Straw is very absorbent and has a high comfort rating. Straw can be relatively dust-free, if carefully selected. However, drawbacks include the high labor requirement in cleaning stables, the large volume of resulting material, and the difficulty in disposal.

The best time to buy straw is at harvest time, but facilities are necessary for storage. Straw is best if it is stored indoors.

Two other highly absorbent materials for bedding are wood shavings and sawdust. Horses will seldom eat these materials and they burn much more slowly than straw. Shavings and sawdust also help keep down odors and require less frequent cleaning than many other materials.

Black walnut, however, should never be used. Severe laminitis (founder) has resulted in horses where black walnut shavings were being used for bedding. Black walnut (*Juglans nigra*) wood contains a number of aromatic chemical agents, some of which are quite toxic to horses. Eating just a few of the fresh shavings will cause severe gastrointestinal irritation and severe founder.

Softwood shavings such as pine are generally a safe and practical material to use. Pine shavings produce considerably less disposable material than straw and are generally disposed of more easily. Although shavings may be more expensive to use than straw, the additional cost can usually be justified in labor saved. Purchasing a large volume rather than buying it by the bale or bag saves money but a considerable storage area may be necessary. The shavings must be as dust-free as possible. Often, shavings are mixed with sawdust, but if too much sawdust is present, it could cause respiratory problems in some horses. The wood-shaving particles should be relatively large.

Sawdust and shavings should be stored indoors. If they are wet, their value as bedding is worthless and they will take a long time to dry.

A good, cheap bedding can be made from corn stalks or ground corncobs if they are readily available. After the corn has been picked, stalks can be chopped with a flail chopper into bedding. The corn stalks should be dry. At times, horses will eat chopped corn stalks.

Ground corncobs can also make an absorbent bedding.

Another comparatively new bedding product is recycled newsprint. This product is pollen-free and has less dust than straw and shavings. As a result, horse owners with allergies or contact lenses, and horses with respiratory conditions may benefit from its use.

The weight of newsprint is less than that of an equal volume of other bedding products. Weekly stripping of stalls with newsprint bedding may not be necessary if the stalls are thoroughly picked daily and existing bedding is fluffed to keep it dry longer. Newsprint is very absorbent, softer and more comfortable than either shavings or straw.

While good ventilation is obviously a part of stable management, proper ventilation can also lower the humidity, keeping most bedding drier and extending its effectiveness.

Basically, the choice of bedding material should be determined by

- availability and price
- absorptive capacity
- ease of handling
- ease of clean-up and disposal
- nonirritability from dust or components causing allergies
- texture and size
- fertility value of the resulting manure

Table 16-1 compares the absorbency of various bedding materials.

Table 16-1 Bedding Materials and Amounts Needed	
Material	**Lbs. of Water/ 100 lbs. Dry Matter**
Wood Products	
Hardwood Chips	150
Hardwood Sawdust	150
Hardwood Shavings	150
Pinewood Chips	300
Pinewood Sawdust	250
Pinewood Shavings	200
Straw	
Barley	210
Oat, long	280
Oat, chopped	375
Wheat, long	220
Wheat, chopped	295
Other	
Corn Stalks, dried	250
Corn Cobs	210
Hay, chopped	200
Peat Moss	1,000
Shredded Newspaper	400

FLY CONTROL

A sound sanitation program is of paramount importance to fly control. All other types of control are doomed to failure without this important first step. Control of stable flies in barnyards, stables, or corral areas usually involves several methods, which also apply for the house fly (see Figure 16-8). Chemical control directed at larval and adult stages of both insects is usually required periodically during the fly season.

Sanitation Around Stable or Corral

The basic aim of a sanitation program is to reduce or eliminate fly larval development sites. A number of areas require attention because the larvae of these flies can develop in varied habitats. Manure management is essential. Timely spreading of manure promotes drying and prevents larvae from developing. In small areas manure provides an ideal breeding site for large numbers of both stable and house flies. Wet areas where manure, mud, and plant debris accumulate also form ideal breeding habitats. Modifications to the drainage around corrals to reduce excess moisture can eliminate fly production sites and make chemical control efforts much more successful.

Chemical Control

A variety of chemical control techniques are available to the horse owner. Generally, control of adult flies using insecticides on surfaces and as sprays to kill existing adult flies are the most effective techniques. In most barnyard situations, a combination of surface treatment and aerosol sprays is used, often on an alternating schedule. Treatments applied directly to horses are not as effective for control of stable or house flies as residual surface treatments. In practice, both techniques usually are needed.

Figure 16-8   Stable fly

Applications of residual insecticides to premises are frequently used to control both house and stable flies. Longer-lasting residual insecticides provide control for an extended period when sprayed onto sites where the adult flies congregate. Flies contact the insecticide when they land on the treated surfaces. Sides of buildings, inside and outside surfaces of stalls, and fences may be potential day or night resting sites for these flies. Observation of the barnyard situation will quickly indicate the favored resting sites for flies.

Knock-down sprays are effective for killing adult flies present at the time of application. The chemicals used for these applications are usually short-residual insecticides that have a quick knock-down and high-contact toxicity. Several types of spray or fogging apparatus may be used. Wind velocities should be low at the time of application and the droplet or particle size should be small (fifty to seventy-five microns) to ensure drift through the corral area. This method requires less application time, but the disadvantage is that it will only kill those flies present at the time of application and thus provides only short-term relief.

Direct applications of sprays and dusts to animals may be used in some situations to protect them. Materials used for direct animal application usually have short-residual activity and this type of application is labor intensive.

Other methods of fly control, such as baits, electric grids, and traps, have some limited use for house fly control but are ineffective for the blood-feeding stable fly. Baits may be used effectively for house fly control in enclosed areas. Fly papers, cords, and strips may also help alleviate fly problems in these areas. Such methods are usually ineffective in open areas.

Control of immature flies (larvae) is sometimes possible. Usually, the best approach is to remove the potential source of fly production with sanitation practices. When this is not possible, a larvicide can kill the developing flies. A larvicidal insecticide may be applied directly to places where eggs are laid and larvae develop.

Biological control has potential for controlling barnyard fly problems. A number of parasites and predators of both house and stable flies exist that help to reduce fly numbers. Some of these natural parasites are available commercially but to date, research has not demonstrated their cost-effectiveness.

Management efforts are needed to control horse and deer flies, black flies, gnats, horn flies, and general nuisance flies like house flies. The following information briefly describes these flies, their life cycles, habits, and control.

Horse and Deer Flies

Horse and deer flies are large biting flies that can inflict painful bites on horses and humans. Several species may become abundant enough to constitute a problem for grazing horses, particularly animals pastured near streams or low, wet areas. Both horse and deer flies have been incriminated in the transmission of equine infectious anemia. Further, because the bite is painful, horses may become restless and unmanageable when they attempt to ward off attacks by these flies. Immature larval horse flies are aquatic or semi-aquatic and the last stage larva overwinters. Life cycles are long. Most species have only one generation per year and some species may have a two-year life cycle. Only female flies feed on blood. Control is difficult. Treating individual animals with repellents or insecticidal sprays may reduce fly bites.

Black Flies

Black flies or buffalo gnats are small ($\frac{1}{12}$ to $\frac{1}{15}$ inch long), humpbacked, biting flies that may reach high populations in the spring and early summer, particularly in pasture areas along streams. The immature stages are found in flowing water. Pupation occurs under water and the adults float to the surface, ready for flight, feeding, and mating. Adult flies feeding on horses and other animals can pose serious animal health problems, and the irritation caused by black fly bites can make horses unmanageable. Anemia, as a consequence of black fly population, is high. Bites may also cause severe reactions such as toxemia and anaphylactic shock. These reactions can result in death. Control is difficult. Species that feed in the ears of horses can be controlled using insecticidal applications or petroleum jelly in the interior of the horses' ears. When possible, horses can be stabled during the day and pastured at night. Black flies only feed during daylight hours and usually do not enter stable areas. Area sprays or general topical applications of insecticides are not very effective.

Biting Gnats

"No-see-ums," "punkies," or biting midges can be a serious pest of horses. Blood loss and irritation associated with the feeding of these very small (usually less than 0.04 inch), blood-feeding flies can be significant. The immature stages of these flies complete development in water in a variety of locations, from tree holes or manufactured containers to lakes and streams. Adults often are unnoticed because of their size and because they are active at night, late evening, or early morning. Direct treatment of horses with wipes or sprays containing insecticides or repellents can provide them relief.

Horn Flies

The horn fly is normally a pest of grazing cattle. However, when cattle and horses are pastured together, this fly will feed on horses. Horn flies are about one-half the size of stable flies and, like stable flies, are biting flies. The horn fly usually remains on the host animal almost continually, both day and night. Females lay eggs on fresh cattle droppings. Sprays or wipes can be used successfully on horses.

Nuisance Flies

Several types of nuisance flies may be associated with horses or their premises. These include the house fly, bottle flies, false stable flies, and other species of barnyard flies. Face flies, usually a pest of cattle, may also affect horses, particularly when cattle are nearby.

Two major pest species that bother horses are the stable fly and the house fly, a nonbiting species. A distinguishing feature, visible to the naked eye, that separates the two species is the distinct stiletto-like proboscis of the stable fly that extends forward beyond the head. This sharply pointed beak is used to pierce the skin and draw blood. The house fly cannot bite since it has sponging mouthparts.

House Fly. Both male and female house flies are grayish-brown with a black- and grey-striped thorax. The house fly is a medium-sized fly ranging from about one-quarter to one-third inch long with sponging mouthparts. House flies do not bite but feed on a variety of plant and animal wastes and garbage, as well as other sources of carbohydrates and proteins (see Figure 16-9).

House fly eggs are about 0.04 inch long, whitish, and slightly curved. The females generally deposit eggs in batches of about 100 eggs at a time. Each female may deposit four to six batches of eggs during an average lifetime of two to four weeks during the summer.

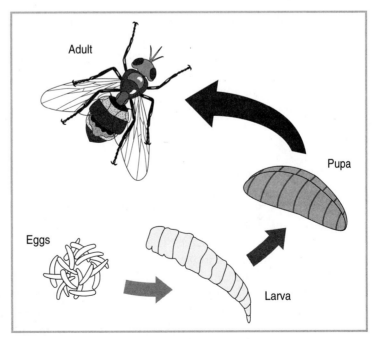

Figure 16-9 Life cycle of the house fly

The three larval stages are similar in appearance to stable fly larvae. The third stage reaches approximately one-half to two-thirds inch in length. Differentiation of the two species is based on the size and shape of the posterior spiracles (or respiratory tract openings).

Pupae are barrel shaped and are of the same approximate size and coloration as stable fly pupae.

House fly females lay their eggs in clusters, preferably in moist decaying organic material. Eggs hatch within eight to forty hours, depending on temperature. Larvae feed on yeast, bacteria, and decomposition products in their development site. Larval development through three stages takes from three to eight days. Larvae crawl to drier areas to pupate when feeding is completed. The pupal stage lasts from three to ten days, depending primarily on temperature. Adults emerge from the puparia and begin feeding within twenty-four hours. Males are ready to mate shortly after emergence and females begin mating by the second or third day. Most females mate once and deposit eggs in batches every two to four days.

The flies feed on carbohydrates and proteins. Females require protein to produce viable eggs. Solid foods are first liquified with saliva and are then ingested using the sponging mouthparts.

The entire life cycle from egg to adult can be completed in as little as ten to fourteen days during warm weather. Like the stable fly, house flies overwinter in sites, such as silage or manure piles, where microbial fermentation heats the larval habitat. House flies may develop throughout the year in heated livestock facilities. They are active near sources of food during daylight hours and generally rest at night on stationary objects both indoors and out. The flies prefer shaded areas during much of the day and commonly move inside structures where livestock are held.

House fly management, like stable fly management, is based on a strong farm sanitation program. The methods for reducing house flies are the same as those discussed for the stable fly.

Face Fly. The face fly is usually a pest of grazing cattle. However, when horses are pastured with or close to cattle or when face flies are numerous, these flies will feed on secretions around the eyes of horses. Adult face flies look much like house flies. The face fly does not bite, but the persistent feeding behavior of the fly makes it a nuisance pest. In addition, the face fly can mechanically transmit parasites or pathogens to the horse. Control of face flies is difficult. Relief can be obtained by stabling horses during the daytime when the face fly feeds. In addition, since the face fly feeds predominantly on cattle, pasturing horses separately from cattle will lessen the incidence of these flies on horses. Topical insecticide applications are usually not effective because face flies spend little time on the vertebrate host.

MANAGEMENT OF PASTURES

Since horses do best when they are allowed to graze, maintaining good pastures is an important management priority. Sound management is essential to keep the desired grass species persistent and productive. Pastures can be improved by the use of lime and fertilizer or by reseeding. The following are management tips for pastures.

Avoid over- or undergrazing. Horses are notorious spot grazers. They will seriously damage desired species in some areas unless they are moved into new pastures frequently. Some form of rotational grazing is desirable. The correct acreage per horse changes with the season as well as with other factors. However, a good rule is to provide at least one acre of good-quality pasture per horse. Five or six **paddocks** should be set up and horses permitted to graze first in one area for about one week and then changed to another. This system helps to keep the legumes and grasses growing better and increases the feed available per acre. In addition, rotating horses from pasture to pasture breaks the life cycle of some parasites (see Figure 16-10).

Clip pastures regularly during the growing season. Clipping to a height of two to three inches after horses are moved to a new paddock helps to control weeds, prevent grasses from heading, and in general keeps the pasture in a more desirable condition.

Figure 16-10 In order to maintain good strong pastures, it is important to rotate horses from pasture to pasture to avoid overgrazing.

Drag pastures with a chain link harrow at least once per year.
Dragging helps to spread manure droppings, which reduces parasite populations by exposing them to air and sunlight. Dragging also helps to smooth over
areas dug up by horses' hoofs on wet soil.

Apply fertilizer as needed. Improved horse pastures must be fertilized
annually if legumes and grasses are to persist and remain productive. The
fertilizer to use depends on the grass species present. A complete **soil test**
every two or three years is the best guide.

Pasture Improvement

Pastures with good stands of desirable grass and legume species need proper
soil fertility combined with good management to assure continuing good horse
pasture. Most permanent bluegrass pastures produce less than 2,000 pounds of
dry matter per acre per year, which is far below their potential. Yields on many
pastures can be doubled simply by applying lime and fertilizer. Liming and top
dressing Kentucky bluegrass pastures with phosphate, potash, and nitrogen
costs much less and is less work than complete pasture renovation.

However, there are times when lime and fertilizer are not enough to
restore a pasture and complete renovation is necessary. When renovating an
old pasture the following points should be considered:

1. Perform a soil test to determine lime and fertilizer requirements. This
    is the only sure way of knowing how much lime and fertilizer are
    needed.

2. Apply required lime several months before doing the actual seed-
    ing. Disking or plowing will help to mix the lime evenly throughout
    the soil.

3. Select a seed mixture that complements the pasture drainage charac-
    teristics.

4. Destroy or suppress the old pasture by plowing or using herbicides.

5. Use the appropriate method of seeding based on extent of tillage.

6. Protect the seeded area until the new plants are well established.
    Where recommended mixtures are seeded without a companion crop
    and weeds are controlled, new seedings can become established in a
    single year.

In heavy traffic areas, along fences and around gates and water troughs,
tall fescue may be used. Fescue is generally considered less palatable than
bluegrass, but tall fescue produces one of the toughest heavy-traffic sods of any
adapted grass. Older stands of fescue often are infested with an endophyte
(within the plant) fungus. Toxins associated with this fungus can cause
lowered reproductive rates, abortion, agalactia (lack of milk), and prolonged

gestation with mares. Thus, whenever establishing new fescue stands for horses, endophyte-free tall fescue seed should be used. Brood mares should be removed from pastures containing endophyte-infested tall fescue at least ninety days prior to foaling.

Managing Health Concerns of Forages

Horses are extremely susceptible to molds, fungi, and other sources of toxic substances in forage. Mold problems generally occur in hay that has been baled at too high a moisture level (20 percent or more) without the use of a preservative. This is especially a problem with first-cutting hay because it is harvested during a period of time when it rains frequently and the weather conditions are less than ideal for hay drying.

Horses should always be fed clean, unmoldy forages. In addition to molds and fungi, some forage species contain chemical compounds that can have negative health effects on horses.

Sudangrass and sorghum-sudangrass hybrids contain compounds that can cause muscle weakness, urinary problems, and death in severe cases. Do not feed these grasses to horses!

Older varieties of tall fescue contained an endophyte fungus that could cause severe health problems if horses have only tall fescue to eat during the summer months. Newer tall fescue varieties that are free of the endophyte fungus are now available.

Another health problem occurs when horses are fed hay that contains blister beetles. When consumed, the beetle causes irritation to the lining of the digestive tract that usually results in death. The danger of blister beetles is discussed in Chapter 14. Alfalfa hay produced in southern areas of the United States is most generally associated with blister-beetle contamination. Do not feed any hay containing blister beetles to horses.

Poisonous plants in pastures or hay can also be fatal to horses. Ornamental shrubs and nightshade are common poisonous plants. Any plant poisonous to other animals is probably poisonous to horses. Some are highly palatable and should be identified and removed from pastures, but many poisonous plants are not palatable and horses will not eat them unless the forage is inadequate to meet their needs. A list of common poisonous plants is found in Table 14-4 in Chapter 14.

WOUND MANAGEMENT

A **wound** is a disruption in the integrity of living tissue caused by physical means. Managing wounds requires recognizing the types and characteristics of wounds and the associated symptoms in the horse. The characteristics of a wound include:

1. The horse's temperature is usually normal, but will be elevated when infection is present and below normal if the horse is going into shock.

2. Pulse is often normal even with severe wounds, but may be increased if blood loss is excessive.

3. Mucous membrane color will range from normal to pale in cases of excessive blood loss.

4. Capillary refill time will be normal except in the case of blood loss and shock, when it may be over two seconds.

Wounds can be classified as clean wounds, contaminated wounds, and infected wounds; or they can be characterized as open or closed wounds.

1. A clean wound is a sterile or noncontaminated wound less than six hours old. After this time, the wound may or may not become infected.

2. A contaminated wound is less than six to eight hours old and, despite the presence of bacteria, the wound is not infected.

3. An infected wound is usually more than six to eight hours old, during which time bacterial activity has infected the wound. The result is pus and dead tissue, and there may be septicemia (the presence of bacteria or their toxins in the blood).

Open wounds include incisions, lacerations, abrasions, punctures, perforations, and penetrating.

Closed wounds include contusions, hematomas, seromas, abscesses, or traumatic hernias.

Veterinarians need to examine, or at least discuss, a wound on an emergency basis during the first six to eight hours after an injury occurs. Suture repair has the best chance for healing if performed within this time period. Once this time has elapsed, wounds that are sutured invariably break down and, for that reason, older wounds are often left to heal by granulation after they have been cleaned, rather than sutured.

Tetanus toxoid booster should be given if the horse has not had one within six months. If the vaccination status of the horse is unknown or uncertain, both tetanus toxoid and tetanus antitoxin should be administered.

Summary

Good management is knowledge in action. Good managers learn to recognize horses under stress and, where possible, take action to alleviate stress. Stress can be caused by mechanical injury, poor nutrition, and disease. Horses indicate stress through behavioral changes.

Positive identification of horses is practiced by a good manager. Horses can be identified by body markings, tattoos, freeze brands, blood typing, and microchips. Positive identification prevents theft, fraud, and is necessary for insurance.

Two of the most critical times in the life of a horse are the neonatal and weaning periods. Good management practices here insure a healthy foal for training and a healthy mare for future foals.

Proper selection of bedding is necessary for stall maintenance. Selection of bedding depends primarily on availability, price, and absorptive capacity.

All good managers will have a sound sanitation program for fly control. This program can also include various chemical methods. An understanding of the types of flies, their life cycles, and their habitats is a necessary part of control.

When possible, horses should be pastured. Sound management keeps pastures productive and improves poor pastures through renovation.

Horses sometimes receive a wound. Wound management requires that the owner recognize the characteristics of the various types of wounds.

REVIEW

Success in any career requires knowledge. Test your knowledge of this chapter by answering these questions or solving these problems.

True or False

1. The Jockey Club was the first organization to set up an accurate identification system for horses.

2. A clean wound is one that has been cleaned properly.

3. A foal's navel cord should be allowed to sever naturally.

4. A foal should be taught to accept basic discipline and handling at weaning time.

5. Horses have only binocular vision.

6. Social order is important in a herd of horses.

Short Answer

7. List the four categories of stress for a horse.

8. List the ten pieces of information every horse owner should keep on file for each horse.

9. What are the three metabolic problems closely associated with nutrition in a horse?

10. Identify four characteristics of a wound.

11. List the five factors that determine the best time for weaning a foal.

12. Name the two substances that could be added to a pasture to double its yield.

13. What is the most popular type of bedding material?

Discussion

14. What are the reasons for marking or identifying a horse?

15. Why is bedding for a horse important, and what are the factors that determine the type of bedding used?

16. Discuss three management practices for pastures.

17. Discuss a fly control program.

18. Why would an enema be used on a foal?

19. Indicate the differences among a relaxed horse, an angry horse, and an interested horse.

20. Define the tying-up syndrome and its signs.

STUDENT ACTIVITIES

1. Using Table 16-1, collect some bedding materials and develop a demonstration to show the water-absorbing ability of each material.

2. Draw a diagram that explains monocular and binocular vision. What type of vision do humans have?

3. Observe behavior of horses within a herd. Record your observations in a written log and with a video camera.

4. Research the tying-up syndrome. Use the Internet or other resources to develop a report that gives the recent physiological explanations for the condition.

5. Draw diagrams of the different types of wounds.

6. Visit several horse pastures and describe their conditions. Identify management practices that are being practiced or those that are missing. Collect, press, mount, and label plant samples from the pastures.

7. Make a set of flash cards to teach someone how to identify the common body markings of horses.

8. Make an insect collection of the flies found around horses. Using an insect key, label the flies with their common and scientific names.

ADDITIONAL RESOURCES

Books

American Youth Horse Council. 1993. *Horse industry handbook: a guide to equine care and management.* Lexington, KY: American Youth Horse Council, Inc.

Blakely, J. 1981. *Horses and horse sense: the practical science of horse husbandry.* Reston, VI: Reston Publishing Company, Inc.

Davidson, B., and Foster, C. 1994. *The complete book of the horse.* New York, NY: Barnes & Noble Books.

Ensminger, M. E. 1990. *Horses and horsemanship.* 6th ed. Danville, IL: Interstate Publishers, Inc.

Evans, J. W. 1989. *Horses: a guide to selection, care, and enjoyment.* 2nd ed. New York, NY: W. H. Freeman and Company.

Hill, C. 1994. *Horsekeeping on a small acerage.* Pownal, VT: Storey Communications, Inc.

Lewis, L. D. 1996. *Feeding and care of the horse.* 2nd ed. Media, PA: Williams & Wilkins.

Price, S. D. 1993. *The whole horse catalog.* New York, NY: Fireside.

Prince, E. R., and Collier, G. M. 1986. *Basic horse care.* New York, NY: Doubleday & Company, Inc.

Stoneridge, M. A. 1983. *Practical horseman's book of horsekeeping.* Garden City, NY: Doubleday & Company, Inc.

University of Missouri-Columbia Extension Division. (n.d.) *Missouri horse care and guide book.* Columbia, MO: Cooperative Extension Service, University of Missouri and Lincoln University.

CHAPTER 17

Shoeing and Hoof Care

Foot and hoof care is essential to the horse because neglect can lead to lameness, a change in the flight of the foot, and a rough gait. The adage "no foot, no horse" is as true as ever. Lameness of feet and legs has many causes. If a problem is not treated immediately, permanent disabilities can result. The foot care a horse receives can hasten or delay permanent unsoundness.

OBJECTIVES

After completing this chapter, you should be able to:

- Describe the internal and external parts of the hoof
- Explain the three main functions of the hoof wall
- Explain why the condition of the frog of the horse's foot is a good indication of the health of the horse
- Discuss how corrective shoeing can help toed-in and toed-out horses
- List steps in picking up a horse's feet
- Explain the importance of inspecting feet daily
- List five tools used in caring for a horse's feet
- Describe how to shoe a horse
- Discuss why it is important to start foot care early in a horse's life
- Describe why trimming should be done carefully
- Explain how the weight of the horse is carried on the foot/hoof
- Name common problems of the feet
- Describe trimming and how trimming can correct minor problems

426

K E Y   T E R M S

Bars	Hoof leveler	Pincher
Clinch cutter	Hoof pick	Plantar cushion
Coffin bone	Hoof wall	Puller
Cold-fitted	Horseshoeing	Rasp
Commisures	Hot-fitted	Rings
Coronet	Laminae	Shod
Deep flexor tendon	Navicular bone	Shoeing apron
Frog	Nippers	
Hammer	Periople	

STRUCTURE OF THE FOOT

To understand proper care of a horse's feet, the structure of the foot and the functions of its various parts must be known and understood. The major parts of a horse's foot are the hoof wall, coronet, sole, frog, and the internal structures such as the bones, cartilages, tendons, and connective tissue (see Figures 17-1 and 17-2).

Hoof Wall

The **hoof wall** is a horny substance made up of parallel fibers. It should be dense, straight, and free from **rings** (ridges) and cracks. Viewed from the side, the wall at the toe should be a continuation of the slope of the pastern.

The main functions of the hoof wall are to:

- Provide a weight-bearing surface not easily worn away

- Protect the internal structure of the foot

- Maintain moisture in the foot

Usually, the hoof wall is thicker at the toe than at the quarter (side) and heel. The hoof wall is protected by the **periople**, a varnish-like coating that also holds moisture in the hoof (see Figures 17-1 and 17-2).

Coronet

The **coronet**, or coronary band, is the source of growth for the hoof wall. It is directly above the hoof wall and is protected by a thick layer of skin and dense hair. A healthy foot will grow about ⅜ inch per month. A change in the rate of growth of the hoof can be caused by a change in the amount of exercise, the ration, illness, and the general state of health and condition of the animal.

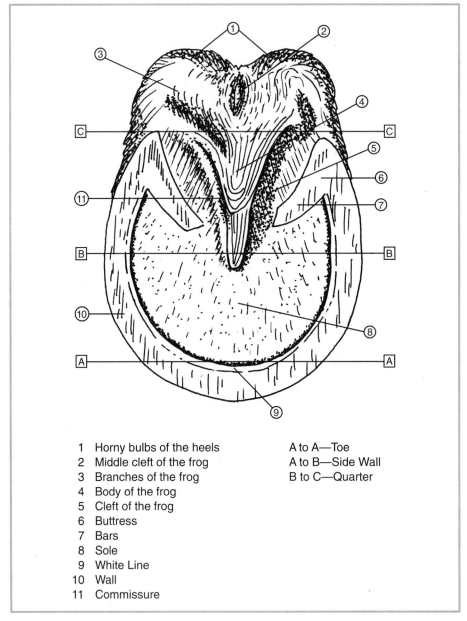

1 Horny bulbs of the heels
2 Middle cleft of the frog
3 Branches of the frog
4 Body of the frog
5 Cleft of the frog
6 Buttress
7 Bars
8 Sole
9 White Line
10 Wall
11 Commissure

A to A—Toe
A to B—Side Wall
B to C—Quarter

Figure 17-1 Parts of the horse's hoof

Injury to the coronary band can result in irregular growth of the hoof wall and can develop into a permanently unsound hoof wall (refer to Figure 17-2).

The hind feet may grow faster than the forefeet, and unshod feet may grow faster than shod feet. The feet of mares and geldings seem to grow faster than those of stallions.

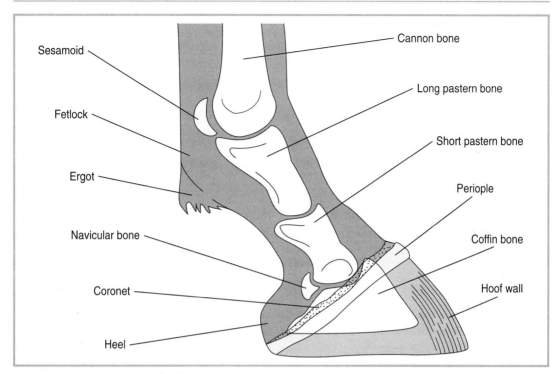

Figure 17-2   The bones in the horse's foot

Sole

The sole of the foot is a horny substance that protects the sensitive inner portions of the foot. It should be firm, slightly concave and of uniform texture. The horse has no feeling at the exterior sole surface.

A flat-footed horse tends to receive more bruises and injuries to the sole. Also, horses that have experienced founder and have developed a dropped sole are more easily bruised at the sole.

Frog

The **frog**, located at the heel of the foot, forms a "V" into the center of the sole (refer to Figure 17-1). The frog is a spongy, flexible pad and is also a weight-bearing surface. It is the intermediate organ between the **plantar cushion** and the source of pressure from the horse's weight. The frog is separated from the sole of the foot by two lines called **commissures**.

The condition of the frog generally is a good indication of the health of the foot. Without proper flexibility, expansion and ground contact, the frog cannot perform its function in complementing the circulation of blood and absorption of shock throughout the foot.

Internal Foot Structure

To be able to provide proper foot care, the owner or handler needs to understand the important internal parts of the horse's foot and their functions (see Figure 17-2).

The **coffin bone** provides the shape of the foot and the rigidity needed to bear weight.

The plantar cushion expands and contracts to absorb shock and pumps blood from the foot back toward the heart.

The **navicular bone** serves as a fulcrum and bearing surface for the **deep flexor tendon**, which is responsible for extension of the foot as it progresses through a stride.

Sensitive **laminae** serve as a means of attachment for the hoof wall and the coffin bone and also as the main area of blood circulation within the foot.

CARE OF THE HOOF

Foot care is one of the most neglected of all horse management practices. Most lameness that impairs the usefulness of a horse can be prevented by proper foot care and reasonable management.

Most foot care practices can be done by the average horse owner. But horse owners should know when to seek the help of a professional, especially for corrective shoeing and disease treatment and control. A **farrier** is a person who cleans, trims, and performs the actual **horseshoeing**. The farrier and the veterinarian should work together to keep the horse sound (see Figure 17-3).

Foot care should be as routine as feeding and watering. It should include:

- Routine cleaning

- Periodic trimming

- Corrections of minor imperfections

- Treatment of foot diseases and injuries

Ideally, a horse's feet should be inspected and cleaned every day. A **hoof pick** or fine-bristled wire brush can be used for cleaning the sole, frog, and hoof wall. This will improve the likelihood that problems will be detected early. A nail or other object stuck in the foot is a serious medical condition and should be treated as soon as possible. Too much pressure from the wire brush can damage the periople. This would disturb the moisture balance of the foot.

The hoof wall grows an average of ¼ inch per month. Most horses' hooves are trimmed and **shod** every six to eight weeks. This depends, of course, on the rate of growth and the wearing of the hoof wall. If the horse spends time on hard surfaces, the hoof wears down faster than it does on a horse in a soft, lush pasture.

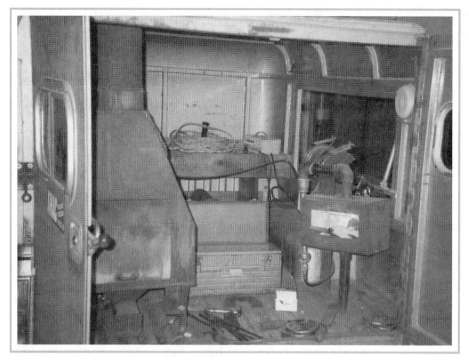

Figure 17-3 A farrier's truck

Sometimes, the sides of the hoof will grow or wear at different rates. This causes the legs to look, and possibly be, crooked. Corrective trimming can level the hoof.

Care of the Foal's Feet

Foot care should begin early by teaching foals to allow handling and cleaning of their feet. If this practice is followed, it will save both the young horse and the farrier considerable trouble later when it is time to trim and shoe. Handling a foal or young horse's feet may be somewhat tricky at the start, but by following proper procedures young horses will soon become very easy to handle and trim. Many foals have crooked legs. Corrective trimming can help straighten their legs by evening the wear on their hooves.

Tools

As with any trade, special tools are used in caring for the horse's feet. The farrier's basic tools include:

- Hoof pick—used to clean any dirt or rocks from hoof crevices
- **Nippers**—used to remove extra hoof wall
- **Clinch cutter** and **pincher** or **puller**—used to remove shoes that have been worn and are ready to be taken off

- **Hammer**—two kinds can be used: one for driving the nails in and the other for shaping or rounding the horseshoe on the anvil
- **Rasp**—needed for leveling the foot
- **Hoof leveler**—used to determine the angle of the hoof wall and if the hoof is level to the ground

Additional equipment often includes a heavy **shoeing apron** to protect the horseshoer and an **anvil** to shape the horseshoes.

Foot Cleaning

The foot should be cleaned from the heel toward the toe with a hoof pick. Special care should be taken to clean the commissures on each side of the frog and the cleft of the frog itself, but the heel should not be opened excessively. This weakens the area and interferes with proper contraction and expansion of the heel (see Figure 17-4).

After riding, the sole must be cleaned and checked for gravel or other foreign objects that could be lodged in the natural depressions of the foot. A nail, gravel, stick or other object can work into the foot and cause lameness for a long time. Objects have been known to exist in a horse's foot for as long as a year before emerging at the heel or along the coronet. When a foreign particle emerges at the coronary area, a sore, called a quittor, usually develops. This problem can easily lead to serious infection.

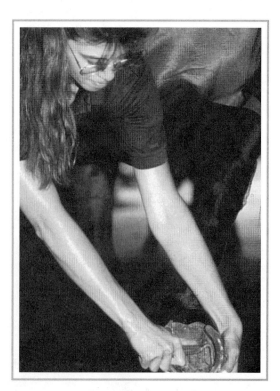

Figure 17-4 Cleaning hooves on a routine basis is good preventive medicine for horses.

Periodic Trimming

Trimming of the feet is important, although it is not needed as frequently as cleaning. Trimming should be done at about four-week intervals on horses kept in stalls or paddocks, or about six-week intervals for horses used heavily or running in pastures.

The main goal in trimming is to retain the proper shape and length of the foot. Most people should feel comfortable pulling shoes and trimming feet while they wait for the farrier.

The bottom of the foot should be kept level and the inside and outside walls should be maintained at equal lengths. The toe of normal feet and pasterns should be three inches long; the quarter, two inches; and the heel, one inch.

The hoof wall should be trimmed with nippers to remove excess length, then a rasp used to smooth and level the bottom of the foot. Each stroke of the rasp needs to run from the heel through the toe to prevent uneven areas in the hoof wall.

A white line is external evidence of the lamination (sensitive laminae) between the hoof wall and the coffin bone. The sole of the foot is usually the same thickness in a normal horse. The sole should not be trimmed to an unnatural shape. To do so would make parts of the sole dangerously thin and tender.

Trimming the sole, referred to as lowering the sole, is done to keep the pressure on the hoof wall rather than on the sensitive inner parts of the foot. The dead, flaky tissue should be trimmed from the sole. Live tissue, elastic when stretched between the fingers, should not be trimmed away.

The frog should not be trimmed excessively because it should contact the ground with each step. It is trimmed only enough to remove dead tissue and to provide a uniform and adequate fissure along the junction of the sole and the frog.

After the bearing surface has been rasped to a level surface of proper length, the edges of the wall should be rounded if the horse will not be shod. This prevents chipping and peeling as the foot contacts rocks, logs or other obstructions.

Trim the heels low enough to promote expansion and prevent contraction of the heels. The main concern is to trim often enough to prevent cracking and uneven wear, which could eventually contribute to the improper set of the feet and legs. With a little practice, most horse owners should be able to routinely trim the feet of horses that do not need corrective work. To prevent harmful mistakes, owners need to seek the help of a professional farrier when trying to correct an improper turn or set of the feet and legs. A more detailed discussion of foot trimming appears later in this chapter.

Maintaining Hoof-Wall Angle

Horse owners should maintain the proper angle of the hoof wall in relation to the ground and the angle of the pastern. Shoes that are left on too long change the angle of the foot relative to the pastern and can cause lameness. The angle of the hoof wall should approximate the angle formed by the shoulder and the pastern—usually 45 to 55 degrees.

Since the hoof wall is narrower at the heel than at the toe, heels wear first, whether the horse is barefoot or on shoes. Low heels put more stress on the tendons of the leg. If a horse is shod at a 50-degree angle, this angle may change. A 50-degree angle might be down to 46 or 47 degrees in four to six weeks. This affects the action of the horse and puts more strain on tendons and ligaments.

As the hoof grows larger, the walls at the heels will overlap the shoe. When a shoe presses on the **bars**, the danger of producing corns in the foot exists. Running a horse with shoes that have been left on too long also can cause bowed tendons. Regular trimming and shoe re-setting are essential in avoiding these problems.

Foot angle varies from breed to breed and much variation is found among horses of the same breed. Generally, the Western breeds have steeper pasterns and a greater angle at the ground than the other breeds. Unless some correction is needed, as in forging and scalping, the foot should be trimmed to its natural angle because any change would result in stress to other areas of the column of bones in the leg.

Corrections of Minor Imperfections

The most common deviations from a normal set of feet and legs are when either front or rear feet toe in (pigeon-toed) or toe out. Other problems commonly corrected by trimming are cocked ankles, buck knees, calf knees, sickle hocks and slight rotations of the cannon bone. Also, some common faults in the movement of feet in a stride—forging, scalping, interfering, and brushing—are corrected by careful trimming.

When trimming feet, conformation of the horse needs to be considered. For example, a splayfooted horse (feet turned out) bears more weight on the inside wall and heel than on the outside. Wear is greatest, both shod and barefoot, where weight is borne. The objective in corrective trimming is to remove more of the outside wall and heel than the inside. This will shift the horse's weight near the center of its feet. A pigeon-toed horse is trimmed exactly the opposite.

Bone structure of adult horses cannot be changed much, but their action can be improved. Corrective trimming of young horses every six weeks or two months up to two years of age will substantially improve bone structure.

Treatment of Foot Diseases and Injuries

Disease organisms concentrate where animals are confined, so cleanliness is important. Horses kept in a stall or small pen should have their feet picked or cleaned daily to reduce the risk of thrush. Thrush is the condition resulting from bacterial penetration into the frog and surrounding area. The bacteria

How To Stay in Good With Your Farrier

Answer these questions about yourself:

- Was the farrier bruised from head to foot by an old spoiled horse while you stroked the horse's neck, proclaiming it wouldn't hurt a fly?

- Did you tip or offer to pay more when the farrier committed extra time and patience to a young horse being shod for the first time?

- Did you call the farrier out to shoe three horses, then decide to shoe only one when he or she got there?

- Did the farrier have to help chase the horse all over the farm to catch it before shoeing?

- Did you pay the farrier promptly and in full?

Answer these questions about the farrier's work:

- Was the shoe shaped to fit the foot?

- Were the foot and shoe both leveled?

- Was the shoe set fully forward, foot not dubbed off?

- Were the frog and bars "opened up" but not pared away?

- Was the experience satisfactory for the horse or were its ribs bruised from blows of the rasp?

Honestly answering these questions will help you build a solid relationship with your farrier.

produce a foul odor and cause the frog to become soft and mushy. If allowed to go untreated, serious lameness can result and extensive treatment will be necessary.

Extremely wet conditions such as a muddy lot or wet stall promote rapid drying of the feet. The natural oils and protective films of the foot are eroded from constant contact with external moisture. Large horses with small feet commonly have hoof dryness problems.

Moisture. Moisture in the horse's feet helps to maintain flexibility and prevent cracking. Most of the moisture needed in a healthy and well-protected foot can come from within.

One way to maintain proper moisture in the foot is to regularly apply a good hoof dressing containing some animal fat such as lanolin. If the dressing is not a petroleum derivative, it can be massaged into the coronet, the frog, and the sole as well as on the hoof wall. The dressing helps to keep the sole pliable and to eliminate dead tissue around the frog and heel. Also, massaging the coronet stimulates growth of a healthy new hoof wall.

Lost Shoes. When a shoe is lost, it is important to promptly cut the hoof wall level with the sole to prevent it from breaking above this point while awaiting the farrier. Removing the opposite shoe and lowering the hoof wall to equal the length of the other hoof will balance the gait of the horse.

Nail Pricks. Much lameness results from nail pricks. Horses should not be ridden in areas littered with trash and boards containing nails. Injury caused by nails can ruin a horse.

As soon as a nail prick is identified, prompt medical attention and packing is needed to prevent infection by ground-borne disease organisms. Horses usually are fitted with a protective boot after being pricked by a nail and may be shod with a pad after the condition has been treated and shows signs of recovery.

Founder. Fat horses tend to have problems with laminitis (founder). This is especially common among horses with some Shetland pony breeding. Grass founder in the spring produces more laminitis than any other single cause. If the horse is fat, grazes abundant grass, and is not exercised, there is great risk of laminitis.

Laminitis commonly causes lameness. Horses with laminitis have extreme pain and soreness, especially in their front feet. They try to bear their weight on their back legs and lighten the front end as much as possible by carrying their front feet forward and their back feet up under their bodies. Therapeutic trimming and shoeing may make a horse with laminitis sound enough for light work and normal reproduction. Chapter 14 provides more details on laminitis.

HORSESHOEING

Many sizes, shapes, and types of shoes are available (see Figure 17-5). Many different types of prefabricated shoes are available that are either **hot-** or **cold-fitted** to the horse. The most important aspect of shoeing is fitting the shoe to the horse and *not* the horse to the shoe.

Learning and practicing safe handling of the horse's feet are important steps in performing routine foot care. In shoeing a horse, farriers follow a number of steps to assure a correct fit.

Figure 17-5 Many different types of horseshoes

Picking Up a Horse's Feet

Pick up the front foot by rubbing the leg up high and gently working down to the ankle. Brace a free hand against the horse's shoulder for more stability. If the horse fails to lift its foot, gentle pressure on the tendon behind the cannon bone with thumb and forefinger usually will persuade the animal to cooperate. Once the foot is raised, allow the horse to hold it in a comfortable position.

If the horse is uncomfortable, it will not stand well. Trying to maintain a relaxed and comfortable position is best for both horse and yourself when holding the horse's leg between your knees. For this reason, shoeing young horses for the first time in fly season can be a challenge unless effective fly repellents are used. Figure 17-6 shows how to pick up a horse's feet.

Picking up the hind foot of foals and young horses is dangerous unless done correctly. Pick up the left foot first, since most horses are accustomed to being handled from this side. Approach the horse from the front and place the left hand on his hip; then run the right hand down the back of the horse's leg to just above the ankle. Pull forward on the cannon until the horse yields its foot. If you feel tense muscles, go more slowly. Step promptly under the raised foot with the inside leg and pull the foot into your lap. Lock it in place with your elbow over the hock and your toes pointed toward each other. Hold the foot in this position so both hands are free to work. If the horse resists, move more slowly.

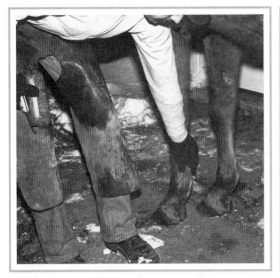 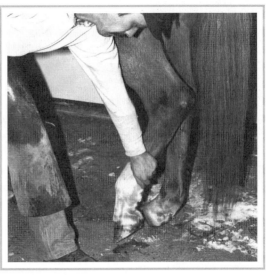

(a)   **(b)**

Figure 17-6 The proper way to pick up a horse's feet: **(a)** Near forefoot: Slide your left hand down the cannon to the fetlock. Lean with your left shoulder against the horse's shoulder. When the horse shifts weight and relaxes on the foot, pick it up. Reverse for picking up the off forefoot. **(b)** Near hind foot: Grasp the back of the cannon just above the fetlock and lift the foot forward. Reverse sides for picking up the off leg.

To lift a hind foot, keep one hand near the hip and go down the leg slowly with the other. Work in close to the horse. If the horse won't yield the foot, squeeze the tendon to get the horse to yield the foot. Move the hand in front of the cannon or fetlock as the foot rises. Position the foot firmly between your knees. If the horse struggles and wishes to regain its foot, let it do so. Repeat the procedure until the horse learns to yield its feet willingly.

Now gently push the horse away with the left hand while pulling its foot toward you with the right hand. Next, step to the rear of the horse with the inside foot, pulling the leg straight behind it, and at the same time drawing the hock up under the left arm. The same procedure is followed for the right hind foot except that it is worked left-handed. The foot, bottom up, can now be rested comfortably on your knees for trimming or shoeing. A satisfactory trimming job can be accomplished with a hoof knife, a rasp, and a nipper.

Removing shoes and cold shoeing, however, require additional tools—including shoepullers, clinchers, shoeing hammer, punch, clinching block, clinch cutters, hoof pick, anvil, shop hammer, and of course shoes and nails. When trimming or shoeing, an apron should always be used. If a shoeing apron is not available, a pair of heavy chaps are good.

Trimming should begin by cleaning the hoof with the hoof pick. Make sure to draw the pick from the heel toward the toe. This method does a good job of cleaning and is safer for the horse. Don't clean from the toe toward the heel. If the horse jerks its foot and the hoof pick from your hand, it can experience severe injury when it steps on the pick in this position.

Removing Old Shoes

Clinches of old nails must be cut or straightened to remove the shoe. If the shoe is pulled without this operation, it will not only be more difficult to remove, but the walls of the hoof may be injured. Clinches may be cut with the clinch cutter or rasped off. A "pull-off" rasp is an old rasp no longer used to level the foot.

Place the blade edge of the clinch cutter under the clinch and straighten it for pulling by light hammer blows. If you have difficulty getting it started, lean the top out and use the back corner nearest your hand.

Most commercial farriers rasp the clinches off with the fine side of their rasp, because it is faster than using a clinch cutter. Place the shoe pullers under the shoe at the heel and push down toward the toe to remove the shoe. This operation is repeated on the opposite heel, always working toward the toe, until the shoe is completely free. Do not pry sidewise because of danger of sprains to the horse's tendons.

Trimming Feet

Begin trimming the foot (see Figure 17-7) by removing the loose, flaky (outermost) part of the bars. Next, trim each side of the frog just enough to open the seams on each side at the heel of the hoof. This helps keep filth from collecting. Do not lower the frog. This structure should touch the ground when the horse stands on the trimmed foot.

After the frog and bars have been trimmed, use your hoof knife to trim out the soft, flaky part of the sole in order to determine how much of the hoof wall should be trimmed away. Observe the juncture of the sole and wall at the toe. Decide how much will need to come off the toe, with a lesser amount at the heels.

Now start the nipper at the heel of the hoof at a depth level with the sole. Often beginners get too deep at the heels and not deep enough at the toe. Proceed around the hoof until the opposite heel is finished. The hoof wall should not be trimmed below the level of the sole. Now the hoof should appear relatively level and both heels should be the same height.

A relatively level hoof that requires a minimum of rasping is not easy for beginners to accomplish. Beginners do not adequately lower the sole, which serves as a guide for the nippers. This results in unevenness or not removing enough of the wall. Such a condition is not serious, but requires unnecessary rasping.

(a)

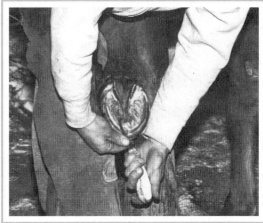

(b)

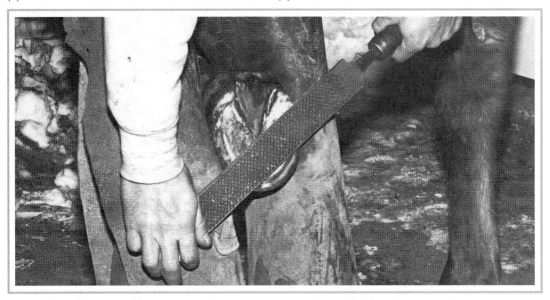

(c)

Figure 17-7 The proper way to trim a horse's hoof: (a) Use of the hoof knife in conjunction with the nippers to pare the dead and flaky tissue from the sole. (b) Use of nippers to cut the horny wall (outer surface) to a proper angle and length to fit the conformation of the animal. (c) The rasp is used to prepare a ground-bearing surface by eliminating jagged and sharp corners on the bottom of the hoof wall. *(Courtesy of Frank Morton, Farrier. Photos by Steven M. Ennis)*

Use the rasp to finish trimming. Draw the rasp from the heel toward the toe, always taking care to keep the pressure equal over the entire foot.

Move the rasp around over the sides and toe, being careful not to get too deep in one spot. Use a sharp rasp for ease and speed in trimming.

The rasp may be reversed and drawn from the toe toward the heel when trimming the inside of the hoof. Too much pressure on the rasp at the heel will lower the heel too much, and the hoof will not have the correct angle.

Checking Foot Levelness

When you think the foot is level, check levelness by sighting down the hoof from heel to toe. Drop the hoof down so it rests in a normal position with the hoof hanging free. Holding the hoof itself may result in its being slightly twisted, and it may therefore appear level when it is not. Be sure neither heel is high and that there are no low spots around the wall.

The hoof should have an angle of 45 to 55 degrees, depending on the conformation of the particular horse—slope of shoulder and length and slope of pastern. If a horse tends to over-reach, trim the hind feet two or three degrees less than the front, for example, if the front feet are fifty-three degrees, the hind feet should be fifty degrees. This will permit the front feet to break over faster, and prevent over-reaching.

Both front and both hind feet must have the same angle. This can be checked by using a hoof level. A hoof level is not absolutely necessary for amateur trimming and/or shoeing, but it helps develop skills and improves accuracy.

If the hoof is to be trimmed and not shod, all that remains is to round the edges of the hoof wall and lower the sole. Round the edges of the wall with the fine side of the rasp. Remove the sharp edges to about one-fourth the thickness of the wall. This reduces the chance of having pieces of the wall break out. If the hoof is to be shod this step is omitted.

Shaping the Shoe

For the beginner, one of the most difficult parts of shoeing is shaping the shoe. The first rule to remember is to shape the shoe to fit the foot. Shaping the shoe may be made easier by marking the heels of the foot and making a paper tracing from heel to heel.

Use a white grease pencil if the hoof is dark, and mark the back of the heel where the heel of the shoe stops on each side of the foot. Trace the outline of the foot on stiff cardboard or a tablet attached to a clipboard.

A well-shaped front foot is uniformly round and wide at both heel and toe. Since horseshoes seldom come in this shape, substantial shaping will be necessary. Shaping is made easier by using the pattern.

Most new shoes are too narrow for the front feet and must be spread and the heels bent in. This is corrected by hammering the shoe on the anvil and comparing it to the pattern or the hoof.

Check the levelness of the shoe on a flat surface. The face of some anvils will do. If the shoe will rock, it has a high spot in it. The ultimate test is a flat board. Raised points caused from hammer blows will have to be beaten down or rasped off before passing this test. Perfect levelness is desirable for both shoe and foot for equal weight distribution. High spots cause a shoe to rock and work loose as well as placing undue strain on that part of the hoof.

Most horses' hind feet are somewhat narrow and pointed at the toe. Shoes are initially round at the toe, so each side will need to be flattened on the back of the anvil. Upon flattening, the shoe will be widened considerably at the heel and must be drawn together.

Nail holes are too small in most manufactured shoes to accommodate the nail. A special punch increases the hole size until the nail head protrudes about one-sixteenth inch. If the nailhead goes flush into the shoe, it cannot be tightened. If it does not go deep enough, it will wear off and the shoe will loosen. Finally, the heel of the shoe should extend to the end of the heel of the front hoof but not beyond. On the hind foot ⅜ inch of the shoe may extend beyond the heel of the hoof.

Nailing the Shoe On

Now the shoe is ready to be nailed on. The white line on the bottom of the hoof marks the outer edge of the junction between the sensitive part of the hoof and the horny hoof. Any nail driven inside this line will cause pain to the horse and will result in lameness. All nails are driven along or just outside this line. If nails are driven very far outside the white line, they will split the hoof out and the shoe will come off prematurely.

Horseshoe nails are beveled on one side, both top and bottom, and are straight on the other side. The beveled side is always put on the inside or nearest the center of the foot. This allows the point of the nail to drift toward the outside of the hoof wall when driven. Most inexperienced people are reluctant to drive a nail. They fear "quicking," or driving the nail into sensitive tissues. This is almost impossible unless driving inside the white line or unless the nail is turned to drift inward. Driving high in the wall does not "quick" the horse. When preparing to nail, the shoe should be positioned so that it is fitted flush with the toe of the hoof. Some farriers prefer to drive a toe nail first because this allows easier positioning of the shoe. If the heel nail is driven first, however, the shoe will move less and will be more stable after driving one nail. The two heel nails and two toe nails should be the first four nails driven. Nails should be at least one inch deep before they come out through the wall. Once through the hoof wall, each nail should immediately be bent over with the claws of the hammer and twisted off flush with the hoof wall. An apron or chaps should always be used when driving nails, even if the horse is completely gentle. Under some conditions even the most gentle of horses will react.

After all nails are driven, they must be set by placing a clinching bar or nipper under the nail stub and striking the head of the nail. This tightens the shoe on the hoof and locks the nail head in the shoe. Excessive hammering will pull the clinches too far down to be rasped under. Many professional shoers use nippers for clinching.

Before clinching, the burs of splintered hoof wall under each nail are rasped off with the fine edge of the rasp. Also, the twisted ends of the nails on top are rasped with the flat, fine side of the rasp before clinching.

Although clinching can be completed with the hammer and clinching bar, it is much more easily accomplished by using clinchers. This tool is placed over the nail and squeezed together, clinching the nail down.

The goal is evenly spaced nails of adequate height and a shoe fitted "full"—that is, out to the edge of the hoof, with no gaps or "daylight" between hoof and shoe (see Figure 17-8). A space under the shoe allowing a knife blade to enter, suggests a poor job.

After all nails have been clinched, excess hoof that may protrude over the shoe can be dressed off. Very little, if any, of this should exist. Rasping above the clinched nail injures the hoof wall and may result in drying out or cracking of the hoof.

After-Shoeing Care

Many horse owners apply a hoof dressing following shoeing, and some even apply it every day or so. This practice is very easily overdone and may reduce both the strength and pliability of the hoof. If the hoof becomes excessively

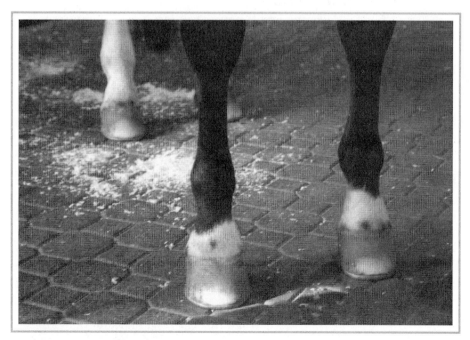

Figure 17-8 Well-shod feet

hard, a small amount of lanolin (wool fat) may be applied to the coronet and the bulbs of the heel.

Additional moisture can be applied by packing the bottom of the hoof with a special type of mud designed for this purpose. One of the simplest and easiest ways to keep a horse's hooves in good condition is to keep the area muddy where the horse goes to drink. This will usually be sufficient moisture to prevent dry cracking, cracked heels, and other problems related to dry hooves.

SUMMARY

Foot care is one of the most neglected of horse management practices, even though it is essential to the horse. The most important aspects of good foot care are regularity, frequency, cleanliness, and the use of proper corrective measures. An experienced farrier can properly clean, trim, and shoe horses for general soundness or for corrective help. Horses should be taught early in life to yield their feet.

REVIEW

Success in any career requires knowledge. Test your knowledge of this chapter by answering these questions or solving these problems.

True or False
1. Permanent lameness can result from improper foot care.
2. Feeding practices can affect the horse's feet.
3. Corrective trimming does not help a foal's feet.
4. Hooves should be trimmed once every three months.
5. Mud around the water source is good for horses' feet.

Short Answer
6. Which grow faster, the hind feet or the forefeet?
7. List four internal and four external parts of the hoof.
8. What part of the horse's foot bears the weight?
9. Name five tools used in horseshoeing.
10. List four problems in the set of the feet and movement of the feet commonly corrected by good horseshoeing.

Discussion
11. Explain why the frog is a good indicator of a horse's health.
12. Discuss the importance of starting hoof care early.

13. Describe simply the process for shoeing a horse.

14. Why is it important to check a horse's feet daily?

15. What are the dangers of trimming incorrectly, cleaning incorrectly, or shoeing incorrectly?

STUDENT ACTIVITIES

1. Visit a farrier about shoeing and hoof care. Find out how farriers are trained, how much their equipment costs, and approximately how much they charge.

2. Make a drawing or collect some of the types of horseshoes. Indicate the use of each type.

3. Develop a checklist describing how to lift each foot; include any precautions.

4. Obtain a prepared specimen of the bones of the foot. Use this to present a report on laminitis (founder).

5. Develop a checklist that stresses the points of good horseshoeing.

6. Diagram the anatomy of a hoof showing a shoe properly nailed.

7. If possible, go with a farrier to observe the shoeing of a horse.

ADDITIONAL RESOURCES

Books

American Youth Horse Council. 1993. *Horse industry handbook: a guide to equine care and management.* Lexington, KY: American Youth Horse Council, Inc.

Blakely, J. 1981. *Horses and horse sense: the practical science of horse husbandry.* Reston, VI: Reston Publishing Company, Inc.

Davidson, B., and Foster, C. 1994. *The complete book of the horse.* New York, NY: Barnes & Noble Books.

Ensminger, M. E. 1990. *Horses and horsemanship.* 6th ed. Danville, IL: Interstate Publishers, Inc.

Evans, J. W. 1989. *Horses: a guide to selection, care, and enjoyment.* 2nd ed. New York, NY: W. H. Freeman and Company.

Frandson, R. D., and Spurgeon, T. L. 1992. *Anatomy and physiology of farm animals.* 5th ed. Philadelphia, PA: Lea & Febiger.

Griffin, J. M., and Gore, T. 1989. *Horse owner's veterinary handbook.* New York, NY: Howell Book House.

Instructional Services Materials. (n.d.) *Curriculum material for agriscience 334: equine science*. College Station, TX: Texas A & M University.

Kainer, R. A., and McCracken, T. O. 1994. *The coloring atlas of horse anatomy*. Loveland, CO: Alpine Publications, Inc.

Kreitler, B. 1995. *50 Careers with horses—from accountant to wrangler*. Ossining, NY: Breakthrough Publications.

Prince, E. R., and Collier, G. M. 1986. *Basic horse care*. New York, NY: Doubleday & Company, Inc.

Self, M. C. 1963. *The horseman's encyclopedia*. New York, NY: A.S. Barnes and Company.

Stashak, T. S. 1996. *Horseowner's guide to lameness*. Media, PA: Williams & Wilkins.

Stoneridge, M. A. 1983. *Practical horseman's book of horsekeeping*. Garden City, NY: Doubleday & Company, Inc.

University of Missouri-Columbia Extension Division. (n.d.) *Missouri horse care and guide book*. Columbia, MO: Cooperative Extension Service, University of Missouri and Lincoln University.

Wood, C. H. ed. 1993. *Art and science of equine production*. Lexington, KY: University of Kentucky, College of Agriculture.

18

Buildings and Equipment

Horse owners use many types and styles of fences, barns, and shelters. Most people have one type or design they will like better than others. But as long as the barn is well-built and meets local building codes, the fence is safe and strong, the shelters are strong, and water is always available, the style and design is not particularly important. This chapter provides some general building guidelines for welfare, safety, health, and cost.

O B J E C T I V E S

After completing this chapter, you should be able to:

- Recommend an environment for horses that addresses welfare, safety, labor, and cost

- List the planning stages of construction

- Identify the space requirements for a horse facility

- Discuss the importance of ventilation in a building housing horses

- Name materials commonly used for stall floors

- Describe the requirements for a horse stall

- Provide guidelines for the selection of feed and water facilities

- Discuss reasons for fencing horses and how to select the right fence

- Name four types of fences

K E Y   T E R M S

Air requirements	R value	Ventilation
Flight	Space requirements	
Polyvinylchloride (PVC)	Stalls	

RECOMMENDED ENVIRONMENT

Horses have lived outdoors with natural windbreaks as their only housing for centuries. So, the simplest housing is the best and healthiest for horses. A large field or paddock along with a simple shelter is adequate housing. When housing is built for horses it should provide for:

- Welfare of the horses

- Safety, health, and comfort of human handlers

- Efficient use of labor

- Cost effectiveness

Providing for the welfare of the horse begins with understanding its environmental needs. The environment involves four main areas:

1. Physical: The physical environment includes such things as temperature, heat-loss factors, stall space, feeder space, and flooring.

2. Social: The social environment involves behavioral considerations related to how horses interact with other horses.

3. Chemical: The chemical environment includes water quality; various gases such as oxygen, carbon dioxide, and ammonia; and air contaminants like dust and molds.

4. Biological: The biological environment primarily includes disease organisms in the air, water, feed, stall materials, and other animals.

Horses use **flight** as a primary defense mechanism. In attempting to flee danger, horses can injure themselves. They are generally nonaggressive, but when threatened, frightened, or in pain, they may strike, bite, kick, or attempt to break out of their stalls or stables. Facilities should provide for the safety of the horses and handlers when these behaviors occur.

Under natural conditions, horses do not spend long periods of time in an enclosed area, such as a stall or stable. In barns, some horses will become bored and develop vices. Providing adequate stall space will tend to minimize vices.

When horses are brought into a building, fresh air needs to be provided and the metabolic products including carbon dioxide, water vapor, and manure need to be removed. Adequate **ventilation** will reduce the presence of air contaminants such as dust, molds, and irritating gases from decomposing manure that can cause respiratory problems.

SPACE REQUIREMENTS FOR HORSES

The first step in building, is knowing the recommended **space requirements** of horses. Table 18-1 provides these recommendations.

BUILDINGS

Horses are housed in buildings primarily for the convenience of the horse owner and handlers. As a result, human environmental needs and wants play a major role in designing horse facilities. Often, human wants may be in conflict with the environmental needs of the horse.

A horse can do well in nearly any temperature if the humidity can be held to a comfortable level and there is enough air movement through the building to keep the air clean and free of condensation. The conditions that are most detrimental to a horse's health are when there is high moisture and the barn is either cold or hot. These are the conditions most likely to harm the horse's respiratory system and to allow the inhalation of pathogens.

Table 18-1 Space Requirements for Horses

Use	Size (feet)	Height of Ceiling	Height of Doors	Width of Doors
Smaller horses	12 × 12	8 to 9 feet	8 feet	4 feet
Broodmare and foaling barn	12 × 12 to 16 × 16	9 feet	8 feet	4 feet
Stallion barn	14 × 14	9 feet	8 feet	4 feet
Barren–mare barn	150 sq. feet per animal	9 feet	8 feet	4 feet
Weanling or Yearling barn	10 × 10	9 feet	8 feet	4 feet
Breeding shed	24 × 24	15 to 20 feet	8 feet	9 feet
Isolation barn	12 × 12	9 feet	8 feet	4 feet
Training, boarding, riding stables	12 × 12	9 feet	8 feet	4 feet

Several items must by considered in the preconstruction planning stage:

- Purpose of the facility
- Number and breed of animals to be housed
- Room for future expansion
- Regulatory requirements
- Budget
- How layout facilitates day-to-day activities

Buildings represent a major cost and consequently, can represent costly mistakes. Valuable information can be obtained by examining buildings designed by others to observe good features and recognize mistakes (see Figure 18-1).

Site Selection

Local zoning requirements should be checked before buying a farm or designing a new building. Some areas place a restriction on the number of acres

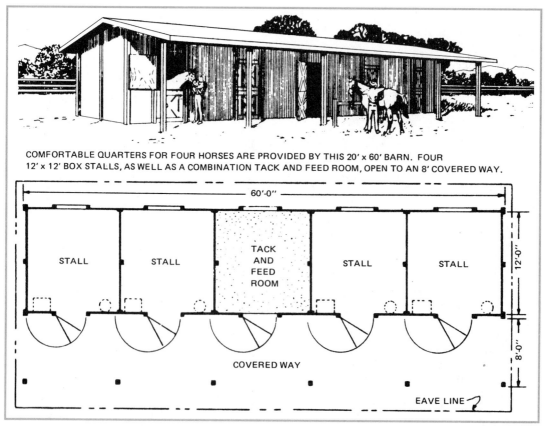

COMFORTABLE QUARTERS FOR FOUR HORSES ARE PROVIDED BY THIS 20' x 60' BARN. FOUR 12' x 12' BOX STALLS, AS WELL AS A COMBINATION TACK AND FEED ROOM, OPEN TO AN 8' COVERED WAY.

Figure 18-1 A four-stall horse facility. *(Courtesy of Clemson University)*

necessary before livestock can be housed. Also, the distance from boundary lines, dwellings, and neighbors may be regulated. If these regulations cannot be met, then it is necessary to apply for a variance and receive approval from the zoning board before building can commence.

The site should allow water to drain away from buildings, working rings, and training tracks. A site with a slope of 2 to 6 percent provides rapid removal of water without causing erosion. A detailed site plan should be developed before making a final decision. The site plan helps ensure that sufficient space is allowed for the buildings, roads, paddocks, working rings, training tracks, and manure storage and use. Manure handling is frequently overlooked and can be a major obstacle to enjoyment and convenient function of the facilities.

The site plan should indicate where water, sewer, and electrical lines enter the building. The building should be situated to take advantage of prevailing winds to effectively use the natural air flow. In the plan, consideration should be given to clients, traffic, impact on neighbors, manure handling, and conditions in the neighborhood that will startle or distract horses.

Site Preparation

Getting a particular location ready for a building involves removing the topsoil, leveling the area and bringing utilities such as water and electricity to the site. The nature of the work usually means that local contractors will be engaged.

Type of Construction

Buildings can be metal frame, pole, or conventional construction (see Figure 18-2). All three have been used with equal success for nearly every type of farm building. There is no general rule as to which type is most economical for any one situation. In fact, it is not unusual to find more variation in price among similar types of construction than among different ones.

Options in the Building

There are many options and alternatives to consider. Even "packaged buildings" offer alternative or optional items. Some of the choices are listed below, along with their advantages and disadvantages.

Windows. Windows are expensive additions to farm buildings and, to keep costs at a minimum, are being used less and less. The only place they are essential is in those structures that must conform to health regulations. When windows are used for light, the window area should equal 8 to 10 percent of the floor area. Plastic roof panels can also be used as a good light source for uninsulated, cold buildings.

Siding. Metal is low-maintenance siding material, and it is available in prepainted finish colors that will last fifteen to twenty years without refin-

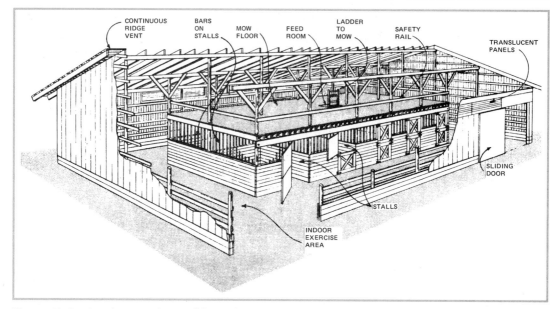

Figure 18-2 A pole-type, nine-stall horse barn with an indoor exercise area. *(Courtesy of USDA)*

ishing. However, metal siding is subject to damage when exposed directly to livestock.

Wood siding will withstand abuse and it has a better insulating value than either metal or masonry, but it requires periodic painting or stain to preserve its appearance and durability.

Masonry requires very little maintenance, but it has a high initial cost and is difficult to insulate. When masonry is used with pole or steel-frame buildings, it requires a separate foundation.

Roofing. Metal roofing can be of aluminum or steel. White-colored roofing has slightly better reflective quality than natural metal. Metal roofing requires less roof framing than shingles and is lower in cost. A roof with a solid deck and shingles has a better insulating value than a metal roof.

Insulation. Insulation is an increasingly important part of modern farm building construction. Even buildings that are considered cold structures are minimally insulated to moderate summer and winter temperature extremes.

Many choices of insulating material are available. To provide a basis of comparison between buildings, insulation should be specified based on its **R value**. General recommended levels are as follows:

- Cold buildings operated at outside temperature
    Ceilings (roof): R2 to R4 for summer heat
    Walls: No insulation

- Buildings where animal heat provides only winter minimum temperatures
    Ceilings: R16
    Walls: R9 to R12
- Buildings with supplemental heating systems
    Ceilings: R24
    Walls: R13

Interior Finish.   Choices of material for the interior finish in farm buildings are almost infinite (see Figure 18-3). A performance specification rather than identification of a specific material usually will provide a better comparison among building manufacturers. Items that should be considered in developing performance specifications include:

- Mechanical strength. If interior finish is exposed to animals, it will have to take considerable abuse.
- Moisture resistance
- Ease of cleaning
- Color

Ventilation.   A good ventilation system must (1) provide fresh air to meet the respiration needs of the animals, (2) control the moisture build-up within

Figure 18-3   Inside facilities at Cutter's in Hailey, Idaho.

the structure, (3) move enough air to dilute any airborne disease organisms produced within the housing unit, and (4) control and/or moderate temperature extremes.

Each of these four provisions requires some optimum rate of air exchange. If respiration and temperature control are provided for, moisture build-up and disease control will be satisfactory.

The basic process that occurs with all successful ventilation systems is as follows:

1. Cool, dry air is drawn into the building.

2. Heat and moisture are added to the air.

3. Warm, wet air is expelled from the building.

Failure to provide for any part of this process will result in failure of the ventilation system.

Air requirements vary with animal size and outside environmental conditions. The ideal ventilation system would be infinitely variable. During extremely cold weather, it should move just enough air to satisfy respiration needs, and in hot weather, the maximum rate should eliminate heat stress.

A ventilation system should be designed to provide at least three levels of air movement. The lowest, or minimum, level provides enough air to meet respiration requirements and operates continuously. This lowest level provides all the air necessary during periods of extremely cold weather or in buildings where a supplemental heating system is in operation. A thermostat may be used to shut off the minimum level when the building temperature drops to near freezing.

The second, or intermediate, level of ventilation provides enough additional air movement to control both temperature and moisture during normal winter conditions. Fans that provide this additional air are usually controlled by thermostats that turn them on whenever the building temperature reaches the desired level.

The high, or maximum, ventilation rate is intended to provide some degree of temperature control during summer months. Maximum-rate fans are controlled by thermostats that turn them on when interior temperature exceeds some set level, usually 75 to 80° F.

Table 18-2 shows the recommended ventilation rates for horses.

Natural ventilation is the most common and cost effective ventilation system for horses. In post-frame construction, the space between the bottom of the roof surface and the top of the girder that supports the roof truss is left open on both sides of the barn. Air enters on the windward side of the barn and exits on the downwind side. Warmer air and moisture that accumulate at the peak of the roof must be allowed to escape. This can be done with cupolas or openings at the ridge. These should be unobstructed air outlets.

Table 18-2 Recommended Ventilation Rate for Horses in a Building at 55°F	
Season	**Cubic Feet per Minute**
Winter, minimum	25
Winter, normal	100
Summer	200

The air inlets and outlets must allow unobstructed airflow with minimal interior obstruction. Roof slopes of at least $\frac{4}{12}$ pitch are most effective in causing air to move from the animal space to the ridge openings or cupolas. Avoiding any overhead storage enhances air movement and reduces the risk of fire. One-half inch to three-quarter inch hardware cloth can be installed to discourage bird entry. Grillwork, rather than solid wall partitions, facilitates air movement through the stalls.

Since wind forces air to move through the building, the barn should be oriented with the long axis perpendicular to the prevailing winds. Other buildings and land features, such as trees, should not block the wind.

Heating. The heating system should be designed to maintain a specified interior temperature when the outside temperature falls. The interior temperature desired will depend on the building's use. An automatic temperature control system should be specified.

Electrical System. The electrical system provides lighting, general outlets, and outlets for special equipment. Adequate lighting can have a positive influence on workers' attitudes, plays a major role in safety, and enhances the management level by increasing people's ability to see potential problems. Horses can sleep in either light or darkness, but they tend to hesitate moving past areas with high contrast. Shadows and sharp differences between light objects and their background should be avoided.

Two types of electrical fixtures are common in the stable area: incandescent bulbs and fluorescent tubes. Fluorescent lighting is four times more energy efficient than incandescent. Protective coverings over tubes or bulbs are essential in stalls, alleys, and anywhere horses could reach the fixtures, and in the feed room where broken glass is undesirable.

Light levels are measured at thirty to thirty-six inches from the floor with a lux meter. For general lighting, such as passageways and recreation areas, ten footcandles of light is considered adequate. For task areas such as grooming stations, tack-care areas, indoor riding arenas, and offices, at least thirty to forty footcandles of light is needed. For reading and fine detail work like veterinary care, seventy footcandles of light is required.

A common mistake in many horse barns is not to provide enough receptacle outlets. At least one double outlet is needed for every two stalls. Outlets should be above the level of the horse's back or recessed into the wall.

Floor. A four-inch-thick concrete floor is sufficient for most farm buildings. Reinforcing is not necessary if floors are placed over a well drained, compacted fill material. Floors should be thickened to eight inches for a distance of two feet in from doors where equipment will be entering the building. A six-sack mix concrete made with air entraining cement should be used, and the floors should slope ⅛ inch to ¼ inch per foot to floor drains.

Stall floors for horses must be made of durable material that is not slippery. It should be absorbent, easy to clean, and resistant to pawing. Floors should require a minimum amount of expense and time to be maintained in a satisfactory condition. Some of the more commonly used materials include clay, a sand and clay mix, limestone dust, wood, concrete, asphalt, and rubber floor mats.

- Good clay is hard to find. Maintaining level, dry clay floors is difficult.

- A mixture of two-thirds clay and one-third sand will allow drainage and it is easy to obtain materials for filling holes and replenishing the surface when necessary.

- Limestone makes a level, hard surface. The thickness of the limestone needs to be four or five inches over six to eight inches of sand or other base material that allows drainage.

- Wooden floors are rough-cut hardwood at least two inches thick that has been treated to retard decay. Wooden floors are slippery when wet, and prone to attract rodents by creating an environment for urine to accumulate and feed to fall through the cracks.

- Stall floors made of concrete are easy to clean and sanitize, but they require more bedding. The use of concrete floors is of general concern because of its association with increased leg problems.

- Asphalt can be used for stall flooring, but many of the problems associated with concrete can occur with asphalt.

- Rubber floor mats need to be placed on a stall floor that is level and packed well. The mat should be a single piece, at least ⅝-inch thick and made of a durable rubber that will withstand pawing. Some bedding may be required.

Some alternate flooring for stalls that could be considered include interlocking rubber paving bricks, fiber-reinforced polyethylene interlocking blocks, and fiber-grade polypropylene.

All of the alternate flooring materials will add to the cost of stalls. The additional cost needs to be weighed against the benefit of the flooring material.

Special Additions. Almost every building will be modified to provide some special feature for the farm it is located on (see Figure 18-4). Some of the more common additions are:

- Bathroom
- Office
- Handling facilities
- Feed storage room
- Tack room
- Special equipment space

Insurability. Farm buildings are becoming extremely complex structures and usually represent a considerable investment that must be protected with insurance. Owners need to make sure the structure will be eligible for any insurance coverage.

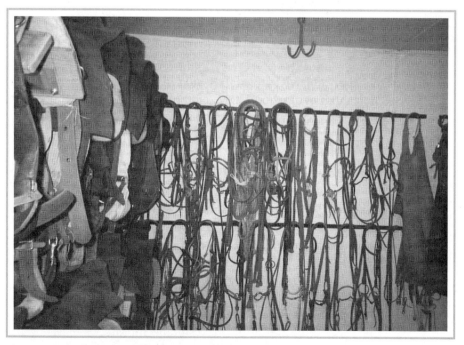

Figure 18-4 Tack should be kept in a room with moderate temperatures where it can be kept dry to prevent damage to the leather.

HORSE BARNS

When building a barn, many important characteristics must be considered. The barn must be able to hold a uniform temperature while being ventilated and maintaining a dry atmosphere. Condensation in the barn can dampen the food, which will become moldy. The floors must be dry and firm, preferably with a nonslip footing. Good drainage is necessary so that ammonia from urine can be washed away and to keep molds from growing. Surfaces should be easily disinfected. Adequate lighting is needed for moving horses around or working in the barn after dark.

Horse barns can be designed for small, medium, or large operations. Some variations on barns include:

- Broodmare and foaling barn
- Barren-mare barn
- Stallion barn and paddock
- Breeding shed and corral
- Weanling and yearling quarters
- Riding stables
- Training stables
- Boarding stables

STALLS

There are many different sizes of **stalls**. A twelve-by-eighteen foot enclosure usually is the largest and a twelve-by-twelve foot stall is the most common and the smallest comfortable size for today's large horses. Smaller stalls often lead to sanitary problems because they must be constantly kept clean. They can also create a high risk of injuries because the horses are constantly bumping into the walls and have little room if they lie down.

For a foaling mare, a twelve-by-sixteen- or twelve-by-eighteen-foot stall is highly recommended. This gives the mare plenty of room to deliver the foal and allows extra room for a veterinarian or other experts, if needed (see Figure 18-5).

Stall Doors

A stall door must hold the horse within the stall in a safe manner. The door should be easy to open and close for the safety of both the horse and the handler and it must be strong and simple to operate. The stall door should be a minimum of four feet wide and at least eight feet in height.

A sliding door is the most suitable for safety and ease of operation. Sliding doors should possess sturdy tracks and rollers, as well as a safe latch. Doors with drop-down bars or latches that protrude can injure a horse. The most

Figure 18-5   To provide enough room for foaling, a stall should measure 12 × 16 or 12 × 18 feet. *(Photo courtesy of Barbara Lee Jensen, After Hours Farms)*

common is the half-wood, half-bar door that allows some ventilation and light into the stall. A full-mesh door allows maximum ventilation and light. Mesh doors are valuable in a foaling barn because they permit foals to receive plenty of fresh air; in a barn that is properly designed and ventilated, they do not allow drafts.

The weight of a full one-piece swinging door can cause hinges and latches to sag, making the door difficult to close properly. Swinging doors also can be a safety hazard when opened into alleyways or other high-traffic areas.

Double doors have two sets of hinges and two latches so that the doors tend to sag and must be reset. Two latches presents a greater risk because one may not be closed. The advantage of double doors is that the top door can be left open, which allows the horse to stick its head outside and take an interest in its surroundings. However, a horse may try to go over the top if it becomes nervous or excited.

Stall Guards

Some horse trainers prefer stalls constructed of webbing or chains or half-metal doors with a neck yoke. Stall guards are the least costly, but they are the least desirable for containing the horse because they are easy for a horse to push out, go over, or break through.

Safety First, and Always

When working with horses and people, the first concern should always be a safe, accident-free environment. To accurately assess the safety of stables, these questions need a truthful answer:

1. Is each buildings' service-entrance equipment located in a dry, dust-free location?

2. Is service-entrance equipment mounted on fire-resistant material?

3. Is service-entrance equipment free of rust and other signs of deterioration?

4. Are electrical fixtures properly covered so they do not fill with cobwebs, dirt, or chaff?

5. Are circuits properly fused with correct-sized breakers?

6. Is all wiring in good condition with no signs of fraying or deterioration?

7. Are all lighting fixtures properly protected?

8. Are stable aisles well lit and at least twelve feet wide?

9. Are stable aisles and walls free of objects that might harm horses?

10. Are all stalls designed to prevent contact with neighboring horses?

11. Are all stall doors equipped with "horse-proof latches" to prevent escape?

12. Are all electrical fixtures and wiring inaccessible to horses or properly protected?

13. Are stalls cleaned and rebedded daily?

14. Is all grain and feed kept in covered containers or bins?

Some general questions needing a truthful evaluation include:

1. Are areas surrounding buildings free of high weeds, grass, and debris?

2. Is hay properly dried and cured prior to inside storage?

3. Are all roofs, walls, windows, and doors weather-tight on hay storage buildings?

4. Are fire extinguishers:
   Located in each building?
   At least five-pound ABC or better?
   Conspicuously hung within fifty feet of any point in the building?
   Protected against freezing?
   Inspected and tagged annually?

5. Are lightning rods properly installed and grounded with conductor cable showing no signs of corrosion?

6. Is a responding fire department within five miles of the farm?

7. Is the telephone number of the fire department conveniently located near telephone?

8. Are NO SMOKING signs posted and enforced?

In the pastures:

1. If post-and-rail fencing is used, are rails secured to the inside of the posts?

2. Are pastures and paddocks free of harmful objects?

3. Are isolated groups of trees fenced off or protected by lightning rods?

4. Are pastures rotated to break the life cycle of parasites?

5. Are shelters provided in pastures and paddocks?

ARENAS AND INDOOR TRAINING FACILITIES

Arenas and indoor training facilities are basically clear-span structures that are part of, attached to, or close to the main horse barn. Arenas should be at least thirty-six feet wide and can be used for exercising and training horses. This width will limit the arena to riding, as it is too narrow to turn a cart around in. Clear-span structures fifty feet wide or wider are used as exercise, training, and riding arenas. Widths of at least sixty feet are best for group riding or driving horses inside a building.

The ceiling height in an arena must be a minimum of fourteen feet for the horse's and rider's safety. The higher the ceiling, the better lit the arena or training area must be to minimize shadows. A sixteen-foot ceiling will allow the training of hunter/jumper horses with ample head room for the rider.

SHELTERS

Shelters in pastures to allow horses to get out of the sun, wind, rain, or other types of weather are common. Some have just a top for shading. Others are enclosed on three sides. In that case, the open side needs to be away from the wind. Metal strips on all edges will prevent the horses from nibbling on the wood and destroying the shelter (see Figure 18-6).

Figure 18-6 An outside shelter at River Grove Farms in Hailey, Idaho.

FEED AND WATER FACILITIES

The design of feed and water facilities is controlled by fads and the likes and dislikes of owners. Design will also be a function of the overall type of facility. For the horse, these facilities need only be simple, safe, and effective. For the care giver, these facilities need to be located so that they can be conveniently filled, checked, and cleaned.

Feeders

Feeders include hayracks, mangers, grain containers, mineral boxes, and self-feeders. The most important point of design should be to keep feed off the ground. Feeding on the ground encourages sand colic resulting from the horses eating every single scrap off the ground.

Locating pipes in front of feeders and waterers helps to keep horses from leaning on them and thus will increase the lifetime of the facility. Concrete aprons under feeders keep the horses off mud in the winter and from eating the dry dirt in summer. A roof over the feeding area keeps the horse, food, and dirt dry.

Rubber tires are excellent feeders as they are easy to eat in and hard to pull food out of. But they are also harder to clean out. Before being filled, feeders should always be checked for any uneaten toxic weeds or moldy hay and to make sure the horses are not off their feed.

Watering Devices

Horses must have constant access to clean, fresh water from either a large waterer that is filled with a hose or from an automatic waterer. Automatic waterers are more convenient, but they must be checked frequently to ensure they are working and that the pipes do not freeze in winter. Also, some horses are frightened by the hissing noise they make while filling up.

A frost-free hydrant, hose, and bucket is the least costly watering system to install. The water hydrant should be recessed in the wall to eliminate the possibility of a horse getting hurt on it or the farm staff hooking it with the wheel of a tractor, manure spreader, or other piece of equipment.

The water hydrant is a trouble-free system that permits the care giver to estimate how much water a horse is drinking. The major concern is keeping it from freezing during the winter. In most cold climates, the waterline will need to be four feet or more below the surface of the ground.

Many farms now use automatic waterers in order to save labor. Special attention should be given to the design of the waterer and its location in the stall. Round waterers that require an angled support brace at the bottom are a hazard to a horse rolling in the stall. The horse or foal can get its leg caught in the angle brace. Automatic waterers are best placed in a back corner of a stall so an overflow tube can be attached and run to the outside of the barn. Automatic waterers must be checked daily to ensure adequate water supply and cleanliness.

FENCES

Sooner or later every horse owner must provide fencing for the pasture, turnout lots, arena, or aisles. The most important considerations are that fences be safe and strong enough to contain the horses and that the price and appearance be acceptable to the owner.

Reasons to Fence in Horses

Horses are much healthier outside in the sun, rain, and even mud, than they are when kept inside. But in order to have horses outside without exposing them to the danger of automobiles, poisonous weeds, and other hazards, safe fences are essential. Fences keep animals away from the property of others and at the same time, fences discourage people from entering the horse's environment. In most areas horses are considered "attractive nuisances" that can be dangerous and, except on the open range, horse owners in most states are required by law to fence in horses.

Fences are also important in making the handling, moving, and sorting of horses easier and less stressful for the horses and less labor-intensive for the handlers. They help separate horses that are not compatible, protect pastures that are not suitable to be grazed, and provide boundaries for other essentials such as exercise paddocks, round pens, riding arenas, and protection from driveways. Fences are a major investment for any horse farm or stable manager.

Selecting the Right Fence

There are many types of fences and fencing materials that can be used for horses (see Figure 18-7). The type of fence needed depends on a number of factors:

- Type of horse being managed
- Intended use of the area
- Density of animals on the fenced area
- Availability of shelter
- Neighbors
- Desired aesthetics
- Projected budget

Obviously, draft horses require taller and stouter fences than those required for miniatures. Mare and foal pastures need to be made safe and solid

(a)

(b)

(c)

(d)

Figure 18-7 Different fencing may be chosen depending on the type of horses being managed, the use of the area, desired appearance, and cost. (**a**) metal piping, (**b**) wire mesh, (**c**) PVC plank, and (**d**) wood.

to protect curious foals from danger. Usually old pleasure horses that are used to fences require less sturdy and visible barriers than young horses, or horses that have never been in pastures with groups before.

Stallions should have taller and stronger fences to keep them in and also to keep children and curious visitors out. When fencing stallions, unacquainted horses, or very valuable horses, the area between paddocks should be double fenced or separated by a twelve-foot empty aisle.

If the pasture provides a significant share of the horses' nutrients, at least two acres per horse should be allowed. If the area is primarily an exercise lot, then the space should be more than 500 square feet per horse. If activity is expected on the inside of the enclosure, then the boards (or other material) should be attached on the inside of the posts. This is primarily for safety reasons in riding arenas where the fence surface protects the cart or the rider's leg from hitting the post. In pastures, it prevents horses from pushing boards away from the posts.

Dividing the total number of horses by the available acreage determines the animal density. Basically, the higher the animal density per acre, the stronger the fence needed. For large pastures with only a few horses an open wire-type fence may be adequate. But when confining several horses in smaller areas, stronger fencing is required.

Either allow the horses access to wooded areas, build a three-sided shed, or fence the area in a manner that provides a stall or lean-to shed to provide shelter from the sun and weather. The arrangement of fences and gates selected should depend on whether it is necessary to allow the horses to get into a building or shelter to access water. Horses tend to congregate near shelter, feed, or water, so the fences in these areas need to be more solid and safe than at the periphery.

If things attractive to horses such as grain crops, other horses, or even the barn are present on the other side of the fence, then the fence should be stronger and possibly taller. Generally, the closer they are to the barn and other horses, the stronger the fences need to be. This situation often calls for a twelve-foot easement or aisle between the fence and the attraction on the other side.

The importance of appearance depends on personal likes and also the priorities in the neighborhood. Fences should be safe, require as little maintenance as possible, and be affordable for the owner. Few people can afford to install and maintain a four-board fence for a large acreage, but this type may be desired for appearance around the barn, entrances, and the front of the property.

Prices for most fences range from less than a dollar to more than four dollars per linear foot. The criteria involved in matching the horses' needs to the attributes of the specific fence should be the most important factors in determining costs. Building fences takes a lot of time. A person's ability to build the fence and the availability of time to do it are factors to consider. Paying someone to install the fence often results in quicker installation and higher-quality fences. Fences using high-tensile wire and polyvinylchloride should be installed by professionals.

Types of Fences

Fences need to be specific for the situation. The types of fences available for horse facilities change and improve every year. Fences discussed here include post and board, woven wire, high-tensile wire, **polyvinylchloride (PVC)**, pipe, diamond wire, electric wire, and various combinations.

Post and Board. This type of fence includes three or four boards hung on wooden posts. Board fences are suitable for line fences, paddocks, and arenas.

The standard design usually includes sixteen-foot, rough-cut, one-by-six-inch hardwood boards fastened on the inside of four-inch (minimum) diameter wooden posts with each staggered board spanning two posts.

Several variations of the post-and-board fence include:

1. Setting the posts on ten-foot centers and using twenty-foot boards

2. Using square five-inch posts instead of the round four-inch posts

3. Using two-by-six-inch milled planks rather than the full one-inch thick rough-cut boards

4. Deleting the fourth board and increasing the space between boards to ten inches (between the bottom board and the ground to sixteen inches). A good practical distance between the ground and the bottom board is just greater that the height of the lawn mower deck. This makes trimming fence rows much easier.

Advantages of the post and board fence include safety, sturdiness, high visibility for the horses, and popular aesthetic appearance. Disadvantages include high maintenance costs, and board replacement.

Woven Wire. Field-livestock woven-wire fence can be purchased in rolls of thirty-nine- to fifty-five-inch widths. The top and bottom wires should be at least 9 gauge with the intermediate wires 11 to 12 gauge. Woven wire with the vertical strands a maximum of six inches apart should be used. The standard design is to hang a forty-seven-inch woven wire five to six inches off the ground on the inside of four- to five-inch-diameter round wooden posts. A one-by-six-inch hardwood board is then nailed above the wire, making the total fence sixty inches tall.

Depending on the topography, the wire may need to be higher off the ground, and the board could be replaced by an electric wire no more than four inches above the woven wire. Posts on eight-foot or ten-foot centers work well with sixteen- or twenty-foot boards. As with any wire fence, strong brace-post sections must be placed in the corners to stretch the wire tight. Brace sections also must be in the lowest points of valleys and at the top of hills to allow straight stretching without the wire pulling the posts out of the ground.

Advantages include high visibility and low maintenance. Disadvantages include stretching and costs.

Diamond Wire. Much like the livestock woven-wire fence, the diamond-wire fence is normally made with forty-eight-inch wire hung six inches off the ground with a six-inch board along the top, making the whole fence about sixty inches tall. Brace post sections are needed to adequately stretch the wire. Wire of 9 to 11 gauge diameter is available in this design. The wire should be all galvanized steel.

The unique interwoven design provides smaller holes than the livestock fence and less danger to horses that might put a foot through the fence or walk up the fence. Standard sixteen-foot one-by-six-inch rough-cut hardwood boards are nailed to the top of five- to six-inch wooden posts with both the wire and the boards fastened to the inside of the posts. Advantages of the diamond-wire fence are safety and low maintenance. Disadvantages are primarily cost and the need for brace sections and ways to stretch the wire tight.

Pipe Fence.   Fences made of pipe are constructed from two-inch to four-inch horizontal pipes welded to four-inch posts. As with wooden posts, the pipe posts should be driven or set thirty to thirty-six inches into the ground. The horizontal pieces should be welded to the inside of the fence or holes cut in the posts so the rails can be slid through the posts. Usually four or five horizontal rails are set six to eight inches apart. The top of the posts must be rounded or capped so sharp edges are not exposed.

Depending on the availability of used well-casing pipe, this type of fence can be economical, sturdy, and require relatively low maintenance costs. The construction requires metal cutting and welding expertise. This fence is more popular in the south where fewer temperature changes reduce the need for repainting.

High-tensile Wire.   High-tensile-wire fences are made with five to seven strands of smooth 12.5 gauge wire spaced eight to twelve inches apart. Rigid brace sections are required at corners, gates, and fence ends. Eight-foot wooden or fiberglass line posts set thirty to thirty-six inches into the ground are placed at fifty- to seventy-five-foot intervals with fiberglass spacers of the same height every twenty to thirty feet. Alternating wires should be electrified, so plastic insulators must be put on the posts on the top and bottom strands and alternating wires in between.

Considering the curiosity of horses, the electrification of this type of fence is essential. Bracing, in-line strainers or tighteners adequate to allow 200 to 250 pounds of tension are needed. The advantage of this fence is that it is sturdy and takes relatively little maintenance, although it does require frequent checks for damage, electrical shortages, and tension loss. Disadvantages are that it has low visibility to the horse, it takes specific expertise and equipment to install, foals can get through the fence, and if a section is damaged, the whole line must be repaired or retightened. Generally, this type of fence should only be used with electricity and in areas of at least five acres.

Polyvinylchloride (PVC).   PVC fence is made of a weather-resistant polyvinylchloride material that can be made in flat or round shapes resembling boards or pipe. Round rail fences consist of five-inch-diameter round posts and three-inch-diameter rails sixteen feet long that span through three posts. Posts with three or four rails are most common. The plank shaped rails, 1½ inches by 5½ inches, usually interlock into slots in the five-inch round or square posts. These single- or co-extrusion polymer products can be made in white, brown, or black and have a UV light protection mixed in to keep the product from fading. The advantages of this type of fence are that it looks great, does not need painting, and requires very little maintenance. The disadvantages are that it is expensive to install and is less sturdy in small areas with high animal density.

Covered Boards. These products consist of a two-foot by six-inch wood plank, usually sixteen feet long covered with polyvinylchloride or plastic. The advantage is that it combines the sturdiness of wood fences with PVC protection that eliminates the need for painting. The board inside can still break and need to be replaced, however.

Cable. Usually these fences are made with pipe posts four- to five-inches in diameter. Twisted-wire cables either run through holes in the posts or are fastened to the inside of the posts. At least six cables should be used with the bottom cable about six inches off the ground and the top strand fifty-four inches high. Cables with a minimum diameter of one-half inch should be used. This fence is not recommended in small areas or with foals. It is difficult to keep horses from entangling themselves in the strands.

Electric and Fiberglass Webbing. Electric-wire fences can have two to four strands of either smooth wire or wire woven into colored fiberglass webbing. This webbing can be one- to four-inches wide and comes in a variety of colors. The key to any electric fence is to make sure that the electrical current is not shorted out by poor insulators, tall weeds, or broken wire. Posts can be wooden with insulators nailed or stapled to the inside of the posts, or they can be metal with plastic clip-on insulators.

The electric material must be kept tight but not with the tension of a high-tensile-wire fence. The posts can be twenty to thirty feet apart and should be set deeper than thirty inches in the ground for permanent fences. The advantages of these fences are their economical features, and the fact that they can be made temporary by using metal posts. The disadvantages include the low visibility to the horse (with smooth wire) and the constant need to check for breaks and shortages so the horses do not become entangled or injured. The fiberglass webbing is safer and more visible.

Nylon or Rubber Fencing. These are fences made of two- to four-inch strips of belting or inner-tube rubber from the tire industry. These strips should be stretched, so the considerations of tension and bracing are the same as they are with wire fences. Rubber and nylon are very durable and safe, but tend to stretch in colder climates and become brittle with time.

Caution should be used in selecting a product that does not have exposed nylon threads, because horses will playfully ingest these and experience colic. This fence can be very safe, but curious foals may weave their way through it. It will need to be replaced regularly in colder climates.

Barbed Wire.   Barbed wire is inexpensive but very dangerous for horses. Whether electrified or not, this type of wire is not recommended for horses.

Other Considerations

Regardless of the material used, horizontal fencing must be fastened onto the inside of the posts so that when horses lean against the fence, they push the boards, pipe, or wire against the posts rather than off the posts. If boards are put on the outside of the posts, they can easily be detached, allowing the horses to escape. When boards are put on the inside, there is no need for vertical face boards.

Round posts generally come from tree stock similar in diameter to the finished post. Square posts, on the other hand, must be milled and wood removed from the original blank; this removes strength as well. Thus a four-inch round post is stronger than a four-inch square post of the same type of lumber.

Posts should be western red cedar, osage orange, western juniper heartwood, or black locust, and hard enough to be useful untreated. Treated softwood posts appear to be more expensive; however, pressure-treated posts last 25 percent longer than untreated posts, and they are often guaranteed for twenty years. The best plan is to buy either treated posts or hardwoods, and then coat the posts with a paint, or other coating before setting them in the ground.

The top of a horse fence should be fifty-four to sixty inches above ground level. Line posts should be set thirty to thirty-six inches deep, requiring either 7½- or 8-foot posts. Corner posts should be 8½- or 9-feet long and set thirty-six to forty-two inches deep. Square posts should be five inches square as line posts and six inches for corner posts. Round posts should be a minimum of four inches top diameter as line posts and six inches as corners.

All fences that rely on tension must have strong corner and brace sections in the line fences. The strength of the brace is based on the cross wires that pull the top of the second and third posts toward the corner and away from the tension. The corner posts should be set in concrete at forty-eight inches; landscape timbers or four-inch-square posts can be used for the horizontal braces. Eight to nine gauge wire should be used with any hardwood piece to twist the post tight.

SUMMARY

In the past, horse housing designs have developed at the whim of humans. Horses are the one species of large animal that many people have tried to fit into the pet category. Building requirements frequently have been established for the comfort and benefit of people and have not taken into consideration the health of the animal. Many horse barns are built with poor or nonexistent ventilation and, especially in colder climates, horses are required to spend many hours each day in a moisture-filled, dust-laden environment.

Minimum air exchanges per hour, size and orientation of buildings, stall sizes, and flooring should be not only comfortable, but safe for equine charges.

The most important consideration in selecting and building fences for horses is that they must safely contain the animals. After determining what type of horse will be fenced and other important criteria, build the strongest fence you can afford. The cost of building the fences will likely be second only to the cost of the property itself and the barn. But for the horse owner, nothing is more comforting than knowing that the horses are outside where they are the most healthy, in a fence that will keep them there safely.

REVIEW

Success in any career requires knowledge. Test your knowledge of this chapter by answering these questions or solving these problems.

True or False

1. The primary defense of the horse is flight.
2. Fencing needs to be on the outside of posts.
3. Natural ventilation is the most common and cost-effective ventilation system for horses.
4. Plastic roof panels should not be used as a light souce in cold buildings.
5. The higher the ceiling, the better lit the arena or training area must be to minimize shadows.

Short Answer

6. List the four main provisions for an equine facility.
7. Name four considerations that are part of the environment for the horse.
8. Identify at least five items to be considered when planning to construct a horse facility.
9. List five variations on barns for different horse operations.

10. Name three types of fencing available.

11. Give one advantage and one disadvantage of the three types of siding—metal, wood, and masonry.

12. What are the stall space requirements for—a stallion, a mare with a foal, and an older horse?

Discussion

13. Describe how the ventilation of an equine facility is different in the winter as compared to the summer.

14. What is the purpose of fencing horses?

15. Name three types of stall flooring and describe the advantages and disadvantages of each.

16. Why is moisture resistance an important feature in a horse building?

17. Describe why adequate lighting is important inside a horse facility.

18. Explain one advantage and one disadvantage each for the following types of fencing—wood, PVC, electric wire, and pipe.

STUDENT ACTIVITIES

1. Make a display showing four types of fencing used for horses. Include the cost, advantages, and disadvantages of each type of fencing.

2. Using a computerized drafting program, design an equine facility.

3. Visit some horse facilities and, using a video camera, document your visit. Use the camera to compare windows, flooring, siding, ventilation, feeders, and waterers.

4. Compare the cost of building materials in your area. For example, compare the cost of siding materials, such as metal, wood, and masonry.

5. Collect samples of at least five types of material used in stall floors.

ADDITIONAL RESOURCES

Books

American Youth Horse Council. 1993. *Horse industry handbook: a guide to equine care and management.* Lexington, KY: American Youth Horse Council, Inc.

Evans, J. W. 1989. *Horses: a guide to selection, care, and enjoyment.* 2nd ed. New York, NY: W. H. Freeman and Company.

Hill, C. 1994. *Horsekeeping on a small acerage.* Pownal, VT: Storey Communications, Inc.

Pelley, L. 1984. *In one barn: efficient livestock housing and management.* Woodstock, VT: The Countryman Press.

Price, S. D. 1993. *The whole horse catalog.* New York, NY: Fireside.

Prince, E. R., and Collier, G. M. 1986. *Basic horse care.* New York, NY: Doubleday & Company, Inc.

Stoneridge, M. A. 1983. *Practical horseman's book of horsekeeping.* Garden City, NY: Doubleday & Company, Inc.

CHAPTER 19

Horse Behavior and Training

Without a clear understanding of horse behavior, individuals cannot become good riders, trainers, or owners. Horse psychology—understanding how a horse perceives the world—is used to encourage horses to respond to the goals of the trainer and rider. This training begins when the foal is still at the mare's side. Those horses that are fit are easier to train.

OBJECTIVES

After completing this chapter, you should be able to:

- Name and describe ten behavioral categories
- Discuss the role of reinforcement in training
- Describe imprinting
- Describe the horse's senses of vision, touch, smell, and hearing
- Identify how to read the emotions of a horse
- Discuss how the gregarious nature of horses can influence their training
- Describe the role of the sense of touch in training
- Characterize longeing and its uses
- Describe the role of aerobic and anaerobic fitness in training horses
- Discuss how a horse is taught during training

K E Y   T E R M S

Agonistic behavior	Flighty	Pig-eyed
Allelomimetic behavior	Grooming behavior	Reactive behavior
Barn-sour	Hard mouth	Reward training
Care-giving behavior	Herd obedience	Sexual behavior
Conditioned response	Imprinting	Sleep and rest behavior
Cue	Ingestive behavior	Unconditioned response
Eliminative behavior	Interval training	
Epimeletic	Mimicry behavior	

BEHAVIOR DEFINED

Some of horse behavior is genetic. This part of their "programming" directs how they interact with their environment to maintain themselves and to survive. The other part of horse behavior is learned as horses respond to their environment. As with all animals, horse behavior can be categorized. Different names are given to these behavioral categories, but most names are similar to these:

1. Reactive behavior
2. Ingestive behavior
3. Eliminative behavior
4. Sexual behavior
5. Care-giving and care-seeking behavior
6. Agonistic behavior
7. Mimicry behavior
8. Investigative behavior
9. Grooming behavior
10. Sleep and rest behavior

Reactive Behavior

Reactive behavior is a classification of activities used by an animal to keep itself in harmony with its environment and to adjust to sudden potentially harmful situations. One form of reactive behavior is a simple reflex, for example, a limb withdrawn in response to a local pain.

Communication and vocalization are also a form of reactive behavior. Some communication is merely body language, for example, as horses exhibit group association by maintaining visual contact. Vocalization is used to exchange signals between mare and foal or between other bonded individuals when they are separated.

Another form of reactive behavior is shelter-seeking. In cold weather, horses augment their shaggy coats by seeking protection from the cold and wind. During a storm horses will turn their backsides to the wind. Following a cold night, they will stand broadside to the sun to expose as much of their bodies as possible to the sun. On the other hand, during hot weather, they will seek shade or a cooling breeze.

Ingestive Behavior

Ingestive behavior includes eating, drinking, food preferences, daily patterns of feeding, the mechanics of obtaining food, and chewing food. The first behavioral trait of all mammals including foals is suckling. Foals start eating solid food when they are only a few days old. They start nibbling on feed and rapidly learn to eat it.

Horses in a pasture eat small amounts of feed all day long (see Figure 19-1). Obviously, horses kept in a stall or corral eat at the convenience of the owner or handler. If food is freely available, horses have a tendency to overeat.

When horses graze they take a bite of grass, move a few steps and take another bite of grass. So they are moving most of the time they are grazing. Horses graze over a large area. If plenty of grass is available, horses will eat the

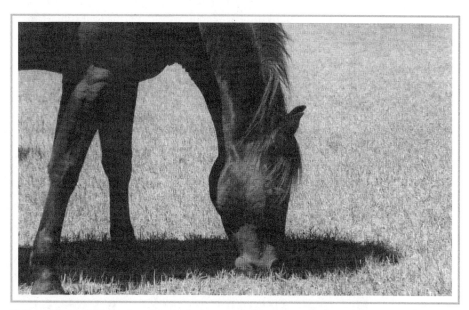

Figure 19-1 Grazing is a form of ingestive behavior. Horses will walk and graze for most of the day. Because they are grazers, horses that are confined to stalls with little or no turnout may begin to exhibit behaviors such as stall walking and cribbing. So it is important that horses be given appropriate exercise and moments of freedom to graze.

top of the stalks and leave the bottom. If pasture is insufficient, horses will overgraze an area and eat the grass down to the surface.

Horses use their molars for chewing before swallowing, whereas cattle ingest large quantities of food with minimal chewing.

Feeding behavior is influenced by learned patterns and preferences, palatability of the feed, the environment, and social associations. Exactly how genetics and environmental contributions affect feeding behavior is not completely clear. Thirst, on the other hand is controlled by centers in the brain that work at maintaining a specific level of body fluid. The control factors regulating thirst are influenced by hormones, salt intake, moisture content of the feed, and environmental factors (see Figure 19-2).

Eliminative Behavior

Eliminative behavior refers to urination and defecation. While urinating, all horses have the same characteristic stance. The neck is lowered and extended, the tail is raised, and the hind legs are spread apart and extended toward the back. Most horses urinate about every four to six hours but they are often reluctant to urinate on a hard surface because the urine splatters on their legs.

Horses defecate every two to three hours, but they will also usually defecate when nervous. To defecate, a horse raises its tail and may hold it off to one side. Horses will defecate while they are moving. Stallions prefer to defecate in a small area and will even back up to a pile of manure to defecate. In a corral, often mares and geldings choose no particular place to defecate and

Figure 19-2 Water must be made available at all times in order to maintain the proper balance of fluid in the body.

will scatter their feces everywhere. In pastures, however, horses tend to deposit their urine and feces in certain areas and graze other areas.

Sexual Behavior

Sexual behavior involves courtship, mating, and maternal behavior. It is controlled by hormones, but some of it may be learned. Stallions find females in heat by sight and smell. Courtship of the stallion prepares him for mating. It is characterized by neighing, smelling, and pinching the mare with his teeth by grasping the folds of the skin in her loin-croup area. Often the stallion will extend his head and upcurled upper lip when around a mare in heat.

Mares in heat show relaxation of the external genitalia, frequent urination, winking of the clitoris, spreading the hindlegs, and lifting the tail sideways. Also, a mare in heat will allow the stallion to bite and smell her.

Care-giving and Care-seeking Behavior

Giving care or attention is very common in horses. Another name for this type of behavior is **epimeletic**. Horses seek attention and care from each other and display this behavior in several ways. During fly season horses stand head to tail and mutually swat flies for each other. Using their incisor teeth they nibble at areas including the base of the neck, withers, back, and croup during mutual grooming. Mutual grooming is frequent during the spring and summer (see Figure 19-3). Licking as a means of care-giving is limited to a mare licking her foal for about the first half hour after parturition.

Figure 19-3 Mutual grooming is very common and very relaxing for both parties. Both *Sidney* and *Berry* look very relaxed and content as they groom each other. *(Photo courtesy of Barbara Jensen, After Hours Farms)*

Horses also signal their desire for care and attention. This type of behavior is also called et-epimeletic behavior. All age groups show this type of behavior, which is most often seen when horses are separated from each other. For example, a young foal when separated from its mother will nicker or whinny for her. Mature horses that are used to being together become upset and call for each other when separated. At the beginning of the separation they will frequently whinny for each other and act very excited. They may also become so nervous and excited that they will try to run through fences.

Agonistic Behavior

Agonistic behavior includes fighting, flight, and other reactions associated with conflict. In all species of farm animals, males are more likely than females to fight. Aggression is used to establish the dominance hierarchy of horses kept together. Most dominant hierarchies are linear but sometimes they can be complex—one horse low in one dominant order may rank above another horse in another dominant order.

Hierarchy is established by some characteristic behaviors. Unacquainted horses approach each other with their heads high; they may also toss their heads. Their necks are arched and ears point forward. The face-to-face encounter is made by smelling or exhaling at each others' nostrils. They may squeal, rear up and threaten to strike during this face-to-face encounter. As the encounter continues the horses may continue smelling each others' necks, withers, rumps, and genitals. At some time during this encounter one horse may decide to turn its hind end around to the other horse and kick with one or both hind legs. Once dominance is established, only threats of aggression are necessary to maintain the hierarchy. To avoid or reduce this type of behavior, a newcomer horse can be penned adjacent to the group. Obviously, this requires strong, high, safe fencing. Horses that are run together from a very young age seldom fight.

Mock fighting is a variation of play. During mock fighting it is common for animals to circle each other, or in a group they will push, nip, and chase each other. Sometimes they will rear on their hind legs and paw at each other. This activity is especially observed in young colts (see Figure 19-4).

Mimicry Behavior

Horses learn to copy the behavior of other horses at a very young age. This is called mimicry or **allelomimetic behavior**. When one member of a group does something, others will do the same thing. For example, horses moving toward water and crossing a pasture display allelomimetic behavior. As one horse starts toward the water others follow. The first horse continues because the rest of the herd is following. Finally, even the most timid horse will follow the group. This type of behavior is closely related to gregarious behavior.

Figure 19-4 Horses that are healthy will often exhibit a wide variety of play. One such play behavior is mock fighting. It is not unusual for a horse that is feeling energetic to explode into a bucking frenzy or race around a pen playfully. *(Photo courtesy of Barbara Jensen, After Hours Farms)*

The close presence of other animals provides companionship and has a quieting effect.

Investigative Behavior

Horses like to explore and investigate a new environment. This curiosity subsides once the environment becomes familiar. If any change or novelty is introduced, investigative behavior reappears. Horses use their senses of sight, hearing, smell, taste, and touch to investigate.

Foals are more curious than older horses. The mare becomes very nervous as she watches her foal investigate. But foals spend much of their time looking and sniffing at objects in their pastures or stalls. Exploratory behavior is sort of a trial-and-error learning activity (see Figure 19-5).

Grooming Behavior

Besides mutual grooming, horses also groom themselves. Horses will paw a dry area and roll on their backs in the dirt. When they get up they shake their whole body. This is a grooming activity.

Horses get rid of annoying insects on their bodies by several methods. Rapidly contracting superficial muscles on the trunk and forelegs causes

Figure 19-5 Foals are very curious and as they get older are more apt to wander further from their mother's side to explore. The mare in this photo is very relaxed and not at all disturbed by her foal's need to wander. Obviously the mare feels comfortable and safe in this environment.

insects to fly away. On their forelegs, shoulders, ribs, flanks, and thigh areas, horses use their heads to remove insects. Insects on the belly are removed with the hind leg, while the tail swats at flies on the hind quarters.

Itching is often relieved by rubbing against some fixed object. The horse will also use its head or a hind foot to scratch an itch. Horses scratch their forequarters, sides, croup and hind quarters with the head, while the hind foot is used on the neck and head.

Sleep and Rest Behavior

Rest and sleep allow the horse to restore its physiological status. During sleep the body makes metabolic recoveries in a short time. During rest, the body conserves energy; the animal may be drowsy but wakeful. The horse rests while standing.

As in humans, sleep in horses occurs in two forms—brain sleep and body sleep (REM). In brain sleep the brain puts out slow electrical waves. In body sleep some electrical currents of the brain are of the same pattern as when the animal is awake. During this time, the eyes move rapidly behind closed eyelids. This form of sleep is known as rapid-eye-movement or REM sleep. During REM sleep the animal needs to be lying down unless it can prop itself up against something. Horses lie down for about four or five periods per day. REM sleep is a deeper sleep than slow-wave sleep. During REM sleep the mind is very active but muscle tone is almost completely lost.

In slow-wave sleep or brain sleep the muscles are not fully relaxed. Horses can be in slow-wave sleep while lying down or while standing up. Because a horse can lock its knees and hocks with a series of tendons controlling leg flexion, the horse can sleep while standing up.

In the course of a twenty-four-hour day the horse is alert and active just a little more than nineteen hours. It spends almost two hours in a drowsy, resting state. So the horse is awake about twenty-one hours a day. The horse actually sleeps—brain sleep and REM—for a total of about three hours a day.

Abnormal Behavior

Horse owners, trainers, and riders need to learn the normal patterns of horse behavior so they can recognize abnormal behavior. For example, abnormal reactive behavior in horses may include such activities as:

- Weaving
- Head nodding and shaking
- Pacing and pawing
- Self-mutilation
- Tail rubbing
- Destructive behavior

Abnormalities of ingestive behavior includes:

- Crib biting
- Tail biting
- Tongue dragging
- Wind sucking (see Figure 19-6)
- Wood chewing
- Eating feces, hair, or soil

LEARNING

Training involves learning new behaviors. The horse learns to make a desired response. Stimuli cause responses. If the response occurs without practice, it is an **unconditioned response**. A response that is learned is called a **conditioned response**. Many of these are used in horse training. The stimulus used to train horses is called a cue. Responses are chained together into maneuvers.

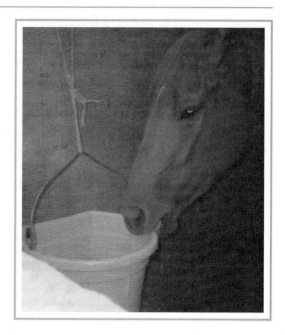

Figure 19-6 Wind sucking is considered an abnormality of ingestive behavior. It is the habit of force-swallowing gulps of air; it is usually accomplished by grasping at an object with the incisor teeth and pulling the neck back in a rigid arch as air is drawn in.

Cues

Horses must learn to recognize **cues**. Trainers start with the cues that are the closest to being natural. A horse is not ridden from hand to leg. It is always from leg to hand. The leg is the most crucial aid to give the horse cues.

As the horse learns the basic cues, the trainer will advance the horse to new cues. The rate of learning depends on the individual horse and the clarity with which the cues are paired. The new cue should always be given first, followed by an old cue that the horse knows. If the horse seems confused then go back to the previous cue until it is clear, then advance. Communication must be clear, so cues must be very specific. Indiscriminate cues will only confuse the horse.

Reinforcement

Reinforcement is something that can strengthen the response to certain stimuli. Primary reinforcers have natural reinforcing properties. Feed, for example, is a primary reinforcer. Very few primary reinforcers are used in training. Secondary reinforcers are learned. Acts of kindness are secondary reinforcers to horses. For example, a soothing voice and rubbing a horse's neck are secondary reinforcers.

All reinforcement is either positive or negative. Positive reinforcement is sometimes called **reward training** and it is effective because the horse wants to give the desired response. Negative reinforcement means the horse will respond to avoid or get rid of the stimulus. The three methods of conditioning with negative reinforcement are punishment, escape, and avoidance. For any reinforcement to be effective, it must be contingent on the response and given immediately.

Three seconds is considered the appropriate time limit. Otherwise, the horse may not understand which behavior is being rewarded or punished.

Young horses are trained using continuous reinforcement. Gradually, this becomes intermittent reinforcement as training progresses. A horse trained on an intermittent schedule will preform longer without reinforcement than a continuously reinforced horse. This is what is referred to as a finished or fully-trained horse.

The more effort required by the horse to make a particular response, the more difficulty the horse will have learning the response. For example, less time is required to train a pleasure horse than to train a jumping horse. This is why it is so important to break each response down into smaller steps. Horses with a greater natural athletic ability have greater potential than those with less ability so it is important to know and understand a horse's conformation and physical limitations when training.

IMPRINTING

Handling and accustoming a foal to human stimulus during the first forty-eight hours after its birth has been shown to psychologically prepare the foal for later handling. This **imprinting** of human contact is most effective if done within the first twenty-four hours of the foal's life. Handling the foal's feet, muzzle, ears, rectum, and girth help prepare it for the future when it becomes necessary to pick up the feet, and clip the muzzle and ears, pass a stomach tube, take temperatures, and tighten a saddle. Time spent handling foals in the first few days of life is time well spent (see Figure 19-7).

Figure 19-7 Proper handling of a newborn makes it easier to handle a horse later for such things as ear clipping. *(Photo courtesy of Cathy Esperti)*

SENSES

How an animal behaves is influenced by its senses of vision, hearing, touch, and smell. Using these senses, the horse interprets and responds to its environment and training.

Vision

A horse has a field of vision that is approximately 220 degrees for each eye, allowing it a panoramic view. Its only real blind spots are directly behind and directly in front; however, a horse can certainly use its sense of smell when an object is directly in front. Because horses are capable of monocular vision (independent viewing) from each eye, they may shy at a bag they just passed when heading in a different direction. The rider may think, "My horse just saw that bag. Why is he spooking now?" He spooked because he may have seen the bag from the eye on the side that it passed first but when he changed directions he was seeing it for the first time with the other eye. A horse has to focus it's attention forward in order to use binocular vision, which is limited to 60 to 70 degrees (see Figure 19-8). A young horse will probably use monocular vision in new situations until it is better experienced.

Generally speaking, horses see poorly. Their eyes have a ramped retina. That is, it does not form a true arc, so parts of the retina are closer to the lens than other parts. The horse adjusts its range of vision by lowering and raising its head, much as a human does with trifocal glasses.

Such a visual arrangement is most convenient for grazing and watching for enemies at the same time, but it is a real handicap in judging height and distance. As a horse approaches a strange jump, it lowers its head, then raises it to appraise the height of the jump. At the point before takeoff it cannot even see the jump since its eyes see separately.

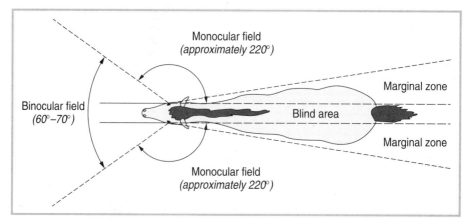

Figure 19-8 The visual fields of a horse. Understanding how a horse sees the world leads to better understanding of horse behavior.

Horses taken from a brightly lit area for loading into a trailer may lower their noses to the floor of the trailer, then raise their heads rather high for loading. In addition to smelling the trailer for identification, they may be trying to find the head position that gives them the best possible vision. They also may be taking time for their eyes to adjust to the light change, a much slower process than for humans.

Young horses that resist trailer loading are doing what saved the lives of their ancestors, who would have regarded a trailer as a dark cave. Horse handlers should allow plenty of time for loading young horses until they are well trained. A good system is to park a trailer in the horse lot and feed young horses in it.

Horses are color blind. They do not perceive blue streams running through green fields, framed in trees with fall-colored leaves. They see a drab mosaic landscape with different amounts of light reflecting from it.

Objects that remain still convey very little information to the horse's brain. A sitting rabbit or bird may be seen readily by the rider, but may remain obscure to the horse until it moves. Horses see movement instantly and react according to temperament, experience, and confidence in the rider. A stall-raised horse may shy sharply at sudden movement.

Young horses need to gain confidence by being gently urged toward objects they fear. If they are concentrating on the fearful object and are punished, they assume the object caused the pain and their suspicions are reinforced. If the rider practices patience in early training, the horse realizes the rider will not ask it to go into dangerous situations and it will lose its fear of strange objects.

Size and position of the eyes and width of the head and body determine front and rear vision. Horses with large, wide-set eyes have more forward and rear vision than others. Even so, there are blind spots at both ends of the horse. This is why you should not approach a horse directly from the rear and why you should speak to the horse when passing behind it.

Frontal vision is affected by width of forehead and how the eyes are set in the head. Most horses probably do not see objects nearer than three feet directly in front of their faces without moving their heads. With their heads in normal position, they do not see the feed they eat nor the ground they step on.

Pig-eyed horses, or those with sunken eyes, see less in front and behind than others. They have often been classified in song and verse as being "mean." Many pig-eyed horses are normal and useful, but one researcher suggests that those growing up in groups of foals may be "picked on" more than others and develop disposition problems.

When being ridden, the horse needs free rein in negotiating obstacles so it has good vision. Horses must be allowed to concentrate when traversing rough terrain, because they must remember their earlier view of the ground now under their feet since they can longer see it. Undoubtedly, some stumbling results from the horse's not watching the area over which it travels and not remembering where the obstacles are.

Hearing

The hearing of most horses is quite good. Rotating ears on movable heads and long necks are advantageous for hearing. Since the horse's sense of hearing is better than its sight, the eyes and ears work together. The ears will point toward a sound so the horse can hear it better. Then the horse will try to see what is making the sound. Horses hear high tones not perceptible to human ears, for instance the blowing of horns in fox hunting. This may cause high-strung Thoroughbred hunters to show anxiety and break out in a sweat.

Fear of parade bands, loud machines, and gunshot noises may result from actual pain to the horse's ears. U.S. cavalry mounts used on pistol ranges would lose their hearing after a few years use in target practice.

Touch

The skin of the horse is a very specialized sense organ. It tells the animal whether something is hot or cold, hard or soft, or whether it causes pain. Some horses will learn to check an electric fence daily with the hairs on their upper lip and will promptly tear it down when the battery fails.

Nerve endings in people are more abundant in the mouth, feet, and hands. Spots of most sensitivity in horses seem to be in the mouth, feet, flanks, neck and shoulders. The mouth is sensitive to pain rather than light pressure. Bitting should be done with care and reins handled with light hands, or else sensitivity in the mouth is lost and a **hard mouth** results.

Some horses are so sensitive to contact in the flank that they may buck when a rider's leg is used too strongly or incorrectly.

Horses vary greatly in skin sensitivity. They love to be groomed and have their backs scratched. Selecting mild grooming equipment is necessary for some thin-skinned horses. Currycombs and "shedding" blades should have fine teeth.

Saddling is a bruising experience for some horses, whereas others seem immune to any feeling when a saddle is placed on them. If a horse humps up and tries to avoid the saddle, flapping cinches and stirrups may be hurting it, it may have back pain, or the saddle may not fit correctly. This reaction needs to be taken seriously and investigated.

Smell

Most animals in the wild state have a good sense of smell. Horses in a research project in England that were frustrated by circling in closed trailers, were able to head directly homeward from a downwind distance of five miles. Domestic stallions can identify mares in heat for great distances downwind.

Colts being saddled for the first few times should be allowed to smell the saddle and the blanket before saddling. This reassures them that the equipment is not dangerous and that it has been used by other horses.

Smell probably dictates grazing habits of horses, although it does not always keep them from eating poisonous plants when forage is abundant.

WORKING WITH HORSE BEHAVIOR

Every time people use horses, they exercise psychology, because their strength is no match for that of horses. If we do not use psychology, a horse may use us to achieve objectives that are not consistent with our intended goal. Such a situation results in owner dissatisfaction and a spoiled or confused horse. Modern horse psychology attempts to anticipate possible behavior of the horse under a variety of conditions and then tries to provide a comfortable condition that will calm and encourage the horse to respond correctly to the handler. Refer also to Chapter 16.

Reading A Horse

Unmanageable situations can often be avoided by correctly reading the emotions of a horse. Ears pinned backward indicate anger or a warning. These signs warn handlers that they may be bitten or kicked. Horses sometimes "fake" anger in an attempt to bluff and scare off a potential intruder. The ear position of horses performing with great resolve, such as hard trotting, pacing, or running a race, should not be misinterpreted, as the ears are sometimes held in a backward position during extreme effort. However, mares with newborn foals are probably not bluffing when their ears are "pinned back," and they should be respected.

Ears forward show interest or suspicion. Some horses show interest in everything they see in new surroundings without finding anything fearful to be avoided. Such horses maintain a good attitude and seem to enjoy their work. Others keep their ears forward and eyes open, afraid of a sudden attack.

Eyes and nostrils show emotion and reflect temperament. Dilated nostrils reflect interest, curiosity, or apprehension. When the eyes flash, nostrils dilate, and muscles tense, the horse is likely about to react. It might be only a slight start, a reverse in direction, or both. But if the cause of fright intensifies, the horse may bolt, rear, or buck. Riders who read their horses' emotions accurately can often steady the horse with reassuring words or control through appropriate hand, leg, and upper body connection (see Figure 19-9).

Memory

Horses usually are considered to have memories second only to elephants. In the wild, if an attack came at a certain place, the herd avoided that spot in the

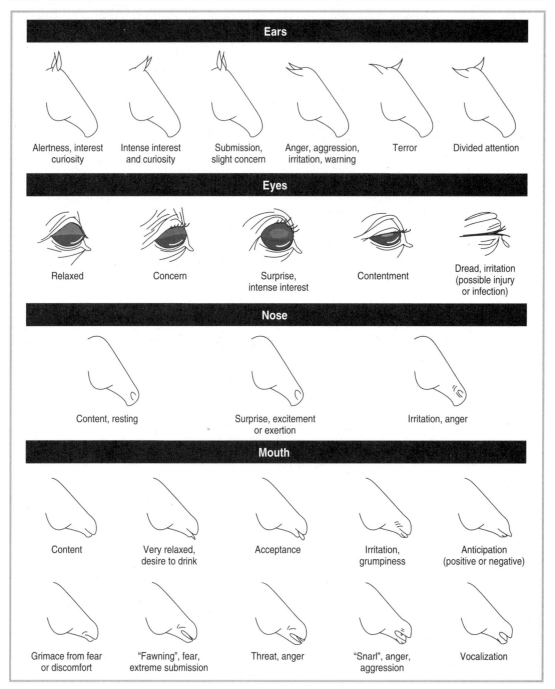

Ears

Alertness, interest curiosity

Intense interest and curiosity

Submission, slight concern

Anger, aggression, irritation, warning

Terror

Divided attention

Eyes

Relaxed

Concern

Surprise, intense interest

Contentment

Dread, irritation (possible injury or infection)

Nose

Content, resting

Surprise, excitement or exertion

Irritation, anger

Mouth

Content

Very relaxed, desire to drink

Acceptance

Irritation, grumpiness

Anticipation (positive or negative)

Grimace from fear or discomfort

"Fawning", fear, extreme submission

Threat, anger

"Snarl", anger, aggression

Vocalization

Figure 19-9 When working with horses it is important to recognize warning signs of aggression. Key features of the horse to watch are the ears, eyes, nose, and mouth. For example, the ears held flat back accompanied with a "snarl" is a sure sign of anger. On the other hand, a soft expression of the mouth along with a soft expression in the eye is usually a good sign of contentment.

future. This caution is still practiced by wild horses in the United States. If it were not for the horse's good memory, it would be considerably less useful to people. A well-trained young horse never forgets its training. Neither does the poorly trained one. For this reason, bad habits should be recognized and corrected before they become fixed.

Horses have not ranked outstandingly well on limited intelligence tests, although they do very complex things routinely when trained. Some horses may be considered highly intelligent because they can open most gates and doors on the farm. But idle horses tend to seek activity, some of which may involve gate latches. Once they succeed, their good memory keeps them trying to open doors. When they get the grain bin open, they remember only the joy of eating. They can't associate overeating with the ensuing colic or loss of hooves from founder.

Gregariousness

Horses are gregarious in nature; they band together. This tendency has practical implications. Wild horses in the center of the herd were safer from attack. This can be seen today with zebras in Africa.

The gregarious tendency can be used to advantage in training young horses. A young horse may be fearful of working alone. Horses walk too slowly and jog too fast until they are well trained. A good training method is to jog them away from the barn and walk them toward it. **Barn-sour** horses result from allowing them to run back to the barn where they can be reunited with the herd and thus rewarded for their behavior. The routine should be changed so this is not expected by the horse.

Situations that produce barn-sour horses need to be avoided. Young horses should be sufficiently trained to be obedient before they are asked to leave the premises with a rider. Ground driving helps. If they show anxiety to get back to the barn, change the routine. A good method is to turn away from the barn each time they try to go to it. A useful technique may be to head the horse away from the barn when bringing it to stop at the end of the training session. After the rider dismounts, the horse is led back to the barn. This method is useful in ring riding.

Group riding brings out the "**herd obedience**" tendency in horses. That is, they all tend to do what others do. For example, if one enters a stream, the others tend to follow (see Figure 19-10).

In the wild state, obedience to leadership meant survival. If the stallion called for silence, every horse stood still. If he commanded flight, they ran at the heels of the lead mare. The stallion ran at the back of the herd to nip those who needed more speed. Horses today are dependent on people for leadership and survival.

Figure 19-10   Trail riding brings out the herd tendency in horses. As the group begins to leave the barn, the less dominant horses will follow the lead horse and the rest of the herd. The lead horse is usually very quick to identify itself. *(Photo courtesy of Barbara Jensen, After Hours Farms)*

Reproductive Behavior

Reproductive hormones are responsible for the behavior of mares in estrus. To check the behavioral traits exhibited at estrus, a mare is "teased" or tested for receptivity to the stallion. A mare in estrus demonstrates receptivity by standing quietly when the stallion approaches, urinating frequently, exposing the clitoris (also known as winking), and raising the tailhead. When the mare is ready for mating she is said to be in standing estrus or in heat.

Mares exhibit a wide range of individuality in the expression of estrus. Some are quite aggressive and demonstrate all these signs of estrus when teased. They may even back up into the stallion. Others show minimal change in their behavior, but allow the stallion to mount. Estrus signs are usually consistent in a single mare from one cycle to the next. Occasionally mares will show signs of estrus when strange new horses are present. But most mares will not show signs of estrus unless a stallion is present. It is nearly impossible to reliably determine estrus if a mare is stabled only with other mares and geldings.

COMMUNICATION AND TRAINING

Training begins while the foal is still on the mare. Handling and teaching it to lead at this young age will help develop a more dependable horse through the years. Halter breaking is not difficult if done appropriately while the foal is young.

Numerous books and videos are written on the subject of breaking and training horses. Many people have their own methods and their own opinions. This is only a start.

Riders should not constantly correct or interrupt the thought train of a horse doing a job that requires deep concentration. The horse can think of but one thing at a time. The rider who continually punishes or corrects the horse distracts attention from the task at hand and the horse can become confused or angry.

Quick reflexes and panic characterized prehistoric horses. Indeed, their life depended on them. They will panic into flight without much consideration of the need for or consequences of such a decision. Young horses fleeing with or without riders may sustain severe injury from running into objects or from total exhaustion.

The runaway horse is simply carrying out the kind of behavior that allowed its ancestors to survive. As some horses get older they tend to become calmer, while others do not.

Speed, quickness and willingness to serve, even at great sacrifice, have made horses most useful to humans. This also poses some dangers and problems.

"Flighty" horses should be handled by experienced riders and must not be hurried into new and strange situations. Even though they are controllable at home, they may not be in strange surroundings and could be dangerous for the novice rider. The object is for the rider or handler to provide the support and will without provoking an unmanageable confrontation with the horse.

Communication of rider to horse is accomplished through voice, legs, and hands—in this order of importance. Voice cues for starting and stopping are easy to give and easily understood by the horse. Rein cues are more complex for both rider and horse, and signify more complicated maneuvers than simple starts and stops. Leg cues are needed for most complex responses, such as rollbacks. Because of the sensitivity of a horse's skin, it can react to the lightest pressure of the rider's leg.

Horses are equally sensitive to insecurity or confidence in their riders, and respond accordingly. If the rider lacks assurance, the horse will feel insecure and perform below its capability.

The horse is a strong, sensitive creature, capable of great speed and quick reactions. It has great ability to adapt to unfamiliar situations. This is why we like horses. Many of the things humans ask them to do are strange to their nature, so we need to understand their reaction to these new situations.

Catching and Haltering

If the mare is gentle, use her to help catch her foal. Lead her into a box stall with the foal following and get the foal in a corner. The mare will help hold the foal while you ease the halter on. Work slowly with a lot of rubbing and quiet talking to calm the foal.

The foal probably will be nervous and scared, but if the mother shows no concern, the job will be easier. After haltering, turn them both loose and let the foal wear the halter about two days. This will give the foal time to get used to the feeling of having a halter on. Then go through the same procedure as before, catching and haltering in the box stall. This is the time to begin leading the foal.

Teaching to Lead

Snap a good lead rope in the foal's halter ring and put a rump rope over its hips with one end coming through the halter. A cotton rope makes a good rump rope. Lead the mare out of the stall and let the foal follow. Stay in front of the foal, pulling forward on the lead rope attached to the halter while also pulling sharply on the rump rope. The foal may jump forward when the rump rope is tightened, so be careful not to get stepped on (see Figure 19-11).

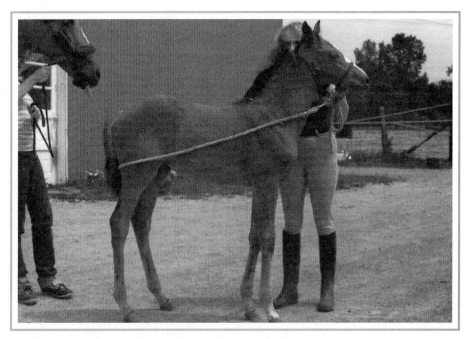

Figure 19-11 The proper use of a rump rope. *(Photo courtesy of Barbara Jensen, After Hours Farms)*

If the mare and foal are led around together while pulling on the lead and rump ropes, soon the foal will be leading. Once the foal begins to lead well, work away from the mare to make sure the foal is leading and not just following the mother. Do this for two or three days for about ten to fifteen minutes each time.

When the mare and foal are brought in for weaning, the foal probably will not be afraid because it will remember the prior experience of not being hurt. Halter and lead it for a few days. Trim its feet and worm it if necessary. Any handling at this age is time well spent. The foal's gentleness and learning to lead will save time during breaking at two years of age when the foal will be saddled, bridled, and mounted for the first time.

LONGEING

Longeing is a procedure in which the horse travels in a large circle around the handler on a long strap or line. It is useful in training young horses and in exercising others. Longeing affords the horse an opportunity to improve balance and develop stride and action. It is also a good way to reduce energy in overactive horses before they are ridden. Longeing can be started after weaning, if the trainer is careful not to let a young horse hurt itself by being jerked off-balance on a longe line (see Figure 19-12).

Figure 19-12 Longeing has a multitude of purposes. It is a great way to allow a horse to release excess energy and warm up prior to being ridden. *(Photo courtesy of Michael Dzaman)*

Before horses are longed, they should be taught to lead from either side, and to stop, stand, and back. They should be gentle and reasonably obedient or easy to control.

The horse should be groomed at the site of training the first time it is longed. This relieves some anxiety and puts it at ease. Also, protecting the horse from flies with repellent allows it to give its undivided attention for the lesson.

Starting

Horses are trained to longe in a small pen. After the horse has circled the ring a few times, the trainer should start to drop away from the shoulder, keeping the horse moving forward by tapping the ground lightly with the whip. The trainer should drop toward the rear of the horse.

To keep the horse moving forward, the trainer should stand by the horse's left leg and hip as the circle is gradually made larger. The trainer should still make a small circle as the line feeds out. The whip will keep the horse from stopping or closing the size of the circle.

The horse should learn to stop and stay on the perimeter of the circle. At the command of "whoa," it may turn and face the trainer. In later lessons, the horse should stop in place on the perimeter until commanded to face inward and come to the trainer. The horse should not be allowed to anticipate commands and make its own decisions. When the horse is stopped in the center of the circle or at the perimeter, it should be taught to stand in place.

Some horses keep a longe line tighter than others. A tight line on a horse is desirable. Short pulls and releases will restrain it. A soft nylon or leather halter, compared to a longeing cavesson, may encourage tight line pulling. When the horse is going in a large circle around the trainer with the right tension on the line, the trainer can stand in one position and give the lesson with minimum effort.

Many horses are definitely one-sided in their preference in longeing. The trainer should change directions to work the weaker side more than the stronger side until the horse will longe in both directions in good form.

Other Uses

One of the best uses of the longe line is to "tune up" an old, well-trained horse. If a horse is suspected of being lame, longeing is a great way to assess its movement. And it is an excellent way for a horse to "learn" self-carriage. Also, a stabled horse can be exercised on a longe line when it is not possible to ride it.

Horses can be trotted across cavalettis or fence posts to regulate length of stride. This is particularly useful in jumping horses and in young horses that do not extend enough in the trot. Gaited horses with a pacing tendency can often be improved by this procedure, as can Western horses that need more length of stride in their extended trots.

Some horses jump well on longe lines. It is good exercise for trained horses and a good way to start young jumping prospects.

Longeing is an important step in preliminary training. A young horse can learn to start, stop, stand, walk, trot, and canter on command. Longeing establishes authority and routines that reduce mounted training time.

Seasoned performers can be exercised, refreshed, or even prepared for new activities from the longe line. Longe line training is a skill that most accomplished horse owners have found worthwhile to develop.

FITNESS

Fit horses are easier to train. Aerobic fitness is probably the simplest and safest type of stamina to train into a horse. Aerobic training requires low intensity, long duration types of work. The heart rate should reach about 120 beats per minute for ten to fifteen minutes. About three to four weeks are required to achieve the training effect. After that time, the intensity or duration must be increased to improve the fitness level (refer to Chapter 6).

Horses need to be able to expend large amounts of energy anaerobically and then replenish that energy aerobically for the next maximum contraction of the muscles. The anaerobic training begins after the aerobic training and this phase is related to specific skill training.

One of the newest concepts in the conditioning of horse athletes is **interval training**. Interval training is simply the use of multiple bouts of work interspersed with a relief interval when partial recovery is allowed. The theory behind this is two-fold. First, it allows more total work to be done, and second, it allows fatigue to be brought on gradually and controlled. While interval training is most applicable to anaerobic types of training, it can be adapted to fit any type of training.

Warm Up and Down

Some general concepts apply to all athletes regardless of species or event. These guidelines improve performance, prevent injury, and minimize the soreness associated with exercise. The first of these is the warm up. This stretches and relaxes the muscles to allow for greater flexibility. Warm up increases the muscle temperature allowing greater use of its energy stores. It also increases the blood flow to the muscles allowing more efficient transfer of oxygen. Trotting, side-passing, two tracking, longeing, and backing are examples of warm up exercises.

After exercise, the horse should be warmed down. Many trainers do this using a mechanical walker or by hand-walking their horses ten to fifteen minutes after the end of exercise (see Figure 19-13). The warm down period should consist of light work decreasing in intensity. This warm down period helps remove

Figure 19-13   A mechanical walker is often used by large horse facilities, such as River Grove Farm in Hailey, Idaho, to ensure proper warming up and cooling down of horses. **Caution:** Horses should never be left unattended on a mechanical walker. It is unsafe.

metabolic by-products such as lactate out of the muscles. Also, it prevents muscles from tightening up after exercise, thereby minimizing soreness.

Fatigue

Trainers must be aware of fatigue during all phases of training. When exercise is intense or of long duration, ATP (adenosine triphosphate) supplies for muscular contraction decline and the by-products of metabolism build up. When this happens, the muscle either runs out of fuel or it is "poisoned" by the harmful by-products. The result is that the muscle can no longer contract efficiently. At that point, other muscle groups start contracting to perform a motion they are not accustomed to. The horse may misstep and injure itself.

Fatigue can be prevented by decreasing the intensity of the exercise and allowing the horse to rest for a period of time. A complete stop is not advisable, just a slow down. Once the heart rate drops below 100 beats per minute, the work can be continued.

Muscles require twenty-six to forty-six hours to replenish their glycogen (energy) stores, depending on the severity of the depletion. Horses need at least one day a week completely off if they are being worked at very high intensity or for extremely long durations.

SUMMARY

How horses interact with their environment is genetic and learned. The behavior of horses can be categorized as follows: reactive, ingestive, eliminative, sexual, care-giving and care-seeking, agonistic, mimicry, investigative, grooming, and sleep and rest. Understanding these ten behavioral categories helps trainers and riders successfully interact with horses. The senses of vision, hearing, smell, and touch influence how a horse interacts with its environment and how a horse learns. Training is a process of teaching the horse to respond to cues. This begins when the foal is still at the mare's side. First the foal learns to lead. Training continues as the horse is saddled, bridled, and mounted for the first time. Longeing also requires training.

Fit horses are easier to train. Fitness includes an aerobic and anaerobic component. Like human athletes, horses in training need a warm up and warm down period. They are also subject to fatigue.

REVIEW

Success in any career requires knowledge. Test your knowledge of this chapter by answering these questions or solving these problems.

True or False

1. For horses, training involves learning.
2. The horse has a very short memory.
3. A horse owner should be able to read the emotions of a horse to avoid unmanageable situations.
4. Idle horses tend to seek activity, such as opening barn doors and latches.
5. During sleep the horse has a period of rapid eye movement.

Short Answer

6. List the ten behavioral categories.
7. Name the senses a horse uses to interpret its environment.
8. When should training begin?
9. What are three ways to communicate with a horse?
10. How are the emotions of a horse read?

Discussion

11. What is the difference between a conditioned and an unconditioned response?
12. Define imprinting.

13. Explain how a horse sees.

14. Define cues, stimuli, response, and reinforcement.

15. Describe the difference between aerobic and anaerobic training in horses.

16. Describe the process of longeing and discuss its uses.

STUDENT ACTIVITIES

1. Develop a report on imprinting. Extend the discussion to animals other than horses, for example, poultry.

2. Observe a group of horses each day at the same time for a period of a week. Document their reactive behavior, ingestive behavior, eliminative behavior, sexual behavior, care-giving and care-seeking behavior, agonistic behavior, mimicry behavior, investigative behavior, grooming behavior, and sleep and rest behavior. Also, note any signs of their emotions, for example, ears pinned back, licking of the lips, tense muscles, etc.

3. Groom a horse and write a description of the process and the tools used.

4. Visit with an experienced rider or trainer. Ask what type of secondary reinforcers can be used with horses. Make a list of these and the expected response.

5. Choose one of the vices horses develop. Describe the vice and any possible solutions.

ADDITIONAL RESOURCES

Books

American Youth Horse Council. 1993. *Horse industry handbook: a guide to equine care and management.* Lexington, KY: American Youth Horse Council, Inc.

Blakely, J. 1981. *Horses and horse sense: the practical science of horse husbandry.* Reston, VI: Reston Publishing Company, Inc.

Ensminger, M. E. 1990. *Horses and horsemanship.* 6th ed. Danville, IL: Interstate Publishers, Inc.

Evans, J. W. 1989. *Horses: a guide to selection, care, and enjoyment.* 2nd ed. New York, NY: W. H. Freeman and Company.

Fraser, C. M., ed. 1991. *The veterinary manual.* 7th ed. Rahway, NJ: Merck & Co.

Miller, R. W. 1974. *Western horse behavior and training.* New York, NY: Doubleday, Inc.

Sizemore, D. M. 1986. *Horsemanship.* Irving, TX: Boy Scouts of America.

Topliff, D. R., and Freeman, D. W. (n.d.) *Physical conditioning of the equine: part I: concepts of equine exercise physiology.* Stillwater, OK: Cooperative Extension Service, Division of Agricultural Sciences and Natural Resources. Oklahoma State University.

Topliff, D. R., and Freeman, D. W. (n.d.) *Physical conditioning of the equine: part II: specific event training programs.* Stillwater, OK: Cooperative Extension Service, Division of Agricultural Sciences and Natural Resources. Oklahoma State University.

CHAPTER 20

Equitation

Equitation is horsemanship, or the art of riding and managing horses. In a sense this whole book is about equitation. This chapter, however, is limited to some necessary skills and information for riding horses in two common ways—English and Western. Methods of mounting, sitting in the the saddle, and dismounting differ slightly but are still basically the same. A rider who gains proficiency in English riding can easily master Western riding, and vice versa. Selecting the proper saddle, safety, hauling, haltering, and tying are other important components of general equitation.

While this chapter ignores many of the other areas of equitation like harness, gaited horses, showing horses, hunters and so on, there are numerous books and videos available for these other areas of equitation.

OBJECTIVES

After completing this chapter, you should be able to:

- Name three styles of saddles and describe their uses
- Indicate the four criteria for selecting a saddle
- Describe the anatomical points on a horse that must be checked when considering a saddle
- Discuss the results of a poorly fitted saddle
- Discuss the effect the rider's being forward or sitting back in the saddle has on the performance of the horse
- Describe the process of saddling and bridling a horse
- Identify guidelines for proper dress around horses, especially for Western riding
- List the steps for proper mounting of a horse
- Give the rules of safe riding
- Describe how to load and haul a horse and how to check the safety of a trailer

- Name three types of halter material
- Describe the process of haltering and adjusting a halter
- Indicate three safe ways of tying a horse

K E Y   T E R M S

Breast collar	Halters	Throatlatch
Cinch	Pommel	Trailer sour
Cross tying	Quick-release knot	Tree
Curb chain	Rigging	

SADDLES

A saddle is one of the first pieces of equipment most people buy after they acquire a horse. It is a major investment; selection and purchase require deliberation and knowledge. The life span of most saddles is several times that of a horse. The selected saddle should fit the needs of the rider and the type of horse. Personal preference should be supplemented with knowledge of the advantages and disadvantages of the many different styles and types of saddles.

Styles of Saddles

The style of riding determines the type of saddle. But a great deal of variation among the saddles within one riding style still exists. Tradition, experience, and exposure to other riders must then be considered. It is also crucial to a rider's success and a horse's physical condition and performance to fit a saddle to both horse and rider.

Table 20-1 summarizes the styles of saddles based on the type of riding.

Western or stock saddles tend to be large and heavy. They are difficult, if not impossible, for youngsters to handle. However, they offer a great deal of

Table 20-1 Styles of Saddles	
Style	**Use**
Stock	Roping, cutting, general purpose, and specialty
Hunt-jumping	Forward seat, balance seat, and polo
Gaited	Lane Fox
Dressage and Miscellaneous	Racing, side saddle, track, and parade

security for a beginner. The thickness of the saddle and the amount of leather under the leg, knee, and seat isolate the horse from the rider which can limit communication. Western saddles are probably more versatile, rugged and durable than other styles. They are available in a wide range of designs and prices. Western saddles can also be purchased in child size. It is important, however, to check the fit on a horse.

Hunt-jump saddles are usually rather light and easily handled. A wide variety of designs and prices are available. In most cases, this type of saddle allows the rider to sit closer to the horse, to feel the horse, and to communicate more readily with seat and legs. As a rule, these saddles require more training of the rider in developing a sure seat than stock saddles do. But this usually leads to much better equitation form (see Figure 20-1). Saddles used to ride and exhibit gaited or park horses, such as the Lane Fox saddle, are rather limited in use. They retain many of the advantages of the hunt-jump saddles. They are lightweight and allow ease of communication. This style of saddle provides minimum security for the rider. As with any style of saddle, proper equitation requires proper training.

Dressage saddles were originally designed for accommodating women's ankle-length skirts. Now they are designed to give the rider maximum ease of communication with the horse and help the rider maintain perfect balance and form, whether the horse is highly collected or mildly extended.

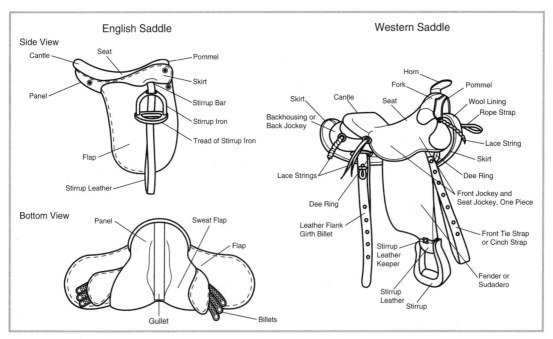

Figure 20-1   The parts of the English saddle and the Western saddle. *(Courtesy of University of Illinois at Urbana-Champaign)*

Many saddles are designed for very specific purposes. These include side-saddle, trick saddle, and special show or display saddle. Using these saddles for anything other than their intended purpose should be discouraged. Safety, comfort of the rider, and ability to maintain soundness of the horse must be considered before beauty or the desire for a unique design.

The selection of a saddle must meet these four basic criteria:

1. It must fit the horse.

2. It should not interfere with the performance or the ability of the horse to perform.

3. It must fit the job or the activities desired.

4. It should fit the rider physically.

Fitting a Saddle to a Horse

Not every saddle fits every horse, just as one size or shape of boot does not fit every person. Some points of the horse's anatomy that must be checked when considering a saddle include:

*   Size and shape of the withers

*   Length of back

*   Slope of shoulder

*   Spring of rib

*   Muscling, especially of the shoulder

To some extent, the rider needs to consider the overall size of the horse, especially with smaller horses and ponies.

Most saddle fitting problems occur at the withers. Ample clearance at the withers is needed to prevent injury, yet not so much space that security is lost. Pressures should not be concentrated on small areas of the back and withers. In a stock saddle with rider mounted, about two inches of clearance should be between the withers and the gullet (underneath front) of the saddle. Insufficient clearance, even with a heavy saddle blanket, means the fork of the saddle is too wide, or the withers of the horse are too high and narrow, or both. Adding a heavy pad or a second or third blanket may help. A narrower saddle is a better solution.

Injury to the withers is usually the result of a poorly fitted saddle. In addition to being painful to the horse, it frequently results in bad habits such as bucking and head slinging, and it may cause the horse to resist saddling. Ill fitting saddles are sometimes a result of the rider's inconsideration, but more often result from a lack of knowledge and attention to the welfare of the horse.

Horses with flat, "mutton" withers often wear saddles that are too narrow. This causes the saddle to sit much too high in front. Additional blankets will help prevent a sore back, but little else can be done to alleviate the problem. To avoid the pain and fatigue that result from this situation, the saddle or the horse needs to be changed. No roping should ever be attempted using an excessively narrow saddle on such a horse.

To fit your horse properly, measure the width of the withers. Width taken at a point two inches below the top of the withers should correspond to the fork width of the saddle. Since blankets and pads will compensate for some misfiting, some variation can be tolerated. Width of the fork of stock saddles varies from 5½ inches to 7 inches. Average saddles are between 6 and 6¾ inches wide. This width accommodates most horses with use of a good blanket or pad. Every secondhand or used saddle should be measured despite claims of size as some spreading occurs with use.

The width of an English saddle **tree** is just as critical as the fork width of a stock saddle, however, it is more difficult to determine as a result of saddle design. A "cut-back" **pommel** may be necessary to prevent damage to the withers. The "cut-back" can range from very slight to over four inches. One major advantage of a "cut-back" pommel is that the saddle can fit a wide variety of horses. The tree of hunt-jump saddles, especially less expensive brands, can spread a great deal without breaking which can drastically change the fit. When considering a used saddle of this type, the width between the points of the tree should be checked regularly. This is true especially if the saddle is to be used in shows. Very little can be done to improve the fit of a wide-fronted hunt-jump saddle except to find a horse with an appropriate anatomy.

A stock saddle should lie directly over the upper end of the horse's shoulder blades. This allows maximum area of contact between horse and saddle, distributing the load and pressures to minimize sore backs.

If the horse is straight-shouldered or if the saddle tends to slip back because of poor riding habits, the sides of the saddle place great pressure on the back edge of the shoulder blades. Even blankets cannot completely eliminate this concentration of pressure. For this condition, a **breast collar** is needed to keep the saddle well forward over the shoulder blades.

Length of a stock saddle should also be considered. A long saddle on a very short-backed horse can cause too much pressure over the loin and kidney area of the horse's back, resulting in injury and soreness. The square cut skirts on some stock saddles may also irritate the flanks of short-backed horses.

Using a Saddle

Performance of any horse can be hindered if the rider does not remain over the horse's center of balance. Since the center of balance changes with different speeds and kinds of activity, a saddle must be selected that provides comfort for both horse and rider so the rider can maintain balance during a specific type of performance. Not only will this aid in achieving maximum perform- ance, it will mean comfort and security to both horse and rider. The center of balance of a horse standing or walking freely lies directly over a point a few inches behind the withers.

As the horse moves forward at speed, the point of balance moves forward. Jockeys provide a good example of weight well forward yet centered over their legs with the ankle, knee, and hip joint acting as shock absorbers so they can balance and move freely with their mounts, permitting full potential performance of the horse. Pleasure riders find that "getting into a half seat" is not only comfortable for themselves, but it also seems to allow freer movement of the horse (see Figure 20-2).

A horse jumping is another rather extreme example of shifting the center of balance. The center of balance shifts as the horse approaches the jump, then the weight drops back, levels, then comes forward. A rider must be in-time with their horse and follow the shifting of balance. Stock seat riders who have attempted to jump in stock saddles can appreciate what the expression "being behind one's horse" means. Not only is this hard on the rider's back and neck, it also is uncomfortable for the horse and usually causes it to refuse a jump.

The more collected a horse is, the farther to the rear the center of balance is displaced. Therefore, the rider of a gaited horse needs to be well back from the withers to free the fore hand legs and put his or her weight more over the horse's hind quarters.

Figure 20-2 The center of balance of horse and rider constantly change and in order to have a balanced ride they must be in harmony with one another as Leslie Burr Howard demonstrates here with her mount. *(Photo courtesy of Barbara Lee Jensen, After Hours Farms)*

Cutting horses work primarily off the hind quarters and are very light on the fore hand legs. Saddles traditionally used have been designed to keep the rider well back from the withers.

The basic design of a saddle usually allows some latitude in placement. The hunt-jump saddle "positions" the rider through the center of the seat. A rider can use various billet strap combinations, however, to change the position by as much as three to four inches. This permits the saddle to be placed properly for different activities or to accommodate a variety of conformation differences. Placement of stock saddles is governed by position of the **rigging**. Rigging can be anywhere from full rigging (directly below the horn) to the center-fire rigging (half-way between the horn and the top of the cantle). The average pleasure rider who does not use a rope will probably find seven-eighths rigging most comfortable and readily available (see Figure 20-1).

The full-rigged saddle was designed especially for roping. It places the horn rigging and **cinch** in a straight line directly over the withers. This permits maximum strength of construction and correctly places the stress from the rope at the withers. Such a design also places the average pleasure rider well behind the center of balance, especially when the horse moves at speed. It does, however, permit the rider to be in balance when the horse is working off its quarters.

Shape of the seat of a saddle is important to both pleasure and equitation riders. Steep seats force the rider to the rear and may offer security, but experienced riders usually find them uncomfortable. This is especially true of pleasure riders with uncollected mounts. Equitation riders must be able to stay in balance with the horse.

Tradition often dictates what type of saddle should be used. Tradition, however, must not replace common sense. It is important to select a saddle designed to permit a specific type of performance.

The stick-forked, flat-seated, low-cantled stock saddle frequently advertised as a roping saddle is not designed for pleasure riding. It is excellent for roping. A roping saddle offers little security in front and little or no support for the hips. Rigging placement also detracts from its usefulness as a pleasure saddle.

The forward seat jumping saddle was designed specifically for jumping. The rider must use relatively short stirrups and ride in the half seat or two-point—that is, the seat is out of the saddle and the rider has two points of contact, the calves. Most saddle makers advertise their saddles using such expressions as roper, cutter, or equitation. Keep in mind that these are advertising claims and should be viewed in the same light as the claims for headache remedies, razor blades, or automobiles.

Fitting a Saddle to a Rider

The saddle should also fit the rider. Saddle size is more critical with English saddles, especially hunt-jump saddles, than with stock saddles. The rider's safety, comfort, and show-ring success all depend on proper saddle size. Length of a hunt-jump saddle is measured from the pommel to the center of the top of the cantle. Standard lengths are sixteen, seventeen, and eighteen inches when the saddle is constructed on a straight-head tree. Lengths on a slope-head tree usually are 1 to 1½ inches less. Probably the most critical test for hunter-seat riders is the position of the knees in the knee pockets. Regardless of length of seat, unless the knees fit into the knee pockets with proper length of stirrup, the saddle does not fit. Although measurements can be made, it is usually advisable to try a hunt-jump saddle for size as it rests on a horse before purchasing it.

Saddle Construction

The tree is the foundation of every saddle. One of the first steps in evaluating a saddle is to check the tree. Until recently, all quality stock saddles were made on a wooden, rawhide-covered tree. Some cheap saddles are made with canvas-covered trees; others are made with the tree only partially covered with rawhide. A relatively recent innovation in saddle-making is the extruded plastic tree. These plastic trees seem to be strong, durable, and free from warping. They reduce weight and cost because they eliminate a great deal of the hand labor of building up a ground seat.

English saddles are usually built on a rigid tree with a straight head or on a spring tree, usually with a sloped head. Slope-head spring trees are relatively new with only a few manufacturers using them, but they seem to be increasing in popularity. Another innovation is the recessed stirrup bar. The combination of slope head and recessed stirrup bar nearly eliminates the hump under the thigh on old models.

Ornate finishes on stock saddles are not always just decorative. The designs serve to hide scratches and to increase the rider's grip. Hand-carved saddles are usually quite expensive. Carving creates a cleaning problem. Embossed saddles are far more common than carved saddles. The high quality of most embossing plates may cause difficulty distinguishing between carving and embossing without careful inspection or looking at the under-side of the leather. Poor-quality embossing is especially noticeable on the swells, where it tends to fade out.

Stirrup adjustments vary considerably. The ease with which adjustments can be made is important if several people use the saddle. The patented Blevins buckle is usually found only on better-quality saddles. It is one of the best and easiest to use. Double-tongued and sometimes single-tongued buckles are normally used on less-expensive saddles with narrow stirrup leathers. Such buckles are satisfactory for adjusting stirrups, yet the overall quality of the saddle may not be acceptable. The quick-change buckle is one of the most common. It usually works well, but it may jam if it is not kept aligned and free of rust. Stirrup pins replace leather laces, which were traditional until recent years.

After final deliberation and selection, the real work of the saddle begins. Remember that the saddle is a device to help the rider maintain a proper and secure seat. In other words, the saddle was meant to be used, not kept on display as a trophy.

Saddling

When saddling a horse for *any* form of equitation, the following rules are crucial:

1. Groom the horse thoroughly to be sure there are no sores on its back or in the cinch area, as this could cause more injury to the horse or cause the horse to wring its tail or buck. If there are saddle sores, consider using extra padding or a girth pad, or give the horse time off until the sores heal. Blankets need to be checked for foreign objects or dirt buildup, and they need to be dry.

2. The blanket or saddle pad should have no wrinkles, and offer adequate padding for the horse. Some horses require more padding than others and some may require extra padding at their withers to prevent binding the shoulders. The saddle cinch (Western) or girth (English) must be clean, as dirty cinches or girths can cause saddle sores.

3. Raise the saddle as high as you can and set it down gently on the horse's back. This helps prevent back soreness and helps assure the horse that the saddling experience is nothing to fear. Throwing the saddle onto the horse's back can cause bruising and may aggravate any existing back problems.

4. Place the saddle properly. This may vary from horse to horse. Do not place the saddle too far forward, which restricts shoulder movement and causes discomfort, or too far back, which can cause kidney damage and sore backs. Until you are proficient at saddling, always have an experienced rider or trainer check your saddle placement prior to riding.

5. For a Western saddle, let the cinch and stirrup down, making sure they do not slam down on the horse's side. For English saddles, hook the girth on one side. Never release the cinch and stirrup by pushing them over the saddle from the left side. This could hurt or startle the horse.

6. Reach under the horse and grasp the cinch or girth with your left hand, facing the rear of the horse. If using a martingale or breast

collar, you may need to thread the cinch or girth through the end of the martingale or breast collar before fastening.

In Western riding, if you use a rear cinch, tighten the front one first. Put the cinch strap through the cinch ring and the rigging ring twice. Then you can either tie a cinch knot to secure the cinch, or you can buckle it if the cinch has holes for it (see Figure 20-3).

7. In Western riding, with the left hand under the buckle to prevent pinching, tighten the cinch slowly, an inch or two at a time. In English, slowly buckle the girth but not too tight. Tightening it too quickly can cause your horse to be "cinchy," or irritable, during saddling. Some horses may even begin biting or rearing when you

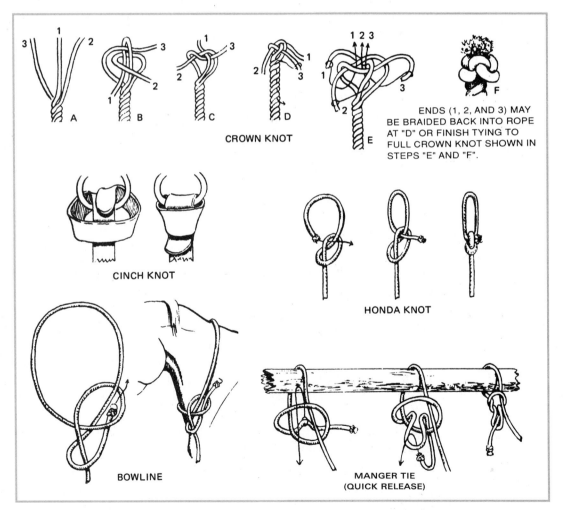

CROWN KNOT

ENDS (1, 2, AND 3) MAY BE BRAIDED BACK INTO ROPE AT "D" OR FINISH TYING TO FULL CROWN KNOT SHOWN IN STEPS "E" AND "F".

CINCH KNOT

HONDA KNOT

BOWLINE

MANGER TIE (QUICK RELEASE)

Figure 20-3 Knots commonly used by horse owners. *(Courtesy of the Appaloosa Horse Club, Inc., ID)*

tighten the cinch if they anticipate discomfort. Tighten the cinch until it is snug enough to hold the saddle on the horse. You can tighten it more before mounting.

If you have a rear cinch, fasten it so that your hand can fit flat between cinch and horse when the rider is mounted. It should not be excessively tight when the horse is first saddled, nor should it be so loose that a back foot could get caught in it. Rear cinches should have a strap connected to the front cinch to prevent them from getting into the flank area.

8. After the horse is walked to the mounting area, recheck the cinch or girth. You probably will be able to take it up another hole or two without getting it too tight. For riding, the cinch should be snug under the heart girth, but not excessively tight. You should be able to fit two fingers under the buckle without much difficulty. Check the cinch again after mounting, as some horses will "blow out" their lungs during saddling, only to relax after you mount, suddenly making the cinch too loose.

9. To unsaddle, simply reverse the above process.

10. If you have had a hard ride, loosen the cinch gradually before taking the saddle off. This allows the blood to flow back under the saddle slowly.

Bridling

Untie your horse before bridling. Working on the horse's left side again, drop the nosepiece of the halter off the nose and refasten the crown strap around the neck. Avoid placing your face too close to the horse's head during bridling and use caution when handling the ears. This helps ensure that you do not get hit in the face should the horse toss its head.

If you have romal reins, or closed reins, place them over the horse's head and neck. If you have split reins, place them over your right shoulder, making sure they do not droop where you or the horse could step on them. Throughout this process, be particularly careful not to wrap any piece of equipment attached to the horse around your hand or arm, as it could cause serious injury.

Spread the crown of the bridle with the right hand and hold the bit in the left. Place your right arm over the horse's head between its ears and approach the horse's mouth with the bit. Be sure to keep the cheek pieces out of its eyes and avoid banging its teeth with the bit.

With the bit pushed lightly against the horse's lips, insert the left thumb in the corner of the mouth. There are no teeth here, so if necessary you can put pressure on the bar of the mouth with your thumb to encourage the horse to open its mouth. Many horses will open their mouths readily as you approach with the bit.

Lift the bridle upward with the right hand as you gently feed the bit over the teeth. Never jerk the bridle, and move with the horse if it moves its head. Place the crown of the bridle over one ear and then the other, bending the ears forward gently as you pull the bridle over them. Rough handling of the ears can cause horses to be head-shy and difficult to bridle. Be careful not to drag the cheek pieces over the horse's eyes. Straighten out the forelock to avoid irritation. Then fasten the **throatlatch**, allowing enough room for you to insert your hand sideways throughout the jaw area (see Figure 20-4).

The bridle should be properly adjusted before you ride. Be sure the browband does not hang down in the horse's eyes and that the bit is neither too high nor too low. The bit should rest on the bars of the mouth. It should be high enough that it creates a small wrinkle at the corners of the mouth. For a snaffle bit, there should be two wrinkles; however, for some other bits it is only one. Be sure you know what is correct for your bit. If the bit hangs so that it comes in contact with the incisor teeth, it is too low.

Also in Western riding, check the **curb chain**, or curb strap. You should be able to fit three fingers sideways between the horse's chin and the chain, but the chain should be tight enough that it places pressure on the chin when you pull back on the reins. This ensures that you have enough control of your horse.

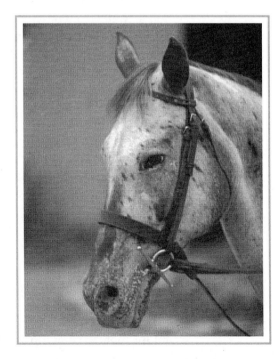

Figure 20-4   Prior to mounting, always make sure the bridle is properly adjusted. The brow band should not hang down in the horses eyes, the bit should be properly adjusted, the nose-band should be two fingers below the check bone (this can vary depending on bridle type), and the excess leather straps should be secured in keepers. *(Photo courtesy of Michael Dzaman)*

WESTERN EQUITATION

The horse and rider need to be suitable for each other. Beginners should ride only calm, dependable horses—preferably older horses—until they are proficient enough to handle more difficult ones.

Equipment must be adequate for the situation and in good repair. Riders should check the rigging, cinches, latigo straps and billets of their saddle to be sure they are strong and not in danger of breaking. Also, riders need to check bridles and reins, especially at stress points, making sure the leather is strong and supple. Leather that is dry and cracked can break easily.

Proper Dress

Wear hard-toed boots with a heel at all times when handling or riding horses. The heel will help prevent your foot from sliding through the stirrup and the hard boot will protect your toes should the horse step on them (see Figure 20-5).

Always wear long jeans, which protect your legs from saddle sores and from hazards on the trail. Avoid shorts and any type of pant made from slick material, such as nylon.

You may want to wear gloves for hand protection, particularly in the winter when they will be exposed to harsh weather. Gloves also may help in the summer because your hands may sweat and make the reins slippery. If you lunge your horse before riding, always wear gloves in case your horse tries to pull away, pulling the line through your hand in the process. Chaps are another option as well. They provide protection for your legs and clothing, and they help to keep you warm in winter.

Avoid dangling jewelry that could get caught on the horse. Loose shirts are a hazard because they can catch on the saddle horn when you dismount. Long hair should be pulled back so your vision is not restricted.

All riders should wear an approved riding helmet to protect their heads in case of a fall.

Wear spurs only when necessary and be sure you have a well-developed leg before attempting to use them. Riders who do not have control of their legs can accidentally gouge or startle their horse. Have an experienced rider or trainer show you how to use them properly, as incorrect use can injure the horse and cause it to buck or run away.

Mounting

Mount your horse in an area away from buildings, trees, fences, and objects on the ground. Pick a spot with good footing and be sure your boots are clean on the bottom. Otherwise, your foot may slip out of the stirrup as you are mounting.

Avoid using deep stirrups or oxbow stirrups for pleasure riding. These are meant for roping and cutting horse riders, and it is difficult to keep the foot in

Riding Attire

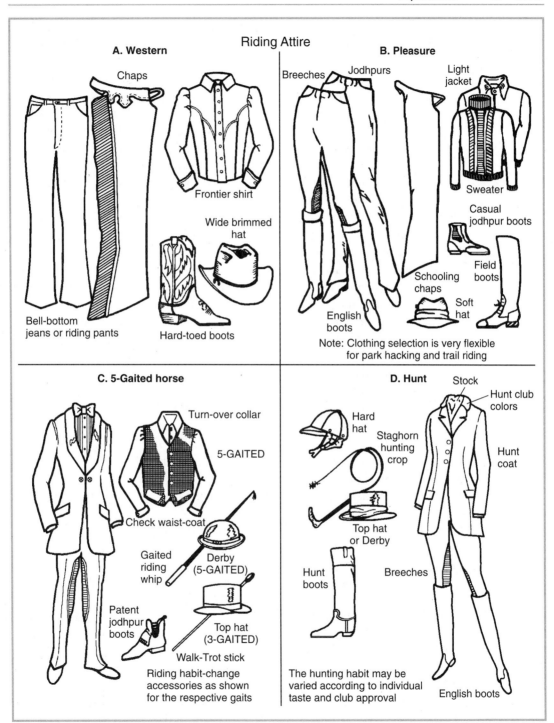

A. Western

Chaps

Frontier shirt

Wide brimmed hat

Bell-bottom jeans or riding pants

Hard-toed boots

B. Pleasure

Breeches Jodhpurs Light jacket

Sweater

Casual jodhpur boots

Schooling chaps Field boots

English boots Soft hat

Note: Clothing selection is very flexible for park hacking and trail riding

C. 5-Gaited horse

Turn-over collar

5-GAITED

Check waist-coat

Gaited riding whip Derby (5-GAITED)

Patent jodhpur boots

Top hat (3-GAITED)

Walk-Trot stick

Riding habit-change accessories as shown for the respective gaits

D. Hunt

Stock

Hunt club colors

Hard hat Staghorn hunting crop

Hunt coat

Top hat or Derby

Hunt boots Breeches

English boots

The hunting habit may be varied according to individual taste and club approval

Figure 20-5 Proper riding attire includes (**A**) Western, (**B**) pleasure, (**C**) 5-gaited horse, and (**D**) hunt. (*Note:* More riders are now wearing helmets as a safety measure.)

the proper position for pleasure riding using these types of stirrups. The depth of a deep stirrup makes it easy for a small foot to go through and get caught.

Before mounting, check the cinch again to make sure it is neither too loose nor too tight. Take one more look at your equipment to be sure everything is adjusted properly. It is proper to mount from the left side, but horses should be trained to allow mounting and dismounting from both sides in case you ever need to use the far side in an emergency. Handling the horse from both sides also helps prevent you and the horse from becoming "one-sided."

Hold the reins in your left hand, positioning your fingers on the reins just as you would when mounted. Take up the slack so that you have light contact with the horse's mouth. Facing the rear of the horse, twist the stirrup to receive your left foot. Make sure your horse stands still during this process. If it tries to walk away, tell it to whoa and pull back on the reins until it stops.

Keep your left hand at the base of the horse's neck and place the right hand on the fork of the saddle on the opposite side. Balance your left hand on the neck to be sure you do not bump the horse's mouth while mounting. If necessary, grab mane or hold on to the bony part of the withers.

Take one or two hops on the right foot and swing yourself up into the saddle, making sure your leg swings clear of the horse's rump. Bumping the horse could startle it, cause it to anticipate discomfort, or prompt it to move off before you are seated. Restrain the horse if it wants to walk off. Be sure your left toe is not pushing into its side.

Sit down softly in the saddle. Flopping down in the saddle could cause a horse to show anxiety or even buck. Even the calmest horse may learn to dislike mounting if you do not show it respect throughout the process. Cold-backed horses usually can be spotted by their tendency to have a "hump" in their backs before riding. The back of the saddle may raise up slightly and the horse may exhibit a stiff walk. Consider lunging such horses before riding to prevent a bucking episode.

If the horse tries to buck, lift your hands and sit deep in the saddle to keep its head up and your body secure, keeping the horse moving forward. The tendency for beginning riders is to lean forward, but this only makes it easier for the horse to buck you off. It is more difficult for the horse to buck with its head up, and you must sit up straight to keep the head up.

Horses should learn to stand after mounting and they should not walk away until asked. Stand quietly for several seconds before moving off so your horse learns that it must be patient and wait for you (see Figure 20-6).

Basic Safe Riding

Start out by riding in an area that is familiar to both horse and rider. Make sure you have the "kinks" out before riding on the trail or in new surroundings. The horse should be quiet and should listen to your cues. Ride with your reins at a comfortable length to encourage the horse to relax and move forward.

(a)

(b)

(c)

(d)

Figure 20-6 (a) Always check the cinch to make sure that the saddle is securely fastened prior to mounting. (b) The horse is mounted from the near (left) side. Begin by placing the left foot in the stirrup. (c) Swing the right leg over the horse. Do not allow it to drag on the horse's hindend. (d) When in the proper seat position, the rider sits relaxed in the center of the saddle. Note that the heels are lower than the toes.

When riding on a road, ride facing oncoming traffic. Riding on roads where there is high-speed traffic can be extremely hazardous and should be avoided. Beginners should never ride on the road unless accompanied by an experienced rider.

Be extremely careful when crossing pavement or hard road surfaces, especially if those surfaces are wet or have oil spots. Ride in these areas at a walk to prevent slipping and to preserve your horse's legs. Give yourself adequate time to cross between cars so you do not have to hurry.

esides being used for transportation, work, sport, recreational activities, and a companion, the horse has become recognized as an integral partner in working with people in therapy and education. In equine facilitated therapy (therapeutic horseback riding) and therapeutic driving activities, the horse is viewed as a tool in therapy, sport, and education for people who are physically or mentally challenged.

For someone who cannot walk, or has difficulty in walking, the horse provides input to the rider, very similar to the motion required in human movement. The three-dimensional movement of the horse at a walk (side to side, up and down, and front and back) is transmitted to the rider. The rider is not only receiving the physical benefits of the horse but is also experiencing fun and mental stimulation.

Recognizing the number of skills and the amount of mental preparation required by a rider suggests this activity is also useful to someone who has difficulty in learning. The value of equine facilitated therapy includes:

- **Exposure to a non-traditional environment.** For someone who is disabled a trip to the farm or horse barn may be quite an excursion and a break from their normal routine. Many people are not accustomed to being around animals, and probably not an animal as big as a horse.

- **Visual experiences.** Many scenes associated with animals are new and exciting when seeing them for the first time.

- **Auditory experiences.** Describing the sounds associated with horses and "life" around the barn is difficult to do without the experience.

- **Olfactory experiences.** The smells of new hay, of mixed sweet feed, a new foal, or just the horse itself stimulate the senses.

- **Tactile experiences.** Physically touching the horse can teach the meaning of coarse, soft, or hard, by the feel of a mane or tail, a short, smooth summer coat, or the feel of a heavy winter coat, the textures of hay and grain, the feel of leather, and also the feel of saddle pads.

- **Physical involvement.** The rider's use and strengthening of muscle groups, reactions, balance, and coordination that occur during equine related activities may be different than at any other time. Also, being mounted on a horse requires the development of spatial awareness skills.

- **Psychological experience.** The horse presents many challenges that, when mastered by the rider, enhance a psychological profile.

Many people benefit by just being able to lead a horse where they want it to go.

- **Expanded vocabulary and identification skills.** The words and terms used when referring to the horse and its surroundings are different and new.

- **The "risk factor."** The ability to work and move around horses with ease takes skill and courage. People learn that they need a certain amount of risk and knowledge in their lives to be healthy and to develop other skills.

- **Eye-hand coordination.** The experienced "horseperson" uses an incredible amount of dexterity and skill to accomplish seemingly simple tasks—locating the horse, catching it, haltering, tying, and grooming.

Therapeutic riding instructors incorporate educational goals into the riding process. They place numbers, letters, shapes, and associated pictures around the arena to use in the lesson. Riders learn all equine management and riding skills to the best of their ability. The horse is a great therapy and teaching tool for people who are physically or mentally challenged, or both.

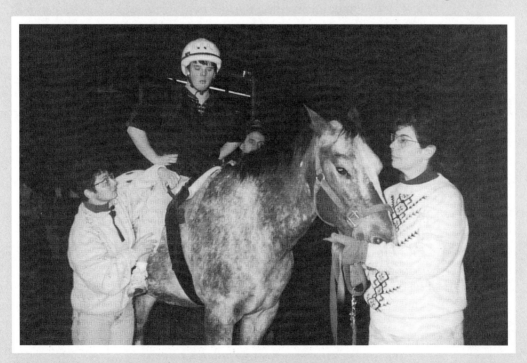

Therapeutic programs provide people with mental, physical, emotional, and learning disabilities the opportunity to enhance the quality of their lives through activities involving horses as shown here at EquAbility, Inc. *(Photo courtesy of Terry Brown)*

Be aware that horses see differently than humans and may spook at strange objects. Keep this in mind as you approach unfamiliar territory so your horse does not jump out into traffic.

If your horse does spook at something new, do not increase its fear by punishing it. Simply keep it moving forward, possibly on a circle, moving back and forth past the object of its fear. Circling in this manner will give the horse an opportunity to see and smell without exaggerating the importance of the object, which will probably reinforce the horse's fear. Allowing the horse to stop and look at the object teaches it that spooking is a way to get out of work. Speak quietly to your horse and give it reassuring pats when it responds properly. Be sure that you remain calm.

When riding with friends, keep a safe distance between horses, whether riding side by side or in a line. When riding single file, keep at least a horse's length between horses. If you tailgate or ride up on the rear of another horse, you may be kicked or your horse may step on the other horse's heels.

When riding side by side, know that some horses do not like this and will try to kick the other horse. Be on the lookout for warning signs, such as pinned ears and one horse swinging its hind end toward the other horse.

If you ride in a group, remember that horses are herd animals and do not like to be left behind. For example, if one rider is left behind to close a gate, the horse may become anxious and want to catch up. This makes mounting difficult and creates a dangerous situation for the rider. It is best to wait until the entire group is ready before moving away. Young horses may become particularly anxious when left behind and some may even panic.

Avoid riding up quickly behind other riders, as it is the horse's nature to join in when other horses start to run. For example, do not lope past another horse at the walk, as this may catch the other rider unaware and cause that horse to take off running with you. It is not uncommon for young, green horses to panic and buck when other riders gallop by if they are not allowed to join in with them.

Riding double is not as safe as riding alone. Not all horses will tolerate two riders, so if you ride double, be sure your mount is suitable. The person riding behind should be a balanced, experienced rider, because if the horse gets nervous, the beginner's tendency is to squeeze with the legs or clench onto the front rider, which will only worsen the situation. Horses are particularly sensitive in the flank area. If the second rider is not careful, he or she can easily clench the horse in this area, causing the horse to buck or try to run away.

Allow your horse plenty of time and plenty of rein when crossing obstacles on the trail. Horses see differently than humans do and they need enough rein to raise and lower their heads to judge height and distance. This also allows them to balance themselves properly. Do not hurry your horse over rough ground. Give the horse time to pick its footing properly.

Always walk back to the barn. If you allow your horse to run home, it will become barn-sour and may become anxious or start trying to take off with you every time you turn toward the barn. A barn-sour horse also may begin misbehaving upon leaving the barn. For this reason, it is a good idea to walk the last quarter-mile of your ride, which also allows the horse to cool down.

Clowning and showing off will increase the likelihood of an accident. Good riders do not need to exhibit their horsemanship skills by showing off. The calmest, safest horse can panic in unusual situations, so always keep this in mind and avoid showing off.

ENGLISH EQUITATION

A correct seat in the saddle is basic to all successful activities with horses. It not only indicates sophistication in horse riding, but affords balance to the rider and aids performance of the horse by correct weight distribution. The accomplished rider does everything possible to divert attention from rider and mount in performance classes, and reduces fatigue of the horse on trail and pleasure rides by sitting balanced in the saddle.

Preparing to Lead

Stand at the left shoulder of the horse, with the reins in your right hand, and the excess, or bight, of reins in your left hand. Lead from this position—not from in front of the horse. Remember, the reins should be over the horse's head and not in riding position.

Mounting

Hold reins with the bight on the right side, your left hand on the horse's withers, while your right hand positions the stirrup over your left foot. If the horse tends to move toward you, keep the right (off) snaffle rein tighter than the left (near) snaffle during mounting, thus reducing the chance of getting stepped on. Whenever possible, use a mounting block.

With your left hand firmly on the withers, grasp the off side of the cantle with your right hand. Take one or two hops on the right foot to attain momentum to mount. Whenever mounting from the ground it is best to have a rider stand on the right side and put weight in the right stirrup to prevent unnecessary torque on the horse's back (see Figure 20-7). Swing your right leg clear of the horse's hips.

Position your right foot in the stirrup before easing your body weight into the seat of the saddle. Avoid dropping heavily into the saddle of strange or "cold-backed" horses (see Figure 20-8).

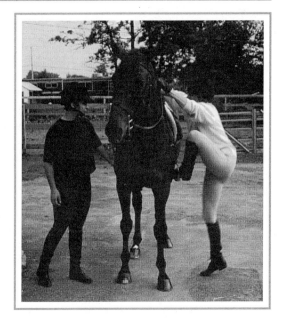

Figure 20-7 Whenever mounting from the ground, it is best to have a rider stand on the right side and put weight in the right stirrup to prevent unnecessary torque on the horse's back. *(Photo courtesy of Michael Dzaman)*

Hand Position

Hands should be held in an easy position, neither perpendicular nor horizontal to the saddle, and should show sympathy, adaptability, and control. Height of the rider's hands above the horse's withers is determined by how and where the horse carries its head. Elbows should be held at the sides in a natural position, neither in too tight nor out too far.

Basic Position in the Saddle

Much of the success of good riders can be attributed to their mastery of this position. To take a basic position, the rider should sit comfortably in the center of the saddle with the feet and legs hanging under the body in a relaxed, natural position. Properly adjusted stirrups will rest between the ankles and insteps of the feet, depending on the build of the rider. The irons should then be placed under the balls of the feet with even pressure on the entire width of the soles. The position of the feet should be natural (neither extremely in nor out). The ankles and insteps should be flexible with heels positioned lower than toes.

When a whip or crop is carried, it should be held in the left hand, butt upward, in mounting (and dismounting). Then it should be transferred quietly to the right hand and held in the same position with its body resting along the rider's right leg.

When the crop is carried as an aid in warm ups or practice, it may be necessary to change it from hand to hand to support the rider's leg when necessary. Unnecessary motion or use of a crop upsets the horse and may result

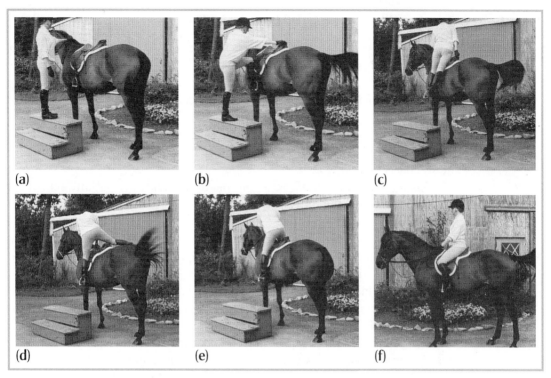

Figure 20-8 **(a)** The horse is mounted from the near (left) side. **(b)** Begin mounting by placing the left foot in the stirrup. **(c)** The rider hops on the right foot to gain momentum and springs up. **(d)** Swing the right leg over the horse. Do not allow it to drag on the horse's rump. **(e)** The right foot is slipped into the stirrup before the body weight is settled into the saddle. **(f)** When in proper seat position, the rider is relaxed in the center of the saddle. *(Photos courtesy of Michael Dzaman)*

in poor performance. Use of the crop as an aid is supported by some trainers, while others do not promote its use.

Preparation to Dismount

Dismounting generally is the reverse of mounting. With your left hand on the withers holding the reins, right hand on the pommel, support yourself in the stirrups in preparation to dismount.

Swing your leg over the horse's back and place your right hand on the cantle of the saddle in preparation for stepping to the ground with your right foot. It is correct either to step down or slide down from this position, depending on the size of the horse and/or the rider.

Proper Dress

An approved safety helmet must be worn at all times. As in Western riding, a good riding boot should be worn. If a riding boot is not worn, a good leather shoe with a reinforced toe and heel is the next best choice (see Figure 20-5).

LOADING AND HAULING

Horse owners will usually find it necessary at some point in time to trailer their horses. Hauling may be necessary at the time of purchase or for horse shows, trail riding, or medical emergencies. Being prepared and maintaining the trailer in road-worthy condition prevents needless delays when the time to haul comes.

Make sure that the trailer is securely and properly hitched to the towing vehicle before loading your horse. Unhitched trailers can easily tip up under the weight of a moving horse.

Loading and unloading must be practiced in advance of any scheduled events. Horses not familiar with being hauled can create an unpleasant beginning to a day's journey. When working with young horses in trailers with partitions, you can boost their confidence if you enter first on the opposite side of the partition. Never go into the same stall you want the horse to go into unless there is an open escape door (see Figure 20-9).

Promptly fasten the bar or chain behind the horse after it loads to prevent it from backing out before you are able to tie its head. When tying it's head, use a **quick-release knot** or a tie with a panic/safety snap. Make sure the horse has enough rope length to permit head movement for balance, but not enough to get its head too low or over to the horse traveling alongside.

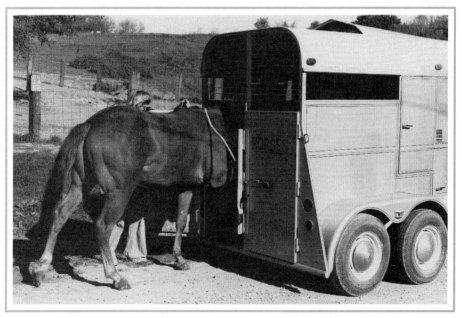

Figure 20-9 Loading a horse into a tandem trailer. *(Photo by Jim Bodine)*

Once the horse is loaded and the gate is closed, check the latches to be sure they are tight and that they cannot bounce up and come loose. There are many types of latches, so be sure that the type you are using cannot come unfastened.

When on the road, stay back from the vehicle in front of you so that you will have adequate room to stop. The extra weight of the trailer will increase the distance normally required to stop your vehicle. Avoid hard stops as they tend to throw horses down. Even if the animals are not injured, they may become fearful and **trailer-sour**, which causes difficulty in future hauling.

When you arrive at your destination, be careful where you unload. Leave enough room behind for unloading and unload on ground that will give good footing for the horse. Be sure you have untied the horse before you release the tail chain or gate. Horses that get unloaded part way and find their heads caught may panic and injure themselves.

Trailer Safety Checklist

Before the horse is loaded, the safety of the trailer should be checked.

- Hitch—Be sure that the hitch is secure and your trailer is properly fastened. Use heavy safety chains to secure the trailer to the towing vehicle.

- Tires—Follow the manufacturer's recommended inflation pressures. A good rule of thumb for safe tire tread is a minimum of $\frac{1}{4}$-inch tread depth. Inspect tires for signs of dry rot. A tire with dry rot is not dependable. Don't forget to have a spare tire that is well maintained.

- Brakes—Replace worn components and test brake operation before beginning the haul.

- Lights—Check for correct and full operation of brake, turning, and marker lights. Interior lights are handy when loading and unloading at night.

- Jacks and safety triangles—Have these available and in good working order in case of roadside breakdowns.

- Floorboards—Horses apply a great deal of pressure on the small area under their hoofs. Floorboards should not be in a rotted or weak condition. Rubber mats on the floor and tailgate provide traction and cushion during loading, unloading, and travel.

- Wheel bearings—These need to be repacked with grease and checked at least every year.

HALTERING

Halters are designed to help catch, hold, lead, and tie horses and ponies. They are nothing else. Every horse should have its own halter, correctly sized and adjusted to fit.

Types of Halters

Some horses are delivered to the new owner in shipping halters. Shipping halters are made of jute fiber (burlap), are light, and usually have a string throatlatch. A shipping halter is inexpensive and adequate for temporary use but is completely unsatisfactory for use as a permanent halter. It cannot be adjusted well (only the throatlatch can be changed) and the fiber lacks strength and durability. This type of halter is also difficult to keep in place on the horse's head and is almost impossible to keep clean.

Rope halters made of braided cotton are very popular. They are strong, relatively inexpensive, and readily adjustable. They are also available in a variety of sizes. The chief disadvantages of rope halters are that they are difficult to keep clean, have a tendency to rot and mildew if not kept dry, and lack the durability found in top quality leather halters.

Another problem with rope halters is that they shrink. Rain, heavy dew, or even high humidity will cause cotton rope halters to shrink. Unless care is taken to frequently readjust rope halters, the shrinking can cause severe pain and even choke the horse.

To eliminate shrinking, a new rope halter should be soaked in water for a few hours or overnight, then thoroughly dried. Clothes dryers, ovens, and other sources of high heat should not be used because they tend to overshrink the halter; heat can also damage the fiber, thus weakening the halter. The type of rope halters used with cattle should not be used with horses. Pulling on the lead rope draws down under the jaw and over the top of the head, much as a lariat rope would. Use these halters only in an emergency. Tie a knot at the point where the lead rope passes through the eye of the halter and the lead rope becomes a halter shank.

Nylon halters have all the advantages of cotton rope halters plus more. They are easily cleaned, not usually affected by dampness, not subject to rotting and mildew, and can be obtained in a variety of colors. Nylon does not shrink; instead, it tends to stretch. In some cases, nylon halters tend to slip at the adjustment points, especially at the crown and under the chin. Therefore, it is necessary to occasionally readjust nylon halters. Nylon halters are more expensive than cotton.

Nylon halters can also be obtained in a flat web design. They look like and are designed like leather halters. They are cheaper than leather, last longer and require less care. However, nylon web halters are difficult to adjust and repair and they do not break. Because of this they can also be dangerous. When using a nylon halter it is best to have a "breakaway" leather crown piece. It is better to have a halter break than a horse in a crisis situation. Like nylon rope, nylon webbing stretches easily. The ends of some pieces of a nylon halter that have been cut with a hot device have a sharp, abrasive edge. These may be removed by cutting with a knife or scissors.

Leather halters are available in a wide variety of types and an even wider variety of prices. Some are adjustable only at the crown piece. These usually must be buckled and unbuckled to be put on and taken off. Some halters have an adjustable chin strap to accommodate various sizes of muzzles, as well as adjustments in the crown piece to fit various lengths of heads. This type of halter is especially well adapted for use on young growing horses or when one halter is used on a number of horses.

Some halters have snaps at the cheek, so unbuckling is not needed when putting on or removing the halter. Leather halters require a great deal of care and attention to keep them in good condition. They must be cleaned regularly and inspected frequently for wear or damage. They are most easily repaired, easiest to individualize with name plates and look dressier than other types of halters. In general, they are also more expensive.

Halters of all types may be purchased in various sizes. Most manufacturers list sizes according to breed, age, type, or weight. Care should be taken when buying halters to save the sales slip and insist on the right of return or exchange if the size selected is incorrect.

Haltering

Putting a halter on a horse is easy if the horse has good manners and has been properly trained.

To halter a horse in a corral, paddock or pasture, the horse first must be caught. The horse should be trained to let you approach from either side.

Carry the halter, unbuckled or unsnapped, in your left hand. The right hand can then grasp the mane at the top of the neck and behind the ears. Or the right arm may be placed under the neck with the fingers extended palm upward, palm toward the neck to grasp the mane from the horse's right side. The left hand can then slip the noseband of the halter over the nose.

At this point the right hand can grasp the crown piece and pull it in place, either pulling it back over the ears or by lifting the crown piece strap over the neck behind the ears. Buckling or snapping completes the job. In the case of halters with snaps at the cheek, it may be easier to use the left hand to push the halter back over the ears and use the right to fold the ears forward under the crown piece. A lead shank can also be used to catch the horse. This is accomplished by placing the lead around the neck and holding both ends as a noose, while the left hand puts the halter in place. This procedure is especially recommended on horses or ponies that resist being haltered.

Haltering Rules

Halters should not be left on horses that will not be watched or inspected at least daily. Young horses especially should not be turned out wearing halters.

Halters may catch on fences, tree branches, or brush. The young horse, unable to free itself, panics—usually with serious consequences. This is why break away halters are used when it is necessary to have a halter on a horse that is turned out. A horse should not be turned out wearing a loose-fitting halter. Horses use their rear feet to scratch their heads, and loose-fitting halters are an open invitation for a back foot to be caught or "hung-up."

TYING THE HORSE

The only "rules" for tying a horse are those dictated by safety and common sense. Tying is only a matter of keeping a horse in one place. Most horses learn to "tie" simply because they find it easier to stand quietly than to fight. All horses should be taught to stand tied and should not be considered fully trained until they do so.

The first requirement in correctly tying a horse is using a knot that can be untied quickly, will not slip, and can be untied even though the horse may be pulling back on the tie rope. The recommended knot for tying a halter rope to a fixed object is a quick-release knot.

Take special care to prevent a horse from breaking loose when tied. Once a horse breaks loose, either from improperly tied knots or breakage of equipment, it is very apt to try harder to break loose the next time it is tied. Halters, tie ropes, and the objects to which they are to be tied should be strong and sound to minimize any chance of the horse breaking free. However, break away snaps can be used for emergency situations where you need to unhook a horse quickly.

Horses should be tied far enough apart so they cannot kick or bite each other. They should be separated by ropes, rails, or distance. A recommended distance between strange horses when tied to a fence or along a picket line is twenty feet. At no time should they be tied closer than ten feet apart.

Any horse that is tied, even in a stall, should not be left unobserved for long periods of time. This is particularly important with young horses. When possible, tie horses where they can watch activities around them. When tied this way, they become less bored and less easily frightened. Horses should never be tied fast with bridle reins. Bridles were not designed to act as halters. Neither were reins intended to be used as tie ropes. A quick pull on a bridle can also cause the bit to cause pain and damage to a horse's mouth.

Problem Horses

Some horses dislike being tied and are known as halter pullers. To help prevent halter pulling or to get around this problem, a lariat rope may be placed around the girth of a horse with the standing part of the rope extending forward to the halter ring from between the front legs of the horse. The end of the lariat is then tied to a fixed object. As the horse backs up, the lariat loop tightens around the horse's middle and the rope through the halter rings pulls the head down, without injuring the neck at the atlas joint. After only a few short sessions the horse learns to stand quietly.

Another method of tying a halter puller is to use a three-fourths-inch or one-inch soft cotton or soft nylon rope. This is tied around the neck. The other end of the rope is threaded through the halter ring and fastened to something solid with a quick release knot. Although very hard pulling could injure the horse, the size of the rope will usually prevent this. This method may not stop a horse from pulling back, but it is a very effective means of keeping it tied. This technique should only be used by a qualified trainer in a unique or controlled situation.

Tying to a Post

To tie a horse to a post, stake, or smooth vertical pole or tree trunk, a knot should be used to prevent the rope from dropping down the pole and from slipping. A much better arrangement, and one that can be untied easily, is to wrap the lead around the post two or three times, then tie a quick release knot and draw out all the slack. This will be apt to slip down the post if not tied tightly, but it is much safer than a hitch, quick-release combination.

The knots should be tied about three and one-half to four feet above the ground, with two or three feet of tie rope between the knot and the halter. It is important to keep the horse from dropping its head down and stepping over the rope. It must, however, be able to get its head up to its normal height.

Tying a horse to a smooth horizontal pole or to a picket line can be safely done in a manner very similar to the procedure used for a vertical pole. In this case, an additional wrap should be made in the hitch, followed by the quick-release knot, to keep everything in place. Just as with the vertical post, the hitch knot may be difficult to untie when the horse pulls back too hard. Therefore, the same procedures as outlined above should be used.

Ground Ties

When there are no suitable objects to which a horse can be tied, it may be possible to use a ground tie. This can be useful on trail rides, when stopping in an open park or pasture. The first step is to dig a small hole about one foot deep. Then tie a long rope such as a lariat to an object such as a large stone, a branch, or even a hammer. Draw the rope tight and place the object in the hole.

Carefully pack the dirt into the hole. The other end of the rope is then attached to the halter ring, with a quick release knot, or it may be placed around the horse's neck and secured with a bowline knot. Unless the horse is especially unruly, there should be no problem.

Before using the ground tie method, or staking a horse out where the rope will lie in similar fashion along the ground, the horse must be trained not to become entangled in the rope. The horse should allow the rope to rub against both the outside and inside of all four legs and should stand quietly if he does become entangled.

Cross Tying

Cross tying restricts movement of the horse more than tying it with a single rope. Two ropes are used to cross tie a horse. Cross tying not only requires special equipment, it requires special training. Most horses object at first to having their heads held with limited movement. To start training, allow lots of slack in both ties. Gradually shorten the ties until the desired control is obtained. The ropes are usually anchored six to eight feet off the ground and they are long enough to allow the horse to stand with its head level.

SUMMARY

After acquiring a horse, saddles are the first piece of tack to be purchased. The style of saddle depends on the type of riding a person intends to do. Saddle styles include the Western or stock, hunt-jump, gaited, dressage, and specialty saddles. Saddles need to be selected to fit the horse and the rider. Saddles range in quality and cost depending on their style and construction.

Saddling a horse is the first step in preparing to ride. This requires time and practice to ensure the comfort of the horse and the safety of the rider. After the horse is saddled, it is bridled, and lead to a safe place for mounting. Proper dress is important for a safe ride and it is important to the type of riding.

In all areas of equitation, safety needs prime consideration. Rules and guidelines of safe riding must always be followed. Loading and hauling of horses presents a safety hazard for the handler and the horse. Tying a horse can be another hazard if not done properly. Halters are designed to help catch, hold, lead, and tie horses. Reins were never intended to be used as tie ropes.

REVIEW

Success in any career requires knowledge. Test your knowledge of this chapter by answering these questions or solving these problems.

True or False

1. Every saddle fits any horse.

2. Every horse should have his own halter correctly sized and adjusted to fit.

3. A recommended distance between strange horses when tied to a fence or along a picket line is 5 feet.

4. In all areas of equitation, safety needs prime consideration.

5. Any horse will take two riders easily.

Short Answer

6. What are the four basic criteria for selecting a saddle?

7. List the five points of a horse's anatomy that should be checked when fitting a saddle.

8. Identity two main differences between Western and English saddles.

9. List five checkpoints for trailers before loading a horse.

10. What are the two main purposes of a saddle?

Discussion

11. Discuss haltering rules and how to halter a horse.

12. Explain ground tying, tying to a post, and cross tying.

13. Discuss why proper clothing is needed while riding a horse.

14. Discuss precautions for safe riding.

15. Briefly describe how to saddle a horse.

STUDENT ACTIVITIES

1. Make price comparisons of different saddles, from low end to high end. Determine their ornateness and what features are included. Find out what kind of materials are used in making the saddles. Present this information in a table.

2. Using a software presentation program, develop a presentation covering safety guidelines when riding, loading and hauling, or tying horses. Or, compare differences between Western and English saddles.

3. Develop a report that compares the differences between English and Western riding. Include a comparison of the tack used by each.

4. Halter and saddle a horse; longe a horse; lead a horse; mount a horse; and tie a horse.

5. Watch videos on Western riders and jumpers and note differences in positions.

6. Make measurements on several horses to see differences in their anatomy and note how important that is when selecting a saddle. Try to use horses of different sizes.

ADDITIONAL RESOURCES

Books

American Youth Horse Council. 1993. *Horse industry handbook: a guide to equine care and management.* Lexington, KY: American Youth Horse Council, Inc.

American Youth Horse Council. 1989. *Basic horse safety manual.* Lexington, KY: American Youth Horse Council, Inc.

Ensminger, M. E. 1990. *Horses and horsemanship.* 6th ed. Danville, IL: Interstate Publishers, Inc.

Miller, R. W. 1974. *Western horse behavior and training.* New York, NY: Doubleday, Inc.

Self, M. C. 1963. *The horseman's encyclopedia.* New York, NY: A. S. Barnes and Company.

Sizemore, D. M. 1986. *Horsemanship.* Irving, TX: Boy Scouts of America.

Topliff, D. R., and Freeman, D. W. (n.d.) *Physical conditioning of the equine: part I: concepts of equine exercise physiology.* Stillwater, OK: Cooperative Extension Service, Division of Agricultural Sciences and Natural Resources. Oklahoma State University.

Topliff, D. R., and Freeman, D. W. (n.d.) *Physical conditioning of the equine: part II: specific event training programs.* Stillwater, OK: Cooperative Extension Service, Division of Agricultural Sciences and Natural Resources. Oklahoma State University.

Videos

University of Idaho. 1995. *Ground handling horses safely.* Ag Communications Center.

CHAPTER 21

Business Aspects

Under the right conditions, and with careful preparation, an equine business can be profitable, both financially and emotionally. For someone poorly prepared and uninformed, a business venture can be a disaster. Beginners should consider starting small and thoroughly understanding the basics of running a business. Beginners should also seek expert legal, tax, and accounting advice before going into business. As experience in the industry is gained, the business manager may expand. Another option is to work with someone who successfully operates a business in the horse industry before going into business for yourself.

OBJECTIVES

After completing this chapter, you should be able to:

- Identify terms related to horse industry business management with their correct definitions
- List reasons for keeping records
- Distinguish between basic kinds of records
- List guidelines for building and maintaining a good credit standing
- List factors that a lender looks for in a borrower
- List factors that a borrower looks for in a lender
- Identify indicators of good loan repayment ability
- List the essential components of all budgets
- Define related management terms
- Describe functions in the management process
- Identify management considerations in planning an equine business
- Explain important skills of managers
- Describe the importance of records and reports

- Explain important human relation skills
- List three types of insurance needed in equine businesses
- Describe the elements of a good boarding contract

K E Y   T E R M S

Accounting	Enterprise	Mortality insurance
Accrual accounting	Enterprise budget	Net worth
Assets	Equity	Partnerships
Balance sheet	Estate	Principal
Boarding contracts	Evaluation	Profitability
Break-even analysis	Fixed costs	Shareholders
Capital	Goals	Sole proprietorship
Cash-basis accounting	Income	Solvency
Cash flow	Interest	Strategic planning
Collateral	Liabilities	Variable costs
Corporations	Liquidity	

COUNTING THE COST

No one should enter an equine business without counting the costs. These costs can include personal or social considerations that may affect the success of the business venture. Or, these costs can be the actual costs of getting into the business and staying in business.

Personal or Social Costs

Some of the personal or social costs to consider before entering an equine business can be gathered from the following checklist:

1. Are you willing to work long, hard, and irregular hours—for example, sixteen hours a day, seven days as week?

2. Do you get along well and communicate effectively with people? Business owners and operators must promote and market themselves and their product.

3. Are you comfortable with mathematical problem-solving and mechanical troubleshooting?

4. Will you seek help when needed?

5. Do you have the technical expertise to manage the operation?

6. Can you afford to hire qualified help?

7. Do you know others in the business who will provide help or information?

8. What related associations or organizations can you join or do you need to join?

9. Are you willing to learn of current practices and new developments?

10. Are you familiar with the legal issues of marketing your product?

11. Do you have the resources to construct and operate a facility?

12. Do you have the right location for the business you wish to conduct?

13. Is the prospective business site located near your markets?

14. Do you live close enough to the business site to visit and monitor it as needed, and to ensure security?

15. What utilities are available at the site of business?

16. Can you control water to, from, and within your system?

17. Can you effectively manage any waste produced by your operation?

18. Will your neighbors and others accept your business operation?

19. Have you discussed your planned operation with the appropriate local, state, or federal agencies?

20. Have you identified the permits and insurance required to construct and operate the business?

21. Do you have the resources—financial, technical, and spacial—needed?

22. Are support services and industries available?

23. Do you have access to a dependable work force for physical labor?

BUDGETS

Estimating the costs and returns for a particular activity is called developing an **enterprise budget**. This procedure reflects the economic value of producing a specific output using a given set of inputs by following specific production practices. **Profitability** can be estimated by subtracting all the costs from the expected revenues.

Two types of costs should be considered in developing enterprise budgets: variable and fixed. **Variable costs** are the expenses that change based on production output, such as feed, veterinary supplies, bedding and fuel. **Fixed costs** are the expenses that do not change, regardless of whether production occurs—expenses such as depreciation, **interest** on investment, insurance, and taxes. Variable costs can often make up the largest portion of the total costs of doing business.

Each business owner faces different situations when trying to analyze the economic feasibility of a business. So, estimates in enterprise budgets should be used only as a starting point for planning.

RECORDS IMPROVE PROFITABILITY

Businesses need a complete and accurate records system in order to make informed management decisions that help maintain or improve business profitability. Records systems serve four functions:

- To assist in reporting to the Internal Revenue Service and other taxing entities, creditors, other asset owners, and others who have a vested interest in the financial position of the business
- To indicate progress
- To diagnose strengths and weaknesses
- To plan

With the proper records, owners determine the actual cost of doing business. Profitability is no longer determined by the money in the bank left to spend.

Records can also help the manager plan and implement business arrangements and do estate and other transfer planning. Managers can use records to determine efficiencies and inefficiencies, measure progress of the business, and plan for the future.

Business owners do not need to be accomplished accountants or experts on taxes and law. They do need to know how to keep the required records for their businesses. They must realize that all business decisions have income tax consequences, and they must be able to evaluate the **accounting** and legal professionals who serve their businesses.

Choosing a Records System

Records systems range from simple, hand-accounting systems using pencil and paper to sophisticated double-entry computer accounting systems. Some require a mix of hand and computer operations.

A system should not only meet the accounting and planning needs of the operation, but it should also satisfy income tax, legal, and other outside reporting requirements. Computer programs should be selected with good detailed instructions for use.

Accounting Methods

Two types of accounting methods are used in agriculture—cash basis and accrual.

Cash-Basis Accounting. This method is used primarily for income tax reporting purposes in service industries. Generally in **cash-basis accounting**, income is recorded as income when it is received and expenses are recorded as expenses when they are paid. Cash-basis accounting is simple and can provide some income tax advantages for businesses that are heavily dependent on inventory changes.

This method also has drawbacks. Cash-basis accounting can grossly distort the financial position, profitability measures, and operational results of the business. Cash-basis accounting needs to be converted to **accrual accounting** for analysis and decision-making purposes.

Accrual Accounting. This method is required for tax purposes for most trading and manufacturing businesses. In accrual accounting, expenses are considered expenses when they are accrued (or committed) and income is counted as income when it is earned. This includes changes in inventories. This method does not depend on how the cash moves in the business. Expenses incurred are matched with related income to determine net income. This approach provides a better continuous picture of profitability. An assessment of cash flow is still needed to determine the financial feasibility of a business.

Basic Record Keeping

Record keeping need not be a complex managerial activity if some simple rules are followed. A well designed record system makes the job easier as well as more efficient. Six suggestions for better record keeping in a business are:

- Always record the gross or total amount. Never, never net it out.
- Always go through all the steps for each transaction.
- Run everything through a checking account.
- Separate business income and expenses from personal income and expenses.
- Do periodic accuracy checks.
- Staple your calculator tape to each page as you total your accounting or ledger book so you can refer back to it.

Tax Records

The Internal Revenue Service requires a set of records to show all taxable income and expenses that are deductible. This can be done in many different formats. The manager or record keeper must maintain accounts to show the three different types of farm income:

1. Sale of items purchased for "resale"
2. Other ordinary income
3. Sale of capital items

According to the IRS, farm income is a whole class of income. Records must also be kept of the two types of expenses—ordinary expenses and capital expenses—along with some expenses that could be classified in either category. Included in the expense category is the annual depreciation record.

The record system chosen should support items on a tax return. The records must provide evidence of the types of income and expenses. This requires sales slips, invoices, receipts, deposit records, and canceled checks. Income and expenses should be clearly identified. Records of loans, debt repayment, and interest expenses must be kept as long as they have any income tax or legal ramifications.

Other required records might include capital item (equipment, physical improvements) records, Social Security records, Occupational Safety and Health Administration (OSHA) records, Federal Unemployment Tax records, worker's compensation, retirement plans, health insurance, operating agreements, carryovers and carrybacks, net operating losses, and income tax credits.

Balance Sheet

The balance sheet shows where the money is invested and how the business is financed. It provides a snapshot of the financial position of the business at a particular point in time. It shows the financial and credit soundness of the business. The balance sheet provides comparative data that can be used for evaluating the business and for developing the farm-earnings statement (see Figure 21-1).

The balance sheet is part of the "big three" in accounting. The other two are the farm-earnings statement and the cash-flow statement (see Figure 21-2). The general accounting equation for the balance sheet is:

$$\text{Assets} = \text{Debt} + \text{Equity} \ or \ \text{Assets} - \text{Debt} = \text{Equity}$$

The balance sheet is divided vertically into two parts—the left part called **assets** (what the business owns) and the right part called **liabilities** (what the business owes). The total of the two parts must be equal. Two kinds of liabilities are included: (1) debt or outside capital and (2) equity (net worth) or inside capital. The debt represents claims lenders have on the assets while equity represents claim owners have on the assets.

Horizontally, the balance sheet can be broken into three categories.

1. **Current Assets**. The first category, current assets, contains those assets that are in cash or are usually turned into cash during the course of the year. For tax purposes, they are assets that would be considered ordinary income if sold or ordinary expenses if purchased.

2. **Intermediate Assets**. The second category includes intermediate assets. They are not true current assets but neither are they true long-term assets. They are assets used in the production of income and are generally viewed as nonreal estate property, such as machinery and productive animals.

BALANCE SHEET

FORM 8A

NAME: Market   Depreciated Cost   DATE:

CURRENT FARM ASSETS	Line No.	Value	CURRENT FARM LIABILITIES	Line No.					
Cash, checking balance	14		Farm accounts payable and accrued expenses						Amount
Prepaid expenses and supplies	15								
Growing crops	16								
Accounts receivable	17								
Hedging accounts	18								
	19								

Crops held for sale or feed	Line No.	Crop Code	Quantity		Judgments and Liens					
	20				Estimated/Accrued Taxes:					
	21				Property					
	22				Income Tax and Social Security					
	23				Accrued Interest: Current					
	24				Intermediate					
	25				Long term					
	26				Subtotal accounts payable and accrued expenses		79			
	27				Current farm notes payable	Due Date	Interest Rate	Annual Installment	Amount Delinquent	Principal Balance
	28									
	29									

Crops under govt. loan	Line No.	Crop Code	Quantity							
	35									
	36									
	37				Total Current Farm Liabilities		80			
	38				INTERMEDIATE FARM LIABILITIES					
	39				Description	Due Date	Interest Rate	Annual Installment	Amount Delinquent	Principal Balance
	40									

Livestock held for sale	Line No.	Lvstk. Code	Quantity							
	45									
	46									
	47									
	48									
	49									
	50									
	51									

Total Current Farm Assets	60	
INTERMEDIATE FARM ASSETS		
Breeding livestock	Number	

Total Intermediate Farm Liabilities   85

LONG TERM FARM LIABILITIES

Description	Due Date	Interest Rate	Annual Installment	Amount Delinquent	Principal Balance

Farm machinery and equipment		

Total Intermediate Farm Assets	65	
LONG TERM FARM ASSETS		
Farm real estate	Acres	

Total Long Term Farm Liabilities   90

TOTAL FARM LIABILITIES

NONFARM LIABILITIES

Nonfarm accounts payable and accrued expenses

FLB stock and co-op equity		
Total Long Term Farm Assets	70	
TOTAL FARM ASSETS		
NONFARM ASSETS		
Vehicles		

Nonfarm notes payable	Due Date	Interest Rate	Annual Installment	Amount Delinquent	Principal Balance

Household goods		
Cash value of life insurance		
Stocks and bonds		

Total Nonfarm Farm Liabilities   95

Total Nonfarm Assets	75	
TOTAL ASSETS		

TOTAL LIABILITIES

NET WORTH

Figure 21-1 The balance sheet

The Big Three

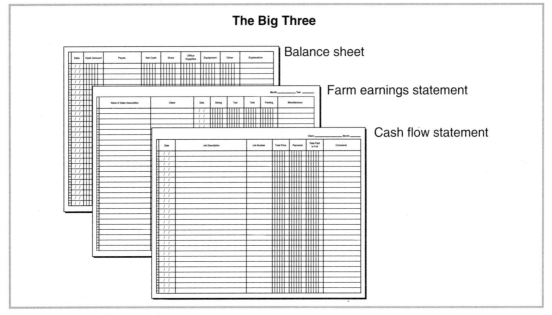

Balance sheet

Farm earnings statement

Cash flow statement

Figure 21-2 The big three

3. **Long-Term Assets**. The third asset category is composed of long-term assets. These generally include real estate property used for producing income.

Asset Value. Determining the appropriate asset values is the biggest challenge when developing a balance sheet. The values selected depend on their use. Two sets of values can be shown for analysis purposes. One value should be the market value, which is what a willing buyer would pay a willing seller (given adequate time and sufficient knowledge). Credit worthiness and loan soundness are measured using this column. Another value sometimes included is the adjusted tax basis, but this data is readily available from properly kept tax records.

Farm-Earnings Statement

The second of the "big three" statements focuses on current activity. It shows the income earned by the business before taxes. The general accounting equation is:

$$\text{Sales} - \text{Cost of Goods Sold} - \text{Operating Expenses} \pm \text{Inventory}$$
$$\text{and Capital Adjustments} = \text{Income Before Taxes } or$$
$$\text{Revenue} - \text{Expenses} = \text{Income Before Taxes}$$

The earnings statement is divided into three sections:

1. Cash-operating statement
2. Adjustments for inventory
3. Adjustments for capital items

The first section shows all cash income and cash expenses and produces a figure called net cash farm income. The second section shows the inventory adjustment, which results in a figure called adjusted net farm operating income. The inventory adjustment is the difference between the ending current assets and beginning current assets, adjusted for changes in accounts payable. The third section shows the capital account adjustment, which results in net farm earnings—or the return to unpaid labor, unpaid management, and equity capital. The capital adjustment is the difference between the intermediate and long-term assets at the end of the year and the intermediate and long-term assets at the beginning of the year.

The earnings statement ties together the information from the balance sheet with cash-basis income tax accounting data. The bottom line is an excellent measure of the profitability of the business.

Cash-Flow Statement

The most action-oriented of the "big three," the cash-flow statement shows how cash moves into and out of the business. The general accounting equation is:

$$\text{Inflows} = \text{Outflows}$$

A complete cash-flow statement can also serve as a cash accuracy check.

Many different formats for developing a cash-flow statement are available. One way is to divide the **cash flow** into four sections:

1. Income—the marketing plan

2. Operating expenses—the production plan

3. Capital purchases—the investment plan

4. Principal, interest, and additional borrowing—the debt service plan

This type of organization gives a better perspective of total cash flow and aids in planning and control.

Three columns are necessary for each accounting period; one set of these columns should be for each month or at least for each quarter. The first column would be called "projected," the second column would be called "actual," and the third column would be called "variance." In this fashion, the cash-flow statement can be used as a financial management control tool. In cash-flow planning for income, operating expenses, and investment, the business manager is asking, "How much am I going to sell or buy? At what unit price am I going to buy or sell?, and, At what time am I going to buy or sell?"

Debt-service information can be obtained from credit records and the balance sheet. A two- to three-year cash-flow history is useful. Then, the manager can find out how this year is going to differ from previous years. This helps make budgeting easier and more accurate (see Figure 21-3).

Cash-Flow Worksheet

FOR YEAR _____	JAN	FEB	MAR	APR	MAY	JUNE	JULY	TOTAL/YR
BALANCE ON HAND								
DESCRIPTION								
INCOME:								
RIDING LESSONS								
HORSE SALES								
HORSE TRAINING								
CAPITAL ITEMS								
STUD FEES								
BREEDING FEES								
INTEREST INCOME								
OTHER INCOME								
TOTAL INCOME								
EXPENSES:								
HIRED LABOR								
TAXES								
INSURANCE								
LEASE RENT								
LOAN PAYMENT								
HORSE PURCHASES								
CHEMICALS								
PESTICIDES								
VACCINATIONS								
FUEL, OIL, ETC.								
FARRIER FEES								
TACK								
STUD FEES								
FEED PURCHASES								
OTHER EXPENSES								
SUPPLIES								
OTHER EXPENSES								
TOTAL EXPENDITURES								
NET INCOME								
LIVING EXPENSES								
ENDING BALANCE								
AMOUNT TO BORROW								

Figure 21-3 A sample cash flow for six months used for projecting or tracking cash, income, and expenditures each month.

The cash-flow statement is useful as an evaluation, control, and planning tool. But used by itself, it can relay false information because it only considers cash. For best results, the cash-flow statement should be used with the balance sheet and earnings statement. Used together, the "big three" provide a complete set of financial statements (see Figure 21-4).

Other Key Accounts

Several other accounts feed into or supplement the "big three" financial statements. These include income accounts, expense accounts, capital item accounts, depreciation records, enterprise accounts, labor records, marketing records, feed records, experimental records, individual machine records, and family records.

Accuracy Checks. Single-entry cash-basis accounting can result in significant errors. It is best to balance the checkbook against the record book on a monthly basis. Then at the end of the year, the manager can make three accuracy checks:

1. Cash flow

2. Profit/net worth

3. Liabilities

When these three accuracy checks balance, the business manager can proceed to file income tax returns and use the records to analyze and manage the business.

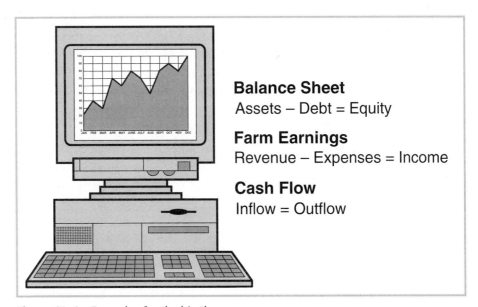

Balance Sheet
Assets – Debt = Equity

Farm Earnings
Revenue – Expenses = Income

Cash Flow
Inflow = Outflow

Figure 21-4 Formulas for the big three

USING AN ACCOUNTING SYSTEM FOR ANALYSIS

Before decisions can be made or analyzed, the necessary information must be available. The primary goal of any accounting system should be to provide business management analysis and control. The accounting system should be geared toward the manager. If the accounting system is not used, it is worthless. The accounting system should supply three types of information:

- Scorekeeping, or evaluating performance (generally, a retrospective view available in the financial statements)

- Attention directing, to flag ongoing operating problems, inefficiencies, and opportunities (identified through analysis of the financial statements)

- Problem solving or analyzing the relative merits of alternative courses of action

The accounting system provides information the manager needs for external reporting for tax and credit purposes. Accounting also provides financial control of routine operations, business management analysis, and reporting to multiple owners.

Tax Requirements

The Internal Revenue Service (IRS) and most state income tax authorities require that enough business records be kept to justify all income and expense claims reported on an income tax return. The lack of standardized requirements for a minimum acceptable set of records has led agribusinesses to store their cash register receipts, invoices, bank statements, and canceled checks in a box or file drawer and do little more. Legally, such records are sufficient. This system can become an extremely expensive one in the event of an IRS examination.

Other Taxes and Investments

Complete and accurate records minimize problems with estate, gift, and property taxes. The ability to participate in investments outside normal business activities can be enhanced by having the information readily available to determine whether a particular investment opportunity is financially feasible.

Credit Applications

Lenders now stress repayment capacity of loans as well as **collateral** security. Most borrowers now need to show that the investment for which the loan is intended will be able to generate enough income to pay back the interest and **principal** owed within the specified time period.

Financial Control of Routine Operations

Astute business managers concern themselves with cash-flow management. Cash budgeting involves all the steps required in the whole business planning process: marketing (including price projections for inputs as well as outputs), yield projections, and enterprise combinations.

Despite the difficulty of preparation, the cash-flow budget helps document managerial abilities and loan repayment capacities. The cash-flow budgeting process can be extended one more step to provide an extremely effective financial control device. If monitored monthly or quarterly, the cash-flow budget can indicate potential problems before they arise. This ability to foresee problems allows the manager to adjust before the fact rather than react afterward.

Business Management Analysis for Strategic Planning

If a business owner is disciplined enough to develop and maintain a records system to meet income tax reporting and credit application needs, then virtually all the needed information will be available to meet what is probably the most important goal of a records system—business management analysis. Good business managers know exactly what their variable and total costs of production are. They know whether they are meeting the goals of their marketing plans or their cash-flow budgets. They have analyzed their strengths and weaknesses, both in physical terms and financial terms. They know where their business has been, where it is now, and where it is going.

Corporations and Partnerships

Multiple-owner forms of business organization require more detailed records because of more intricate tax reporting requirements, state corporation laws, and additional documentation needs of lenders. Perhaps the most important need for more detailed records in **partnerships** and **corporations** comes from the likelihood of problems and potential conflicts among the individuals involved.

Lease and Family Distributions

Individuals involved in informal family partnerships, joint ventures, and share leases need to rely on a detailed records system to ensure fairness in distribution of profits and contributions.

Uses and Interpretations of the Statements

Just doing the scorekeeping—producing the financial statements—is not enough. It takes interpretation and analysis of the financial information to meet the attention-directing and problem-solving needs.

Interpretation begins by evaluating **net worth**—a key measure of financial wealth. On a market-value basis, net worth shows what would be left if all assets were converted to cash and all liabilities paid. Next, the net income should be sufficient to meet withdrawals for family consumption. More cannot be taken out of the business than is earned. On a cost basis, the change in net worth from one year-end balance sheet to the next equals net income minus withdrawals. The next step is to use data from the financial statements for a systematic financial analysis of the operation.

Financial Analysis. The first step in financial analysis is to identify appropriate criteria that will facilitate a comprehensive analysis; then measures for each criterion must be established. For each of the following five criteria, one or more measures are suggested (see Figure 21-5).

Liquidity. **Liquidity** is a short-run concept describing a firm's ability to meet short-run obligations when due without disrupting the normal operation of the business. The ratio of current assets to current liabilities is a common measure.

Solvency. A longer-run concept, **solvency** relates to capital structure and a firm's ability to pay all obligations if assets were liquidated. The focus is on total debt in relation to equity. It is a financial risk measure because the risk of not being able to repay borrowed capital and interest increases as the proportion of debt to net worth increases. Another equally useful measure is debt as a percentage of total assets.

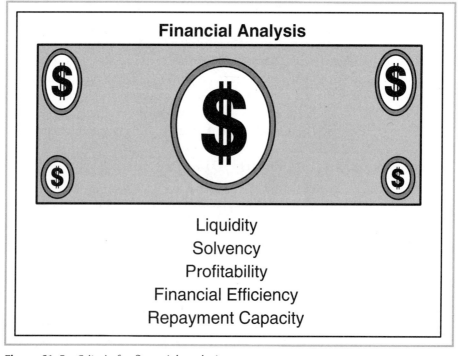

Figure 21-5 Criteria for financial analysis

Profitability. Net income, or profitability, relates to revenue less expenses. But a dollar measure of net income is not sufficient, because the size of the business is not considered. Furthermore, net income is typically a return to unpaid labor, management, and capital, in contrast to other businesses where it is a return only to capital. Return on assets and return on equity are two common measures of profitability. Net income is typically adjusted to get a return to capital expressed as a ratio to total assets.

Financial Efficiency. Financial efficiency is a measure of the efficiency of a business in generating profit out of gross production. The secret of a successful business is to maximize the dollar value of profit out of each $1,000 value of farm production—a measure of gross production. Net farm income divided by the value of farm production is one useful measure. Similarly, operating expenses, interest, and depreciation can individually be evaluated as a proportion of the value of farm production.

Repayment Capacity. This is an assessment of the firm's ability to repay debt. Ability to repay capital debt and interest is a major concern. One measure is all interest plus principal payments on capital debts, expressed as a percentage of the value of farm production. A nonratio method—capital debt repayment capacity—is calculated as net income plus depreciation less withdrawals.

How Are We Doing?

Financial measures and ratios can be interpreted three ways: (1) comparative analysis, (2) trend analysis, and (3) actual vs. budgeted. Comparative analysis is a comparison of one's operations results with those of operations of comparable size and type. For example, if an operation's debt to asset ratio is 40 percent, how does this compare with the debt level of other successful operators? Trend analysis compares results in one year with results achieved in past years. A trend analysis shows strengths and weaknesses and helps focus attention on areas where further strengthening is needed. Comparison of actual performance with the cash-flow budget requires developing an operational plan for the year ahead and then comparing monthly or quarterly performance with projections. Management should focus on variances, or the differences between budgeted and actual performance.

INCOME TAX

A business run by a **sole proprietor** pays no federal income tax. Instead, the taxable income of the business is included in the proprietor's personal income, and taxes are paid at the individual tax rates. Federal income taxes for a partnership are treated in a similar manner. The partnership files an information return showing the business's income and expenses, the names of the partners, and how the partnership earnings will be divided among the partners. The profits, losses, capital gains and losses, and tax credits are allocated to partners according to the terms of the partnership agreement. The partners pay taxes on their respective shares of partnership income as individuals.

Federal income tax savings may occur if a business incorporates and becomes subject to federal income taxation under Subchapter "C" of the Internal Revenue Code. Because a corporation is considered a separate taxpayer, the corporation can divide its income among the corporation, owner-operator employees, and **shareholders**. The corporation pays individuals associated with the corporation for their contributions—owner-employees receive a salary for their labor, and management and shareholders receive dividends for their capital investment. Residual income after all expenses are paid is taxed to the corporation at corporate income tax rates. Whether federal income taxes will be lower after incorporation depends upon the corporation's earnings level, the tax rates for individuals versus that for corporations, and the allocation of earnings.

When the corporation is owned primarily by a family, the tax objective is to minimize the family's total annual income tax burden. This means that the total taxes paid by the corporation, in addition to the personal income taxes paid on the stockholder-employee's salary, and any other personal income should be less than the total personal income taxes paid by the owners before incorporation.

Another tax advantage of incorporation is the increased business deductions available because the owners who work for the corporation become employees of the corporation. In addition to the employee's salary, the corporation can take a deduction for fringe benefits such as group life-insurance plans, medical and hospital plans, pension and profit-sharing plans, and others. It permits the corporation to use pre-tax dollars to pay for benefits received by a stockholder, which the same individual not in a corporation would acquire by using after-tax dollars. This results in more after-tax total income available to the stockholder-employees.

A disadvantage of Subchapter "C" corporations is that double taxation is possible. It occurs when corporations pay dividends to their shareholders. Dividends are distributed from the corporation's after-tax income and shareholders must include dividends in their taxable income. Thus, shareholders are in effect paying taxes a second time on the same profits.

If a corporation elects to be taxed under the special tax option or Subchapter "S" method, the corporation is not a taxpayer for income tax purposes. That is, the corporation itself is not taxed on an income. The income of the corporation "flows through" to the shareholders and each shareholder pays a tax on the individual's prorated share of the corporation's earnings when filing an individual income tax return. All income is taxed the year it is earned whether or not it is retained or distributed. Subchapter "S" rules are similar to partnership rules in that an information return is filed annually on behalf of the corporation.

Thus, corporate earnings in a Subchapter "S" corporation are taxed only once—to the shareholder. This avoids the double taxation possibility present with Subchapter "C" corporations.

Just because Federal income taxes may be reduced by incorporation, not all taxes and costs will necessarily be reduced. Rather, a number of increased costs and taxes exist with corporations. All of these must be examined in arriving at the total savings possible by incorporation.

Payroll Taxes

After incorporation, the sole proprietor or partner changes status from employer to employee. The business has at least one additional employee, if not more, which results in increased payroll taxes.

Social Security taxes are increased since the combined employee and employer rates under the corporate structure are higher than for self-employed individuals—partners or sole proprietors.

Stockholders-employees of corporations are also subject to Worker's Compensation charges on their salaries and are entitled to benefits under the act. This is not true of sole proprietors or partners in a partnership. A stockholder-employee's salary may also be subject to the unemployment compensation tax.

Another disadvantage to owner-operators of incorporated businesses is that personal income taxes must be paid through quarterly estimates or withholding, rather than as a lump sum.

Business Structure Must Fit Objectives

A small-scale family business begins usually as a sole proprietorship. When circumstances surrounding the operation suggest a partnership or corporation, an in-depth analysis needs to be made. An analysis of the organizational characteristics and the objectives of the family is perhaps the most important, but still the most neglected, phase of the process.

Usually, the decision does not need to be rushed. It is a relatively easy and inexpensive process to incorporate or form a partnership, but it may not be so easy and inexpensive to dissolve the corporation or partnership. Those thinking about changing business structure, should take enough time to weigh the advantages and disadvantages of each type for their particular situation.

ESTATE PLANNING

Capital transfer among common property owners in a partnership or corporation is a significant consideration when family members decide to continue the enterprise as an operating unit beyond the retirement of the present owners. With proper planning, the partnership and corporate structure can be used to reserve resources for retirement, transfer property to family members, and minimize expenses and transfer taxes.

Regardless of the business structure—be it sole proprietorship, partnership, or corporation—it is possible to develop a sound **estate** plan. The capital transfer through the estate can be handled with jointly held property ownership, wills, and trust arrangements. Although partnership or corporate structures do not in themselves solve estate-transfer problems, they can make capital transfer somewhat easier.

THE COMPUTER IN MANAGEMENT DECISIONS

Agribusiness has never been more dynamic and competitive than it has today. Each decision a manager makes, or fails to make, can significantly impact the business. In some cases, a decision can simply affect a single production cycle of an enterprise. In others, a decision can change the direction of an entire operation.

In this fast-paced, high-risk climate, computers can play an important part in helping managers make crucial decisions about their operations (see Figure 21-6). Some programs are designed for strategic management, which is concerned with positioning the organization for success by matching its long-range direction with its resources, management capabilities, and the economic environment of the industry. Other programs address tactical management, which focuses on the day-to-day, season-to-season activities needed to carry out long-range strategic plans.

Figure 21-6 Once a luxury, computers are now essential to successful business management.

The success of strategic or tactical decision making depends to a large degree on managers' access to relevant information and their ability to use that information effectively in making decisions. Today's computer programs can help gather important data, provide a framework for analyzing options, and perform calculations thoroughly and accurately at a fraction of the time it would take to do the same thing with pencil and paper.

Strategic Decisions

An owner's most important strategic management decisions deal with deciding the long-range direction of their businesses. Each must decide what **enterprise**, or combination of enterprises, offers the best long-term potential, how big the business should be, the type of financing needed, the amount of debt that can be handled, and how to ensure adequate profit from the business now and in the future. The most effective way to approach these and other strategic questions is to identify a wide range of options and then narrow the field to the most feasible plans. The decisions made must be consistent with both business and family goals, available resources, the management ability available, and the risk-bearing capacity of the business and the people involved.

Tactical Decisions

While strategic management addresses long-range plans and objectives, tactical management focuses on the activities that move the organization toward those goals. Computer programs are available to help business managers monitor or analyze production practices, develop financing plans, establish labor schedules, and create marketing plans.

Computer software can help managers develop marketing plans, keep production records, track breeding programs, and many other tactical decisions. As with strategic planning, computers can speed up the decision-making process, reduce mathematical errors, and help the manager think critically about alternative courses of action.

Once a manager develops annual and long-range plans and answers, the "how" and "when" questions, the next step is to compare what is actually happening in the business with expectations for production, marketing, and finances. Computers can help carry out the important and time-consuming task of monitoring operations on a daily basis. They can also help managers make adjustments when performance fails to meet expectations.

Having access to such a system depends on what records the manager is willing and able to keep on a day-to-day or week-to-week basis. Whether they are kept in a hand or a computerized system, they must be such that they can be summarized and analyzed at any point in time. The computer clearly has the advantage here in terms of quickly recalling structured data, calculating the desired measures and detailing comparisons of plans to the actual outcomes in a timely fashion.

Computers for Decision Making

Of course, the manager must have the right kind of information for each decision, whether strategic or tactical, and the information must be current and accurate. This includes information about the world-at-large as well as about the business itself. For strategic decision making, the manager must keep abreast of general economic conditions, world supply and demand, credit policy, and so forth. In the area of tactical planning, the manager must keep up to date on current and future market prices, weather information, and other factors. While much of this information can be gleaned from the general news media, managers can often get information tailored specifically for their concerns through commercial computerized agricultural information networks.

When it comes to information about operations, many managers have been content to keep only the data necessary for income tax preparation. But in today's dynamic and competitive world, that approach is no longer sufficient. At a minimum, the manager must maintain production data in a useful format as well as information on assets, liabilities, and all credit transactions.

As with any technology, the value of the new computerized management tools depends on how conscientiously and wisely they are used. For the system to reach its potential, the manager must be willing to record vital data on a regular basis and use the resulting information and analyses in making crucial business decisions.

CREDIT

Sound use of agricultural credit is a two-way street affecting both borrower and lender. The individual seeking credit must be prepared to demonstrate to the lending institution that the proposed financing is feasible.

In any borrower-lender relationship, the borrower supplies up-to-date financial and production records to give the lender an understanding of the business. Financial records include a balance sheet and an income statement, as well as historical and projected cash flows. If possible, three to five years of financial and production data are desirable. Many lenders today are asking for income tax returns for the past three years.

On the other hand, it is the lender's responsibility to analyze these documents in a logical and systematic manner. This results in a timely decision on the borrower's credit worthiness. While good financial management is the primary responsibility of the borrower, both lender and borrower must use sound credit practices.

Selecting a Lender

Selecting a lender or lenders is a critical aspect of financial management. An owner/operator should shop for credit and investigate several sources before making a final decision. The borrower, must be prepared to make judgments as well as be judged. Five guidelines to use in rating the quality of a credit service are:

- Select a knowledgeable lender who understands the equine industry today.

- Select a lender who has experience in equine industry credit and a commitment to agriculture.

- Choose a lender who is willing to discuss lending policies and terms and provide prompt action to credit requests.

- Choose a lender who has the capacity to meet anticipated credit needs.

- Select a lender who has a reputation for honesty and integrity.

A lender with a reputation for honesty will judge potential borrowers on the same basis. A strong borrower-lender relationship is one of mutual confidence. Maintaining confidentiality of information and objectively evaluating a situation—being able to say "yes" or "no" to a credit request and backing the decision with facts—are strong attributes to seek in selecting a lender.

Preparing for the Lender

As a borrower, you must provide current, accurate financial statements and supporting records. The following tips will help make negotiating a financial package go more smoothly and ensure a good borrower-lender relationship.

- Arrange credit in advance. Lenders do not like surprises. Do not inform the lender of a major decision after the fact. This can destroy trust and credibility and make future credit more difficult or impossible to obtain.

- Allow your lender time to review your plans and make suggestions. Major purchase decisions are sometimes made on the basis of emotion rather than profitability. A lender can provide objectivity and counsel in reviewing your credit request. Explaining your goals and plans builds the lender's confidence and trust in you, which strengthens the working relationship.

- Keep your lender informed. Even the best of businesses face adversity that reduces the ability to repay. Inform your lender as soon as possible of changes in plans or unforeseen problems that will interfere with making loan payments. Communication is the key element in the initial request and throughout the credit process.

A potential lender or investor wants to see a written business plan that can be used as a guide in developing the business. Business plans can have different formats depending on the type and source of funding being sought or the general purpose of the plan. If funding is already in place, a business plan can easily be used as a guide for strategy. Successful business plans contain certain elements.

Title Page. The title page should minimally include four pieces of information: the name of the proposed project, the name of the business, the people involved, and the address and phone number of the primary contact.

Table of Contents. The table of contents should include the major topics of the body of the business plan and critical tables or figures. The main function of the table of contents is to guide the reader to the critical areas of the business plan.

Statement of Purpose. This section is a brief mission statement of the equine venture: what is to be accomplished, why this project was chosen, and how it is to be done. The statement should include outside funding requirements and describe the repayment plan and the source of repayment.

Executive Summary. This section presents the key elements of the business plan to prospective lenders or investors. The length should be kept under five pages to increase its likelihood of being read. The summary should begin with a brief restatement of the purpose of the project. The financial aspects of the venture should be summarized showing projected returns, outside financing requirements and basic timing of cash flows.

The Business. The next section of the business plan provides details of the venture. The following points should be addressed:

- *History.* You should describe your present organization including when it was founded, progress made to date, present structure, past financing, and prior successes and experiences.

- *Description.* The proposed venture should be laid out concisely. The location, size, products, facility design, and installation procedures need to be described. This section answers in detail what the venture intends to do.

- *Market.* The business plan should include a description of the market in which the product or service will be sold.

- *Marketing.* This section describes how the venture will specifically tailor its product or service to the market. Also included here, is a product or service description and how you will be taking your product to the market.

- *Competition.* An analysis of significant competitors should be presented. Possible reactions from competitors and the effects your product or service will have on the market should be addressed. Strategic moves by your venture in relation to competitors should be discussed and explained.

- *Operations*. A detailed explanation of operating processes and limitations should be presented. Managerial techniques of the processes should be discussed. A suggested technique for describing operations is to divide the project into its separate functions. Each function can be explained in detail along with its importance. The functions can then be brought together to form the whole system of operations.

- *Management*. This section discusses how the venture will run. The organizational structure is described to show lines of authority and responsibility. An organizational chart should be presented depicting clear functional duties and communication lines. Each critical functional area requires managerial control and feedback descriptions. The type and style of management can be discussed.

- *Research and Development* (Optional). If the venture is involved with new techniques or new products or services, this section should be included. The discussion should include whether expertise is to be hired or developed from within. Details such as cost, expected accomplishments, timing, and so on should be described.

- *Personnel*. The complete personnel plan is presented in this section to show the expected hiring needs and personnel policies. The expected number of full- and part-time employees, the amount of overtime, and any seasonal trends should be included. Also, fringe benefits and control policies need to be listed as well as whether it is critical to keep employees. A demonstration of the knowledge of any laws that may effect employees should be included.

- *Loan Application and Effects*. This section presents any source of external financing and describes why it is necessary and what benefits are expected. Borrowing for equipment, land, and feed should be listed with an explanation of the terms, purchase options, the collateral required, and why borrowing makes sense. Any assumptions concerning financing should be spelled out.

- *Development Schedule*. The timing of the venture development can be presented by chart or in some other form showing critical dates. This section should include the decision milestones and guide the reader through each stage of the venture. An example is a calendar chart showing various stages of the venture and any completion dates. Any information such as dependence on governmental agencies, weather, or equipment manufacturers should be included.

- *Summary*. This section should present in abbreviated form all prior sections and information about the business. It highlights the most important facts and assumptions and excludes the details.

Financial Plan. This section includes sources and applications for capital, equipment list break-even analysis, pro forma balance sheet, pro forma income statement, pro forma cash-flow budget, historical financial statements, equity capitalization, debt capitalization, and supporting documents.

- Maintain a high level of integrity. If you expect a lender to be honest and aboveboard at all times, then the same will be expected of you. Inaccurate information and failure to honor commitments will jeopardize the borrower-lender relationship.

Analyzing Credit Use

Once credit is obtained, properly managing the credit becomes a major challenge in the business. Three basic financial statements—the balance sheet, income statement, and cash-flow statement—are tools used to monitor the financial strength of the business. When compiled and supported by accurate financial information, these tools can provide the support needed for many of the strategies and financial decisions faced.

Any business—agribusiness firm, farm, corporation, or small business—must meet certain criteria to be successful, particularly if credit is used. A successful business must exhibit strong repayment ability, liquidity and solvency, and profitability and financial efficiency. The lender's cornerstones of sound credit, the five Cs, include the same qualities (see Figure 21-7). Both producer and lender can determine the financial status of the business with these criteria:

- Character (honesty, integrity, and management ability)
- Capacity (repayment ability and profitability)
- Capital (liquidity and solvency)
- Collateral (minimizing risk to the lender)
- Conditions (for granting and repaying the loan)

Any analysis of the use of credit is only as strong as the quality of financial and other information provided. Circumstances such as size and mix of enterprises, costs, values, commodity prices, collateral values, type of business entity, and time of year can all affect analysis. Do not base final decision on any one

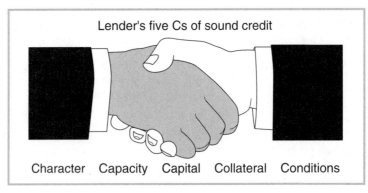

Figure 21-7 The five Cs

factor, but rather, on a balanced, comprehensive approach. Comprehensiveness is the number one factor in developing any valid analytical process.

The Lender's Viewpoint

Many lenders use a systematic approach to analyzing credit. They use some or all of the following guidelines and yardsticks:

- Annual earnings summary
- Earnings-coverage ratio
- Debt-payment ratio
- Business operating efficiency
- Current ratio
- Percentage equity
- Collateral position

Credit Management. Once the debt and repayment structure is in place, constant monitoring by the lender and management of credit is essential. Debt structure and repayment terms, tracking of security, and marking progress of repayment are frequent problems if numerous creditors are involved.

Sound credit analysis may include periodic review of open accounts with merchants, dealers, and suppliers. A check on personal credit-card balances can be used to analyze personal accounts. A sign of strong cash flow and credit management is when accounts payable, after initial billing, average less than five percent of revenue. If unpaid bills average more than ten percent of revenue, it is a sign of pending credit problems. Any sharp increase in accounts payable or a general trend upward will be carefully scrutinized by the lender.

Individual and Business Resources. Evaluating the financial situation and management of an agricultural business frequently involves more than analysis of the basic financial statements. A lender will look at the health and age of the individual requesting credit as well as that of the entire family. The stability of family relationships and evidence of estate planning or transfer of farm assets and short- and long-term goal setting are prime considerations. Education and practical experience should be observed, as well as how management techniques are applied to the operation. A good credit analysis will include on-site investigation of the overall resources—land, buildings, improvements, horses and equipment—and personal living habits.

Lenders will critically evaluate the effects that economic and market trends have on the business. Forecasts and other projections of costs and expenses related to various enterprises can help determine the overall health of the business and the borrower's needs, desires, and strategies for success.

MANAGEMENT

Expert management is critical to the success of any business venture. That is why time must be set aside for management, even if the principal laborer is also the manager.

Management involves overlapping activities such as:

- Setting goals and objectives
- Identifying and responding to problems
- Gathering, compiling, analyzing, and applying relevant information
- Making, carrying out, and evaluating results of specific decisions
- Training, directing, and evaluating employees
- Controlling financial decisions and operations
- Monitoring all aspects of the operation

Management Functions

At the core of good management is a set of **goals** and objectives for the business, developed and understood with clarity by the owner, by management, and by labor. Expectations about levels of annual earnings and production, maintenance of buildings and grounds, tradeoffs between **capital** appreciation and current earnings, long-term growth, and achievements must be established. While these goals and objectives are not always formalized in writing, they need to be reasoned and discussed.

Figure 21-8 shows five basic management functions or activities used to achieve the goals and objectives of a business.

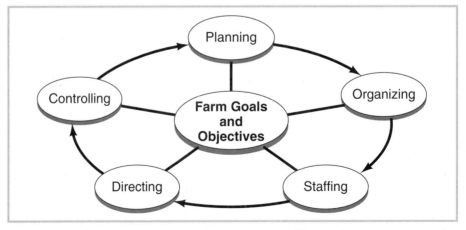

Figure 21-8   Five basic management functions or activities used to achieve the goals and objectives of a business

Planning. While all five of the basic functions are important, planning is crucial because a good plan involves all the other functions. Planning involves:

- Setting daily priorities and schedules: What should be included in today's "To Do" list? Who should complete each priority activity?

- Recognizing problem areas and looking for alternative solutions.

- Making a financial plan and cash-flow statement for the year, knowing when and how much credit must be obtained, and where the cash will come from to meet the regular obligations.

- Looking at alternative marketing plans.

- Establishing the overall enterprises for the business.

- Developing the business: How fast should the business grow? Is new staff needed? What professional development is needed for each manager?

Planning cannot be done just once a year. It is an ongoing process. Plans need to be revised when established checks or measures determine that goals are not being reached. Most planning deserves undivided attention. An uninterrupted hour with a banker, at the computer, or in discussion with a trusted neighbor or partner may save a lot of money, time, and energy.

Organizing. Organizing is establishing an internal structure for the roles and activities required to meet the organization's goals. The manager must decide the positions to be filled and the duties, responsibilities, and authority attached to each. Organizing also includes the coordination of efforts among people.

Organizing includes:

- Deciding who reports to whom. This is often referred to as the chain of command.

- Determining the functions of each position (job design), including the degree of authority.

- Establishing the work routines and standard operating procedures for each production enterprise.

Staffing. Staffing is as crucial to a small or part-time business as it is to a much larger one. Often, the need to figure out how to get all the jobs done on time is even more critical for a small business. No business should try to operate without the possibility of hiring assistance when needed. Assistance can range from hiring a teenager after school to help with a few operations to contracting with an accountant to prepare tax records. Staffing activities include:

- Recruiting and hiring workers

- Training and evaluating workers

Directing. Directing is closely related to staffing. The smaller the business, the more the two are interlocked. Delegation of authority is often one of the most difficult things for the manager of a small business to accomplish. All workers need to know their responsibilities and have a sense of when they can

What is Your Management Philosophy?

Management is the art of successfully pursuing desired results with the resources available to the organization. Management is:

- People-oriented

- An art, not a science

- An ability to establish and meet prescribed goals

- Working with available resources, stressing efficiency

Some describe management as a division of areas of responsibility, for example, finance, marketing, production, and personnel. Others view it as coordinating a series of resource inputs, for example, money, markets, material, machinery, methods, manpower—the Six M concept. Other concepts of management divide it into industrial engineering, organizational, behavioral, or functional approaches.

Industrial Engineering Approach. This approach scientifically analyzes work processes and strives for increased productivity. Frederick Taylor, the "father of scientific management," was a major pro-

ponent of this management concept. Job descriptions, time-and-motion studies, and production standards form the basis of the industrial engineering approach to management. This school of management asserts that if relationships and tasks are carefully designed, productivity is the natural result. The use of power and organizational authority will result in maximum effectiveness.

Organizational Approach. This approach studies the ways in which power and authority may be distributed in order to increase productivity. The organizational approach focuses on such areas as specialization, division of labor, ways in which power and authority are distributed throughout the organization, staff relationships, span of control (for example, how many employees can be controlled by one supervisor), and span of attention (for example, how many different operations can be controlled by one manager). This school of management asserts that if the tasks are carefully designed, productivity is the natural result.

Behavioral Approach. This approach stresses managing human resources in

make decisions and when the boss must be involved. The lines of authority become more crucial with more employees.

Motivation is part of directing. Knowing what is going on and listening to employee concerns helps build communication and confidence. Creating a team spirit where every worker feels some responsibility for the success or failure of the operation is desirable. Openness and understanding by a manager are respected in close working relationships.

order to better the working environment for both the employer and employee, which, in turn, increases overall productivity. The behavioral approach urges the manager to enlarge and enrich jobs to give individual workers more responsibility and authority and to provide a working environment in which employees can satisfy their own needs to be recognized, accepted, and fulfilled. Douglas McGregor, Abraham Maslow, and Frederick Herzberg were leaders in developing this approach to management.

Abraham Maslow developed one of the most useful and widely used models for human needs. Maslow's hierarchy is based on the idea that different kinds of needs have different levels of importance to individuals, depending on the individual's current level of satisfaction. Needs basic to human survival take priority over other needs, but only until survival has been assured. After that point, other needs form the basis for the individual's behavior:

- Survival—The most basic human concern is for physical survival, such things as food, water, warmth, and shelter.

- Safety—Once immediate survival has been assured, humans are concerned about the security of their future physical survival. To-day, this may take the form of income guarantees, insurance, retirement plans, etc.

- Belongingness—After safety is assured, people become concerned with their social acceptance and belonging.

- Ego status—With a comfortable degree of social acceptance, most individuals become concerned with status in their group. Group respect and the need to feel important depend heavily on the responses of other group members.

- Self-actualization—The highest level of need, the feeling of self-worth, may be achieved through creative activities such as art, music, helping others in community activities, or building a business.

Functional Approach. Finally, another popular concept views management as a series of functions. This school of thought describes management as PODCC—planning, organizing, directing, coordinating, and controlling. Two other functions should be added—communicating and motivating—since these functions underlie the success or failure of the first five functions. In this approach, the best of all schools of management philosophy can be combined.

Controlling. Controlling is another key function. Control is the part of business management that determines what new methods are needed to turn out positive results when an investment decision is proven to be less profitable than planned. Control requires keeping track of expenses and **income**. It forces a manager to monitor what is happening every day.

PLANNING—THE SECRET OF BUSINESS SUCCESS

What separates a successful business from an unsuccessful one? Numerous factors—quality of the land, location, managerial skill, and sufficient **equity** capital—are all important. Yet, some businesses that seem to have these basics are less successful than other businesses that are not so well endowed.

An important attribute of good business management is to be able to step away from the immediate concerns to see the future. **Strategic planning** is analyzing the business and the environment in which it operates in order to create a broad plan for the future.

For smaller businesses, the most effective planning may take place at the kitchen table. To establish an appropriate atmosphere for strategic planning requires setting aside time away from the day-to-day problems and interruptions so that the key participants—owners, managers, family members—can reach a common understanding about what they want to do in the next three to five years, and how they want to do it.

Management needs to take a broad overview of the economy and the industry to determine the major opportunities and threats. Tactical planning is concerned with day-to-day and week-to-week decisions. The results of strategic planning could lead to new enterprises, major capital investments, or perhaps even an exit from the business. This broader focus over a longer time distinguishes strategic planning from tactical planning.

Why Do Strategic Planning?

Strategic planning permits more profits, in the long run, by:

- Establishing a clear direction for management and employees to follow
- Defining in measurable terms what is most important for the firm
- Anticipating problems and taking steps to eliminate them
- Allocating resources (labor, machinery and equipment, buildings, and capital) more efficiently
- Establishing a basis for evaluating the performance of management and key employees

- Providing a management framework that can be used to facilitate quick response to changed conditions, unplanned events, and deviations from plans.

Steps in Strategic Planning

Strategic planning involves the first seven steps shown in Figure 21-9. The eighth step—implementation—is strategic management (see Figure 21-9).

Step 1. Define the Mission. The mission statement defines the purposes of the organization and answers the question "What business or businesses are we in?" Defining the business's mission forces the owner-operator or manager to carefully identify the products, enterprises, and/or services toward which the organization's production is oriented. This statement answers the question: What do we do and why do we do it?

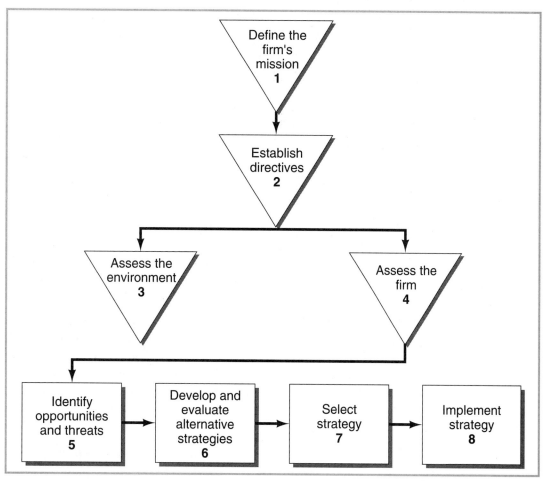

Figure 21-9 Strategic planning involves the first seven steps.

A mission statement is not necessarily a long document. In fact, it should contain fewer than 100 words, and two or three sentences may be sufficient.

Answering this question will suggest goals that will help to clarify objectives in the next step.

Step 2. Establish Objectives. Goals, which are the general, long-term desires of owner-operators or managers, clarify the business's purpose. For example, a goal could be to gain regional recognition. Objectives should translate the goal into concrete terms. Objectives should be measurable and straightforward statements such as the following:

- Increase sales by 100 percent in the next five years

- Increase advertising budget by 25 percent in the next three years

- Increase foal crop by 30 percent in the next five years

- Add 10 more stables in the next two years

These objectives should be chosen in such a way that they contribute to reaching the goals identified in Step 1. Each objective has two characteristics—it is measurable, and it is time limited. This allows management to evaluate progress in implementing the plan.

Step 3. Assess the External Environment. Every organization faces uncertainties, threats, and opportunities that are beyond its control. Market forces may cause prices to plunge, either in the long-run or short-run. Over production, declining consumer demand, a strong dollar, high interest rates, changes in government policies, and regulation of labor and pesticides are external threats that can cut profits or make business more difficult. New market opportunities are created by demographic changes, changing consumer lifestyles, population growth in selected regions, and technological breakthroughs.

The operator or manager must understand the economic, social, and technological forces that will affect the firm. Then reasonable expectations may be formulated about what will happen to product prices, interest rates, the rate of inflation, labor markets, and input prices over the next three to five years.

Step 4. Assess the Organization's Strengths and Weaknesses. The quality and quantity of resources within the control of the operator or manager is the first part of this assessment. What are the abilities and limitations of the operator or manager? What skills and abilities do the employees have? How modern and efficient are the facilities? How large is the resource base? How much water is available? What is the cash position of the business? These resources need to be compared to those of competitors.

Step 5. Identify Opportunities and Threats. This step combines the data gathered in Steps 3 and 4 to determine the threats and opportunities the business might encounter in the planning period. Difficulties in the external environment can present opportunities in another segment of agriculture. For example, concern about exercise creates a need for horseback riding. Companion animals may provide help for troubled teens. Some firms have avoided problems by creatively turning an external threat into an opportunity.

Step 6. Develop and Evaluate Alternative Strategies.

Step 6, along with Step 7, is at the heart of the strategic planning process. This is the point at which the business develops the alternative plans that describe the methods for achieving objectives and obtaining greater long-run profits. In what ways can the business gain a competitive advantage?

Some types of strategies operators or managers use to gain a competitive advantage include:

- Become more efficient. Increase profits by:
    - — Reducing input use
    - — Using more, or higher quality, inputs to increase revenue more than costs
- Seek out alternative enterprises
- Exploit quality differences
- Integrate horizontally
- Integrate vertically
- Reduce risks through diversification and hedging
- Identify new markets or narrow markets—a niche

Organizations that have some degree of control in the market, because of fewer competitors or the possibility for differentiation of their products and services, have the potential for additional strategies to gain a competitive advantage.

Once alternatives are developed, Step 6 is only half completed. These alternative strategies must be evaluated. In practice, management may develop a long list of possible alternatives. These can usually be whittled down with reasoning and logic. Once the obvious losers are eliminated, "pencil-pushing" is in order. No single or preferred method is used for evaluating alternatives, but some combination of the following may be used:

- Budgeting alternatives—both profitability and cash flow
- **Break-even analysis**
- Projections of income, cash flow, and **balance sheet** statements
- Computerized decision aids

Step 7. Select a Strategy.

From the analysis in Step 6, the firm selects a strategy—an alternative or a combination of alternatives—that will enable the operator or manager to achieve the desired objectives. Evaluating alternatives may show that the original objectives are not feasible. The operator or manager may have to move back to Step 2 and select new objectives or reformulate combinations of alternatives. Selection of a final strategy may involve trade-offs among goals. One alternative is seldom superior to all the other alternatives for attaining each of the goals of the operator or manager and his or her family. The process of strategic planning should be recognized more as an art than a science.

Step 8. Implementation.   The eighth step—implementation—is a crucial link in the strategic management chain. Management must periodically look back on the plan and determine how well the business is doing toward reaching its objectives. Assessing implementation may point to mid-course corrections. Assessment enables planners to understand the planning process. Perhaps objectives were set too optimistically or perhaps critical threats or opportunities were not recognized. Recognizing and correcting the plan's weaknesses will improve strategic planning the next time the process is undertaken.

Strategic planning should not be viewed as a formidable task resulting in detailed plans. It should be written, but a few pages will suffice. The process should include all the key players participating in the strategic management discussion. All individuals involved in managing the operation must understand where the business is going, how it plans to get there, and what problems or opportunities lie ahead.

SETTING GOALS FOR BUSINESS MANAGEMENT DECISIONS

Almost everyone is enthusiastic about goals. Most people like to discuss goals and some boast of having goals. People who teach management stress the importance of setting goals. Listen to almost any management guru to hear ideas like these:

- Identify your goals. Manage to reach them.

- Management is goal-directed.

- Take charge of your life and work—set goals and reach them.

- Without goals, you cannot be a manager because you will not know what you want to achieve through your management decisions.

Almost everyone agrees on the importance of goals. The paradox of goals is this: many people will publicly affirm that they have identified their goals—and that goals are important. But, most cannot or will not record and communicate their vision of desired outcomes in the form of a goal statement that can be communicated to others and used to guide their own management decisions.

Identifying goals has both immediate and long-term payoffs—the quality of daily management outcomes and focus of long-term decisions are improved. Those who regularly set and write down goals report benefits like:

- Communication among family members improved.

- Management decisions and work activities effectively focused on priority concerns.

- Cash-flow management in the production unit and household improved as impulse buying of production inputs and household items declined.

- Borrowing, risk, and interest expense declined.

- Conflict was reduced, and working relationships improved.

- Expenses were kept under control, and profits increased.

- Anxiety and concern over the present and future reduced.

- A better balance between production activities and family life was achieved.

Goals and commitment—this is a combination that cannot be beaten. It is the combination that ensures that the business will grow, change, and remain profitable.

HUMAN RESOURCES AND THE BUSINESS

Effective human resource management begins with planning. Using a plan requires that personnel be recruited and then managed effectively.

Personnel Planning

Effective personnel planning starts with a self-assessment by managers. Their personal characteristics, attitudes, strengths and weaknesses, and supervisory skills directly affect the working relationships among employees and others in the business.

Personnel needs depend on the work (tasks) to be done, the types of products produced, and the machinery and technology of each operation. An analysis of personnel needs should result in a statement of the kind and amount of work to be done, which, in turn, provides a basis for determining the number and types of workers needed.

Matching current personnel—family and nonfamily—with tentative job descriptions is a critical step in developing job descriptions for new employees. Identifying mismatches between job descriptions and current responsibilities may help point out training needs, adjustments in job descriptions, shifts in responsibilities and, most important, tasks that cannot be adequately handled with existing personnel.

Hiring Employees

For a team of family and hired workers to function efficiently and effectively, one or more supervisors must carry out the following five personnel management functions: work scheduling, training, motivation, evaluation, and discipline.

Work Scheduling. The reason for work planning and scheduling is to increase labor efficiency. Waiting for instructions, searching for a supervisor, duplicating the work of another employee, waiting for equipment to be available, and doing maintenance work during critical production periods are examples of inefficiencies caused by poor work scheduling.

Work scheduling should be based on a list of tasks to be accomplished, the machinery and equipment needed for the tasks, the people available to do the tasks, and the time in which the work must be done. A task list identifies what needs to be done within the next period or periods of time. The work schedule accompanying the task list identifies the workers and equipment for the tasks. Providing instructions to workers about the tasks they are to do and when and where they are to do them is the final element of the work schedule. The instructions do not have to be given every day if employees are well trained and well supervised.

Training. Managers who hire workers with little equine work experience must provide extensive training to new employees. The complexity of many farm tasks, the risk of injury to untrained workers, and the labor inefficiencies that result from undirected, on-the-job stumbling make training essential.

Hiring experienced workers is sometimes considered an alternative to carefully planned and implemented training programs. In fact, all employees require training. Experienced employees may require considerable training to change poor work habits, inefficient practices, and lax attitudes toward safety that can endanger themselves and fellow workers. Some employers even prefer to hire inexperienced workers for some tasks because training can focus on the skills that are needed and not on retraining or changing old habits.

Motivation. Employees—family members included—do not change their behavior simply because someone tells them to do so. In fact, threats, bribery, and other types of manipulation may make little difference in an employee's work habits or attitude. The challenge for the manager is to balance workers' needs for job satisfaction with the overall business goals. To do this, the manager must identify employees' most important unsatisfied needs and then determine the feasibility of satisfying those needs through work itself or conditions at the workplace.

A person working primarily to satisfy a need for social interaction may care little about labor productivity or sales. Can that person satisfy social needs at break times, before and after work, or through casual conversation during work? Or must the worker be disciplined for wasting time on the job?

Evaluation. A formal **evaluation** program lets employees know where they stand on a regular basis and includes guidelines for wage increases. The evaluation should tell employees how they are doing, identify areas where improvement occurred, and offer constructive suggestions for future work improvement. Specific plans for training and job improvement should be discussed. Workers should also have the opportunity to make suggestions, raise questions, and air frustrations and complaints.

In addition to ongoing daily or other regular communication with workers, at least one formal evaluation meeting should be conducted with each employee every year. This meeting provides opportunities to review performance and progress during the past year and to establish performance goals for the coming year.

Compensation should be discussed during the evaluation meeting. Any changes in compensation should be consistent with the strengths and weaknesses discussed in the evaluation meeting. Merit increases should go only to those who have earned them, and employees should understand why they are or are not getting a raise.

Discipline. Workers function best when the rules are clear and they know the consequences of breaking them. Discipline problems can be minimized through careful employee recruitment and training, clear communication of work rules, and proper attention to human needs. When discipline is necessary, the supervisor should not sidestep the responsibility. Failure to provide discipline sends wrong and confusing messages to workers.

MARKETING

Successful marketing helps ensure the success of the business. Owners should know the market and how to market their product. Because of good records they should know exactly what it costs to produce the services or animals they are marketing. Astute owners know their position in the market.

Once the market position is defined, business owners identify potential customers, who may include professional breeders, trainers, and 4-H and other equine groups. Aggressive businesses explore new markets and develop visible campaigns to interest customers. This can be done by writing articles, conducting clinics and demonstrations, or perhaps placing video tapes in high-traffic areas such as malls.

The market is determined by supply and demand. Location of a business is important to supply and demand in some operations. Advertising can be used to increase demand and capitalize on or offset some of the effects of location. New business owners should determine the importance of location to the success of the business.

The next step is to place a value or price on products or services. This price should be based on the value of the horse, or the cost to provide the service.

Once the market position is examined, the market strategy should be designed. Market strategy includes advertising and building a positive image.

Marketing Strategy

The marketing challenge is probably at the forefront for many business owners. The boom times when the horse industry was growing at a phenomenal rate appear to be over. Market strategy should include advertising and the building of a positive image.

Advertising. The tendency during pressing times is to stop advertising, cut down on promotion and eliminate the cost associated with these activities. The problem is that this also eliminates visibility and may even hurt creditability in some instances. Companies that do not cut back their advertising budgets achieve greater increases in profit than companies that do cut back. Business owners should consider limiting advertising during boom times and set aside money for advertising during recessions. Advertising and promotion should not be considered a temporary activity. People talk about the importance of timing in the horse industry. People who effectively advertise appear to have good timing more often than those who do not.

Advertising takes on many forms and there is no definite pattern that must be followed to be effective. Some tested techniques can serve as guidelines. For example, the advertising of valuable or expensive horses should be targeted at the broadest possible market. Breed journals and other institutional advertising such as specialty or regional publications may be appropriate.

If national advertising is used, brochures and fliers can be designed for a direct-mail campaign. The use of personal letters is often a powerful marketing tool. Video tapes may also be prepared to promote a service, product, or horse.

The Advertisement. A good guideline for planning an advertisement is to plan for the cost to be based on a proportion of anticipated sales. The advertisement should be attractive, direct, and professional. If pictures are included they should be portrait quality and present the product, service, or horse at its best overall advantage.

All advertisements should have some distinctive style or personalized signature. The name of the farm, owner, manager, and trainer, the phone number and mailing address should all be included. Directions or a small map may also be provided.

Capture the reader's attention with a simple, attractive advertisement. Consistent advertising builds confidence in the stability of a business. Advertising budgets should be planned on a yearly basis with a series of advertisements placed in appropriate magazines or journals.

Preplanning for a year will avoid the last minute panic to meet publication deadlines. Continually evaluate your advertising campaign for effectiveness and be prepared to modify your marketing strategy.

BOARDING AGREEMENTS

Boarding contracts will vary, but they should contain at least these five key elements:

1. Emergencies. Emergencies ranging from cuts to colic are a part of horse ownership. The boarding contract should address how the facility will handle emergencies, especially if the owner is unavailable. For example, the facility might request a broad authorization to procure veterinary attention should an emergency arise when the owner cannot be reached.

    The owner might want to limit the stable's authorization, give the stable special instructions, or set a dollar limit on emergency veterinary care. The owner might also want to designate someone as a contact person who is authorized to make decisions regarding the horse in the owner's absence. In either situation, the stable would be wise to have the owner acknowledge that he or she will pay the veterinarian's bill.

2. Insurance. The horse boarding facility should know that the horse has **mortality insurance**. Equine mortality insurance companies give an emergency telephone number that the person in possession of a horse must call when the horse becomes injured or ill. Insurance policies typically require that the company be notified promptly of serious health problems while the horse is still alive. With proper notice, the company can evaluate each problem, and it may want to do any number of things such as consult with the attending veterinarian, order an investigation or new course of treatment, get a second opinion, consent to have the horse put down, and/or order a post-mortem examination.

3. Equine Activity Liability Act Language. As of October 1995, thirty-five states across the nation have laws that protect their horse industries. About twenty of these laws require that contracts used by equine professionals (such as boarding facility operators) include a specific "warning" or other language limiting liability for injuries received due to the inherent nature of equine activities. Form contracts sold in stores and found in books usually will not provide this very important information. Check your state's law to determine whether your boarding contract should include this language.

4. Facility-wide Equine Health Programs. The boarding contract presents a good opportunity to list schedules or disclose the facility's health program and have all boarders consent to it. These provisions will promote the general well-being of all horses on the premises.

5. Release of Liability (if allowed under governing law). Many states legally permit parties to sign liability releases. In those states, the releases are well worth the paper they are written on. Boarding facilities that avoid releases are missing out on a good opportunity to try to limit their liability. Keep in mind that releases should be drafted with the assistance of a knowledgeable attorney. Having a release does not eliminate the need for proper insurance.

For the protection of the facility and its customers, the horse boarding relationship deserves a carefully written contract (see Figure 21-10). Details are nothing to be afraid of, and they can benefit everyone. To protect everyone involved, boarding contracts should be reviewed by a knowledgeable attorney.

INSURANCE

Insurance protects an individual, business, or organization against unexpected losses. Common insurance programs in the equine industry include coverage for farm, ranch, and stable operations, commercial equine liability, care/custody control, equine event sponsors, horse club liability, pleasure and show horse owners' liability, and equine mortality insurance.

Farm, Ranch, and Stable Insurance

Farms, ranches, and stables require specialized knowledge and equipment. All require a specialized insurance provider and agent to protect their investments from emotional and financial impact of fires, theft or litigation.

Coverage options include:

- Dwellings, stables, barns, riding arenas, and other farm buildings
- Guaranteed replacement cost on owner-occupied dwellings
- Replacement cost—actual cash value on other farm structures
- Tack and equipment
- Computer hardware and software
- Named perils on horses
- Spoilage coverage for medicines and vitamins
- Machinery coverage (unspecified, replacement, and newly acquired)
- Personal liability

AFTER HOURS FARMS BOARDING CONTRACT

This Boarding Contract is made and entered into this _____ day of _____ 19 ___ , by and between Barbara Lee Jensen d/b/a After Hours Farm, hereinafter designated "Manager," and _____ hereinafter designated "Owner," and if Owner is a minor, Owner's parent or guardian _____. Manager agrees to accept Owner's horse _____ for boarding and it is the plan and intention of the Owner to board this horse. For and in consideration of the agreements hereinafter set forth, the Owner and the Manager mutually agree as follows:

1) Owner agrees that Manager and employees are not liable for death, sickness and/or accident including consequential damages caused to the horse, except if caused by the willful and wanton negligence of the Manager; in addition, Owner agrees to hold Manager completely harmless and not liable for any injury whatsoever caused to the Owner, and/or any loss or damage to any personal property.

2) It is the responsibility of the Owner to carry full insurance including coverage on his/her horse and all personal property.

3) Owner shall pay the Manager for boarding services the fee of $_____ per month plus the applicable sales tax. This shall include the following: stall, bedding and cleaning, 11 bags of shavings/month, up to 8 pounds of grain/day, hay, regular feedings, daily turn-out, use of facilities.

4) The boarding fee is due on the first of the current month. In the event that payment is overdue by 15 days, Manager is entitled to a lien against the horse for the amount due and shall be entitled to enforce lien and sell the horse for the amount due according to the appropriate laws of New York State.

5) The horse shall be free of infections, contagious, or transmissible disease. A negative Coggins and proof of current worming and immunizations are required.

6) Manager reserves the right to notify the Owner within seven (7) days of horse's arrival if horse, in Manager's opinion, is deemed dangerous or undesirable for a boarding stable. In such case, Owner is responsible for removing horse within seven (7) days and for all fees incurred during horse's stay. After all fees have been paid, this Contract is concluded.

7) In the event of sickness and/or accident to the horse, after reasonable efforts have failed to contact Owner, Manager has permission to contact a veterinarian for treatment.

8) This contract will be concluded when the Manager or Owner has given thirty (30) days notice to conclude the contract.

9) This Contract is nonassignable and nontransferrable.

10) This Contract represents the entire agreement between the parties. No other agreements or promises, verbal or implied, are included unless specifically stated in this written agreement. This Contract is made and entered into the State of New York and shall be enforced and interpreted under the laws of this state. Should any clause be in conflict with State Law, then that clause is null and void. When the Manager and Owner and Owner's parent or guardian, if Owner is a minor, sign this contract, it will then be binding on both parties, subject to the above terms and conditions.

_____   _____

Manager's Signature   Owner's Signature (or Owner's Parent or Guardian)

Address and Telephone of Owner _____

Description of Horse _____

Figure 21-10 An example of a boarding agreement. *(Courtesy of Barbara Lee Jensen, After Hours Farms)*

- Premises/operations liability (boarding, breeding, training, showing, riding instruction)
- Track liability for incidental horse racing
- Care, custody, and control coverage
- Farm/horse operations continuation expense, for example, loss of income or the expenses necessary to carry on normal business after a loss

Commercial Equine Liability Insurance

Commercial horse trainers and instructors should carry liability insurance. This policy provides comprehensive general liability coverage for bodily injury and property damage claims as a result of business activities such as training, instruction, clinics, horse sales, breeding, and boarding.

Under the terms of this type of policy, an individual is protected from a variety of claims that might be brought against them as a result of their equine business activities.

Care/Custody Control

Individuals caring for or boarding horses can be held responsible if:

- A boarded animal is injured attempting to jump a fence
- A horse being trained dies in a barn fire
- An animal ingests a foreign substance in its feed and dies
- An employee forgets to lock a gate and a broodmare gets loose, is injured, and loses her foal

Care/custody control insurance coverage protects an individual against liability resulting from death or injury to animals in their custody, not only from fire but also from other causes. The protection provides that if an animal owner makes a negligence claim against a policyholder covered by this type of insurance, the insurance company will defend the policyholder and will be responsible for making payment if it is determined that the loss is due to negligence and the policyholder is legally liable.

Equine Event Insurance

Equine event insurance is for sponsors of horse shows. Any number of show days can be insured. The coverage is designed to provide protection from liability claims that may result from bodily injury or property damage to a spectator attending a show. This insurance can cover:

- Spectator liability
- Products liability, for example, concession-stand sales

- Show judges and officials
- Premises owner

In most cases, this policy can extend coverage for the preparation and dismantling of the show, limited to one day prior and one day after the show. The premium is based on the number of show days and the limit of liability selected.

Horse Club Liability Insurance

Horse club liability insurance covers the premises at which meetings, shows, and other activities are held by the club. Meetings, trail rides (noncompetitive), gymkhanas, and other events are automatically covered when they are conducted for the sole benefit of the members.

Liability programs commonly available include, spectator liability, personal injury coverage (libel/slander), premises owner's protection, products liability coverage (concession stand), and coverage for show judges and officials. Most basic policy premiums include two days of events in which nonmembers participate and up to 100 members.

Pleasure and Show Horse Owners Liability

Pleasure and show horse liability insurance programs are designed to provide complete liability protection to a person who owns a horse or horses used exclusively for pleasure or show where a homeowner's or tenant's policy will not provide coverage. To qualify, an individual cannot be personally involved in a professional training, breeding, or boarding operation.

A pleasure and show horse owner's liability plan covers an individual against bodily injury claims that may result from the use of the horse and includes property damage claims.

In many instances, this policy can be extended to cover premises owned or leased by an individual when a homeowner's policy will not provide personal liability where the horse is boarded.

Equine Full Mortality Insurance

Equine full mortality insurance includes theft, transit, and limited emergency colic surgery. Subject to the exclusions and conditions contained in the policy, each insured animal is covered against loss by death only. The policy does not cover minor injuries, depreciation in value, failure of the animal to perform the functions for which it is kept, or veterinarian or similar expenses to preserve the animal's life.

The policy also insures against loss resulting from the intentional and voluntary destruction of an insured animal for humane reasons to terminate incurable and excessive suffering arising out of a peril insured against if:

1. The insurance company has agreed to the destruction of the animal or

2. A qualified veterinary surgeon appointed by the insurance company certified that the suffering was so intense that immediate destruction was imperative for humane reasons.

Coverage is limited to specific named perils, for example, fire and/or windstorm, tornado, cyclone and hail; explosion or earthquakes; flood (meaning the rising of natural bodies of water), drowning; accidental shooting by a person or other than the insured or employees of the insured, and transit.

Coverage includes the direct of damage caused by theft or attempted theft (but not mysterious disappearance or escape); attack by dogs or wild animals; and collision of an animal with vehicles except those owned by the insured.

Optional coverage may include:

-   Equine major medical and surgical, which provides coverage for veterinarian fees that are a direct result of medical and surgical treatment for specified animals.

-   Equine loss of use, which provides coverage for specified animals that, during the policy term, become permanently unfit for use specified in the schedule. The animal must first be insured for full mortality and medical/surgical.

-   Importing costs or international transit

-   Gelding surgical coverage, for stallions gelded over the age of two

-   Stallion infertility

-   Barrenness

Insurance for Buying or Selling on Contract. Mortality insurance can be purchased for the total sales/purchase or lease/purchase price of a horse and protect both the buyer's and seller's financial interests in the animal. Both buyer and seller receive a copy of the policy and are paid their entire financial interest in the animal should it die before the contract is paid in full.

Rates are based on the breed, use, and age of each horse. All animals must be checked by a veterinarian before coverage is put into effect.

SUMMARY

No one should enter an equine business without first counting the personal, social, and financial costs. Developing an enterprise budget and record-keeping system is essential to the success of a new business. Accounting systems can be simple or complex, hand-kept or computerized. Besides meeting tax requirements, records provide an effective tool for planning and evaluating the business. When an equine business needs credit, records support the need and the ability to repay the debt.

Obtaining credit is a two-way street. Borrowers look for fair, understanding, and knowledgeable lenders. Lenders lend money to honest, knowledgeable, borrowers who can demonstrate a plan and ability to repay the loan. Both the borrower and the lender manage credit.

Management skills are necessary to operate an equine business successfully. These skills are gaining importance as the economy changes and the workplace changes. Management involves the best use of human resources and the best use of financial resources. Records are essential to good financial management.

Finally, any equine business must consider marketing, the need for insurance, and contracts.

REVIEW

Success in any career requires knowledge. Test your knowledge of this chapter by answering these questions or solving these problems.

True or False

1. A balance sheet is a type of accounting method.

2. Cost of feed represents a fixed cost.

3. Most businesses increase advertising when the business slows down.

4. Equine full mortality insurance insures a horse only against a trailering accident.

5. Cash basis is one accounting method.

Short Answer

6. Give an example of a fixed cost and a variable cost.

7. List five basic management functions used to achieve the goals and objectives of an equine business.

8. List four functions for records.

9. Name the three categories on a balance sheet.

10. List the five Cs that are the cornerstone of sound credit.

Discussion

11. Describe strategic planning.

12. Discuss the components of a good boarding contract.

13. Give three tips for negotiating a loan.

14. Explain how a business is marketed.

15. Describe how insurance can protect an equine business.

STUDENT ACTIVITIES

1. Visit with a representative of a local small business association. Discuss the types of ownerships for a business—sole proprietorship, corporation, or partnership. Develop a list of advantages and disadvantages of each.

2. Collect case studies of business problems, either financial or managerial. Suggest solutions for these problems. Match your skills against those of local business owners and managers.

3. Visit with several local business owners and discuss the problems encountered in human relations and human resources in the workplace. Discuss the roles of leadership training and managing of people.

4. Collect a variety of marketing material from some equine businesses. Evaluate its effectiveness. For example, compare the quality of photographs, paper, and clarity of writing.

5. Obtain a copy of a boarding contract. Read the contract for clarity. Check it for the five elements of a good contract that are listed in this chapter.

6. Check with local insurance agents and find out the type of insurance coverage they offer for horses and equine businesses. If possible, obtain an insurance policy. Read it and put the major points in your own words.

ADDITIONAL RESOURCES

Books

American Youth Horse Council. 1993. *Horse industry handbook: a guide to equine care and management.* Lexington, KY: American Youth Horse Council, Inc.

Bay, C. C. 1990. *Horse farm mangement workbook.* East Lansing, MI: Cooperative Extension Service, Michigan State University.

English, J. E. 1995. *Complete guide for horse business success.* Grand Prairie, TX: Equine Research, Inc.

Kreitler, B. 1995. *50 careers with horses—from accountant to wrangler.* Ossining, NY: Breakthrough Publications.

Pelley, L. 1984. *In one barn: efficient livestock housing and management.* Woodstock, VT: The Countryman Press.

CHAPTER 22

Career Opportunities

The purpose of education and learning is to become employable and stay employable—to get and keep a job. People look for careers and careers look for people. Two broad categories of career opportunities in the equine industry are working for someone else or working for yourself. Success in any career requires some general skills and knowledge as well as some very specific skills and knowledge unique to a chosen occupation in the horse industry.

O B J E C T I V E S

After completing this chapter, you should be able to:

- List the basic skills and knowledge needed for successful employment and job advancement
- Describe the thinking skills needed for the workplace of today
- Identify the traits of an entrepreneur
- List six occupational areas of the horse industry
- Identify the careers that require a science background
- Describe the general duties of the occupations in six areas of the horse industry
- Describe the education and experience needed to enter six areas of the horse industry
- List six general competencies needed in the workplace
- Describe five ways to identify potential jobs
- List eight guidelines for choosing a job
- List ten guidelines for filling out an application form
- Describe a letter of inquiry or application
- List the elements of a resume or data sheet
- Describe ten reasons an interview may fail
- Discuss what research studies indicate about basic skills and thinking skills for the workplace

K E Y   T E R M S

Competencies	Demographic	Letter of application
Creative thinking	Entrepreneur	Letter of inquiry
Cultural diversity	Follow-up letters	Resumes
Data sheet	Forecasts	Risks

GENERAL SKILLS AND KNOWLEDGE

Over the past few years, research study after study has indicated that potential employees never acquire some very basic skills and knowledge. Without these basic skills and knowledge, the specific skills and knowledge for employment in the horse industry is of little value. The new workplace demands an even better prepared individual than in the past. Finally, those individuals working for themselves must develop a trait called entrepreneurship. This may also be a good trait for any employee.

Basic Skills

Success in the workplace requires that individuals possess skills in reading, writing, arithmetic and mathematics, listening, and speaking, at levels identified by employers nationwide.

Reading. An individual ready for the workplace of today and the future demonstrates reading with the following **competencies**:

- Locates, understands, and interprets written information, including manuals, graphs, and schedules to perform job tasks
- Learns from text by determining the main idea or essential message
- Identifies relevant details, facts, and specifications
- Infers or locates the meaning of unknown or technical vocabulary
- Judges the accuracy, appropriateness, style, and plausibility of reports, proposals, or theories of other writers

Reading skills in the horse industry are necessary to keep up with new information, to understand directions for feeding or treating horses, or to understand the language of a contract (see Figure 22-1).

Writing. Individuals ready for the workplace of today and the future demonstrate writing abilities with the following competencies:

- Communicates thoughts, ideas, information, and messages
- Records information completely and accurately

Figure 22-1   Reading skills are important to success.

- Composes and creates documents such as letters, directions, manuals, reports, proposals, graphs, and flow charts with the appropriate language, style, organization, and format
- Checks, edits, and revises for correct information, emphasis, form, grammar, spelling, and punctuation

In the equine industry, writing skills are necessary for such tasks as keeping stable records, taking a message, describing disease conditions, or requesting a test.

Arithmetic and Mathematics.   The workplace of today and the future requires individuals with competencies in arithmetic and mathematics. Arithmetic is the science of computing with numbers by the operation of addition, subtraction, multiplication, and division. Mathematics is the application of arithmetic. These important competencies are:

- Perform basic computations
- Use numerical concepts such as whole numbers, fractions, and percentages in practical situations
- Make reasonable estimates of arithmetic results without a calculator
- Use tables, graphs, diagrams, and charts to obtain or convey information
- Approach practical problems by choosing from a variety of mathematical techniques

- Use quantitative data to construct logical explanations of real-world situations

- Express mathematical ideas and concepts verbally and in writing

- Understand the role of chance in the occurrence and prediction of events

Anyone not convinced of the value of arithmetic and mathematics to the horse industry should consider the skills required to figure feed conversion ratios or growth rates.

Listening. Individuals working today and in the future must demonstrate an ability to really listen. This means to receive, attend to, and interpret verbal messages and other cues such as body language. Real listening means the individual comprehends, learns, evaluates, appreciates, or supports the speaker.

Speaking. Finally, individuals successful in the workplace of today and tomorrow, demonstrate these speaking competencies:

- Organize ideas and communicate oral messages appropriate to listeners and situations

- Participate in conversation, discussion, and group presentations

- Use verbal language, body language, style, tone, and level of complexity appropriate for audience and occasion

- Speak clearly and communicate the message

- Understand and respond to listener feedback

- Ask questions when needed

Speaking skills are used when obtaining and maintaining a job and communicating with coworkers. Career opportunities in sales rely on good speaking and communication skills.

Thinking Skills

Contrary to the old workplace, many research studies indicate that employers in the new workplace want workers who can think. Employers search for individuals showing competencies in these areas: **creative thinking**, decision making, problem solving, mental visualization, knowing how to learn, and reasoning (see Figure 22-2).

Creative Thinking. Creative thinkers generate new ideas by making non-linear or unusual connections or by changing or reshaping goals to imagine new possibilities. These individuals use imagination freely, combining ideas and information in new ways.

Figure 22-2 Thinking skills are important for successful employment.

Decision Making. Individuals who use thinking skills to make decisions are able to specify goals and limitations to a problem. Next, they generate alternatives and consider the risks before choosing the best alternative.

Problem Solving. As silly as it sounds, the first step to problem solving is recognizing that a problem exists. After this, individuals with problem-solving skills identify possible reasons for the problem and then devise and begin a plan of action to resolve it. As the problem is being solved, problem solvers monitor the progress and fine tune the plan. Being able to recognize a disease condition and look for solutions is a good example of problem solving in equine science.

Mental Visualization. This thinking skill requires an individual to see things in the mind's eye by organizing and processing symbols, pictures, graphs, objects, or other information. For example, this type of individual sees a stable and corral from a diagram or understands wiring and plumbing from a schematic.

Knowing How to Learn. Perhaps of all the thinking skills, this is most important with the rapid changes in available technology. This type of individual recognizes and can use learning techniques to apply and adjust existing and new knowledge and skills in familiar and changing situations. Knowing how to learn means awareness of personal learning styles—formal and informal learning strategies.

Reasoning. The individual who uses reasoning discovers the rule or principle connecting two or more objects and applies this to solving a problem. For example, physics teaches the theory of mechanical advantage, but the reasoning individual is able to use this information in understanding how the bones, joints, and muscles of the horse work individually and together.

General Workplace Competencies

Besides the basic skills and the thinking skills, the workplace of today and tomorrow demands general competencies in the use of resources, interpersonal skills, information use, systems, and technology.

Resources. Resources of a business include time, money, materials, facilities, and people. Individuals in the workplace must know how to manage:

- Time through goals, priorities, and schedules
- Money with budgets and **forecasts**
- Material and facility resources such as parts, equipment, space, and products
- Human resources by determining knowledge, skills, and performance levels

Interpersonal. More than ever, people cannot act in a vacuum. Most people are members of a team where they contribute to the group. They teach others in their workplace when new knowledge or skills are needed. More than ever, and at all levels, individuals must remember to serve customers and satisfy their expectations. Through teams, individuals frequently exercise leadership to communicate, justify, encourage, persuade, or motivate individuals or groups. As part of employment teams, individuals negotiate resources or interests to arrive at a decision. Finally, all interpersonal skills require individuals to work with and use **cultural diversity**.

Information. The information age is here. Individuals in the workplace must cope with and use information. Successful individuals will identify the need for information and evaluate the information as it relates to a specific job. With the computer, individuals in the workplace must organize and process information in a systematic way. Also, with all this information available, individuals must interpret and communicate information to others using oral, written, or graphic methods. For example, breeding and performance data must be summarized. To manage production information, computer skills are the key (see Figure 22-3).

Systems. No longer can any aspect of a business or industry be viewed as a part that stands alone. Every part is part of a system and individuals now seek to understand systems, whether they are social, organizational, technological, or biological. With an understanding of the systems in a business, trends, predictions, and diagnoses can be made. Individuals can then modify the system to improve the product or service. For example, a person who breeds horses must understand the biological systems of reproductive genetics and animal behavior.

Technology. Technology makes life easier only for those who know how to select it, use it, maintain it, and troubleshoot it. Technology is complicated. Successful individuals learn to apply appropriate new technology through all the basic skills, the thinking skills, and general workplace competencies.

Personal Qualities

Even with all the training in basic skills, thinking skills, and general workplace competencies, some individuals still fail for lack of personal qualities. These include responsibility, self-esteem, sociability, self-management, and integrity or honesty.

Responsible individuals work hard at tasks even when the task is unpleasant. Responsibility shows in high standards of attendance, punctuality, enthusiasm, vitality, and optimism in starting and finishing tasks.

Figure 22-3 Computers open access to more information than ever before. *(Photo courtesy of Michael Dzaman)*

Those possessing self-esteem believe in themselves and maintain a positive view of themselves. These individuals know their skills, abilities, and emotional capacity. They feel good about themselves.

Successful individuals demonstrate understanding, friendliness, adaptability, empathy, and politeness to other people. These skills are demonstrated in familiar and unfamiliar social situations. The best examples are sincere individuals who take an interest in what others say and do.

Along with self-esteem is self-management. Individuals successful in business accurately assess their own knowledge, skills, and abilities while setting well-defined and realistic personal goals. Then, once goals are set, those who manage themselves, monitor their progress and motivate themselves through the achievement of goals. Self-management also implies a person who exhibits self-control and responds to feedback unemotionally and nondefensively.

Finally, to be successful in the equine industry, an employee or entrepreneur requires good old-fashioned honesty and integrity. Good ethics are still a part of good business.

ENTREPRENEURSHIP

The most common view of an **entrepreneur** is one who takes **risks** and starts a new business. But some traits of entrepreneurship are desirable at many levels of employment. Within any organization, an entrepreneur may:

- Find a better or higher use for resources
- Apply technology in a new way
- Develop a new market for an existing product
- Use technology to develop a new approach to serving an existing market
- Develop a new idea that creates a new business or diversifies an existing business

Almost anyone can be an entrepreneur. It is an attitude more than anything else. Yet, an attitude that incorporates many desired traits. The attitude of an entrepreneur includes:

- Taking risks with clear appreciation for the odds
- Focusing on opportunities and not problems
- Placing primary focus on the customer
- Seeking constant improvement
- Being impressed with productivity and not appearances
- Recognizing the importance of example
- Keeping things simple

- Using open-door and personal-contact leadership
- Encouraging flexibility
- Communicating purpose and vision

Entrepreneurs are ready for the unexpected, differences, new needs, change, **demographic** shifts, changes in perception, and new knowledge. Entrepreneurs are good employees and good employers. Entrepreneurs keep the equine industry growing.

EMPLOYMENT IN THE HORSE INDUSTRY

Some people consider the only jobs available in the equine industry to be those in the actual raising or training of horses. But the industry as a whole requires a large number of people to support the infrastructure of suppliers, producers, and marketers.

Specific jobs or employment opportunities in the equine industry can be grouped into general categories. These include primary careers; horse shows and rodeos; the racehorse industry; recreation; equine supplies, support, and services; education; marketing; and research and development. Each area requires some unique skills.

Primary Careers

Primary careers are for those who want daily contact with horses and the horse industry, outside of horse shows, rodeos, and the racehorse industry (see Figure 22-4).

Primary Careers in the Horse Industry

Artificial inseminator	Horse auctioneer
Breed registry officer	Horse club operator
Breeder	Horse camp operator
Broodmare manager	Horse buyer
Equine consultant	Mounted police officer
Equine nutritionist	Ranch or feedlot cowhand
Equine researcher	Rehabilitation therapist
Equine veterinarian	Riding instructor
Exercise rider	Stable manager
Extension horse specialist	Stallion manager
Farm/ranch manager	Trainer
Farrier	Veterinary technician
Groom	

Figure 22-4 An exercise rider can exercise racehorses or simply help busy horse owners keep their horses in shape. *(Photo courtesy of Michael Dzaman)*

Supplies, Support, and Services

Occupations in the supplies, support, and services area include those that support the horse industry by providing the inputs necessary for an operation to be productive. These careers have contact with horses and the people in the horse industry but not necessarily on a daily basis (see Figures 22-5 and 22-6).

Careers in Equine Supplies, Support, and Services

Accountant	Horse trailer sales and design
Advertising copywriter	Insurance agent
Architect	Land consultant
Attorney	Photographer
Author/writer	Reporter/Journalist
Commercial artist	Saddle maker
Feed store operator	Tack and clothing retailer
Feed manufacturer	Tack and equipment maker
Film production and distribution	Transportation specialist

Horse Shows and Rodeos

For individuals interested in working with horse show or rodeos a variety of jobs are available. Some offer permanent employment and a stable income. Some require travel to attend horse shows or rodeos in different places (see Figure 22-7).

Figure 22-5   A small shop that sells tack and related supplies is just one place to work for those who are interested in equine supplies, sales, and service areas.

Figure 22-6   The saddle making shop of Bob Severe, Burley, Idaho

Figure 22-7 Being a course designer is just one of the many jobs available in the horse show industry.

Careers in Horse Shows and Rodeos

Announcer	Rodeo cowboy
Course/jump designer	Rodeo pick-up rider
Drug inspector	Show groom
Fair or exposition manager	Show veterinarian
Jockey	Show manager
Judge	Show secretary
Jump builder	Show receptionist
Publicity director	Steward
Ring master	Stock contractor
Rodeo clown	Ticket seller

EDUCATION AND EXPERIENCE

Requirements to begin working in the horse industry vary depending upon the level of work. One requirement common to all is practical work experience in the industry. Often to gain this practical experience, the new employee begins at an entry-level job and then is advanced through the organization. Advancement depends on the skills and knowledge the employee brings to the job, the skills and knowledge gained on the job, and productivity on the job.

Entry-level educational requirements vary, but all of the basic skills, thinking skills, and general workplace competencies discussed earlier are

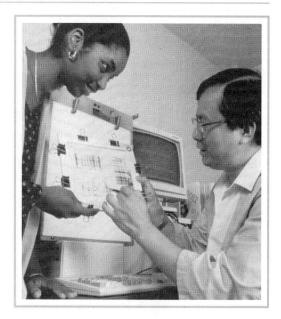

Figure 22-8 This student is receiving invaluable work experience through a Specialized Adult Education program at her high school.

important. These skills should be obtained in high school, and reinforced during additional training and schooling. More specialized education in the equine industry is offered in some high schools, community colleges, and universities.

Many high school agriscience programs provide the education necessary for lower-level entry positions. Often high school programs in agriscience provide students with supervised work experience in some aspect of agriculture. This is invaluable for getting a job and helping individuals determine if they wish to pursue additional education (see Figure 22-8).

Some community colleges and other postsecondary schools provide specialized equine programs with practical experience as a part of the schooling. Programs at community colleges focus on entry-level technician jobs.

Universities and colleges offering bachelor's degrees, master's degrees, and doctoral programs provide some highly specialized education in equine science.

IDENTIFYING A JOB

Finding that first job or finding a different job can be difficult. Whole books, videos, and seminars are available on finding jobs. What follows are some suggestions. The Additional Resources section at the end of this chapter contains more information.

Sources for locating jobs include:

- Classified advertisements in newspapers

- Magazines or trade journals and publications
- Personal contacts
- Placement offices
- Employment or personnel office of a company
- Public notices
- Computerized on-line services

Newspapers, magazines, trade journals, and publications can be good resources for locating a job. By reading the advertisements in these publications, the potential employee can determine the demand for his or her job skills. Also, the potential employee can compare his or her skills and training with those listed in the advertisement.

A different twist on the classified advertisement is the "situation wanted" section of newspapers, magazines, and trade journals. Many people secure excellent jobs by advertising their skills.

Personal contacts are still the top source of jobs. Employers do not like to make mistakes. Some feel that if a trusted acquaintance makes a recommendation this lessens the chances of making a mistake in hiring. Also, personal contacts may know of jobs opening up before they are publicly announced. This gives the potential employee more time to prepare and research the job. Personal contacts include friends, relatives, teachers, guidance counselors, and employees of the hiring company.

Placement offices provide vocational counseling, give aptitude and ability/interest tests, locate jobs, and arrange job interviews. There are three types of placement offices: public, private, and school. These agencies work to match employers with prospective employees. Often too, an agency knows how to help potential employees prepare and present themselves.

Public placement offices are supported by federal and state funds. Their services are free. Private placement offices charge for the services they provide. This usually is a percentage of the beginning salary. Individuals using private placement services sign a contract before services are provided. High schools, trade schools, and colleges may maintain a placement service for their students. They also help individuals identify their aptitude or interest for a job and help in preparation for job interviews.

Many companies support their own employment or personnel office. Individuals seeking employment can fill out application forms and/or leave **resumes** in case a job becomes available.

Finally, some companies seeking new employees may issue a public notice of some kind. This includes posters or fliers on bulletin boards around a community. Posters or fliers are sent to related businesses and are posted on their bulletin boards. Schools and colleges often receive public announcements of jobs.

Computerized posting of jobs is another kind of a bulletin board. Some on-line information services maintain computerized databases of jobs. Interested individuals use a phone, computer, and modem to contact the computerized database to search for jobs that match their qualifications and desires. This type of job listing opens the door wide to potential jobs but often not to local jobs.

GETTING A JOB

Once some job possibilities are identified, the work begins. Getting a job is difficult and requires some preparation. Again whole books, videos, and seminars teach how to get a job. A few tips follow. Before applying for a position, it is a good idea to do a little research on the company and the job. These things should be learned about the job and the company:

- Name of the company
- Name of personnel manager or person who will conduct the interview
- Company address and phone number
- Position available: minimum requirements and job responsibilities
- Geographic scope of the company—local, county, state, regional, national
- Company's product(s) and demand for the product(s)
- Recent company developments

Application Forms

If the company requires an application form, remember you are trying to sell yourself with the information you give. Review the entire application form before you begin. Pay particular attention to any special instructions to print or write in your own handwriting. When answering ads that require potential employees to apply in person, be prepared to complete an application form on the spot. Take a pen and a list with the information you will need to complete the application form. This information may include your social security number; the addresses of schools you have attended; names, phone numbers, and addresses of previous employers and supervisors; names, phone numbers, and addresses of references. The following guidelines will provide you with some direction when completing application forms.

- Follow all instructions carefully and exactly.
- If your application is handwritten, rather than typed, write neatly and legibly. Handwritten answers should be printed unless otherwise directed.

- Application forms should be written in ink unless otherwise requested. If you make a mistake, mark through it with one neat line.
- Be honest and realistic. Give all the facts for each question but keep your answers brief.
- Fill in all the blanks. If the question does not pertain to you, write "not applicable" or "N/A." If there is no answer, write "none" or draw a short line through the blank.
- Many application forms ask what salary you expect. If you are not sure what is appropriate, write "negotiable," "open," or "scale" in the blank. Before applying, try to find out what the going rate for similar work is at other locations. Give a salary range rather than exact figure.

Letters of Inquiry and Application

The purpose of a **letter of inquiry** is to obtain information about possible job vacancies. The purpose of a **letter of application** is to apply for a specific position that has been publicly advertised. Both letters indicate your interest in working for a particular company, acquaint employers with your qualifications, and encourage the employer to invite you for a job interview.

Letters of inquiry and application represent you. They should be accurate, informative, and attractive. Your written communications should present a strong, positive, professional image both as a job seeker and future employee.

The following list should be used as a guide when writing both letters of inquiry and letters of application.

1. Use 8½ × 11 inch white typing paper, not personal or fancy paper and use an attractive, simple format. The letter and envelope should be neatly typed, error free, and without smudges.
2. Write to a specific person. Use "To Whom It May Concern" if you are answering a blind ad.
3. Make your letter short and specific, one or two pages at most (leave details to the resume).
4. Set a positive tone and use logical, organized paragraphs with ideas that are expressed in a clear, concise, direct manner.
5. Use carefully constructed sentences that are free of spelling or grammatical errors. Avoid slang words and expressions and excessive use of the word "I."
6. Avoid mentioning salary and fringe benefits.
7. Write a first draft, then make revisions.
8. Proofread the final letter yourself, and also have someone else proofread it. Be sure you have addressed it and signed it correctly.

A letter of inquiry should:

1. Specify the reasons why you are interested in working for the company and ask if there are any positions available now or anticipated in the near future.

2. Explain how your personal qualifications and work experience would help meet the needs of the company. (Since you are not applying for a particular position, you cannot relate your qualifications directly to job requirements.) Mention and include your resume.

3. Express your interest in being considered a candidate for a position when one becomes available and state your willingness to meet with a company representative to discuss your background and qualifications. (Include your address and a phone number where you can be reached.)

4. Address letters of inquiry to "Personnel Manager" unless you know his or her name.

A letter of application should:

1. Indicate your source of the job lead (newspaper ad, etc.). Specify the particular job you are applying for and the reason for your interest in the position and the company.

2. Explain how your personal qualifications meet the needs of the employer and explain how your work experience relates to the job requirements. Mention and include your resume.

3. Request an interview and state your willingness to meet with a company representative to discuss your background and qualifications. (Include your address and a phone number where you can be reached.)

4. Address your letter to "Personnel Manager" unless you have the name of a specific person to whom applications should be sent.

Resume or Data Sheet

Many jobs require a resume or **data sheet** (see Figure 22-9). The following information should be included:

- Name, address, and phone number
- Brief, specific statement of career objective
- Educational background—names of schools, dates, major field of study, degrees or diplomas—listed in reverse chronological order
- Leadership activities, honors, and accomplishments
- Work experience, listed in reverse chronological order
- Special technical skills and interests related to job
- References

RESUME
Roger Brown

PO Box 1238
Anywhere, ID 00000

Tel: 000/888-8888; e-mail address: rbrown@cyberhighway.net

Career Objectives

Obtain satisfying job in the equine industry that provides advancement opportunities during my career.

Education

Local High School, Anywhere, ID: Graduated 1990.
College of Southern Idaho, Twin Falls, ID, 1991–1993: A.A.S, Equine Science.

Activities and Honors

- Active in 4-H Club with equine emphasis. Learned judging and equitation.
- Member FFA for three years and was elected president during senior year.
- Member Block and Bridle Club 1991–1993.
- Advisor to local 4-H Club.

Employment and Work Experiences

January 1995 to Present: Sun Valley Stables, Sun Valley, ID; general help; cleaned stalls and exercised horses.

July 1993 to December 1994: ABC Grocery, Anywhere, ID; restocked shelves; boxed groceries; worked into checker position.

References

Available on request.

Figure 22-9 A good resume shows information quickly and clearly.

Limit your resume to one page if possible. Make sure it is neatly typed, error free, and logically organized. Be honest when listing qualifications and experiences and make sure your strong points will stand out clearly at a glance. Employers look for a quick overview of who you are and how you might fit into their business. On the first reading, the employer will spend only ten to fifteen seconds reading a resume, so be sure to present relevant information clearly and concisely in an eye-catching format.

The Interview

The next step in the job-hunting process is the interview. There are many sources of good information on how to do well in an interview. Perhaps the best advice comes from the interviewer's side of the desk. This is a list of reasons interviewers give for **not** placing applicants in a job.

1. Poor attitude; not really anxious to work or interested only in the salary and benefits of the job
2. Unstable work record; not having any direction or goals
3. Lack of confidence and self-selling ability
4. Lack of skill and experience; bad references
5. "Bad mouthing" former employers
6. Too demanding (wanting too much money or to work only under certain conditions)
7. Unavailable for interviews or canceling an appointment
8. Poor appearance, poor grammar, use of slang
9. Lack of manners and personal courtesy, chewing gum, smoking, fidgeting
10. No attempt to establish rapport; not looking the interviewer in the eye; being evasive

Follow-up Letters

Follow-up letters are sent immediately after an interview. The follow-up letter demonstrates your knowledge of business etiquette and protocol. Always send a follow-up letter regardless of whether or not you had a good interviewing experience and even if you are no longer interested in the position. When employers do not receive follow-up letters from job candidates, they often assume the candidate is not aware of the professional courtesy they will need to demonstrate on the job.

The major purpose of a follow-up letter is to thank those individuals who participated in your interview. In addition, a follow-up letter reinforces your name, application, and qualifications to the employer and indicates if you are still interested in the job position. It also offers you an opportunity to restate your reasons for wanting the job and why you think you are a strong candidate.

A Job is More Than Money

Before taking a job be certain that it is what you want. While the salary or the wage is important, job satisfaction is something quite different and very important. Jobs quickly become routine and mundane. For example, a job with little fulfillment and challenge for some people can easily become a chore. Before taking a job or even while looking for a job, answer these questions.

1. Does the job description fit your interests?
2. Is this the level of occupation at which you wish to work?
3. Does this type of work appeal to your interests?
4. Are the working conditions suitable to you?
5. Will you be satisfied with the salary and benefits offered?
6. What are the advancement opportunities? Can you advance in this occupation as rapidly as you would like?
7. Does the future outlook satisfy you?
8. Is the occupation in demand now and in the foreseeable future?
9. Do you have or can you get the education needed for the occupation?
10. What type of training is available after taking the job?
11. Can you get the finances needed to get into the occupation?
12. Can you meet the health and physical requirements?
13. Will you be able to meet the entry requirements?
14. Do you know of any reasons you might not be able to enter this occupation?
15. Is the occupation available locally or are you willing to move to a part of the country where it is available?

Also, before taking a job or looking for a job do a little personality inventory of yourself. Consider the following:

1. Do I like to be alone or with people?
2. Am I mechanical or artistic?
3. Would I rather work independently or work under supervision?
4. Would I prefer to think or be active?
5. Can I take authority and responsibility for others?
6. Must I have freedom to express creativity?
7. What things do I like to do? Make a list.
8. At what time of day can I work best?
9. Can I work under pressure or stress?
10. Make a list of your strong points. Consider skills, hobbies, and leisure-time activities you can offer an employer.

Do your research and your job will be more rewarding and you will feel better about yourself.

SUMMARY

The goal of education and training is primarily to become employable and stay employable—to get and keep a job or run a successful business. The world of work requires people who can read, write, do math, and communicate. Rapidly advancing technology has made this even more critical. The modern workplace now looks for people who possess thinking skills. With a solid set of basic skills, future employees also need to relate well to other people; be able to use information; understand the concept of systems, and use technology. Old-fashioned ideas, like responsibility, self-esteem, sociability, self-management, and integrity are not out-of-date.

Jobs in the equine industry range from those very closely tied to the industry to those that support the equine industry. In general, potential job areas include supplies and services, training, production, marketing, research, and development. Education and training for jobs in the equine industry vary from on-the-job training to high school and college degrees.

After training and education, finding and getting the right job may still be a challenge. Several good resources exist for locating a job. Still, the best one is personal contact. Well-written letters of inquiry and application, a clear, eye-catching resume, and being prepared for the job interview help secure a job.

REVIEW

Success in any career requires knowledge. Test your knowledge of this chapter by answering these questions or solving these problems.

True or False
1. Reading skills are not important for feeding horses.
2. Good ethics are still a part of conducting a horse business.
3. A veterinary technician does not require a science background.
4. General appearance is important in a job interview.
5. The horse industry includes much more than just riding and training horses.

Short Answer
6. List three traits of an entrepreneur.
7. Name four general competencies needed in the workplace.
8. Give three sources of information for finding a job.
9. List five types of work available in the horse show industry and five types in the support and services area of the horse business.
10. Name ten primary careers in the horse industry.

Discussion

11. What is the purpose of knowledge and education?

12. Explain the difference between a letter of application, a letter of inquiry, and a follow-up letter.

13. What kinds of information are listed in a resume?

14. Describe five reasons why an interview may fail.

15. Discuss two reasons why computers are important in the horse industry.

STUDENT ACTIVITIES

1. Gather sample resumes from local sources. Develop your own resume or data sheet.

2. Collect position announcements and classified ads for jobs in the horse industry. Write a letter of job inquiry and a letter of job application for their selected job using this information.

3. Develop a list of questions frequently asked during an interview. Use the questions in role-playing job interviews and videotape the interviews.

4. Visit a public or private placement office. Following the field trip, discuss the office's policies and how they affect job searchers and employers. Alternatively, invite a representative from a state employment agency to explain how employment agencies can help students gain employment.

5. Attend a career field day. Locate individuals currently employed in the horse industry to discuss career opportunities.

6. Select one career in the horse industry of interest and prepare a research paper on the career using a computer and word processing software. The paper should identify the knowledge and skills required and the employment opportunities.

7. Collect pictures or photographs of people engaged in various careers with horses and prepare a bulletin board collage.

8. Meet with a resource person such as a business owner or personnel manager to discuss what he or she looks for in resumes, application letters and forms, and during interviews.

9. Visit with local agribusiness people to discuss the importance of employee work habits, basic skills and attitudes and how they affect the entire business.

ADDITIONAL RESOURCES

Books

Aslett, D. 1993. *Everything I needed to know about business I learned in the barnyard.* Pocatello, ID: Marsh Creek Press.

Business Council for Effective Literacy. 1987. *Job-related basic skills: a guide for planners of employee programs.* New York, NY: Business Council for Effective Literacy.

English, J. E. 1995. *Complete guide for horse business success.* Grand Prairie, TX: Equine Research, Inc.

Kreitler, B. 1995. *50 careers with horses—from accountant to wrangler.* Ossining, NY: Breakthrough Publications.

Mandino, O. 1970. *The greatest salesman in the world.* New York, NY: Fredrick Fell Publishers, Inc.

Peters, T. 1994. *The pursuit of wow!* New York, NY: Vintage Books.

Smith, M., Underwood, J. M., and Bultmann, M. 1991. *Careers in agribusiness and industry.* 4th ed. Danville, IL: Interstate Publishers.

U.S. Department of Education. 1991. *America 2000: an education strategy.* Sourcebook. Washington, D.C.: U.S. Department of Education.

U.S. Department of Labor, Employment, and Training Administration. 1978. *Dictionary of occupational titles.* Washington, D.C.: American Association for Vocational Instructional Materials.

Videos

Makin' Headlines and Canadian Trotting Association. 1995. *Harness racing careers.* London, Ontario, Canada: Makin' Headlines and Canadian Trotting Association.

American Horse Council. n.d. *Unbridled opportunities: careers in the horse industry.* Washington, D.C.: American Horse Council.

A P P E N D I X

Due to its location in a book, and because of its name, an appendix is often ignored by the reader. But an appendix contains valuable information that can enhance a reader's understanding and learning. Further, information in an appendix is quick and easy to find.

The information in this appendix includes a variety of useful conversion factors, feed standards and facts, a feeding worksheet, a gestation calculator, several health checklists and schedules, and the addresses for breed registries and equine organizations. Armed with this information, the reader can understand more, plan more, and learn more.

Table A-1 Conversion Tables for Common Weights and Measures	
Metric Conversions	**Equaled Amount**
1 pound	454 grams
2.2 pounds	1 kilogram
1 quart	1 liter
1 gram	15.43 grains
1 metric ton	2.205 bands
1 inch	2.54 centimeters
1 centimeter	10 millimeters or .39 inches
1 meter	39.37 inches
1 acre	.406 hectare

Table A-2 Weight Conversions	
Measurements	**Equaled Amount**
8 tablespoons	¼ lb.
3 teaspoons	1 tablespoon
1 pint	1 lb.
2 pints	1 qt.
4 quarts	1 gallon or 8 lb.
2,000 lb.	1 ton
16 ounces	1 lb.
27 cu. ft.	1 cu. yd.
1 peck	8 qts.
1 bushel	4 pecks
Other Conversions	
1%	.01
1%	10,000 ppm
1 megacalorie (mcal.)	1,000 calories
1 calorie (big calorie)	1,000 calories (small calorie)
1 mcal.	1 therm

Table A-3 Standard Weights of Farm Products Per Bushel	
Product	**lb.**
Alfalfa	60
Apples (average)	42
Barley (common)	48
Beans	60
Bluegrass (Kentucky)	14–28
Bromegrass, orchardgrass	14
Buckwheat	50
Clover	60
Corn (dry ear)	70
Corn & cob meal	45
Corn (shelled)	56
Corn kernel meal	50
Corn (sweet)	50
Cowpeas	60
Flax	56
Millet (grain)	50
Oats	32
Onions	52
Peas	60
Potatoes	60
Ryegrass	24
Rye	56
Soybeans	60
Spelt	30–40
Sorghum	56
Sudangrass	40
Sunflower	24
Timothy	45
Wheat	60
Milk, per gallon	8.6

Table A-4   Storage and Feeding Dry Matter—Losses of Alfalfa

Storage Method	Storage Loss	Feeding Loss
Small bales, stored inside	.04	.05
Round bales, stored inside	.04	.14
Hay stacks, stored inside	.04	.16
Round bales, stored outside	.12	.14
Hay stacks, stored outside	.16	.16
Haylage, vertical silo	.07	.11
Haylage, bunk silo	.13	.11

Table A-5   Bushel Weights and Volumes

Item	lb./cubic ft.	cubic ft./ton
Oats = 32 lb./bu.	26	77
Barley = 48 lb./bu.	38.4	53
Shelled corn = 56 lb./bu.	44.8	45
Wheat = 60 lb./bu.	48	42
Corn & cob meal = 70 lb./bu.	28	72
Soybeans = 60 lb./bu.	48	42
Rye = 56 lb./bu.	44.8	45
Soybean oil meal = 54 lb.	—	37
Dairy feed = 35 lb.	—	57

Table A-6	Measurement Standards, Hay and Straw	
Item	Average cu. ft./ton	Range cu. ft./ton
Hay, baled	275	250–300
Hay, chopped–field cured	425	400–450
Hay, chopped–mow cured	325	300–350
Hay, long	500	475–525
Straw, baled	450	400–500
Straw, chopped	600	575–625
Hay, loose	480	370–390
Straw, loose	800	750–850

Table A-7 Fahrenheit to Centigrade Temperature Conversions[1]					
°F	°C	°F	°C	°F	°C
100	37.8	77	25.0	54	12.2
99	37.2	76	24.4	53	11.7
98	36.7	75	23.9	52	11.1
97	36.1	74	23.3	51	10.6
96	35.6	73	22.8	50	10.0
95	35.0	72	22.2	49	9.4
94	34.4	71	21.7	48	8.9
93	33.9	70	21.1	47	8.3
92	33.3	69	20.6	46	7.8
91	32.8	68	20.0	45	7.2
90	32.2	67	19.4	44	6.7
89	31.7	66	18.9	43	6.1
88	31.1	65	18.3	42	5.6
87	30.6	64	17.8	41	5.0
86	30.0	63	17.2	40	4.4
85	29.4	62	16.7	39	3.9
84	28.9	61	16.1	38	3.3
83	28.3	60	15.6	37	2.8
82	27.8	59	15.0	36	2.2
81	27.2	58	14.4	35	1.7
80	26.7	57	13.9	34	1.1
79	26.1	56	13.3	33	0.6
78	25.6	55	12.8	32	0.0

1. Formulas used: $°C = (°F - 32) \times \frac{5}{9}$ *or* $°F = (°C \times \frac{9}{5}) + 32$

Table A-8	Conversion Factors for English and Metric Measurements			
To Convert the English	**To the Metric Multiply by**	**To Convert Metric**	**Multiply by**	**To get English**
acres	0.4047	hectares	2.47	acres
acres	4047	m.^{2}	0.000247	acres
BTU	1055	joules	0.000948	BTU
BTU	0.0002928	kwh	3415.301	BTU
BTU/hr.	0.2931	watts	3.411805	BTU/hr.
bu.	0.03524	m.^{3}	28.37684	bu.
bu.	35.24	L	0.028377	bu.
ft.^{3}	0.02832	m.^{3}	35.31073	ft.^{3}
ft.^{3}	28.32	L	0.035311	ft.^{3}
in.^{3}	16.39	cm.^{3}	0.061013	in.^{3}
in.^{3}	1.639×10^{-5}	m.^{3}	61012.81	in.^{3}
in.^{3}	0.01639	L	61.01281	in.^{3}
yd.^{3}	0.7646	m.^{3}	1.307873	yd.^{3}
yd.^{3}	764.6	L	0.001308	yd.^{3}
ft.	30.48	cm.	0.032808	ft.
ft.	0.3048	m.	3.28084	ft.
ft./min.	0.508	cm./sec.	1.968504	ft./min.
ft./sec.	30.48	cm./sec.	0.032808	ft./sec.
gal.	3785	cm.^{3}	0.000264	gal.
gal.	0.003785	m.^{3}	264.2008	gal.
gal.	3.785	L	0.264201	gal.
gal./min.	0.06308	L/sec.	15.85289	gal./min.
in.	2.54	cm.	0.393701	in.
in.	0.0254	m.	39.37008	in.
mi.	1.609	km.	0.621504	mi.

(continued on next page)

Table A-8 Conversion Factors for English and Metric Measurements *(concluded)*

To Convert the English	To the Metric Multiply by	To Convert Metric	Multiply by	To get English
mph	26.82	m./min.	0.037286	mph
oz.	28.349	gm.	0.035275	oz.
fl. oz.	0.02947	L	33.93281	fl. oz.
liq. pt.	0.4732	L	2.113271	liq. pt.
lb.	453.59	gm.	0.002205	lb.
qt.	0.9463	L	1.056747	qt.
ft.^{2}	0.0929	m.^{2}	10.76426	ft.^{2}
yd.^{2}	0.8361	m.^{2}	1.196029	yd.^{2}
tons	0.9078	tonnes	1.101564	tons
yd.	0.0009144	km.	1093.613	yd.
yd.	0.9144	m.	1.093613	yd.

m.^{2} = square meters; kwh = kilowatt hours; bu. = bushel; m.^{3} = cubic meters; L = liter; cm.^{3} = cubic centimeters; in.^{3} = cubic inches; m. = meter; km. = kilometer; yd.^{3} = cubic yard; cm. = centimeter

Date of Service	Estimated Date of Birth	Date of Service	Estimated Date of Birth	Date of Service	Estimated Date of Birth
01-Jan	06-Dec	30-May	05-May	27-Oct	02-Oct
06-Jan	11-Dec	04-Jun	10-May	01-Nov	07-Oct
11-Jan	16-Dec	09-Jun	15-May	06-Nov	12-Oct
16-Jan	21-Dec	14-Jun	20-May	11-Nov	17-Oct
21-Jan	26-Dec	19-Jun	25-May	16-Nov	22-Oct
26-Jan	31-Dec	24-Jun	30-May	21-Nov	27-Oct
31-Jan	05-Jan	29-Jun	04-Jun	26-Nov	01-Nov
05-Feb	10-Jan	04-Jul	09-Jun	01-Dec	06-Nov
10-Feb	15-Jan	09-Jul	14-Jun	06-Dec	11-Nov
15-Feb	20-Jan	14-Jul	19-Jun	11-Dec	16-Nov
20-Feb	25-Jan	19-Jul	24-Jun	16-Dec	21-Nov
25-Feb	30-Jan	24-Jul	29-Jun	21-Dec	26-Nov
01-Mar	04-Feb	29-Jul	04-Jul	26-Dec	01-Dec
06-Mar	09-Feb	03-Aug	09-Jul	31-Dec	06-Dec
11-Mar	14-Feb	08-Aug	14-Jul	05-Jan	11-Dec
16-Mar	19-Feb	13-Aug	19-Jul	10-Jan	16-Dec
21-Mar	24-Feb	18-Aug	24-Jul	15-Jan	21-Dec
26-Mar	01-Mar	23-Aug	29-Jul	20-Jan	26-Dec
31-Mar	06-Mar	28-Aug	03-Aug	25-Jan	31-Dec
05-Apr	11-Mar	02-Sep	08-Aug	30-Jan	05-Jan
10-Apr	16-Mar	07-Sep	13-Aug	04-Feb	10-Jan
15-Apr	21-Mar	12-Sep	18-Aug	09-Feb	15-Jan
20-Apr	26-Mar	17-Sep	23-Aug	14-Feb	20-Jan
25-Apr	31-Mar	22-Sep	28-Aug	19-Feb	25-Jan
30-Apr	05-Apr	27-Sep	02-Sep	24-Feb	30-Jan
05-May	10-Apr	02-Oct	07-Sep	01-Mar	04-Feb
10-May	15-Apr	07-Oct	12-Sep	06-Mar	09-Feb
15-May	20-Apr	12-Oct	17-Sep	11-Mar	14-Feb
20-May	25-Apr	17-Oct	22-Sep	16-Mar	19-Feb
25-May	30-Apr	22-Oct	27-Sep	21-Mar	24-Feb

Table A-9 Gestation Calculator for Horses[1]

1. Assumes a 340-day gestation.

Table A-10 Vaccination Schedules for Different Classes of Horses

Pregnant Mares

Month of Gestation	1	2	3	4	5	6	7	8	9	10	11
Pneumabort-K (rhinopneumonitis)			X		X		X		X		
Flu (influenza)			(X)							X	
E & W (encephalitis: Eastern & Western)			(X)							X	
Tetanus toxoid			(X)								
Strangles*											
Potomac horse fever (PHF)*											
E-Se (Great Lakes states)**											
Equine viral arteritis (EVA)*											

  * Depends on region, incidence, past history of disease.
  ** 1 cc/100 lb. of mare, ex., 10 cc/1,000-lb-mare.
  (X) = optional

Foals

Age	At birth	2 months	3 months	4 months	5 months	6 months
Rhinopneumonitis		X* then every 2 months				
Flu		X* then every 2 months				
E & W		X*				
Tetanus toxoid	X					
Strangles						X

(continued on next page)

Table A-10 Vaccination Schedules for Different Classes of Horses *(concluded)*						
Pregnant Mares						
Nonvaccination injection of Vitamin E-Selenium (equine product)	X**					

* Initial 4-way, booster 3 weeks later (the time when passive, colostral immunity decreases).
** 1 cc/100 lb. (For Great Lakes, selenium-deficient states), most foals: 1 ccIM at birth.

Barren & Maiden Mares	
4-way	1 to 2 times per year (depends on traffic, exposure to a larger population)
Rhino-pneumonitis	Rhinomune if separated from pregnant population; otherwise Pneumabort-K (More frequent administration).
Strangles	As above
PHF	As above
EVA	3 weeks prior to breeding to a known shedding stallion if the mare has a negative titer, i.e., no prior exposure to the virus

Stallions	
Rhino-pneumonitis	Rhinomune 2x/year, before breeding and at the end of the season
Flu	Contained in 4-way, 2x/year, then alone at alternate two-month periods
E & W	Within 4-way, 2x/year
Tetanus toxoid	Within 4-way, or once a year
Strangles	If endemic
PHF	If endemic
EVA	If endemic, prior history, outbreak on farm. One month prior to breeding season.

Yearlings
Assuming prior boosters as foals/weanlings, continue with 4-way 2x/year and strangles vaccine. Rhinomune 2x/year.

Table A-11 Vaccination Schedules	
Type of Vaccination	**Suggested Schedule**
Tetanus toxoid	1 booster per year or when lacerated (cut) or injured.
Influenza	1 booster per year unless there is a special situation.
Rhinopneumonitis	1 booster per year unless there is a special situation. (Pregnant mares require a separate vaccination schedule.)
(EEE) Eastern equine encephalomyelitis	1 booster per year in endemic areas after initial vaccination series.
Potomac horse fever	1 booster per year after initial vaccination series in endemic areas.
Strangles	1 booster per year. This vaccine should be used only if necessary, based on past history of the horse or farm.
Equine viral arteritis	Special situation.
Rabies	Special situation.
Leptospirosis	Special situation.

Notes: Horses one year and older will be considered adults for scheduling vaccinations.

Special situations, such as erratic outbreaks of a given disease, heavy exposure to a variety of horses at shows or racetracks, or increased traffic into the farm or stable, will require a different vaccination schedule from the average program.

Consult your veterinarian about special situations. Clear combination vaccines (3- and 4-way products) that vaccinate against several diseases simultaneously with your veterinarian before use.

Never vaccinate a sick horse until you check with your veterinarian—severe complications can result.

This schedule is based on proper use of vaccinations (foals given the initial injections in a vaccination series) up to one year of age.

Some vaccine products use one injection as a booster while others require a two-injection series repeated every year. Check with your veterinarian.

Table A-12   Checklist for Internal Parasites

Controlling internal parasites (worming) requires treating at certain intervals during the year with appropriate worming agents. This program should be designed in consultation with your veterinarian who will consider:

- ☐ Number and type of horses
- ☐ Arrangement of pastures
- ☐ Traffic of horses
- ☐ Resistance of parasites to worming agents
- ☐ Use of tube worming vs. paste wormers
- ☐ Breeding program—pregnant mares and foals
- ☐ Use of fecal tests to determine effectiveness of treatment

The average adult horse should be wormed at least four times a year (every three months) or more frequently if your veterinarian advises. Even an effective program will not eliminate all parasites, but it will keep them at a tolerable level so they will not affect the health of the horse.

Table A-13   Checklist for External Parasite Control

External parasites include flies, mosquitoes, lice, etc. The three primary reasons to control external parasites are for the comfort of the horse, to prevent certain skin diseases, and to lower the incidence of some diseases that are transmitted by insects.

The two most effective methods of external parasite control are cleanliness and appropriate use of chemical repellents.

Caution: Check with your veterinarian if the horse is exhibiting any of the following symptoms:

- ☐ 1. Hair loss other than normal shedding
- ☐ 2. Excessive itching
- ☐ 3. Sores on the skin

Always follow the label directions on insecticides carefully.

Table A-14   Checklist for Equine Dental Program

Each horse that is a yearling or older should have a dental examination by a veterinarian at least once per year. Other examinations might be necessary if symptoms occur.

Symptoms that dictate examination:

- ☐ 1. Difficulty chewing
- ☐ 2. Reluctance to drink cold water
- ☐ 3. "Quidding"—dropping food out of the mouth
- ☐ 4. Excessive unchewed grain in the manure
- ☐ 5. Constipation colics
- ☐ 6. Weight loss
- ☐ 7. Swelling or tenderness in jaw region
- ☐ 8. Reluctance to accept a bit

During the examination, the horse's teeth can be "floated"—filed down to remove any sharp edges that can interfere with proper chewing. In the younger horse, other procedures might also be required, such as removal of "wolf teeth" or "dental caps."

Table A-15	Feed Requirement Worksheet

Class of Horse _____

Owner's Name _____	_____ Mature horse at rest
Address _____	_____ Mature horse at moderate work
Breed of Horse _____	_____ Mare in last 90 days of pregnancy
Weight of Horse _____ Age _____	_____ Mare in peak of lactation (first 3 months)
	_____ Growing foal _____ Mature weight

		Dig. Energy	Dig. Protein	Ca	P	Vit. A 1,000
A. Your horse's daily requirements		_____ mcal.	_____ lb.	_____ g.	_____ g.	_____ I.U.s
	I	II	III	IV	V	VI
B. Ration lbs. of		X(mcal./lb.) =mcal.	X(C P/lb.) =lb. C P	X(Ca/lb.) = g. Ca	X(P/lb.) = g. P	X(1,000 I.U.s/lb.) = 1,000 I.U.
Feedstuff	each					
_____	_____	X(__)=__	X(__)=__	X(__)=__	X(__)=__	X(__)=__
_____	_____	X(__)=__	X(__)=__	X(__)=__	X(__)=__	X(__)=__
_____	_____	X(__)=__	X(__)=__	X(__)=__	X(__)=__	X(__)=__
_____	_____	X(__)=__	X(__)=__	X(__)=__	X(__)=__	X(__)=__
_____	_____	X(__)=__	X(__)=__	X(__)=__	X(__)=__	X(__)=__
C. Total __lb. supplied by ration		__ mcal.	__ lb.	__ g.	__ g.	__1,000 I.U.s
D. Horse's requirement		__ mcal.	__ lb.	__ g.	__ g.	__1,000 I.U.s
E. Needed nutrients		__ mcal.	__ g.	__ g.	__ g.	__1,000 I.U.s

F. Instructions

1. Determine the class of horse and record requirements from Tables 1 and 2 in line A.
2. List ration ingredients and pounds of each in the appropriate columns.
3. Be sure ration ingredients in B do not exceed 2 to 2.5% of body weight.
4. Obtain feed compositions from Table 4 and record in cols. II, III, IV, V, and VI.
5. Multiply lbs. of each feed (col. I) times each value in cols. II, III, IV, V, and VI.
6. Total the nutrients from each source to get total in ration (line C values).
7. Copy nutrient requirements from line A to line D.
8. Subtract line C values from line D values and record any deficiencies in line E.

Table A-16 Breed Registry Associations

Akhal Teke Registry of America
Rt. 5, Box 110
Staunton, VA 22401-8906

American Andulasian Horse Assn.
6990 Manning Road
Economy, IN 47339-9736

American Baskir Curly Registry
P.O. Box 246
Ely, NV 89301-0246

American Buckskin Registry Assn.
P.O. Box 3850
Redding, CA 96049-3850

American Connemara Pony Society
2630 Hunting Ridge
Winchester, VA 22603

American Cream Draft Horse Assn.
P.O. Box 2065 Noble Avenue
Charles City, IA 50616-9108

American Dartmoor Pony Assn.
15870 Paseo Mantra Road
Anna, OH 45302

American Dominant Gray Registry
10980 "8" Mile Road
Battle Creek, MI 49017-9560

American Exmoor Pony Registry
c/o American Livestock Breeds Conservancy
P.O. Box 477
Pittsboro, NC 27312-0477

American Hackney Horse Society
#A 4059 Iron Works Road
Lexington, KY 40511-8462

American Hanoverian Society
4059 Iron Works Pike
Lexington, KY 40511

American Holsteiner Horse Assn.
#1 222 East Main Street
Georgetown, KY 40324-1712

American Horizon Horse Registry
P.O. Box 564
Belen, NM 87002-0564

American Indian Horse Registry
Rt. 3, Box 64
Lockart, TX 78644

American Miniature Horse Assn.
2908 SE Loop 820
Fort Worth, TX 76140-1073

American Miniature Horse Registry
P.O. Box 3415
Peoria, IL 61614-3415

American Morgan Horse Assn.
P.O. Box 960
Shelburne, VT 05482-0960

American Mustang and Burro Assn.
P.O. Box 788
Lincoln, CA 95648

American Mustang Assn.
P.O. Box 338
Yucaipa, CA 92399

American Paint Horse Assn.
P.O. Box 961023
Fort Worth, TX 76161-0023

American Quarter Horse Assn.
P.O. Box 200
Amarillo, TX 79168

American Quarter Pony Assn.
P.O. Box 30
New Sharon, IA 50207

American Saddlebred Horse Assn.
4093 Iron Works Pike
Lexington, KY 40511-8434

American Shetland Pony Club
P.O. Box 3415
Peoria, IL 61614-3415

Table A-16   Breed Registry Associations *(continued)*

American Shire Horse Assn.
2354 315 Court
Adel, IA 50003

American Suffolk Horse Assn.
4240 Goehring Road
Ledbetter, TX 78946-9707

American Tarpan Studbook Assn.
1658 Coleman Avenue
Macon, GA 31201-6602

American Trakehner Assn.
1520 West Church Street
Newark, OH 43055

American Walking Pony Registry
P.O. Box 5282
Macon, GA 31208-5282

American Warmblood Registry
(also American Warmblood Society)
6801 West Romley Avenue
Phoenix, AZ 85043

American Welara Pony Society
P.O. Box 401
Yucca Valley, CA 92286-0401

Appaloosa Horse Club
P.O. Box 8403
Moscow, ID 83843-0903

Arabian Horse Registry of America
12000 Zuni Street
Westminster, CO 80234-2300

Belgian Draft Horse Corporation of America
P.O. Box 335
Wabash, IN 46992-0335

Caspian Horse Society of America
Rt. 7, Box 7504
Brenham, TX 77833

Chilean Corralero Registry International
230 East North Avenue
Antigo, WI 54409

Cleveland Bay Horse Society of
North America
P.O. Box 221
South Windham, CT 06266

Clydesdale Breeders of the U.S.A.
17378 Kelley Road
Pecatonia, IL 61063

Falabella Miniature Horse Assn. of
America
P.O. Box 3036
125 Glenwood Drive
Gettysburg, PA 17325

Florida Cracker Horse Assn.
P.O. Box 186
Newberry, FL 32669-0186

Friesian Horse Assn. of North America
4127 Kentridge Drive SE
Grand Rapids, MI 49508-3705

Galiceno Horse Breeders Assn.
Box 219
Godley, TX 76044-0219

Golden American Saddlebred Horse
Assn.
4237 30th Avenue
Oxford Junction, IA 52323-9724

Haflinger Assn. of America
14570 Gratiot Road
Hemlock, MI 48626-9416

Half Quarter Horse Registry of America
29264 Bouquet Canyon Road
Sangus, CA 91350

Half Saddlebred Registry
319 South Sixth Street
Coshocton, OH 43812-2119

Table A-16 Breed Registry Associations *(continued)*

International Arabian Horse Assn.
   (also includes Half-Arab and Anglo-
   Arabian registries)
P.O. Box 33696
Denver, CO 80233-0696

International Arabian Horse Registry of
   North America
P.O. Box 325
Delphi Falls, NY 13501-0325

International Buckskin Horse Assn.
P.O. Box 268
Shelby, IN 46377-0268

International Colored Appaloosa Assn.
P.O. Box 4424
Springfield, MO 65808-4424

International Morab Breeders Assn.
   (up to 75% Arabian or Morgan)
South 101 West 34628 Highway 99
Eagle, WI 53119-1857

International Plantation Walking Horse Assn.
P.O. Box 510
Haymarket, VA 22069-0510

International Sporthorse Registry and
   Oldenburg Verband N.A.
P.O. Box 849
Streamwood, IL 60107

International Trotting and Pacing Assn.
575 Broadway
Hanover, PA 17331-2007

Jockey Club, The
821 Corporate Drive
Lexington, KY 40503-2794

Lippizan Assn. of North America
P.O. Box 1133
Anderson, IN 46015-1133

Missouri Fox Trotting Horse Breed Assn.
P.O. Box 1027
Ava, MO 65608-1027

Mountain Pleasure Horse Assn.
P.O. Box 670
Paris, KY 40362-0670

National Pinto Arabian Registry
942 Kathryn Lane
Royse, TX 75189

National Pinto Horse Registry
P.O. Box 486
Oxford, NY 12820-0486

National Spotted Saddle Horse Assn.
P.O. Box 898
Murfreesboro, TN 37133-0898

New Forest Pony Assn.
P.O. Box 206
Pascoag, RI 02859

North American District of the Belgian
   Warmblood Breeding Assn.
General Hunton Road
Broad Run, VA 22014-4877

North American Draft Cross Assn.
742 Rebecca Avenue
Westerville, OH 43081

North American Exmoors
RR 4 Box 273
Amherst, Nova Scotia, B4H 3Y2
CANADA

North American Morab Horse Assn.
W. 3174 Faro Springs Road
Hilbert, WI 54129

North American Mustang Assn. and Registry
P.O. Box 850906
Mesquite, TX 75185-0906

Table A-16   Breed Registry Associations *(continued)*

North American Selle Francais Horse Assn.
P.O. Box 646
Winchester, VA 22604-0646

North America Shagya (Arabian) Society
2520 60th Ave., SW
Rochester, MN 55902

North American Single-Footing Horse Assn.
P.O. Box 1079
Three Forks, MT 59752-1079

North American Trakehner Assn.
1660 Collier Road
Akron, OH 44320

Norwegian Fjord Assn. of North America
24570 W Chardon Road
Grayslake, IL 60030

Palomino Horse Assn.
HC63, Box 24
Dornsife, PA 17623

Palomino Horse Breeders of America
15253 E. Skelly Drive
Tulsa, OK 74116-2637

Palomino Ponies of America
160 Warbasse Junction Road
Lafayette, NJ 07848-9408

Paso Fino Horse Assn.
P.O. Box 600
Bowling Green, FL 33834-0600

Percheron Horse Assn. of America
P.O. Box 141
Fredericktown, OH 43019-0141

Performance Horse Registry
P.O. Box 24710
Lexington, KY 40524-4710

Peruvain Part Blood Registry
2027 Cribbens Street
Boise, ID 83704

Peruvian Paso Horse Registry of North
    America
    (and Peruvian Paso Part-Blood Registry)
#4 1038 4th Street
Santa Rosa, CA 95404-4319

Pintabian Horse Registry
P.O. Box A
Karlstad, MN 56732

Pinto Horse Assn. of America
1900 Samuels Avenue
Fort Worth, TX 76102-1141

Pony of the Americas Club
5240 Elmwood Avenue
Indianapolis, IN 46203-5990

Quarter Sport Horse Registry
1463 Country Lane
Bellingham, WA 98225-8515

Racking Horse Breeders Assn. of America
Rt. 2, Box 72-A
Decatur, AL 35603

Ridden Standardbred Assn.
1578 Fleet Road
Troy, OH 49373

Rocky Mountain Horse Assn.
1140 McCalls Mill Road
Lexington, KY 40515

Royal Warmblood Studbook of the
    Netherlands
North American Department
P.O. Box 828
Winchester, OR 97495-0828

Spanish-Barb Breeders Assn. International
12284 Springridge Road
Terry, MS 39170

Spanish Mustang Registry
Rt. 3, Box 7670
Wilcox, AZ 85643

Table A-16   Breed Registry Associations *(concluded)*

Spanish-Norman Registry, Inc.
P.O. Box 985
Woodbury, CT 06798

Standardbred Pleasure Horse
   Organization
31930 Lambson Forest Road
Galena, MD 21653

Swedish Gotland Breeders' Society
Rt. 3, Box 134
Corinth, KY 41010-9010

Swedish Warmblood Assn. of North America
P.O. Box 1587
Coupeville, WA 98239-1587

Tennessee Walking Horse Breeders' and
   Exhibitors' Assn.
P.O. Box 286
Lewisburg, TN 37091-0286

Thoroughbred Horses for Sport
P.O. Box 160
Great Falls, VA 22066

Thoroughbred in Sport Assn.
964 Gale Drive
Wisconsin Dells, WI 53965

United Quarab Registry
31100 NE Fernwood Road
Newbury, OR 97132-7012

United States Icelandic Horse Federation
38 Park Street
Montclair, NY 07042

United States Lippizan Registry
13351 DP Chula Road
Amelia, VA 23002

United States Trotting Assn.
750 Michigan Avenue
Columbus, OH 43215-1191

Universal Perkehner Society
P.O. Box 1874
Cave Creek, AZ 85311-1874

Walkaloosa Horse Assn.
3815 North Campbell Road
Otis Orchards, WA 99027

Walking Horse Owners' Assn. of America
#3A 1535 West Northfield Blvd.
Murfreesboro, TN 37129

Welsh Pony and Cob Society of America
P.O. Box 2977
Winchester, VA 22604-2977

Westfalen Warmblood Assn. of America
18432 Biladeau Lane
Penn Valley, CA 95946

Table A-17   National and International Horse Organizations

American Driving Society
P.O. Box 160
Metamora, MI 48455-0160

American Granprix Assn.
3104 Cherry Palm Drive
Suite 220
Tampa, FL 33619

American Horse Shows Assn., Inc.
220 E. 42nd Street, #409
New York, NY 10017-5876

American Hunter and Jumper Foundation
340 E. Hillendake Road
Kennett Square, PA 19348

American Paint Horse Assn.
P.O. Box 961023
Fort Worth, TX 76161-0023

American Polocrosse Assn.
250 Morning Glory Lane
Durango, CO 81301

American Quarter Horse Assn.
P.O. Box 200
Amarillo, TX 79168-0001

American Royal Assn.
1701 American Royal Court
Kansas City, MO 64102

American Saddlebred Grand National
4093 Iron Works Pike
Lexington, KY 40511-8434

American Team Penning Assn.
1776 Montano Road, NW, Building #3
Albuquerque, NM 87107

American Vaulting Assn.
642 Alford Place
Bainbridge Island, WA 98110-4608

American Warmblood Society
6801 W. Romley Avenue
Phoenix, AZ 85043

Appaloosa Horse Club of Canada
Box 940
Claresholm, AB T0L 0T0
CANADA

Appaloosa Horse Club, Inc.
P.O. Box 8403
Moscow, ID 83843-0903

Arabian Breeders' Marketing Network
    Augusta Futurity
Atlantic Coast Cutting Horse Association
P.O. Box 936
Augusta, GA 30903-0936

Azteca Association of Canada
R.R. 2
Paris, ON N3L 3E2
CANADA

Barrel Futurities of America, Inc.
4701 Parsons Road
Springdale, AR 72764

Canadian Belgian Horse Assn.
R.R. 3
Schomberg, ON L0G 1T0
CANADA

Canadian Buckskin Assn.
P.O. Box 135
Okotoks, AB T0L 1T0
CANADA

Canadian Cutting Horse Assn.
234 17th Ave., NE
Calgary, Alberta T2E 1L8
CANADA

Canadian Dressage Owners & Riders
    Association
R.R. 2
Millbrook, ON L0A 1G0
CANADA

Table A-17 National and International Horse Organizations *(continued)*

Canadian Driving Society
40774 Taylor Road
De Roche BC V0M 1G0
CANADA

Canadian Fjord Horse Assn.
Box 411
Dauphin, MB R7N 2V2
CANADA

Canadian Hackney Society
R.R. 1
Linsay, ON K9V 4R1
CANADA

Canadian Haflinger Assn.
General Delivery
Shagulandah, ON P0P 1W0
CANADA

Canadian Horse, The
Upper Canada District
1289 Pilon Road
Clarence Creek, ON K0A 1N0
CANADA

Canadian Icelandic Horse Federation
R.R. 1
Vernon, BC V1T 6L4
CANADA

Canadian Percheron Assn.
Box 200
Crossfield, AB T0M 0S0
CANADA

Canadian Shire Horse Assn.
1882 Conc. Road 10
Blackstock, ON L0B 1B0
CANADA

Carriage Assn. of America, Inc., The
RD 1, Box 115
Salem, NJ 08079

Clydesdale Horse Assn. of Canada
R.R. 2
Thomton, ON L0L 2N0
CANADA

Del Mar National Horse Show
2260 Jimmy Durante Blvd.
Del Mar, CA 92014

Gladstone Equestrian Assn.
P.O. Box 119
Gladstone, NJ 07934

Golden American Saddlebred Horse Assn.
4237 30th Ave.
Oxford Junction, IA 52323-9724

Hanoverian Horse Society
R.R. 2
Elora, ON N0B 1S0
CANADA

Intercollegiate Horse Show Assn.
P.O. Box 741
Stoney Brook, NY 11790-0741

International Arabian Horse Assn.
P.O. Box 33696
Denver, CO 80233-0696

International Buckskin Horse Assn.
P.O. Box 268
Shelby, IN 46377-0268

International Halter-Pleasure Horse Assn.
256 N. Highway 377
Pilot Point, TX 76258-9624

International Hunter Futurity
P.O. Box 13244
Lexington, KY 40583-3244

International Jumper Futurity
P.O. Box 2830
Roseville, CA, 95746-2830

Table A-17 National and International Horse Organizations *(continued)*

International Side-Saddle Org., The
P.O. Box 282
Albany Bay, NH 03810-0282

Japan Racing Assn.
New York Office
399 Park Avenue
27th Floor
New York, NY 10022

Malayan Racing Assn.
Paddock Block
Bukit Timah Racecourse
Singapore 1128

Masters of Foxhounds Assn. of America
Route 3, Box 51
Morven Park
Leesburg, VA 22075

National Barrel Horse Assn.
725 Broad Street
Augusta, GA 30901-1305

National Cutting Horse Association
4704 Hwy. 377 S.
Fort Worth, TX 76116-8805

National Grand Prix League
2508 Keller Pkwy.
St. Paul, MN 55109

National Horse Show Commission
Route 1, Box 257
Graham, AL 36263-9519

National Hunter and Jumper Assn., The
P.O. Box 1015
Riverside, CT 06878-1015

National Reining Horse Assn.
448 Main Street, #204
Coshocton, OH 43812-1200

National Snaffle Bit Assn.
1 Indiana Square, #2540
Indianapolis, IN 46204

North American Riding for the Handicapped
 Assn.
P.O. Box 33150
Denver, CO 80233

Palomino Horse Assn.
Box 24, Star Route
Dornsife, PA 17823

Palomino Horse Breeders of America
15253 E. Skelly Drive
Tulsa, OK 74116-2637

Professional Horsemen's Assn. of America
20 Blue Ridge Lane
Wilton, CT 06897-4127

Pyramid Society, The
P.O. Box 11941
Lexington, KY 40579

Ride and Tie Assn., The
1865 Indian Valley Road
Novato, CA 94947

Societe Des Eleveurs De Checaux Canadiens
68 rue Deslauriers
Pierrefonds, PQ H8Y 2E4
CANADA

Special Olympics International
1350 New York Ave., NW, #500
Washington, DC 20005-4709

Tennessee Walking Horse National
 Celebration
P.O. Box 1010
Shelbyville, TN 37160-1010

United Professional Horsemens Assn.
4059 Iron Works Pike
Lexington, KY 40511-8434

United States Combined Training Assn.
P.O. Box 2247
Leesburg, VA 22075-2247

Table A-17 National and International Horse Organizations *(concluded)*

United States Dressage Federation P.O. Box 6669 Lincoln, NE 68506-0669	United States Team Penning Assn. P.O. Box 161848 Fort Worth, TX 76161-1848
United States Equestrian Team Pottersville Road Gladstone, NJ 07934	United States Team Roping Championships P.O. Box 7651 Albuquerque, NM 87194
United States Olympic Committee One Olympic Plaza Colorado Springs, CO 80909	United States Vaulting Federation RD 1, Box 235 Pittsown, NJ 08867-9722
United States Polo Association 4059 Iron Works Pike Lexington, KY 40511-8434	

GLOSSARY

Like a foreign language, terms unique to equine science can be baffling to the newcomer. When individuals travel to a foreign country and want to do business, they are expected to know the language of the country. The same is true for the individual wanting to learn about horses. Indeed, the term glossary *means obscure or foreign words of a field. Successful individuals use the glossary and learn the language. Words not found in the glossary may be listed in the index or defined within a chapter of the book.*

Abdomen—The part of the body between the thorax and the pelvis (the belly region).

Abdominal worm—Nematode parasite that lives in the abdominal cavity.

Abductors—Muscles that move a limb away from the center plane of the horse.

Abortion—Premature termination of a pregnancy.

Abscess—A localized collection of pus in a cavity formed by disintegration of tissues.

Absorption—To take in by various means.

Accounting—A system of recording, classifying, and summarizing commercial transactions in terms of money.

Accrual accounting—An accounting system in which expenses are considered expenses when they are committed and income is counted as income when it is earned. This includes changes in inventory.

Acid solution—A solution with pH less than 7 (for example, a mixture of equal parts of vinegar and water).

Actin—A protein present in muscle tissue. Acting with myosin, actin produces a muscle contraction.

Active immunity—A long-lasting immunity that is achieved when an animal is challenged and stimulated to produce its own antibodies.

Acute—Refers to a disease that runs a short, severe course.

Additive gene—The members of a gene pair that have equal ability to be expressed.

Adductors—Muscles that pull a limb toward the center plane on the horse.

Adenosine triphosphate (ATP)—The universal energy-transfer molecule.

Adhesion—The abnormal union of surfaces normally separated by the formation of new fibrous tissue resulting from inflammation.

Adipose—Fat tissue.

Adrenal cortex—Outer portion of the adrenal gland producing corticosteroids.

Adrenal medulla—Center portion of the adrenal gland producing epinephrine and norepinephrine.

Aerobic—Occurring only in the presence with oxygen.

Afferent—Nerves that carry impulses towards the central nervous system.

Agglutination—A clumping together of living cells caused by an antibody.

Aglactic mare—A mare not producing milk in adequate quantities.

Agonistic behavior—Combative behavior.

Aids—The means by which a rider communicates with a horse (for example, hands, legs, voice, and seat).

Air requirements—Refers to the ventilation necessary for the size and number of animals in a building.

Albino—A horse with the dominant allele W, which lacks pigment in skin and hair at birth. The skin is pink, the eyes brown (sometimes blue), and the hair white. Such a horse is termed white.

Alkaline solution—A solution with pH greater than 7 (for example, a spoonful of baking soda in a pint of warm water).

Allantois—Embryonic/fetal membrane.

Allele—The alternative form of a gene having the same place in a homologous chromosome. Or genes on the same location of a pair of homologous chromosomes.

Allergy—Heightened sensitivity to a particular substance that does not affect the majority of the group.

Alveoli—The sac in the lung where the exchange of oxygen and carbon dioxide occurs.

Amble—This is a lateral gait distinguished from the pace by being slower and more broken in cadence. It is not a show gait.

Amino acid—The building blocks which make up the body's protein.

Amniotic fluid—Fluid contained within the innermost of the fetal membranes just outside the fetus.

Anaerobic—Occurring without oxygen.

Anaphylactic shock—An extreme antigen-antibody reaction.

Anatomical—Refers to the structural parts of the body and the relation of its parts.

Androgens—Hormones that maintain and control masculine characteristics.

Anemia—A condition in which blood is deficient in red blood cells.

Anestrus—Occurs usually in the winter, and is the time when a mare does not cycle or have a heat period.

Aneurysm—When a blood vessel is dilated then fills with blood.

Angle of incidence—Refers to the angle at which the incisor teeth meet.

Angulation—The amount of angle.

Ankle—The joint connecting foot and the leg.

Annuals—Plants that complete their life cycle from seed in one growing season.

Anoplocephaliasis—A disease of yearlings at pasture caused by tapeworms.

Anterior—Forward (in space) or toward the head.

Anthelmintics—Drugs used to treat worms in horses.

Antibiotic—Substance with the capacity to inhibit the growth of or kill microorganisms.

Antibodies—Large protein molecules that destroy bacteria, yeast, some viruses, and toxins.

Antigens—Substances that, when introduced into an organism, induce an immune response consisting of the production of a circulating antibody.

Anti-inflammatory—Drugs that can be used to lessen pain and decrease inflammation.

Antiseptic—An agent used in the treatment of wounds or disease to prevent the growth and development of germs.

Anus—The exterior posterior opening of the digestive tract.

Anvil—A heavy block of iron or steel on which metal may be forged.

Aorta—The main vessel that carries blood to all bodily organs except the lungs.

Apprenticeship—A job that involves working under the supervision of a professional for a variable amount of time. This type of job may or may not include a salary.

Archeohippus—One of the early ancestors of the modern horse.

Artery—A vessel that carries blood from the heart.

Arthritic—Inflammation of a joint.

Articulation—Where the joints come together.

Artificial insemination—Introducing sperm cells into the female reproductive tract by means other than natural service.

Ascarids—Part of the large groups of parasites known as roundworms.

As-fed basis—Indicates that the amount of nutrients in a feed or diet is expressed in the form in which it is fed.

Assets—The property or resources owned and controlled by a business.

Assimilation—The transforming of digested foods into an integral and homogenous part of the solids or fluids of the organism.

Atrophy—A wasting away or shrinking of muscle.

Atropine—An alkaloid compound used as an antispasmodic.

Auditory—The sense of sound.

Autonomic nervous system—The system that is concerned with control over the digestive system, eyes, blood vessels, glandular products, and other automatic functions.

Axon—A nerve cell that conducts impulses away from the cell body.

Azoturia—A disease common to draft horses characterized by the passing of red or brown urine.

Babesia—Small protozoan parasites that occur in red blood cells. Ticks are the intermediate hosts.

Baby teeth—Temporary teeth.

Back—Trotting in reverse.

Bacterial spore—A microscopic form of a bacterium that is very resistant to damage.

Bad mouth—A malocclusion where the top and bottom teeth do not meet.

Bag up—A term used to describe the development of the mammary glands near the time of parturition.

Balance—The ability of a horse to coordinate action, go composed, and in form.

Balance sheet—Statement of the assets owned and liabilities owed in dollars. It shows equity or net worth at a specific point in time.

Bald face—A wide white marking that extends beyond both eyes and nostrils.

Banding—A style of manes seen in western show horses. Manes are sectioned and fastened with rubber bands.

Bang's—Another name for Brucellosis.

Barn-sour—A horse that will run back to the barn.

Barren mares—Mares that have not had foals.

Bars—(1) The structure that keeps the hoof wall from overexpanding. It is a support structure that angles forward from the hoof wall. (2) The gap between a horse's incisors and molars. (3) Side points on the tree of a saddle.

Basal metabolism—Minimal energy requirements to maintain vital body processes.

Base narrow toe-in—Narrow at the feet.

Base wide toe-out—Wide at the feet.

Beat—Refers to the time when a foot—or two feet simultaneously—strike the ground. Beats may or may not be evenly spaced in time.

Bedding—A cushioning material for an animal.

Bench knees—Lateral deviation of the cannon bone.

Benzimidazole—A classification of antiparasitic drug.

Birth date of a foal—For racing or showing events, the foal's birthday is considered as January 1, regardless of the actual month it was born.

Bishoping—Artificial altering of the teeth of an older horse to make it sell as a young horse.

Bit—The part of the bridle that is put in the horse's mouth and used to control the animal.

Biting—Horses have several reasons for biting; for example, when too much pressure is applied in grooming or during cinching the saddle girth. They may also bite in self defense.

Blacksmith—A person who trims and put shoes on horses' feet.

Blastula—A hollow ball of cells, one of the early stages in embryological development.

Blaze—A type of coloring on the face of the horse.

Blemish—A blemish differs from an unsoundness in that it is unattractive, but does not and is not apt to interfere with the horse's performance. It is usually an acquired physical problem that may not make the horse lame but may interfere with the action of the horse. A blemish does not have to be an unsoundness.

Blindness—This is characterized by cloudiness of the cornea or complete change of color to white. Pale blue, watery eyes may indicate periodic ophthalmia (moon blindness).

Blistering—Application of an irritating substance as treatment for a blemish or unsoundness. Blistering increases the blood supply to the site of the blister and induces more rapid healing.

Bloom—A shiny coat for show horses.

Boarding contract—Agreement between the owner of a horse and the owner of the stable.

Bobtailed hackney—A large horse used to pull carriages.

Bog spavin—A soft fluctuating enlargement located at the upper part of the hock and due to a distention of the joint capsule.

Bolt—To eat rapidly; to startle.

Bone spavin (jack spavin)—A bony enlargement at the base and inside back border of the hock. It may fuse bones and render joints inarticulate.

Boot—A device that can be applied to the foot to prevent it from injuring the elbow.

Bots—The larvae of an insect, the bot fly.

Bowed tendon—A serious discrimination involving any or all of a group of tendons and ligaments, but usually the superflexor tendon, the deep flexor tendon, and the suspensory ligament. It is caused by severe strain and wear and shows up as a thickened enlargement of the tendon that occupies the posterior space in the cannon region between the knee and ankle or between the hock and ankle.

Bow-line knot—A type of knot that will not slip.

Braiding—A style of manes seen in hunters and jumpers. Manes are sectioned and braided into small, neat braids.

Braid puller—A piece of baling wire bent to form a long, narrow loop.

Break-even analysis—Determining where income is equal to the total of the fixed costs and variable costs of doing business.

Breast collar—A collar sometimes used to keep the saddle in place.

Breeching—The part of a harness that passes around the rump of a draft horse.

Breed—(1) To produce young; a particular sort or kind of animal. (2) A group of horses selected for their common ancestry and common characteristics.

Breeding true—This means the offspring will almost always possess the same characteristics as the parents.

Breed registry—An organization that tracks horses breeding true or with a common ancestry.

Bridle path—A trail or path designated for use by horses and riders. Also, a space clipped in the mane just behind the ears for the crownpiece of a bridle or halter. The bridle path should be two-to-eight inches long, depending on the breed of the horse.

Bronchi—The two main branches of the trachea going to the lungs.

Bucked shins—Temporary unsoundness characterized by inflammation of the bone covering along the front surface of the cannon bone.

Buck-kneed—Standing with knees too far forward.

Bulk—Excessive amounts of fiber or water.

Calcification—Replacement of the original hard parts of an animal by calcium carbonate.

Calf-kneed—When the knees tend to bow inward.

Calippus—One of the early ancestors of the modern horse.

Calks—Grips on the heels and the outside of the front shoes of horses, designed to give the horse better footing and prevent slipping.

Camped out—A condition where the leg is too far back and behind the plumb line. Usually the whole leg is involved and the plumb line is at or in front of the toe instead of behind the heel.

Camped under—This is the opposite of camped out.

Canines—The pointed teeth beside the in-cisors.

Cannon bone—The bone extending between the knee or hock and the fetlock joint in horses.

Canter—A slow, restrained, three-beat gait in which the two diagonal legs are paired, thereby producing a single beat that falls between the successive beats of the other unpaired legs.

Capillaries—The smallest blood vessel that connect the arteries and veins.

Capillary refill—The number of seconds it takes for the color to return to an area of the gum of the horse that has been pressed with the thumb once the thumb is removed. One to two seconds is normal.

Capital—The amount of money that can be obtained through borrowing or selling assets that is used to promote the production of other goods.

Capped elbow—A blemish at the point of the elbow, also called shoe boil. It is caused by injury from the shoe when the front leg is folded under the body while the horse is lying down.

Capped hocks—An enlargement at the point of the hock; it is usually caused by bruising.

Carbohydrates—Any of a group of neutral compounds composed of carbon, hydrogen, and oxygen including the sugars and starches. They are used immediately for growth or stored for future use.

Carnivores—Animals feeding or preying on animals, eating only animal food.

Carotene—A compound from which vitamin A is synthesized.

Carriage traces—Straps, chains, or ropes of a harness, extending from the collar (specifically the hames) to the vehicle or load.

Cartilage—A translucent elastic tissue that composes most of the skeleton of embryos and very young vertebrates. Most cartilage is replaced by bone.

Cash-basis accounting—An accounting system in which income is recorded as income when it is received and expenses are recorded as expenses when they are paid.

Cash flow—Actual cash levels for a business.

Caslick—A procedure in which the vulva is sutured to prevent infection.

Catalyzed—When the chemical reaction rate is increased.

Catheter—A slender tube inserted into a body cavity for drawing off or administering fluids.

Caudal—Posterior.

Cecum—The blind pouch that forms the beginning of the large intestine.

Cell—The smallest unit of life.

Centaur—Mythical Greek race that were imagined to be men with the bodies of horses.

Center of gravity—The centered mass of the horse. The center is most commonly located in the middle of the rib cage just caudal to the line separating the cranial and middle thirds of the body. The forelimbs bear 60 to 65 percent of the body's weight, because the center of gravity is located more cranially.

Centers—Another name for centrals.

Centrals—First incisor teeth.

Centrioles—Two cylindrical bodies, located near the nucleus that play a part in cell division.

Cerebellum—The part of the brain responsible for control of voluntary muscular movement.

Cerebrum—The part of the brain, anterior to the brain stem, responsible for memory, intelligence, and emotional responses.

Cervix—The outer end of the uterus.

Chestnuts—Horny, irregular growths on the inside of the horse's legs. On the front legs, they are just above the knee. On the rear legs, they are toward the back of the hock. Chestnuts are like human fingerprints because no two are alike, and they do not change in size or shape throughout the horse's adult life.

Chiggers—The larval stage of harvest mites; they affect horses' feet and muzzles.

Chip fractures—Occur in several different places but are most common at the knee. They are small fractures that break off one of the bones in the knee. They are usually caused by high amounts of concussion and stress on the knee and are seen most frequently in racing horses.

Chromatids—As a result of the syntheses during the interphase stage of mitosis, each chromosome consists of two sister chromosomes, chromatids, that are identical in their structural and genetic organization. They become visible when mitosis begins.

Chromosome—Microscopic structure found in the nucleus of cells that carries the genes.

Chronic—Continuing a long time.

Cinch—The part of a western saddle used to hold it onto the horse under the girth area.

Classes of horses—The classification of horses according to their use.

Clinch cutter—Tool used to remove used horseshoes.

Clinching block—A tool used to remove old horseshoes.

Club foot—In this condition, the foot axis is too straight and the hoof is too upright.

Cob—A close-knit horse, heavily boned, short coupled and muscular, but with quality; not so heavy or coarse as to be a draft animal. A cob is usually small, standing under 15 hands.

Cocked ankles—These may appear in front but are more common in hind legs. Severe strain or usage may result in inflammation or shortening of the tendons and a subsequent forward position of the ankle joints.

Coffin bone—The bone of the foot of a horse, enclosed within the hoof.

Coldblood—A horse of draft horse breeding.

Cold-fitted—Horseshoes that are applied to the feet without heat.

Colic—A broad term that describes a horse showing abdominal pain. There are a number of causes but generally indicates pain in the digestive tract.

Collar—Part of a harness of a draft horse fitted over the shoulders that helps to take the strain when a load is pulled.

Collateral—Property, savings, stocks, and so forth deposited as security additional to one's personal or contractual obligations.

Collected—Term used to describe a horse that has full control over its legs at all gaits and is responsive to the cues of its rider.

Color breed—Color breeds do not breed true colors, for example, albinos, paints, Appaloosas, buckskins, white, creme, or spotted.

Colostrum—The first milk secreted by a mare just before and the first day after foaling. It is high in antibodies that protect the newborn against infectious diseases.

Colt—A male horse up to three years of age.

Combined immunodeficiency disease—A condition in which an animal is deficient in cells of the immune system.

Commissures—A band of nerve fiber connecting the two halves of the brain or spinal cord or paired ganglia.

Competencies—Abilities or capabilities of employees.

Compost—Piling organic matter in a way that encourages decay and decomposition.

Concentrates—Classification of feedstuffs that are high in energy and low in crude fiber.

Conception—The act of becoming pregnant.

Conditioned response—Response to a stimulus that is learned.

Condition score—A subjective score given to a horse based on its overall body fat.

Confinement—Refers to a situation in which a horse lives in an enclosed stall, versus free in a paddock or pasture.

Conformation—How the horse is shaped according to type and/or breed.

Congenital—Condition that exists at birth; acquired during the prenatal period.

Conquistadors—Early Spanish explorers and warriors.

Conservative treatment—A more reserved treatment used for bone fractures in horses. The treatment consists of stall confining, hand walking, and anti-inflammatory drugs to reduce pain and swelling.

Constrict—To draw together or render narrower.

Contouring—Being able to shape something to fit the contours of something else.

Contracted feet—This is caused by continued improper shoeing, prolonged lameness, or excessive dryness. The heels lose their ability to contract and expand when the horse is in motion.

Contracted heels—An abnormal contraction of the hoof wall at the heels.

Contraction—A complex interaction of many parts of the nervous system and the muscular system. It is controlled by the nervous impulses received by the muscle cells.

Cooler—A large square of wool or acrylic material used to cover a horse from head to tail. A cooler is useful for cooling out a horse.

Cornea—The transparent part of the coat of the eyeball that covers the iris and pupil.

Corners—The third incisor.

Corns—Reddish spots in the horny sole, usually on the inside of the front feet, near the bars. Advanced cases may ulcerate and cause severe lameness.

Coronary band—Area where the hoof meets the leg; it produces the hoof wall.

Coronary vessels—The blood vessels that provide nourishment to and encircle the heart like a crown at the juncture of the atria and the ventricles and send branches to both structures.

Coronet—The dividing line between the hoof and the leg of a horse.

Corporations—A body of people recognized by law as an individual person having a name, rights, privileges, and liabilities distinct from the individual members.

Corpus hemorrhagicum—The bloody spot on the ovary immediately after ovulation. It becomes a corpus luteum.

Corpus luteum—The gland formed on the ovary following ovulation. It produces the hormone progesterone.

Corticosteroids—Hormones from the adrenal glands.

Cowlicks—Permanent hair whorls that cannot be brushed or clipped out. They are located mainly on the forehead and neck.

Cracks (quarter, toe, heel)—Associated with a hoof wall that is too long and has not been trimmed frequently enough. They can also de-velop with horses that are in rain and mud for long periods of time. The mud draws water out of the hoof wall and when the hoof dries it often cracks.

Cradle—A device useful in preventing an animal from licking or biting an injured area.

Cranial—The part of the skull that encloses the brain.

Creative thinking—Ability to generate new ideas by making nonlinear or unusual connections or by changing or reshaping goals to imagine new possibilities; using imagination freely, combining ideas and information in new ways.

Creep feeding—Feed supplied to the foal in an area unavailable to adult horses.

Crescent—Shaped like the moon in its first quarter.

Crestfallen—A heavy neck which breaks over and falls to one side.

Cribbing—When horses grasp an object (feed box edge or manger) between their teeth and apply pressure, gradually gnawing the object away if it is not metal.

Cross-firing—The same as forging in a pacer in which the inside of the near fore and hindleg (or the reverse) strike in the air as the stride of the hindleg is about completed and the stride of the foreleg is just beginning.

Cross tying—A method of using two ropes to secure a horse so the head is level.

Croup—The rump of a horse.

Crude protein (CP)—Total amount of protein in a feed.

Crupper strap—A leather strap with a padded semicircular loop. The loop end goes under the tail and the strap end is affixed at the center of the back band of a harness or the cantle of a saddle to prevent the saddle from slipping over the withers.

Cryptorchid—A horse in which one or both testes are maintained in the body cavity; it is sterile if both tests are undescended, but fertile in a suspended testis.

Cue—The stimulus used to train horses.

Cultural diversity—Term used to describe the American workplace representing people from different backgrounds.

Cups—The deep indentures in the center of the surfaces in young permanent teeth.

Curb—An enlargement on the back of the leg, just below the hock. It is caused by trauma to the plantar ligament that causes the ligament to become inflamed and then thickened.

Curb bit—A bit that works with leverage action on a horse's mouth. A curb bit must have shanks and a curb strap or chain.

Curb chain—Used in combination with a curb bit. It acts against the chin groove to produce a painful pressure.

Cut-back—A saddle cut back at the withers.

Cytoplasm—The material that lies within the cell and contains several organelles and granules in suspension.

Dam—The female parent of a horse.

Dam's produce—Offspring from a particular dam.

Data sheet—Similar to a resume; contains pertinent information about a potential employee.

Deciduous—Temporary teeth.

Deep flexor tendon—The inner part of the leg responsible for extension of the foot as it progresses through a stride.

Defecation—Release of feces from the bowel.

Degree of finesse—Determined with gaited and parade horses by how well they "move" off their hocks.

Dehydration—An abnormal depletion of body fluids.

Demographic—Having to do with vital and social statistics.

Dental stars—Marks on the incisor teeth appearing first as narrow, yellow lines in front of the central enamel ring, then as dark circles near the center of the tooth in advanced age.

Dentition—The process of cutting teeth.

Deoxyribonucleic acid (DNA)—A nuclein acid of complex molecular structure that is the major component of genes; it plays an important role in the gene action of chromosomes.

Depressed—A state when the horse is extremely relaxed and muscles are flaccid.

Dermatitis—Inflammation of the skin.

Detoxification—The act of removing poison or of the effect of poison.

Diaphragm—A body partition of muscle and connective tissue. Separates the thoracic cavity from the abdominal cavity.

Diestrus—A period of sexual inactivity between two estrus cycles.

Diffusion—The spreading out of molecules in a given space.

Digesta—The food, fluids, and nutrients moving through the digestive system.

Digestible energy (DE)—That portion of the gross energy in a feed that is not excreted in the feces.

Digestion—The breakdown of foods in the digestive tract to simple substances that may be absorbed by the body.

Digestive disturbances—Any abnormal digestive activity that causes discomfort in the animal.

Dilation—Expansion of an opening, becoming wider or larger.

Diluters—Any of several types of fluids used to dilute and increase the volume of semen.

Directness—Also known as trueness is the line in which the foot is carried forward during the stride.

Discrimination—A prejudice or partiality of unsoundness.

Disease—Any condition of a horse that impairs normal physiological functions.

Disinfectant—A chemical that destroys harmful microorganisms.

Distal spots—Dark spots on a white coronet band.

Distended—Enlarged or swollen.

Domesticated—Tamed or gentled for use by man.

Dominant gene—A gene that is expressed.

Dorsal—Back.

Draft horse—A large breed of horses used for work.

Dressage—A method in which, through body movements, and without using hands, feet, or legs, a rider can guide a trained horse through natural maneuvers.

Driving—Horses harnessed and controlled from behind.

Dry-matter basis—Method of expressing the concentration of a nutrient. Dry matter indicates the part of the feed that is not water.

Dun—Body color yellowish or gold with a mane and tail that may be black, brown, red, yellow, white, or mixed. Usually dun has a dorsal stripe, zebra stripes on the legs, and transverse stripes over the withers.

Duodenum—The first part of the small intestine.

Dwelling—This is a perceptible pause in the flight of the foot as though the stride had been completed before the foot strikes the ground. It may occur either front or rear and is particularly common in heavy harness horses, heavy show ponies, and some saddlers.

Dystocia—A retained placenta in the mare.

Easy keepers—Horses that are easy to feed.

Ectoparasites—Parasites that live on the outside of their hosts.

Edema—Excessive accumulation of fluid in tissue spaces.

Efferent—Nerves that carry impulses away from the central nervous system.

Ejaculation—When the semen is expelled through the urethra.

Electrolytes—Any molecular substance that, in solution, will dissolve into its electronically charged components called ions.

Eliminative behavior—Behavior demonstrated by horses during defecation or urination.

Embryo—The earliest stages in the development of an organism before it has assumed its distinctive form.

Embryo transfer (ET)—Removal of developing embryos from one mare and their transfer to the uterus of another.

Endurance fibers—These are type IIa muscle fibers used during periods of aerobic work such as jogging or long-distance riding. These fibers can use carbohydrates, fat, or protein for energy.

English—A type of riding dictated by the saddle, tack, clothing, and riding method used.

Enterprise—A specific process or activity that requires a certain amount of risk to make a profit.

Enterprise budget—A look at the costs and risks involved with producing one commodity or making one product.

Entrepreneur—One who starts and conducts a business assuming full control and risk.

Enzyme—A substance that increases the rate of a chemical reaction.

Eocene epoch—A division of geologic time following the Paleocene and ending about thirty-seven million years ago.

Eohippus—The earliest known ancestor of the modern horse.

Epiphysitis—Swelling or inflammation around the growing points of the long bones.

Epithelial—The tissues that form one or more layers of cells that cover most internal and external surfaces of the body.

Epona—An ancient Gaul goddess of horses who lovingly protected the horse and stable and also kept watch over the grooms and carters.

Equine encephalomyelitis—An inflammatory disease of the brain and spinal cord.

Equitate—The act or art of riding horseback.

Equity—The value remaining in a business in excess of any liability or mortgage.

Esophagus—A muscular tube extending from the pharynx down the left side of the neck and through the thoracic cavity and diaphragm to the stomach.

Estate—One's entire property or possessions.

Estrogen—Steroid hormone produced by the ovary responsible for estrus behavior.

Estrus (Estrous)—The period of sexual excitement (heat) during which the female will accept the male in the act of mating.

Euthanasia—The act of painlessly putting to death animals suffering from incurable conditions or diseases.

Evaluation—Determining worth, performance, value, or conformation; appraisal.

Event—A competition including dressage, stadium jumping, and cross-country jumping.

Evolution—Refers to a process of continuous change from a lower, simpler state to a higher more complex state.

Expiration—Expulsion of air effected by a relaxation of muscles and a contraction of rib and abdominal muscles.

Exploratory behavior—Learning from, investigating, and being attentive to the environment.

Extenders—Any diluter with additives to extend the lifespan of sperm cells.

Extensor—To extend out straight. The extensor tendon is attached to the front of the coffin bone.

External respiration—Consists of two movements—inspiration and expiration.

Extracellular—Outside the cell.

Eye worm—A parasite that lives in the tear duct and conjunctival sac of the horse's eye.

Farrier (horseshoer)—A person who cares for horses' feet, including trimming and nailing on horseshoes.

Fatigue—Exhausted, wearied with labor or exertion.

Fat-soluble vitamin—Vitamins found in the fat portion of the feed and stored in the fatty tissues of the horse.

Feather fetlocks—Long hair growth on the fetlocks of a horse.

Feathering—A fringe of hair around the horse's foot just above the hoof. Some breeds naturally have more feathering or a heavier fetlock than others.

Feral—Horses that were once domesticated and have become wild.

Fermentation—Decomposition of organic substances especially carbohydrates under anaerobic conditions. These conditions are often created by the enzymes produced by microorganisms such as yeast, molds, and bacteria.

Fertilization—When a sperm is fused with an egg.

Fetlock—The joint above the hoof of a horse.

Fetus—The later stage of foal development within the uterus.

Fibrosis—Thickening of affected skin.

Filly—The name for a female horse until the age of three.

Firing—Making a series of skin blisters with a hot needle over an area of lameness.

Firing marks—Where one leg strikes another.

Fistulous withers—An inflammation of the withers affecting this region in much the same way as poll evil affects the poll. Source may be a bruising or bacterial infection.

Fixed costs—Costs that usually do not fluctuate with an increase or decrease in production.

Flanks—The fleshy part of the side between the ribs and the hip.

Flat foot—Conformation that lacks the natural concave curve to the sole. Instead, the sole is flat and predisposed to more contact with the ground. Flat foot increases the chance for sole bruises and resulting lameness.

Flexing at the poll—To give its head to the bit.

Flexion tests—Helps to determine the extent and location of a fracture or other problem of the leg.

Flexor—The tendons that cause the fetlock joint to flex or bend; located behind the cannon bone.

Flight—The horse's primary defense mechanism.

Flighty—A nervous horse.

Floating—Filing off the sharp edges of a horse's teeth.

Flukes—Trematode parasitic worms that are flat and leaf-shaped.

Flushing—The process of removing the embryo from a mare in preparation for embryo transfer.

Foal—A young, unweaned horse of either sex.

Foal heat—The heat that occurs directly after parturition. It is often not fertile.

Follicle—A saclike structure within the ovary that gradually enlarges and, with one burst, releases an egg into the oviduct. Following this rupture, the corpus luteum forms.

Follicle stimulating hormone (FSH)—A hormone secreted by the anterior pituitary and responsible for follicular growth and ovulation.

Follow-up letter—Letter written immediately after a job interview.

Food and Agricultural Organization (FAO)—An agency of the United Nations that conducts research, provides technical assistance, conducts education programs, maintains statistics on world food, and publishes reports with the World Health Organization.

Forage—Feedstuffs from the leaves and stocks of plants.

Foramen magnum—The skull is attached to the first vertebra of the spine and has a large opening, the foramen magnum, through which the spinal cord passes.

Forbs—Any nongrasslike plant that is relatively free of woody tissue that an animal consumes.

Forecasts—Calculations done beforehand.

Forging—Forging is striking the end of the branches or the undersurface of the shoe of the forefoot with the toe of the hind foot. This is the diagonal foot in pacers, and the lateral foot in trotters.

Foundation sires—All registered foals must have their ancestry traced back to the founding stallions.

Foundation stock—Refers to the original animals of the breed.

Founder (laminitis)—An inflammation of the sensitive laminae under the horny wall of the hoof. All feet may be affected, but the front feet are most susceptible.

Fox trot—A slow, short, broken type of trot in which the head usually nods. In executing the fox trot, the horse brings each hind foot to the ground an instant before the diagonal forefoot.

Fracture—A break or crack in a bone.

Free fatty acids—Major components of lipids and fats.

Free radicals—A molecule with one or two unpaired electrons the do not interact with each other; they are often very reactive and unstable.

Freeze brand—An identifying mark made with copper stamps or marking rods that are cooled in liquid nitrogen or dry ice. This is an unalterable system of angular symbols developed by Dr. Keith Farrell, a veterinary medical officer with the USDA.

Frog—A triangular-shaped formation in the sole of the horse's foot. It should be full and elastic and help to bear the weight of the horse.

Fulcrum—The support about which a lever turns.

Full mouthed—Refers to a horse having all of the permanent teeth and cups present.

Gait—Defined as a horse's way of going or the way of moving its legs during progression. The horse is more versatile in selecting gaits than any other four-legged animal and it uses several gaits unique to the species in a distinctive rhythmic movement of the feet and legs. A gait is characterized by distinctive features, regularly executed.

Gaited horses—Horses that perform gaits other than the four natural gaits (walk, trot, canter, and gallop).

Gallop or run—The run, or gallop is a fast, four-beat gait where the feet strike the ground separately, first one hind foot, then the other hind foot, then the front foot on the same side as the first hind foot, then the other front foot, which decides the lead.

Galvayne's groove—A mark on the tooth used to determine the age of the horse.

Gametes—Sex cells.

Ganglia—Secondary nerve centers located chiefly along the spinal cord. They receive and dispatch nerve impulses that do not have to reach the brain (including such stimuli as heat, pain, excessive pressure, and others), but are immediately switched over to motor filaments and cause certain muscles to react instantaneously.

Gaskin—The area of the horse's rear leg just below the thigh and stifle area and above the hock. They are usually heavily muscled.

Gastric digestion—Chemical breakdown of foodstuffs by the stomach.

Gastric juice—The digestive fluid secreted by the glands in the mucous membrane of the stomach.

Gastrulation—This is the beginning process of cell differentiation when the ectoderm, mesoderm, and endoderm are being formed.

Gelding—A castrated male horse.

Genes—The fundamental units of genetics that determine all the hereditary characteristics of animals.

Genetic—The interaction of the genes in producing similarities and differences in individuals related by descent.

Genetic variation—Differences in genetic makeup.

Genome—A complete set of chromosomes.

Genotype—The genetic makeup of an animal.

Gestation—The period during which a female is pregnant.

Get of sire—Offspring from a particular sire.

Gingivitis—Inflammation of the gums.

Glomeruli—Ball-shaped tiny filters, of which each kidney has several million, located in the outer portion or cortex that filter approximately 200 gallons of liquid a day, rejecting blood cells and proteins but permitting fluid salts and other chemicals, including nitrogenous wastes, to pass through.

Goals—The end objectives or terminal points of a business.

Goiter—An enlargement of the thyroid area.

Golgi apparatus—A special type of membrane mixture found near the nucleus. In cells that synthesize and secrete products, the Golgi apparatus is the site of the material that is accumulated.

Grade—An animal that is not registered with a specific breed registry.

Granules—One of the small bodies in the cytoplasm of cells.

Grass founder—Founder caused by lush pasture.

Grease heel—See scratches.

Grooming behavior—Behavior exhibited when horses care for their hair coat.

Grullo—Body color that is smoky or mouse-colored. Each hair is mouse-colored. The mane and tail are black and usually the lower legs are black and often there is a dorsal stripe.

Gum disease—See gingivitis.

Gymkhanas—A meet for various athletic contests or games for horses, or the place where they are held.

Hack—A hack is an enjoyable, good riding or driving horse, sometimes considered a small Thoroughbred in Europe or a Saddlebred in America.

Hackamore—A bitless bridle used in the West for training horses.

Half-stocking—A white marking from the coronet to the middle of the cannon.

Halter pulling—A habit that develops when a horse pulls at whatever it is being tied to.

Halters—Sometimes called head collars. They are used for leading and tying a horse.

Hammer—One of the tools used for horseshoeing.

Hand—The height of a horse. The measurement is taken from the top of the withers to the ground; a hand is four inches.

Hand mating—Breeding that is monitored closely by the handler.

Hard at the heels—When elastic cartilages under the skin that serve as part of the shock-absorbing mechanism ossify they are firm but movable inward and outward by the fingers.

Hard keepers—Horses which need more feed per unit of body weight.

Hard-keeping—A horse that has a very hard time putting or keeping weight on their body.

Hard mouth—Term used when the membrane of the mouth where the bit rests becomes toughened and the nerves deadened because of continued pressure from the bit.

Heart girth or girth—The circumference of the chest just behind the withers and in front of the back.

Heart rate—The number of times the heart beats in a minute.

Heat—Another word for estrus, or when a mare is receptive to a stallion. This time period is also when follicles develop and a mare ovulates.

Heaves—An incurable respiratory disease of horses.

Heaving—Caused by a loss of elasticity in the lungs, resulting from a breakdown in the walls of a portion of the air cells. There is an extra contraction of the flank muscles during expiration. It is often heard.

Heel calk—Grips on the heels of the front shoes of horses, designed to give the horse better footing and prevent slipping.

Herbivores—Animals that subsist primarily on the available vegetation and decayed organic material in the environment.

Herd obedience—The tendency of horses to do what the group does.

Heredity—Passing, or capable of passing, genetically from parents to offspring.

Hernia—A protrusion of the internal organs through an opening in the body wall.

Heterozygous—Different genes for the same trait.

Hierarchy—The dominance hierarchy requires that each horse recognize the other horse and determine through some initial aggressive acts (biting or kicking) and submissive acts (running away) which horse is dominant and which is subordinate.

High ringbone—A bony enlargement on the pastern bones. It occurs at the pastern joint.

Hinny—A cross between a male horse, or stallion, and a female donkey (called a jennet or jenny). A hinny is similar to the mule in appearance but smaller and more horselike, with shorter ears and a longer head.

Hippocamp—One of the early ancestors of the modern horse.

Hippomane—Soft, dark brown body of tissue that may be floating among the membranes of the passed placenta.

Histogenesis—The process of tissue formation.

Hitching post—A rail to which horses are tied.

Hobble—A type of restraint used on horses in which either the front feet or hind feet (at the pastern or fetlock joints) are placed in straps to keep them from kicking or walking or wandering too far.

Hock—The large joint halfway up the hind leg of a horse. Analogous to the heel of a human.

Homeostasis—Maintaining a balance between all the parts.

Homozygous—Possessing identical genes.

Honda—A ring of rope, rawhide, or metal on a lasso through which the loop slides.

Hoofhead—The top of the hoof.

Hoof leveler—The tool used to determine the angle of the hoof wall and if the hoof is level to the ground.

Hoof pick—A tool used for cleaning the sole, frog, and hoof wall.

Hoof testers—A tool that picks up increased sensitivity, commonly over the toe, in the horse's foot.

Hoof wall—A horny substance made up of parallel fibers that should be dense, straight, and free from ridges and cracks.

Hormone—A product of living cells that affects the activity of cells.

Horseshoeing—The process of putting shoes on horses.

Hosts—The animals from which a parasite obtains food. Different parasites require different numbers and types of hosts.

Hot-fitted—Horseshoes which are applied with heat.

Hunters—Horses that are subjectively judged while jumping fences, or horses ridden in fox hunts.

Hybrids—Animals produced from the mating of two different breeds.

Hydrolyzes—Splitting a compound with the introduction of water.

Hydrotherapy—The application of cold water to the affected area, usually hosing the leg.

Hypercalcemia—High levels of calcium in the blood.

Hyperparathyroidism—Overactive parathyroid.

Hypochloremia—Low levels of chloride in the blood.

Hypothalamus—An area in the brain that controls visceral activities, regulates body temperature and many metabolic process, and influences certain emotional states.

Hypothyroid—An underactive thyroid gland.

Hyracotherium—Another name for *Eohippus*.

Idle—Nonworking.

Ileum—The part of the small intestine that connects the jejunum to the cecum.

Immune system—The system in the body that protects and fights diseases.

Immunity—A condition in which an animal is resistant to a disease.

Immunoglobulins—Antibodies that are members of a related group of gamma globulin molecules.

Immunological—Having to do with the immune system or disease resistance.

Impaction—Constipation.

Implants—Devices used to repair a fracture in a long bone.

Imprinting—The imposition of a behavior pattern in a young animal by exposure to stimuli; for example, exposure to humans.

Incisors—Any one of the twelve front teeth.

Income—Amount of money received periodically in return for goods, labor, or services.

Infectious—Diseases caused by pathogenic organisms present in the environment or carried by other animals.

Infective—Capable of producing an infection.

Inflammation—Redness, swelling, pain, heat, and disturbed function of an area of the body.

Influenza—A contagious viral disease characterized by respiratory inflammation, fever, muscle soreness, and often a loss of appetite.

Infundibulum—(1) The end of the oviduct nearest the ovary; (2) The funnel-shaped inside of the tooth.

Inherited—Received genetically from parents.

Ingestive behavior—Behavior exhibited by a horse during feeding.

Inorganic—Something composed of substances of other than plant or animal origin.

Input costs—Money required to begin production.

Inspiration—Inhallation brought about by a contraction of the diaphragm and an outward rotation of the ribs.

Insulin—Lowers the blood glucose and gets glucose across the cell membrane where it can be metabolized.

Interdental space—The space behind the incisors and ahead of the six lower molars in each branch of the mandible.

Interest—Payment for the use of money or credit.

Interfering—Striking the supporting leg, usually at the fetlock, with the foot of the striding leg. Interference commonly occurs between the supporting front leg and a striding front leg or between a supporting hind leg and a striding hind leg.

Intermediate host—Host of a parasite before the last host.

Intermediates—Second incisor teeth.

Internal respiration—The interchange of gases between the blood and the body tissues.

Interphase—A long phase in the process of mitosis.

Interval training—The use of multiple bouts of work interspersed with relief interval when partial recovery is allowed.

Involution—The process during which the uterus returns to normal following parturition.

Ivermectin—A drug used to control parasites.

Jacks—Male donkeys.

Jejunum—The middle part of the small intestine.

Jennet—A female donkey. Also called a jenny.

Jockeys—Professional riders of horses in races.

Joint—A union of two bones.

Joint capsule—The fibrous sac that encloses the entire joint.

Jumpers—Horses that jump at shows and compete for height and time.

Kicking—Horses may kick for a variety of reasons, for example, to get out of a stall, impatience, during feeding time, or from pain.

Knocked-down hip—A fracture of the external angle of the hip bone (ilium) resulting in a lowering of the point of the hip.

Laceration—A cut with jagged edges.

Lactation—Producing milk.

Lactic acid—The chemical produced in the body when glucose or glycogen are used for energy in the absence of oxygen.

Lactose—The sugar found in milk.

Laminae—The flat tissue in the sole or base of the hoof.

Laminitis—An inflammation of the sensitive laminae under the horny wall of the hoof. All feet may be affected, but the front feet are most susceptible. It is characterized by ridges running around the hoof. (Also known as founder.)

Large colon—The part of the digestive tract (large intestine) that is enlarged in the horse to allow time for the digestion of cellulose.

Large intestine—Includes the cecum, large colon, small colon, rectum, and anus.

Larvae—Newly hatched, wormlike stage in the life cycle of an insect.

Larynx—The area of the respiratory tract between the pharynx and the trachea.

Lateral cartilages—Elastic cartilages just under the skin and extending above the hoof on each side of the heel that serve as part of the shock-absorbing mechanism.

Laterally—Toward the side.

Laxative—Medicine administered to the horse to produce evacuation of the bowels.

Leg cues—Signals given through the rider's legs to the horse.

Legumes—Plants with the characteristic of forming nitrogen-fixing nodules on their roots, in this way making the use of atmospheric nitrogen possible.

Lesions—Abnormal changes in the structure of an organ due to disease or injury. Can be internal or external.

Letter of application—Sent with resume or data sheet when applying for a job.

Letter of inquiry—Sent to a potential employer inquiring about possibility of employment.

Liabilities—Just or legal responsibilities.

Libido—Sexual drive.

Lice—External parasites; they may be biting or sucking. Very host specific, they can cause serious skin irritation or anemia.

Ligament—Tough, fibrous tissue that connects bones or cartilages at a joint. Ligaments can also support an organ.

Light horse—This horse is used primarily for riding, driving, showing, racing, or utility on a farm or ranch. A light horse is capable of more action and greater speed than a draft horse.

Liniment—A preparation (mostly alcohol-based) used in treatment of mild strains, sprains, etc., as a counter-irritant to increase blood flow.

Lipizzan Stallions—Stallions of the Lipizzan breed, located at the Spanish Riding School, formerly in Vienna, now located in Wels, Austria.

Liquidity—A business's ability to meet short-run obligations when due without disrupting the normal operation of the business.

Livery stable—A boarding stable for horses and other animals.

Lochia—A brown fluid found in the uterus during uterine involution following pregnancy.

Longe—The act of exercising a horse on the end of a long rope, usually in a circle.

Longitudinally—Running or placed lengthwise.

Low ringbone—A bony enlargement on the pastern bones. It occurs at the pastern-coffin bone joint at about the level of the coronet band.

Lumbar—The portion of the lower back near the loin area.

Lungworm—A roundworm parasite that lives in the lung. More common in donkeys than in horses.

Luteal phase—The period of time during the estrus cycle when the corpus luteum is producing progesterone.

Lymph—Fluid that assists in carrying food from the digestive tract to the tissues and waste products back to the blood stream.

Lymph nodes—The glandlike bodies found in the lymphatic vessels that produce lymphocytes and monocytes.

Lymph vessels—Ducts that transport lymph.

Lysosomes—Small bodies where large numbers of enzymes are stored.

Macrominerals—Minerals found in the body in large quantities for example, calcium and phosphorus.

Maiden mare—A female horse that has not been bred or had a foal.

Malignant—Tending to produce death or deterioration.

Malocclusion—Where the top and bottom teeth do not meet.

Maltose—A disaccharide composed of two molecules of glucose.

Mandible—Lower jaw.

Mange—An itching skin disease caused by parasitic mites.

Mare—The name for a female horse after the age of three.

Market value—Refers to the price for which an animal might sell for at auction.

Mastication—The processing of chewing.

Mastitis—Inflammation of the mammary glands.

Maxilla—Upper jaw.

Meconium—The soft, dark greenish-brown accumulation of digested amniotic fluid, glandular secretions, mucous, bile, and epithelial cells in the digestive tract during the development of the foal.

Megacalorie—One thousand kilocalories, or one million calories.

Meiotic cycle or meiosis—Meiosis is the division of egg and sperm cells. The meiotic cycle produces the gametes or sex cells. The steps in the cycle are prometaphase, metaphase, anaphase, and interphase.

Merychippus—One of the early ancestors of the modern horse.

Mesentery—Where the small intestine lies in folds and coils near the left flank, being suspended from the region of the loin by an extensive fan-shaped membrane.

Mesohippus—One of the early ancestors of the modern horse.

Messenger ribonucleic acid (mRNA)—The protein-coding instructions from the genes are transmitted indirectly through mRNA.

Metabolic—Pertaining to the normal biochemical processes of the body.

Metabolic alkalosis—Condition in the body when the pH increases above the normal levels.

Metabolic disorder—Any abnormalities in normal body functions.

Metabolism—The physical and chemical processes in an organism by which living matter is produced, maintained, and destroyed, and by means of which energy is made available.

Metabolites—The products of metabolism.

Metabolizable energy (ME)—Energy in the feed that is useful to the animal for growth, production, and reproduction. It represents that portion of the gross energy that is not lost in the feces, urine, and gas.

Metritis—Inflammation of the uterus.

Microbial action—Digestion by very minute organisms.

Microchip—A small silicon chip the size of a grain of rice containing the horse's registration number or identification number. A specially designed needle and syringe are used to implant the microchip.

Microfilaments—Long, thin, contractile rods that appear to be responsible for the movement of cells, both external and internal.

Microminerals—Minerals in the body in small quantities, for example, iron and zinc.

Microtubules—Hollow, cylindrical groupings of tubelike structures that help give the cell shape and form. They are also involved in other cell processes.

Midges—Bloodsucking arthropods used as an intermediate host to the nematode onchocerciasis. This parasitic infection usually occurs in the connective tissue, flexor tendons, and suspensory ligaments of the horse.

Milk teeth—The first teeth that an animal develops.

Mimicry—Imitating the behavior of another animal.

Minerals—Essential nutrients for horses, for example, calcium and phosphorus.

Miniature—A very small horse.

Miocene epoch—A division of geologic time beginning about twenty-three million years ago.

Mites—Various small parasitic arachnids; closely related to ticks. They are a secondary host for tapeworms.

Mitochondria—They are composed of an outer membrane and a winding inner membrane. A series of chemical reactions that occur on the inner membrane convert the energy of oxidation into the chemical energy of ATP. Almost all of the energy passes through this molecule before being used in cell function.

Mitosis—Cell division that occurs in plants. The steps in mitosis are early prophase, late prophase, metaphase, anaphase, telophase, and interphase.

Molar—A permanent tooth.

Monday morning disease—Another term for azoturia.

Monkey mouth—Where the lower jaw and tooth structure extend beyond the top teeth.

Monodactyl—Having a single toe.

Moon blindness—Periodic ophthalmia or moon blindness is an inflammation of the inner eye. It is due in part to a vitamin B deficiency. It usually impairs vision and treatment is usually unsuccessful.

Morphogenesis—The process of development during which cells differentiate into specialized types of cells.

Morrill Land Grant Act—Federal law that established land grand institutions in each state.

Mortality insurance—Insurance covering financial losses due to the death of a horse.

Motile—Exhibiting or capable of movement.

Motion behavior—Predictable behavior exhibited by an animal in normal movement.

Mouth speculum—An instrument used to hold a horse's mouth open.

Mucous membranes—A membrane that lines the cavities in the body and connects the inside of the cavity to the outside.

Mucous secretions—Viscous, slippery secretions produced by mucous glands.

Mule—A cross between a female horse and a male donkey.

Muscle tone—Development of strength and firmness of muscles.

Mustang—A wild horse; native horse of the western plain.

Mutagen—An abnormal cell.

Myelin—The substance that covers certain axons and nerve fibers.

Myofibrils—Long, thin tissues that are the contractile elements within muscle cells.

Myosin—A protein present in muscle tissue.

Myxoviruses—A type of virus that attacks the entire respiratory track.

Nasal—Pertaining to the nose.

National Research Council (NRC)—Examines literature and current practices in the nutrition and feeding of horses and publishes recommendations on horse nutrition.

Navicular bone—A small bone located in the horse's foot.

Navicular disease—Inflammation of the navicular bone. It causes horses to go lame.

Neck—The joining of the root and gum.

Necrosis—Death of tissues.

Nematodes—Parasitic worms called roundworms with unsegmented cylindrical bodies. They have complete digestive systems.

Neonatal—Relating to or affecting the newborn shortly after birth.

Nephrons—Part of the kidney; a renal tubule.

Nerves—Bands of white tissue emanating from the central nervous system and ganglia and extending to all parts of the body.

Net energy (NE)—That energy fraction of the feed that is left after the fecal, urinary, gas and heat losses are subtracted from the gross energy. Net energy more precisely measures the real value of feed.

Net worth—Financial condition of a business listing all assets, values of assets, and liabilities of a business.

Neurotransmitters—Biochemicals used to transmit a nerve impulse at a synapse. For example, actecholine or epinephrine.

Nippers—First incisors; also called centrals. In horseshoeing, a tool used to remove extra hoof wall.

Nomenclature—A set or system of names.

Nonadditive gene—The members of the gene pairs will not be equally expressed when nonadditive gene pairs control a trait.

Noninfectious—Diseases caused by environmental problems, nutritional deficiencies, or genetic defects.

Nonruminant—Monogastric or without a functional rumen.

Nonruminant herbivore—A single-stomached animal that eats primarily plant material.

Nucleotides—Basic building blocks of DNA that are each composed of one sugar, one phosphate, and a nitrogenous base.

Nucleus—Each cell contains one which directs the activity of the cell.

Nutrients—A substance that provides nourishment for the body.

Nutritional deficiencies—Deficiency of any necessary substance that provides nourishment for the body.

Off side—Right side of the horse.

Olfactory—Sense of smell.

Oligocene epoch—A division of geologic time beginning about thirty-seven million years ago.

Omnivores—Eating both plant or vegetable and animal food.

Onagers—Relatives of the horse.

Open mare—A mare that was either not bred or did not conceive in the previous season.

Ophthalmia—Inflammation of the eyeball or conjunctiva. See moon blindness.

Orbital—The eye socket.

Ordinances—Local laws or regulations enacted by a city council or other similar body.

Organ—Any part of an animal that performs a specific function.

Organelles—The inside parts of a cell such as the Glogi apparatus, nucleus, ribosomes, centrioles, microfilaments, microtubles, lysosomes, and storage particles.

Organic—Chemical compounds of carbon combined with other chemical elements and generally manufactured in the life processes of plants and animals. Most organic compounds are a source of food for bacteria and are usually combustible.

Osmotic pressure—The pressure needed to prevent water from flowing across a semipermeable membrane into a more concentrated solution from a less concentrated one.

Osselets—Soft, warm swellings over the front and sometimes sides of the fetlock joint.

Ossify—Cartilage being made into bone.

Osteochondrosis—A metabolic disease of cartilage resulting in bone and joint defects.

Osteomalacia—A disease in which the bones become softer. Occurs in adult animals.

Ovaries—Endocrine glands in the female that produce the egg (ovum) at ovulation.

Over-at-the-knee—The same as calf-kneed or buck-kneed.

Overo—Basically either dark or white in color, with no white crossing the back. The spotting is usually roan and extends downwards from the back. The tail is usually of one color. Overo horses usually have bald faces, and glassy eyes are not uncommon.

Overreach—The hitting of the forefoot with the hind foot.

Oviducts—The tubes through which the ovum passes to the uterus.

Ovulation—The release of egg from the mature follicle on the ovary.

Ovum—Egg.

Oxidative phosphorylation—A series of chemical reactions occurring on the inner membrane that convert the energy of oxidation into the chemical energy of ATP. In this process the predominant energy transfer molecule is ATP.

Oxidize—To combine with oxygen to release energy.

Pace—A fast, two-beat gait in which the front and hind feet on the same side start and stop at the same time.

Paddle—To throw the front feet outward as they are picked up.

Paddock—A small fenced area.

Palatability—Acceptability of taste.

Palatable—Appealing to the palate or taste.

Paleocene epoch—A division of geologic time beginning about sixty-five million years ago.

Palpate—To examine by touch.

Parasites—Organisms that live in or on another organism of a different species for the purpose of obtaining food.

Parotid—The largest saliva gland.

Parrot mouth—A result of the upper and lower incisors not meeting because the lower jaw is too short.

Partnership—A form of business organization with multiple owners.

Parturition—The act of foaling or giving birth.

Passive transfer—The process by which antibodies are passed from mare to foal.

Pastern—That part of the horse's leg between the fetlock and the coronet.

Pasture—(1) Area for grazing horses; or (2) grass or other forage that grazing horses eat.

Pasture mating—Natural mating.

Patella—The flat, moveable bone at the front of the stifle joint of a horse.

Patella (upward fixation of)—This is when the patella is moved above its normal position and locks into place. It will prevent the horse from flexing its stifle and the stifle and hock will be extended. The horse will drag its leg since it cannot flex it and bring it back under the body.

Pathogenic—Disease-causing.

Pedal osteitis—Caused by chronic inflammation to the coffin bone, usually of the front feet. Persistent pounding of the feet, chronic sole bruising, or laminitis are causes.

Pedigree—A record of the ancestry of an animal.

Pegasus—A mythological flying horse.

Penis—The stallion's reproductive organ that is used to deposit semen or sperm into a mare's vagina.

Perennials—Plants that normally continue to grow for three or more seasons.

Performance record—The record tracking the actual ability and production of an animal or its offspring.

Pericardium—The sac that encloses the heart.

Periople—A varnish-like coating that holds moisture in the hoof and protects the hoof wall.

Periosteum—A dense connective tissue that covers the surface of each bone.

Peristalsis—Successive waves of involuntary muscle contraction passing along the walls of the intestine or other hollow muscular structures that force the contents onward.

Permanent pastures—Pastures on which horses graze all the time.

Perpendicular—Being or set at right angles to a given line or plane.

Phantom—An object used for a stallion to mount instead of a mare.

Pharynx—A short, somewhat funnel-shaped muscular tube between the mouth and the esophagus that also serves as an air passage between the nasal cavities and the larynx.

Phenotype—The visual characteristics of an animal.

Photoperiod—Refers to the number of hours of light in a day.

Photorefractory—A time when an animal fails to respond to changes in light.

Pig-eyed—Those horses with sunken eyes, seeing less in front and behind than others.

Pincers—Also called centrals.

Pincher—A tool used to remove horseshoes.

Pineal gland—The gland in horses and most other mammals that is responsible for melatonin synthesis.

Pinworms—Parasites that cause intense itching around the anus, from the females laying eggs there, causing the horse to rub its hindquarters, resulting in the hair being rubbed off over the tailhead.

Pitch—Jumping action of a horse in its attempt to unseat its rider.

Pituitary gland—The pituitary and the hypothalamus work together as a functional unit to coordinate the endocrine and nervous systems in their actions, with the hypothalamus being the "center" of the autonomic nervous system and master of the pituitary.

Placenta—The organ that develops in the female during pregnancy lining the uterus and holding the fetus and attached by the umbilical cord.

Plantar cushion—The part of the foot that expands and contracts to absorb shock and pumps blood from the foot back toward the heart.

Plantar ligament—The ligament that runs behind the rear cannon bone in a horse.

Plasma—The fluid portion of blood, as distinguished from corpuscles. Plasma contains dissolved salts and proteins.

Plasma membrane—An extremely thin membrane of lipid (fat) and protein separating the cell from the environment and from other cells. It controls the transport of molecules in and out of the cell.

Plates—Referring to a type of horseshoe.

Pleistocene epoch—A division of geologic time beginning about two million years ago.

Pliocene epoch—A division of geologic time beginning about five million years ago.

Pliohippus—One of the early ancestors of the modern horse; the first true monodactyl.

Plumb line—When a weight is placed on the end of a string to measure the perpendicularity of

something, such as the straightness of the leg of a horse.

Pointing—A stride in which extension is more pronounced than flexion. A horse guilty of a pointed stride breaks or folds its knees very slightly and is low-gaited in front. Also, used to indicate the standing position a horse frequently takes when afflicted with navicular bone disease or injury to the foot or leg, standing on three legs and pointing with the fourth.

Points—Black coloration from the knees and hocks down as in bays and browns. Sometimes includes tips of ears.

Poll—Having to do with the head area of the horse.

Poll evil—A fistula—lesion or sore—on the poll; it is difficult to heal and caused by injury. This is an acquired unsoundness resulting from a bruise or persistent irritation in the region of the poll.

Polyestrous—Having many heat cycles during the year. Mares have more regular cycles at the peak of breeding season when there is more light and no cycles at all during the winter months.

Polyvinylchloride (PVC)—A type of plastic used in building materials.

Pony—A breed of very small horse that is not over fourteen hands high.

Posterior—Backward (in space).

Postpartum—The time immediately after birth.

Poultice—A moist, mealy mass applied hot to a sore or inflamed part of the body.

Pounding—A heavy contact with the ground, usually accompanying a high, laboring stride. Faults in conformation that shift the horse's center of gravity forward tend to create pounding.

Precipitates—To separate out from a solution.

Premaxilla—The area of the jaw that contains cavities for the six upper incisor teeth.

Premolar—Temporary teeth.

Prepuce—Foreskin.

Principal—Property or capital.

Profit—The money that remains after all fixed and variable costs are deducted from income.

Progeny—Offspring or descendants of one or both parents.

Progesterone—A hormone that is released by the ovary before the fertilized egg is implanted.

Proglottids—The segments of a worm.

Prolactin—A hormone involved in lactation produced by the anterior pituitary.

Proprioceptors—The joint proprioceptors give the horse a sense of the positions of its limbs. A sensory receptor situated within the body that is responsive to internal stimuli.

Prostaglandins—A group of hormones that are unsaturated fatty acids and responsible for control of the estrus cycle and timing of parturition.

Protozoal—Any of a subkingdom or phylum of animals containing microscopic, single-celled organisms that reproduce typically by binary fission.

Protozoans—Single-celled animals that occur in the bloodstream and intestinal tract of horses.

Proud flesh—Excess scar tissue on an injured area.

Proximal—Situated closer to the origin.

Przewalski's horse—The oldest species of horse still in existence. This horse was discovered only in the last century.

Puberty—Sexual maturity. The age when an animal becomes capable of reproduction.

Puff—See road puffs.

Puller—Tool used to remove worn horseshoes.

Pulse—The expansion and contraction of the arteries.

Pupae, pupa—Stage of insect development between larvae and adult during which the insect is quiescent.

Pyrantel pamoate—A deworming compound.

Qualitative traits—Traits usually controlled by a few genes showing sharp distinction between phenotypes. For example, coat color.

Quantitative traits—Traits controlled by many pairs of genes with no sharp distinction between phenotypes. For example, growth rate and speed.

Quick-release knot—A knot that unties quickly or breaks loose for the safety of the animal or the handler.

Quidding—A condition in which horses drop food from the mouth while in the process of chewing. It is usually caused by bad teeth or bad gums (stomatitis or gingivitis). It can also be caused by paralysis of the tongue.

Quittor—A festering of the foot anywhere along the border of the coronet. It may result from a calk wound, neglected corn, gravel, or nail puncture.

Rack (single foot)—A fast, flashy, unnatural, four-beat gait in which each foot meets the ground separately at equal intervals; originally known as the "single-foot," a designation now largely discarded.

Radius—The shorter and thicker of the two bones of the forearm.

Random segregation—The random transfer of chromosomes and their genes to form gametes.

Rasp—A tool used for leveling the foot.

Rate of passage—The time required for something to move through an area.

Ration—The feed allowed an animal during a twenty-four-hour period.

Reactive behavior—Activities horses use to maintain themselves in harmony with their environment and to adjust to sudden potentially harmful changes.

Receptors—The part of the neuron that receives internal and external stimuli such as sight, taste, smell, or hearing.

Recessive gene—A gene that is not expressed.

Rectum—The terminal part of the intestine.

Red blood cells—Originate in the red bone marrow, liver, and spleen, and carry oxygen from the lungs to the tissues.

Reflex—Signals traveling on the afferent nerves that never reach the brain. They go directly to the spinal cord and then back to the efferent nerves and the muscles. An example would be a kick in response to a surprise.

Regenerate—To restore or produce anew.

Registered—A horse whose name, along with the name and number of its sire and dam, has been recorded in the records of its breed association.

Relaxation—Controlled by the nervous impulses received by the muscle cells, a muscle either contracts or relaxes.

Relaxin—A hormone from the ovaries that causes relaxation of the pelvic ligaments and possibly relaxation of the cervix at parturition.

Respiration rate—The number of times an organism breathes in a minute.

Resume—Summary of an individual's employment and educational history.

Reward training—Positive reinforcement.

Rhythm—Regularity of each stride.

Ribonucleic acid (RNA)—A substance found in the cytoplasm and nuclei of cells that promotes the synthesis of cell proteins.

Ribosomes—Tiny particles within the cell made of RNA and protein; present in large numbers in most cells. They are the site of protein synthesis.

Rickets—A bone disease in young horses caused by a deficiency of calcium.

Ridgling (cryptorchid)—A horse in which one or both testes are maintained in the body cavity; it is sterile if both testes are undescended, but fertile in a suspended testis.

Rigging—The part of the saddle involved in securing the cinch around the horse.

Ringbone—A bony enlargement on the pastern bones, front or rear. It is caused by bony development around these joints due to tearing and damage of the ligaments and tendons at these bones.

Rings—Ridges.

Risks—A chance of encountering harm or loss.

Road puffs—Soft enlargements located at the ankle joints and due to enlargement of the synovial (lubricating) sacs. Also called windgalls.

Roadster—This is a horse used for driving and includes the heavy and fine harness horses and ponies.

Roan—More or less uniform mixture of white with black hairs on the body, but usually darker on head and lower legs; can have a few red hairs in mixture.

Roaring—The sound made when air is inhaled into the lungs. Also called whistling.

Rodeo—A public event in which the more exciting features of a roundup are presented, as the riding of broncos, branding, lariat throwing, etc.

Rolling—This is excessive side-to-side shoulder motion. Horses wide between the forelegs and lacking muscle development in that area tend to roll their shoulders. The toe-narrow fault in conformation can also cause rolling.

Rotational grazing—The practice of changing the pasture areas of horses for better utilization.

Roughage—Feedstuffs with a high fiber content.

Roundworms—Nematode worms having no segments.

Run—See gallop.

Running martingale—A strap that terminates with two rings, that when properly adjusted, has the effect of preventing the elevation of the head beyond a certain level.

Running walk—A slow, four-beat gait, intermediate in speed between the walk and rack. The hind foot oversteps the front foot from a few to as many as eighteen inches, giving the motion a smooth gliding effect. It is characterized by a bobbing or nodding of the head, a flopping of the ears and a snapping of the teeth in rhythm with the movement of the legs.

Run up—The process of sliding an English stirrup iron up the inside of the stirrup leather so that it does not bounce on the horse's side or get caught on a projection.

R value—The level of insulating ability of a material. The higher the number the better the insulative factor.

Sacking it out—Rubbing a horse that has not been saddled with a saddle pad.

Saline solution—A solution of salt and water (for example, one teaspoonful of salt in one pint of boiled water for a 0.84 percent salt solution).

Saliva—Moistens and lubricates the mass of food for swallowing and, as a digestive juice, acts on the starches and sugars.

Sarcolemma—The outer envelope of skeletal muscles.

Scalping—Occurs when the hind foot hits above or at the line of the hair (coronet) against the toe of a breaking over (beginning the next stride) forefoot.

Scapula—The shoulder blade.

Scars—Marks left on the skin after the healing of a wound or sore. They may appear on any part of the body.

Scratches (grease heel)—A low-grade infection or scab in the skin follicles around the fetlock.

Scrotum—The pouch containing the testicles.

Secretory—Products produced by glands in the body that aid in digestion.

Selenium—A mineral required in the diet.

Semen—Sperm cells plus fluid from the accessory glands.

Seminiferous tubules—Small coiled tubules in the testes where spermatozoa are produced.

Sensitization—Allergic.

Serviceability—The usefulness of a horse for its intended function.

Sesamoid bones—The two pyramidlike bones that form a part of the fetlock or ankle joints on the front and rear legs of a horse.

Sesamoiditis—A condition that consists of a fracture of one or both of the pyramid-like bones that form a part of the fetlock or ankle joints and join with the posterior part of the lower end of the cannon bone.

Settle—Breeding successfully.

Settling percentage—The percent of mares bred that conceive.

Sex determination—Where females carry the XX chromosome and the males carry the XY chromosome.

Shareholders—Owners of a share of a company; stockholders.

Sheath—The double fold of skin that covers the free portion of a male horse's penis.

Shock—A state of profound depression of the vital processes of the body.

Shod—Refers to a horse with horseshoes.

Shoe boil—Also called capped elbow. A soft, flabby swelling at the point of the elbow; it is usually caused by contact with the shoe when the horse is lying down.

Shoe boil roll—Same as a boot.

Shoeing apron—A piece of heavy clothing used to protect the horseshoer.

Short cycle—A cycle that runs shorter than a normal estrus cycle.

Sidebones—When the lateral cartilages ossify making the horse hard-at-the-heels.

Silage—Fermented roughage.

Silent heat—An estrus period with no outward signs of receptivity to the male.

Singeing lamp—An apparatus used to burn the long hairs off a horse's body.

Sire—The male parent of a horse.

Small colon—Part of the large intestine where the moisture in the food is reabsorbed.

Small intestine—The site of most nutrient absorption.

Smooth mouth—Refers to a horse that has no cups present in the permanent teeth.

Snaffle bit—A bit that works with direct action on a horse's mouth. A snaffle bit may have a jointed or a solid mouthpiece.

Snip—A white or beige mark over the muzzle between the nostrils.

Soft—Sensitive, relaxed hands that follow a horse's mouth without becoming rough or inflexible are termed "soft" hands.

Soil test—Used to determine contents of soil and whether fertilizer or some other substance is needed to increase yield.

Sole—The bottom of the hoof.

Sole proprietorship—Form of business organization where one individual owns the business.

Soluble—Able to be dissolved.

Solvency—Having sufficient means to pay all debts.

Space requirements—Refers to the area needed for the size, type of operation, and number of animals in a building.

Speedy-cutting—Occurs when a trotter or pacer traveling at speed hits the hindleg above the scalping mark and against the shoe of a breaking over forefoot.

Spermatozoa—Male reproductive cells.

Spinal cord—One of the most important parts of the central nervous system. This system supplies the body with information about its internal and external environment. It conveys sensation impulses to the brain or spinal cord and other parts of the body.

Splint—An inflammation of the interosseous ligament that holds the splint bones to the cannon bone causing swelling. Usually associated with conformation problems.

Splint boots—Protective covering for the front legs, extend from below the knee to just above the fetlock.

Spooky—A horse that is easily frightened.

Sprain—Any injury to a ligament usually occurring when a joint is carried through an abnormal range of motion.

Spring—The manner in which weight settles back on the supporting leg at the completion of the stride.

Sprung—The appearance of the ribs when the mare develops a wider stride to compensate for the increased weight she is carrying during pregnancy.

Stabling—Refers to the overall facility where a horse lives.

Stagnant—Stale, motionless.

Stallion—A male horse over four years of age.

Stall walking—When a horse walks too much in its stall. It reduces condition and induces fatigue.

Stamina—Ability to endure.

Star—A solid white mark on the forehead. The shape may range from oval to diamond to a narrow vertical, diagonal, or horizontal star.

Statistical analysis—Numerical comparisons that indicate the relative standing compared to other horses.

Step—The distance between imprints of the two front legs or the two back legs.

Stepping pace—This is a slow gait and it is a show gait. It is a lateral, four-beat gait done under restraint in showy, animated fashion with front

foot on the right followed by hind foot on the right.

Stepping pace or slow pace—A modified pace in which the objectionable side or rolling motion of the true pace is eliminated because the two feet on each side do not move exactly together. Instead, it is a four-beat gait with each of the four feet striking the ground separately.

Sterile—Free from living microorganisms; unable to reproduce.

Sternum—Breast bone.

Steroids—Fat-soluble organic compounds. These inactivated hormonal substances are water soluble and are readily eliminated through the urine.

Stethoscope—An instrument for the detection and study of sounds within the body.

Stifled—Refers to a displaced patella of the stifle joint sometimes crippling the horse permanently.

Stillbirth—When a foal is born dead.

Stimuli—Any factors or environmental changes producing activity or response.

Stock—Livestock domesticated for farm use.

Stocking—A white marking from the coronet to the knee.

Stocking plus—A white marking like the stocking, but the white extends onto the knee or hock.

Stocking rate—Number of animals per acre of pasture.

Stomach hair worm—A very small nematode parasite, very thin and hairlike.

Stomach worm—A parasite that produces fibrous tumors, or numerous nodules, which, if close together, form a tumor. They cause gastric and cutaneous habronemiasis. Source of summer sores.

Stomatitis—Inflammation of the mouth.

Strategic planning—Analyzing the business and the environment in which it operates to create a broad plan for the future.

Stress—A demand for adaptation. There are four categories of stress—behavioral, immunological, metabolic, and mechanical.

Stride—The distance between successive imprints of the same foot.

Stride stance—The weight-bearing phase of the stride.

Stride suspension—The nonweight-bearing phase of the stride.

Stringhalt (stringiness, crampiness)—An ill-defined disease of the nervous system characterized by sudden lifting or jerking upward of one or both of the hind legs; most obvious when the horse takes the first step or two.

Strip or stripe—A white mark starting at eye level or below and ending on or above the upper lip. The size and shape of a stripe may vary widely and must be described in detail as to width, length, and its relationship (whether it is connected or unconnected) to a star.

Strongyles—Part of the large groups of parasites known as roundworms.

Stud books—Permanent books of breeding records.

Sucklings—Foals that are not weaned from their mothers.

Sucrose—A disaccharide composed of one molecule of glucose and one molecule of fructose.

Summer sores—Parasitic infections caused by stomach worms where lesions ooze serum, are very itchy, and disappear in cold weather. They occur where horses cannot reach them.

Supplement—A feed or mixture richer in a specific nutrient than the basic feedstuffs in a ration.

Sweat—In the horse, heat is dissipated through sweating or evaporation and by air movement across the body. To do this, the blood transports the heat from the working muscles and the core to the skin.

Sweat scraper—A grooming tool useful for scraping sweat or water off a horse's hide.

Sweeney—A wasting away of the shoulder muscle overlying the scapula of the horse.

Swing—The nonweight-bearing phase of the stride.

Synapses—The connections between nerve cells.

Synovial membrane—A thin secreting membrane that lines joint capsules; it secretes synovia, a clear, slightly yellowish fluid with the appearance and consistency of the white of a watery egg that lubricates the joints.

Synovitis—Inflammation of a synovial membrane.

Systemic—Affecting the whole.

Tack—Equipment used in riding and driving horses, such as saddles, bridles, etc.

Tail board—Used in the prevention of tail rubbing.

Tail rubbing—Constant rubbing of the tail from irritation of parasites. Sometimes develops into a habit.

Tandem—A type of trailer in which the horses ride side by side.

Tapeworms—Large worms that have a head and proglottids or segmented bodies that attach to the intestine of the horse.

Tattoo—An identifying mark on the mucous membrane on the upper lip made by rubbing ink into perforations made by a tattoo gun.

Temporary pastures—Pastures on which horses graze part of the time.

Tendon—Part of the bands of tough, fibrous connective tissue forming the end of a muscle that serves to transmit its force to another part.

Territorialism—Gives the horse space for its basic functions and for care of its home and feeding.

Testes—The male reproductive organs producing sperm cells and hormones.

Tetanus shot—An injection that helps protect an animal from developing the disease tetanus (lockjaw).

Thoracic—Of or pertaining to the thorax (the area between the neck and the abdomen).

Thoroughpins—A distention of the synovial bursa, and considered a discrimination. They can be pressed from side to side.

Threadworm—An intestinal parasitic worm that is unique because only the adult female is parasite; it can exist outside the host; also contributes to foal heat diarrhea.

Throatlatch—The part of the bridle under the horse's throat that connects the bridle to the head.

Thrombin—A biochemical in the blood partially responsible for the process of clotting.

Thrush—An inflammation of the fleshy frog of the foot caused by a fungus. It is blackish in color, foul smelling, and associated with filthy stalls. It may cause lameness.

Thyroid—The gland that secretes thyroxin, the hormone that controls the rate of metabolism.

Ticks—Bloodsucking parasites; often an intermediate host; can cause death in foals.

Tie down—A strap that, when properly adjusted, has the effect of preventing the elevation of the head beyond a certain level without cramping the horse.

Tie stall—A stall in which a horse is fastened by a halter.

Tissue—The structured groupings of cells specialized to perform a common function necessary to the survival of the horse.

Titer—The strength of a solution as determined by titration.

Tobiano—Basically, a white horse in which regular, distinct spots extend down over the neck and chest. All four legs are white, and the face is usually marked the same as in other color patterns found in horses.

Total digestible nutrients (TDN)—The term that indicates the energy density of a feedstuff. It takes into account the amount of fat, protein, and carbohydrate in the feed.

Tourniquet—An instrument used to stop localized circulation of blood by pressure on the local blood vessels.

Toxemia—An abnormal condition associated with toxic substances in the blood.

Toxic—Poisonous.

Toxicity—The state of being poisonous.

Toxin—Poison.

Trace mineralized salt—Salt containing a mixture of the microminerals.

Trachea—The windpipe.

Trailer-sour—Horses that become fearful of trailers and are difficult to haul.

Tranquilizers—Drugs that have a calming effect on animals.

Trappy—A gait that is a short, quick, choppy stride. Horses with short and steep pasterns and straight shoulders tend to have a trappy gait.

Trauma—An injury or wound to a living body.

Traverse or side step—The traverse or side step is a lateral movement of the animal to the right or left as desired without moving forward or backward.

Tree—Basic unit of a saddle; the tree determines the shape of the saddle and may be made of plastic, fiberglass, aluminum, or wood covered with rawhide.

Triple Crown winners—Three-year-old horses winning all three of these races: Kentucky Derby at Churchill Downs, Preakness at Pimlico in Baltimore, and Belmont Stakes at Elmont, New York, in one season.

Trojan Horse—A classic legendary horse in which Greek soldiers hid to gain entry into the ancient city of Troy.

Troponin—A special protein involved in muscular contraction.

Trot—A natural, rapid, two-beat, diagonal gait in which the front foot and the opposite hind foot take off at the same split second and strike the ground at the same time.

Turbinates—Soft, bony structures in the head. These structures are supplied with a great deal of blood.

Two-year-olds—Yearlings.

Tying-up syndrome—A metabolic disorder of the muscles that is associated with forced exercise after a period of rest during which the animal has access to feed; thought to be a mild form of azoturia.

Ulcerations—Breaks in skin or mucous membranes.

Umbilicus—The navel.

Unconditioned response—A response that occurs without practice.

Unicorn—A mythological horselike animal which had a single large horn in the middle of its forehead.

Unnerve—Cutting or removing a nerve from the body.

Unsoundness—Any condition that interferes or is apt to interfere with the function and performance of the horse.

Unthrifty—Unhealthy or inefficient; lacking vigor or bloom.

Ureters—Part of the urinary system which passes urine to the bladder.

Urethra—The canal through which urine, and semen in the male, are discharged.

Urine—The fluid that contains nitrogenous waste and any excess salts or sugars not required by the body.

Uterus—Part of the female reproductive tract; it consists of a body, cervix, and two horns, one of which receives the fertilized ovum for development.

Vaccine—A substance that is either inactivated killed organisms or modified live organisms prepared for inoculation. It stimulates the immune response and/or produces durable immunity with a single dose.

Vacuoles—Large liquid-filled areas in cells.

Vagina—The part of the female reproductive tract that receives the sperm during mating and functions as a passageway during parturition.

Variable costs—Costs that increase or decrease in relation to an increase or decrease in production.

Vein—A vessel that carries blood from the tissues to the heart.

Venae cavae—The large veins entering the heart.

Venereal disease—A disease transmitted by breeding.

Ventilation—Air movement in and out of a building.

Ventral—Surface opposite the backbone.

Vertebrae—Any of the bones or segments in the spinal column.

Vertebral column—The backbone.

Vesicle—A small, thin-walled cavity.

Veterinarian—An animal doctor.

Vices—Habits acquired by some horses that are subjected to long periods of idleness. Hard work and freedom from close confinement are distinct preventives.

Viral—Of, pertaining to, caused by, or the nature of a virus.

Visceral—Smooth muscle is sometime called visceral muscle.

Vital signs—Indications of an animal's health: heart rate, temperature, and respiration rate.

Volatile fatty acids—Acetic, butyric, and propionic acids that are produced by microbial digestion of carbohydrates.

Vulva—The external opening to the vagina.

Walk—A natural, slow, flat-footed, four-beat gait; each foot takes off from and strikes the ground at a separate interval.

Warmblood—Refers to the overall temperament of light to medium horse breeds. Warmblood horses are fine-boned and suitable for riding.

Water-soluble vitamin—Vitamins that are available in feedstuffs or synthesized by microorganisms in the intestine.

Waxed teats—When drops of sticky, clear, or amber-colored fluid excreted prior to parturition become dried and hard, coating the ends of the teats and giving them a waxy appearance.

Way of going—How a horse moves.

Weanlings—Foals that are weaned from their mothers.

Weaving—A rhythmical shifting of the weight from one front foot to the other. It is not a common vice, but when carried to extremes, it renders a horse almost useless.

Wet mare—A mare that has foaled during the current breeding season and is nursing the foal.

Whistling—A paralysis or partial paralysis of the nerves that control the muscles of the vocal cords. Also called roaring.

White blood cells—The active agents in combating disease germs in the body.

White on knee or hock—A separate white mark on the knee or hock.

White spots—White spots on the front of the coronet band or on the heel.

Windgalls—Soft enlargements located at the ankle joints due to enlargement of the synovial (lubricating) sacs. Also called road puffs.

Winding—Twisting the front leg around in front of the supporting leg as each stride is taken. Sometimes it is called threading, plaiting, or rope-walking. Wide-chested horses tend to walk in this manner. Winding increases the likelihood of interference and stumbling.

Wind puffs—Soft enlargements located at the ankle joints and is due to the enlargement of the synovial (lubricating) sacs.

Wind sucking—When a horse identifies an object on which it can press its upper front teeth while pulling backward and sucking air into the stomach, usually accompanied by a prolonged grunting sound. The habit is practiced while eating, thus causing loss of food.

Winging—An outward deviation in the direction of the stride of the foreleg. It is the result of a narrow or pigeon-toed standing position. Winging is exaggerated paddling and very noticeable in high stepping horses.

Winking—When the mare raises her tail to urinate, and the labia opens to expose the clitoris while she assumes a mating position.

Withers—The highest part of the back located at the base of the neck in a horse.

Wolf teeth—The first permanent premolars.

Wound—A disruption in the integrity of living tissue, caused by physical means.

Wry muzzle—A distortion of the muzzle.

Yearlings—Horses between one and two years of age.

Zygote—Cell formed by the union of the male and female gametes (the sperm and egg) and the individual developing from this cell.

INDEX

A